BIOMECHANICS

A CASE-BASED APPROACH

Sean P. Flanagan, PhD, ATC, CSCS

Department of Kinesiology
California State University, Northridge

JONES & BARTLETT
LEARNING

World Headquarters
Jones & Bartlett Learning
5 Wall Street
Burlington, MA 01803
978-443-5000
info@jblearning.com
www.jblearning.com

Jones & Bartlett Learning books and products are available through most bookstores and online booksellers. To contact Jones & Bartlett Learning directly, call 800-832-0034, fax 978-443-8000, or visit our website, www.jblearning.com.

Biomechanics: A Case-Based Approach is an independent publication and has not been authorized, sponsored, or otherwise approved by the owners of the trademarks or service marks referenced in this product.

Some images in this book feature models. These models do not necessarily endorse, represent, or participate in the activities represented in the images.

The author, editor, and publisher have made every effort to provide accurate information. However, they are not responsible for errors, omissions, or for any outcomes related to the use of the contents of this book and take no responsibility for the use of the products and procedures described. Treatments and side effects described in this book may not be applicable to all people; likewise, some people may require a dose or experience a side effect that is not described herein. Drugs and medical devices are discussed that may have limited availability controlled by the Food and Drug Administration (FDA) for use only in a research study or clinical trial. Research, clinical practice, and government regulations often change the accepted standard in this field. When consideration is being given to use of any drug in the clinical setting, the health care provider or reader is responsible for determining FDA status of the drug, reading the package insert, and reviewing prescribing information for the most up-to-date recommendations on dose, precautions, and contraindications, and determining the appropriate usage for the product. This is especially important in the case of drugs that are new or seldom used.

Production Credits

Publisher: William Brottmiller
Associate Acquisitions Editor: Megan R. Turner
Editorial Assistant: Agnes Burt
Production Manager: Julie Champagne Bolduc
Production Editor: Joanna Lundeen
Senior Marketing Manager: Jennifer Stiles
V.P., Manufacturing and Inventory Control: Therese Connell

Composition: Aptara®, Inc.
Director of Photo Research and Permissions: Amy Wrynn
Cover and Title Page Image: © Maria Teijeiro/OJO Images/ Getty Images
Printing and Binding: Edwards Brothers Malloy
Cover Printing: Edwards Brothers Malloy

To order this product, use ISBN: 978-1-4496-9792-1

Library of Congress Cataloging-in-Publication Data
Flanagan, Sean, 1968-
    Biomechanics : a case-based approach / by Sean Flanagan.
        p. ; cm.
    Includes bibliographical references and index.
    ISBN 978-0-7637-8377-8—ISBN 0-7637-8377-3
    I. Title.
    [DNLM: 1. Biomechanics. 2. Movement. 3. Musculoskeletal Physiological Phenomena. WE 103]
    612.7'6—dc23
                                                                                         2012034313

6048
Printed in the United States of America
17 16 15 14 13   10 9 8 7 6 5 4 3 2 1

This book is dedicated to anyone who has ever taught me anything, which is pretty much everyone I have ever met.

Brief Contents

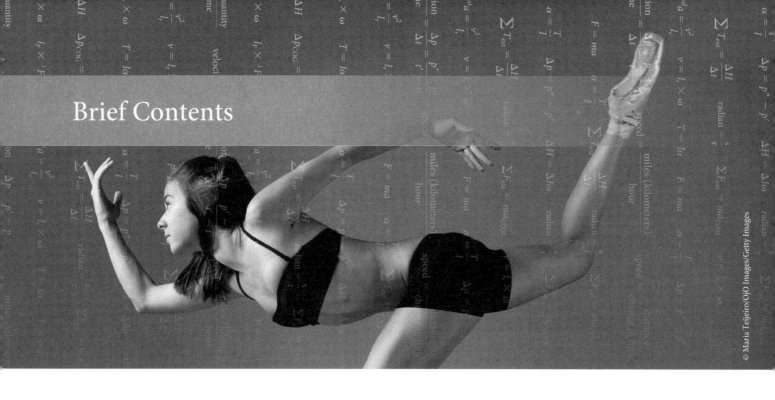

Contents

© María Teijeiro/OjO Images/Getty Images

Preface

This book is designed to be a first course in human biomechanics. Although it was written with an undergraduate kinesiology student audience in mind, I believe it is equally well suited for students in a graduate-level clinical curriculum, such as athletic training, physical therapy, and chiropractic medicine. This is more of an "ideas" book than a "methods" book, and it is written under the assumption that students have a rudimentary knowledge of anatomy and algebra. Trigonometry and geometry are used throughout the book, but "refreshers" appear at the appropriate places. I do not make use of calculus.

Personally, I think many students have a hard time with biomechanics because it is taught in an intimidating manner, with an emphasis on getting the "right numbers" without an understanding of what the numbers actually mean. I have chosen to take a different approach in this book. First, I have used a conversational writing style because I believe that information presented this way is easier to understand without sacrificing rigor. Second, I have tried to make the material less daunting and more meaningful by presenting a Section Question before each major section. Tying new concepts to everyday experience and highlighting research to show how information obtained in the lab can be applied in practice allows the student to better relate to the content. Third, I have placed an emphasis on concepts over computation and expressing these concepts physically, mathematically, and graphically. My hope is that students get an intuitive feel for which way the data should "go" before ever attempting to calculate a number. It might seem that my extensive use of equations contradicts this goal, but I wanted to introduce the symbolic logic behind the equations, and then draw a link between the concepts and the equations. Graph interpretation allows students to visualize this link.

Each lesson opens with a set of Learning Objectives. Marginal Key Terms, Tables, Figures, Boxes, and Important Point boxed features are used throughout the text. Competency Checks are found after every major section and follow the first three areas of Bloom's taxonomy: remember, understand, and apply. An alphabetized Glossary has been placed at the end of the book for optimum review and study. My goal in organizing the content in such a fashion is to lead students to better comprehension and optimal retention.

As for the material itself, I have organized the book into 17 lessons that cover the three levels of biomechanical analysis: whole body, joint, and tissue (bone, cartilage, ligament, tendon, and muscle). I chose not to move sequentially from one level to the next but to use a "whole-part-whole" organization. I begin with elucidating mechanical principles using the whole body level (point-mass, center of mass, and rigid body models) and then discuss the basic material mechanics of biological tissues and unique properties of the muscle-tendon complex. Throughout my career, I have been influenced by a systems science perspective, which states that you cannot get a complete understanding of a system by examining the parts in isolation. For this reason, the muscle–tendon complex is then put into a joint system. After reviewing some mechanical properties of the individual joints of the musculoskeletal system, the mechanics of multi-joint systems is then introduced. In Lesson 17, the three levels are integrated in the context of analyzing movement to improve performance and/or reduce the risk of injury.

I hope that this book provides you with an alternative perspective for teaching and learning the science of biomechanics. Comments and criticisms are welcomed and appreciated.

PEDAGOGICAL FEATURES

Biomechanics: A Case-Based Approach incorporates a number of engaging pedagogical features to aid the student's understanding and retention of the material.

Each lesson starts with Learning Objectives, which highlight the critical points of each lesson.

LEARNING OBJECTIVES

After finishing this lesson, you should be able to:

- Define the following terms: abscissa, absolute value, acceleration, average value, axis, body, cadence, direction, displacement, distance, frame of reference, gait, instantaneous value, kinematics, net value, ordinate, orientation, origin, point, position, relative speed, sense, scalar, slope, speed, step, stride, system, vector, and velocity.
- Explain the difference between speed and velocity.
- Write equations for the following concepts: distance, displacement, speed, velocity, and acceleration.
- Identify speed on a position–time curve.
- Identify velocity on a position–time curve.
- Identify acceleration on a velocity–time curve.
- Explain the difference between instantaneous and average kinematic measures.
- Describe situations in which velocity is more important than acceleration.
- Describe situations in which acceleration is more important than velocity.
- List the determinants of gait velocity.

Important Point! boxes clarify essential math concepts relevant to the content within the specific section.

> **Important Point!**   A net acceleration causes a change in velocity, and a change in velocity causes a displacement.

Section Questions present salient questions to address the point of focus for each section.

Section Question

According to the Disability Statistics Center at the University of California, San Francisco,[1] 1.6 million Americans use wheelchairs (**Figure 4.1**). Understanding the motion of the wheel is essential in understanding the mobility of wheelchair users. Is the motion of the wheel the same as the motion of a body previously explored?

Section Question Answers provide contextual responses to the each section question.

Section Question Answer

Angular motion is similar to linear motion in that there are positions, displacements, velocities, and accelerations. The relation between position, velocity, and acceleration are also the same. The main difference is that you are replacing a linear term with its angular equivalent.

Using the first three levels of Bloom's taxonomy, Competency Checks ask students conceptual and quantitative questions to assist in gauging their understanding of the material.

COMPETENCY CHECK

Remember:

1. Define the following terms: work, positive work, negative work, and mechanical energy expenditure.
2. List the assumptions associated with mechanical energy expenditure.

Understand:

1. How much work is required to do the following? Is the energy entering or leaving the system?
   a. Raise a 10 kilogram mass 0.5 meters.
   b. Lower a 5 kilogram mass 0.1 meters.
   c. Raise a 15 kilogram mass by 0.5 meters, and then lower it back to its original starting position.
2. How much mechanical energy expenditure is required to do the following?
   a. Raise a 10 kilogram mass 0.5 meters.
   b. Lower a 5 kilogram mass 0.1 meters.
   c. Raise a 15 kilogram mass by 0.5 meters, and then lower it back to its original starting position.

Apply:

1. List several activities where:
   a. Work would be an appropriate measure for analysis.
   b. Mechanical energy expenditure would be an appropriate measure for analysis.
   c. Neither work nor MEE would provide useful insights.

Equations are numbered throughout the lesson for easy referral.

---

frictional (or other external) forces, and the time of impact is extremely small. Because the change in momentum of the system is zero in each case, you know:

$$\Delta L = L' - L = 0 \qquad (9.2)$$

where the prime sign will indicate the time immediately after impact, and the momentum without the prime sign is the momentum immediately before impact. You know that momentum is the product of mass and velocity:

$$L = mv \qquad (9.3)$$

It is also important to note that the momentum of any system is the sum of the momentum of each body in a system:

$$L = \sum (m_i \times v_i) \qquad (9.4)$$

where the symbol i represents the number of bodies in the system. If there are two bodies (a and b) in your system, then:

$$L = m_a v_a + m_b v_b \qquad (9.5)$$

If you substitute Equation 9.5 into Equation 9.4, you get:

$$\Delta L = (m_a' v_a' + m_b' v_b') - (m_a v_a + m_b v_b) = 0 \qquad (9.6)$$

Essential Math boxed features provide a review of mathematical material crucial to the understanding of biomechanics.

| Box 2.1 | Essential Math: Ratios and Rates |

A **ratio** is simply one number divided by another number:

$$\text{ratio} = \frac{\text{one quantity}}{\text{another quantity}}$$

A **rate** is a ratio between the change in one quantity and the change in time:

$$\text{rate} = \frac{\Delta \text{one quantity}}{\Delta \text{time}}$$

The delta symbol (Δ) is shorthand for "change in." Think of the dividing line as "per," so we can think of a rate as a change in one quantity (position, velocity, force, work) per a change in a unit of time (seconds, minutes, hours). Rates are going to be very important in biomechanics. From algebra, you should be able to recognize that the rate will be larger if the change in the quantity is increased and/or the change in time is decreased.

$$\frac{\Delta \text{one quantity}}{\Delta \text{time}} \quad \text{or} \quad = \text{Larger ratio}$$

Increase this → (numerator)
Decrease this → (denominator)

Applied Research boxed features provide examples that are helpful in illustrating biomechanical concepts and present evidence of the practical value of biomechanics.

| Box 9.1 | Applied Research: Effective Mass and Head Injuries in American Football |

Head injuries in American football are a serious problem, particularly those resulting from helmet-to-helmet contact. In many cases, the injury to the offensive player receiving the impact is greater than the injury sustained by the striking, defensive player. In this investigation, the researchers provide an explanation for why this is the case. Reconstructing actual, recorded game-time head injuries using instrumented dummies in the laboratory, they found that the striking player aligned their head, neck, and torso (called spearing), increasing the effective mass of the striking player to 1.67 times that of the player being hit. In a follow-up investigation, they compared these impacts to punches to the head delivered by Olympic-caliber boxers. They found these impacts did not transfer as much linear momentum as the football head strikes due to the lower effective mass of the fist.

Data from: Viano DC, Pellman EJ. Concussion in professional football: biomechanics of the striking player—Part 8. *Neurosurgery*. Feb 2005;56(2):266–278.

Viano DC, Casson IR, Pellman EJ, Bir CA, Zhang LY, Boitano MA. Concussion in professional football: comparison with boxing head impacts—Part 10. *Neurosurgery*. Dec 2005;57(6):1154–1170.

Key Terms are highlighted and defined in the margins throughout the lesson and compiled into a **Glossary** at the end of the book.

Plastic deformation   A deformation in which the object does not return to its original dimensions after the deformation

original shape. With a **plastic deformation**, the object has been "stretched out of shape" and will not ever return to its original dimensions. Think of the little plastic thingy (the scientific term) that holds a six-pack (of soda) together. Have you ever tried to put a can back in the plastic thingy after you have taken it out? Chances are you were not very successful because the thingy underwent a "plastic" deformation.

When will an object undergo a plastic versus an elastic deformation? If an object is deformed too much, there is actually a microtearing of the material. The point where it is deformed too much is called the **yield point**: Any deformations beyond the yield point result in permanent (plastic) deformations.

Yield point   The amount of deformation that marks the transition from elastic to plastic deformations, and deformation beyond this point results in a permanent deformation

A comprehensive and instructional art package includes photographs and illustrations throughout the book to encourage learning with a unique visual appeal.

Pushing this way

Weight acts in the downward direction only

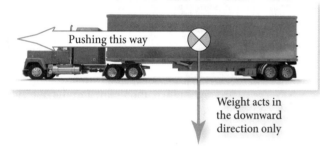

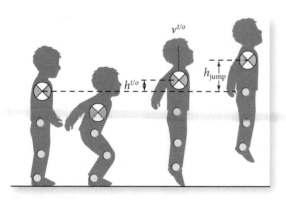

$v^{t/o}$

$h^{t/o}$

h_{jump}

At the end of each lesson, a Summary reinforces key ideas and helps students recall and connect the concepts discussed. Key Concepts are presented in table format for review. Review Questions test comprehension of the concepts discussed within the lesson. References used in the lesson are also listed.

Table 8.1	Key Concepts

- Energy
- Work
- Mechanical energy expenditure
- Efficiency
- Power

SUMMARY

In this lesson, you learned about an alternative to Newton's laws for analyzing human movement. This method involved the concepts of work, energy, and power (Table 8.1). Because of some issues with using these concepts with biological systems, mechanical energy expenditure was introduced. The first law of thermodynamics was compared to the center of mass equation, and efficiency and economy were introduced. Impulse–momentum and work–energy methods provide complementary information and a more complete analysis of movement for several different tasks.

REVIEW QUESTIONS

1. Define the following terms: energy, kinetic energy, potential energy, gravitational potential energy, strain potential energy, work, mechanical energy expenditure, efficiency, economy, and power.
2. State the conservation of energy and the first law of thermodynamics.
3. What is meant by the term *negative* work or power?

REFERENCES

1. Feynman RP. *Six Easy Pieces*. New York: Basic Books, 1995.
2. Watson D. Energy Definition. 2011. Internet Communication.
3. Zatsiorsky VM. *Kinetics of Human Motion*. Champaign, IL: Human Kinetics, 2002.
4. Aleshinsky SY. An energy-sources and fractions approach to the mechanical energy-expenditure problem. 1. Basic concepts, description of the model, analysis of a one-link system movement. *J Biomech*. 1986;19:287–293.

Instructor Resources available for download to adopters of the book include PowerPoint Lecture Presentations, Image and Table Bank, Test Bank, and Instructor's Manual. For access, contact your Representative at www.jblearning.com.

The Companion Website for *Biomechanics: A Case-Based Approach*, **go.jblearning.com/biomechanics**, offers students and instructors an unprecedented degree of integration between their text and the online world through many useful study tools, activities, and supplementary information. Study tools include Student Practice Problems, Weblinks, Flashcards, an Interactive Glossary, and Crossword Puzzles. This interactive and informative website is accessible to students through the redeemable access code provided in every new text.

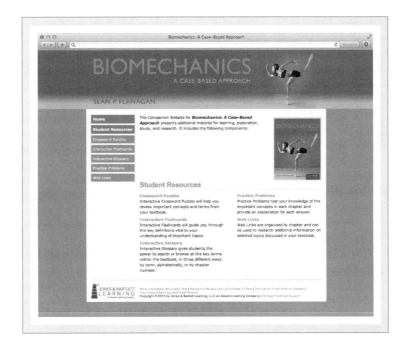

Acknowledgments

Rarely are endeavors of this magnitude undertaken alone. There are many people who have helped shape me and my view of biomechanics, and therefore, they deserve at least some acknowledgment, if not credit, for this final product. Although words cannot adequately express my gratitude, I hope they will accept my deepest and heartfelt thanks. At the risk of omitting some very important people, I would like to specifically mention a few.

First, I wish to thank my family: Jennifer, for continually pushing me outside of my comfort zone and for putting up with all of the lost nights and weekends I spent researching and writing; my mother-in-law, Sally Montelongo, for all of the support over the years; and my parents, Joe and Mary Flanagan, who made sure I did my homework (and checked it), giving me the proper foundations to become a scholar. The pyramid can be built only as high as its foundation is wide.

Next, I wish to thank some of the people who have taught, inspired, and guided me over the years: Dr. George Salem (University of Southern California), for everything; Dr. Kornelia Kulig (University of Southern California), for being a mentor, collaborator, and friend; Dr. Paul Vanderburgh (University of Dayton), for showing me how to be a scientist and ingraining in me that you must first understand what the numbers mean before you learn how to calculate them; Dr. Christopher Powers (University of Southern California), for imparting his knowledge of clinical biomechanics; Dr. James Gordon (University of Southern California), for challenging me to be "more theoretical"; Dr. Jill McNitt-Gray (University of Southern California), for some of the most enjoyable and valuable biomechanics classes ever; Dr. Lloyd Laubach (University of Dayton), for giving me the encouragement and support I needed to pursue doctoral studies; Dr. Karl Stoedefalke and Dr. Bob Christina (Pennsylvania State University), for being true inspirations; Dr. George DeMarco (University of Dayton), for showing me how to be "edutaining"; Dr. Carolee Winstein, Dr. Cesar Blanco, and Dr. Lucinda Baker (University of Southern California), for teaching me how science can be used to further our understanding of normal and pathological motion; Dr. Erik Johnson and Dr. Hung Leung Wong (University of Southern California), for advancing my understanding of mechanics a thousand-fold; Dr. Bill Whiting and Dr. Shane Stecyk (California State University, Northridge), for helping make my transition from student to faculty a seamless one; Dr. Paul Lee (California State University, Northridge), for explaining that whole Lagrangian thing to me; and my fellow doctoral students at the Jacqueline Perry Musculoskeletal Biomechanics Research Laboratory, for all of the thought-provoking discussions.

I also wish to thank the many professionals at Jones & Bartlett Learning for their support, encouragement, and patience: Shoshanna Goldberg, Megan Turner, Joanna Lundeen, Agnes Burt, Prima Bartlett, Kyle Hoover, and anyone else who I may not have worked with directly but who helped make this book what it is. Additionally, I wish to thank all of the reviewers whose comments strengthened this book.

Finally, I want to thank all of my students (past and present) who either directly or indirectly helped shape this book. Special thanks to: Shawn Sorensen; Stephen Cho; Shin-Di Lai; James Kohler; Josh Phillips; Lloyd Magpantay; Matt Hank; Matt McCann; Janelle Kulik; and Lulu Silveyra. This book is the culmination of the teachings, ideas, and concepts of the professors, colleagues, and students whom I have had the pleasure of associating with over the years, but any errors contained within are solely mine.

Reviewers

The author and publisher would like to thank the following reviewers for their feedback in the development of this textbook.

William R. Barfield
Health and Human Performance College of Charleston
and Department of Orthopaedic Surgery
Medical University of South Carolina

Elizabeth C. Davis-Berg
Columbia College, Chicago

Barry A. Frishberg
Professor of Health Sciences
South Carolina State University

John C. Garner, III, PhD, CSCS
Applied Biomechanics Laboratory
University of Mississippi

Dr. Carrie Hendrick
Lenoir-Rhyne University

Don Hoover, PT, PhD, CSCS
Rockhurst University

ChengTu Hsieh
California State University, Chico

Hsin-Yi Liu
North Carolina Central University

G. William Lyerly, PhD
Coastal Carolina University

Paula Maxwell
James Madison University

Richard Robinson
University of Indianapolis

Brian K. Schilling
University of Memphis

Jeremy D. Smith, PhD
School of Sport and Exercise Science
University of Northern Colorado

Henry Wang
School of Physical Education, Sport, and Exercise
Science
Ball State University

Lesson 1

Introduction

LEARNING OBJECTIVES

After finishing this lesson, you should be able to:

- Define the following terms: biomechanics, kinematics, kinetics, and mechanics.
- List the four parts that are used for a symbol in this book.
- Explain why you should study biomechanics.
- Explain how understanding biomechanics can achieve this purpose.
- Describe the three sets of principles that are used in biomechanics.
- Explain the difference between kinematics and kinetics.
- Describe the rules for hierarchical modeling.

1.1 BIOMECHANICS: UNDERSTANDING THE RULES GOVERNING MOVEMENT

In the 1999 Warner Brothers movie *The Matrix*, Neo (played by Keanu Reeves) learns from Morpheus (played by Laurence Fishburne) that he has been living in a sort of virtual reality of generated computer code (Figure 1.1). While in the Matrix, Morpheus is capable of doing incredible things, and he tells Neo that he can, too, as long as he understands the rules under which the Matrix operates:

> *What you must learn is that these rules are no different than the rules of a computer system. Some of them can be bent, others broken.*

Although we do not live in a virtual reality programmed by machines (at least I do not think we do), we do live in a world that is governed by rules. And although we may not

Figure 1.1 Neo is capable of doing incredible things once he understands the rules of the Matrix.

© Moviestore Collection Ltd/Alamy

be able to bend or break these rules, we are capable of doing some pretty amazing things with our bodies if we understand them. **Biomechanics** is a branch of science that looks to uncover these rules by applying the methods of mechanics to the study of the structure and function of biological systems.[1]

> **Biomechanics** The study of the structure and function of biological systems by means of the methods of mechanics

With this impressive-sounding definition out of the way, two questions may immediately come to mind:

1. Why should I study these rules?
2. Who needs to know them?

The answer to the first question is to help people to move better. To be fair, many biomechanicians study more than just people, and some do not necessarily study movement. However, that is the focus of this book. And by move better, I mean to improve performance or reduce the risk of injury, which may have more expansive meanings than what immediately come to mind.

Important Point!   You study biomechanics so you can help someone improve their performance or reduce their risk of injury.

Performing better can have several different connotations, depending on the task. For example, you may wish to have someone jump higher or throw farther. These are obvious examples of a better performance. But you may also wish to decrease the amount of energy necessary to walk across a room or up a flight of stairs, or simply be able to accomplish a task such as buttoning a shirt or combing hair. Do not think that performance is limited to high-achieving athletic competitions. Performance occurs during any human activity, including those that are part of your everyday life.

Important Point!   Performance occurs during any human activity, including those that are part of everyday life.

A lot of human activities are inherently risky, and you will never be able to eliminate all injuries that can occur as a result of participating in that activity. But there are a lot of ways that you can decrease the *potential* for injury. For example, certain ways of moving can place loads on the body that it was not designed to handle (called mechanopathology). Alternately, an injury or disease can change the way a person moves as she attempts to "work around" the condition (called pathomechanics), placing inappropriate loads on different structures (and/or degrading performance). In addition, environmental (e.g., a slippery floor) and other external factors (such as being hit by an opponent) can be potentially injurious. Understanding injury mechanics requires you to know something about the person, the environment, other people in that environment, any equipment that is being used, and the complex ways these factors interact.

Mechanopathology   The mechanics that result in an injury

Pathomechanics   The mechanics that are a result of an injury

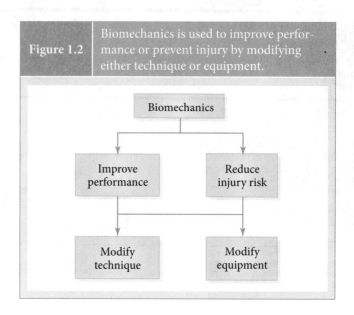

Figure 1.2   Biomechanics is used to improve performance or prevent injury by modifying either technique or equipment.

In general, these two objectives (improving performance and decreasing injury risk) are achieved by either modifying a person's technique (the way the person moves) or the equipment that is used as part of the activity (Figure 1.2). It stands to reason that if the way a person moves limits performance or exposes that person to potentially injurious forces, then changing the way the person moves can improve performance or decrease risk of injury. Similarly, if there are environmental or external factors that are impeding performance or increasing the potential for injury, then equipment may help. As with performance, you should not use too narrow a definition for equipment. Biomechanicians have certainly been involved in the modification of athletic equipment such as helmets and ski poles, but they also study shoes and even the characteristics of floors or machines in industrial settings. So consider performance and equipment in the broadest sense of the terms.

If the answer to the first question is that understanding biomechanics is important in helping people to move better, then the answer to the second question should be obvious: Anyone who is involved with the movement of people needs to understand the rules governing those movements. If you are (or are going to be) involved with teaching skills (such as a physical educator, personal trainer, dance instructor, or coach), preventing or rehabilitating from injury (such as an athletic trainer, physical therapist, chiropractor, or physician), designing equipment to be used by people (such as an ergonomist or engineer), or modifying the structure of the body (such as an orthopedic surgeon), you need to study biomechanics. Biomechanics is a rich field with a broad number of applications. This book's focus is on understanding the core concepts—the rules governing human movement. You will learn more about the applications in discipline-specific texts.

1.1.1   Mechanical, Multisegment, and Biological Principles

Now that you understand why you should know the rules of human movement, you can begin to study them. The rules can be roughly grouped into three sets of principles: mechanical, multisegment, and biological (Figure 1.3).[2] Each will have a different focus in this book.

To understand the rules under which we move, you must first have an understanding of physics or, more specifically, classical mechan-

> **Mechanics**   The study of forces and their effects

ics. Classical mechanics is interested in the motion of bodies under the action of a system of forces. It basically deals with the motion on a physical (size and speed) scale, of everyday things that you can potentially see. It is called "classical" because it was developed based on the work of Sir Isaac Newton and those that followed him, but it excludes such "modern" topics as quantum mechanics (which deals with physics on an extremely small size scale) and the work of Einstein and relativity (which deals with physics on an extremely large speed scale).

Classical mechanics (Figure 1.4) is usually divided into two areas: either things that are moving or things that are

> **Kinematics**   The study of motion without consideration for what is causing the motion

not. *Dynamics* deals with things that are moving and can be further broken down into kinematics and kinetics. Kinematics is the study

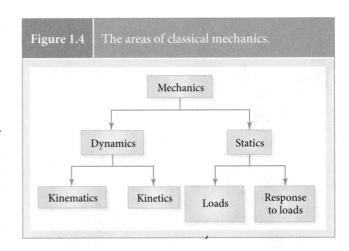

Figure 1.4   The areas of classical mechanics.

of motion without consideration of the causes of that motion. Kinetics deals with the causes of that motion. *Statics* deals with loads on things

> **Kinetics**   The study of the forces that cause motion

that are not moving. An offshoot, if you will, of statics is materials science. Whereas statics deals with the loads applied to a body, materials science examines the material properties of that body and their response to a load. In biomechanics, this is often referred to as *tissue mechanics*.

If you have ever taken a high-school physics class, you have probably learned some basic principles of classical mechanics (such as Newton's laws of motion), or what I just referred to as the first set of rules. The second set of rules is a bit more complicated, but they are an extension of the first. When you learned Newton's laws, you probably learned them for a single body, such as a ball or pendulum. The human body is not a single element (although it can sometimes be modeled that way), but is made up of many connected segments. For example, the "simple" act of reaching for something requires movement of both the upper arm and forearm, and throwing will require the coordinated activity of the lower extremities, trunk, and upper extremities. Some unique properties emerge when the body is looked at as a system of interacting elements rather than a single body (or by looking at the parts in isolation). The second set of rules acknowledges the multi-segmented nature of the human body.

The third set of rules is based on the fact that we are not inanimate objects or machines. As a biological being, you will not violate the laws of physics, but you do influence them in a particular way. This is where the "bio" portion of biomechanics comes in. For example, you may already know Newton's Second Law ($F = ma$). Do not worry too much about it if you do not; you will learn all about it in the lesson on linear kinetics. For now, suffice it to say that this law is one of the foundations

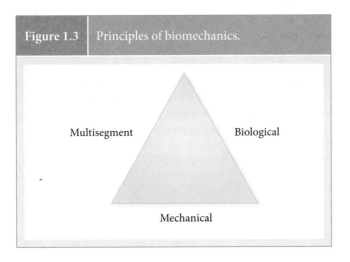

Figure 1.3   Principles of biomechanics.

Figure 1.5 | The code of the Matrix.

© Remy Behear Merriex/ShutterStock, Inc.

of classical mechanics, and you are as affected by it as any other object on Earth. However, in many instances, the source of the "F" in Newton's Second Law are your muscles.[3] Although the law will always hold up, a rather large number of factors based on the anatomy and physiology of your muscles influences the production of force.

1.1.2 Mathematics: The Code

The "code" of the Matrix was depicted in the film as "falling green rain," a combination of mirror images of half-width kana characters, Latin letters, and numbers (Figure 1.5). The program could be read because information was concisely represented by the code. By reading the code, people outside the Matrix could determine what was happening inside the program, but they had to understand the code.

Our world, too, is represented by a "code," but it is not digital rain. Our code is written in the language of mathematics, for many of the same reasons: It allows a relatively large amount of information to be concisely represented. All you have to do is know how to read the code.

Many people are scared of math, but it does not have to be *that* bad. Let us take a look at how it works. Consider the following equation:

$$c = a \times b^2 \qquad (1.1)$$

Looking at the code, what does it tell you? Well, there are several useful pieces of information. First, some variable, c, is completely determined by two other variables, a and b. Second, increasing either a or b will increase c, so if you wanted to maximize c, you would want to increase a and b. Conversely, if you wanted to minimize c, you would have to decrease a and b. Third, the two variables are not "weighted" equally: b is squared. If you doubled a, then you increased c by a factor of 2; but you will increase c by a factor of 4 if you double b because it is squared.

Now say that a was your variable of interest. You could rearrange Equation 1.1 as follows:

$$a = \frac{c}{b^2} \qquad (1.2)$$

Now a is determined by c and b. The variable c has the same effect on a as a had on c. Increasing one will increase the other by the same amount. Variable b is now the denominator. This means that if you increase b, you will decrease a. Doubling b will decrease a by one quarter. In a similar way, you can rearrange the equation to solve for b and then make the appropriate analyses. It was a lot easier to learn this one line of code than memorizing various facts about the relations between a, b, and c. Was it not?

This book assumes that you have a rudimentary level of math skills, and there will be a number of refreshers, depicted as "Essential Math" boxes, throughout the lessons when I think they are needed. In fact, your first one is presented in Box 1.1. One thing that I think puts people off about math is that they spend so much time trying to get the "right number" that they lose sight of what goes into the equation or what the number actually means. In this book, you will focus on the equations and the interpretations. The numbers can come later.

Box 1.1 | Essential Math: Algebra

When manipulating equations, the general rule is to perform the same operation on both sides of the equal sign to maintain the equality. Here are some specific examples to help you manipulate equations.

If two variables are added (or one subtracted from another) and you want to move a variable from one side of the equal sign to the other, then subtract it from both sides if it is positive and add it to both sides if it is negative. Example:

$$a = b + c$$
$$a - c = b + c - c$$
$$a - c = b$$

If two variables are multiplied (or one is divided by another) and you want to move a variable from one side of the equal sign to the other, then divide (if it is a multiple) or multiply (if it is a divisor). Example:

$$a = \frac{b}{c}$$

$$a \times c = \frac{b}{c} \times c$$

$$a \times c = b$$

Sometimes you may have to perform successive operations to get to your variable of interest. Just pay close attention to the order in which you perform the operations. If a divisor is only under the variable, then you perform your addition and subtraction first. Example:

$$a = \frac{b}{d} + c$$

$$a - c = \frac{b}{d} + c - c$$

$$a - c = \frac{b}{d}$$

$$(a - c) \times d = b$$

If the divisor is under several variables, you must multiply first. Example:

$$a = \frac{b + c}{d}$$

$$ad = \left(\frac{b + c}{d}\right)d$$

$$ad = b + c$$

$$ad - c = b$$

And usually, by convention, you want to isolate your variable of interest on the left-hand side of the equation. So the last equation is rewritten as:

$$b = ad - c$$

Notice one equation had a multiplication sign and another did not. If two variables are next to each other with no signs between them, then multiplication is assumed. For example:

$$ad = a \times d$$

A Note on Symbols

The language of math is a language of symbols. It is just more concise to abbreviate something like force with an F or acceleration with an a. Unfortunately, although some symbols are universal, others are not. Physics, engineering, and biomechanics may all use different symbols for the same thing (physicists like to use p for momentum, whereas engineers

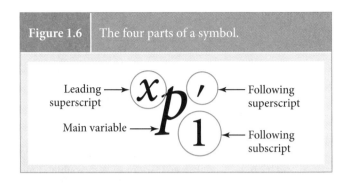

Figure 1.6 The four parts of a symbol.

like to use L). What is even worse is that sometimes the same symbol will have two different meanings in the same book or field (W is both weight and work; s is both seconds and speed). In addition, a capital letter may represent something different than a lowercase one. So you must be careful when looking at different books and papers.

In this book, great care was taken to ensure that symbols only have one meaning, although a lowercase letter may have a different meaning than a capital one. You will be introduced to them as they appear. For now, it is important to know about some of the conventions that will be used. A symbol can have up to four parts (Figure 1.6). The main part is the variable itself. The leading superscript indicates direction. It could be the direction in a particular reference frame, such as x, y, or z (you will learn all about reference frames in the next lesson) or a direction that is perpendicular ($\perp$) or parallel ($\parallel$) to some body of interest. The following subscript refers to the body, which is particularly important if you are keeping track of more than one. The following superscript will indicate a change in time for the same variable. Time "zero," or the start, does not have a following superscript. Time point 1, which is some change from time zero, will have a single prime ($'$). Time point 2, which is some change from time 1, will have a double-prime ($''$), etc. Rarely will there be a need in this book for more than 3 primes.

Do not let all these symbols scare you off. You will be gradually introduced to them, and they make a lot of sense once you get the hang of them. In addition, super- and subscripts will only be used when necessary for clarity. For example, if you are only dealing with one body, there will not be a need for a following subscript, and if the direction is obvious, there will not be a need for a leading superscript. In this way, things will be kept as simple as possible.

1.1.3   Hierarchical Modeling: Keeping Track of the Variables

Equation 1.1 and its various manipulations are fairly easy to keep straight because you only have to keep track of three variables. What if you also had access to even more lines of code? Consider the following equations:

$$a = d - e \tag{1.3}$$

$$e = \frac{g^2}{h + i} \tag{1.4}$$

Although c is still completely determined by a and b, each of those variables is determined by other variables, some of which are determined by still others. Things can get a little messy, and it can be hard to keep track of all the variables. One aid to assist you with keeping track of things is called a deterministic model,[4] although the term *hierarchical model*[5] seems to be more appropriate and will be used throughout this book.

The basic idea of a hierarchical model is presented in Figure 1.7. At the very top, you list your performance criterion, preferably in mechanical terms. In the level below, you include all the factors that determine that variable. It is that simple. The only real rules of hierarchical modeling are: (1) the factors included in the model should be mechanical quantities, and (2) each of the factors included in the model should be completely determined by those factors that are linked to it from below.[4] Once you have identified all the factors, you can annotate those that you cannot control (usually by striking through the box). For example, you can't control gravity. So although you want to include it in your model and keep track of its effects, you are not going to worry about trying to change it.

To give you an example, the hierarchical model for variable c using Equations 1.1 to 1.4 is presented in Figure 1.8. Your interpretation of the model is that c is determined by the following variables: b, d, g, h, and i. Notice that these boxes have nothing below them. Also note that I did not include variables a and e in my interpretation because they were determined by other factors, and it would be redundant to include them.

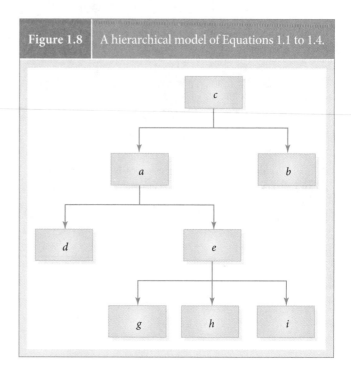

Figure 1.8 | A hierarchical model of Equations 1.1 to 1.4.

The equations were used to develop the model, and the model was used to help determine appropriate equations. They work together. You will see hierarchical models used throughout the text, even when there are only a few variables. This was done to get you used to working with them. In addition, in many instances the model will be added to in subsequent lessons as your understanding becomes more complex. Although some of the limitations of hierarchical models will be discussed in Lesson 17, they are a good way to keep track of variables and assist you in your thinking.

1.2   HOW TO USE THIS BOOK

You may have been clued in by the last two sections that this book will make extensive use of equations (but not necessarily numbers) and hierarchical models to help you understand important concepts. Again, do not get too scared by these. You will find that they are helpful aids in presenting information in a concise way and to organize your thinking. Look for the boxes on "Essential Math" to go over topics where you are rusty. If you have a strong math background, you can safely skip them.

In many ways, this book follows a "whole–part–whole" structure. The first part (Lessons 2–9) is organized around the principles of classical mechanics. In many ways, it is applied physics. And although the information is similar to what you would get in an introductory physics text, the examples used relate specifically to human movement. Because the connections to applied biomechanics may not always be evident,

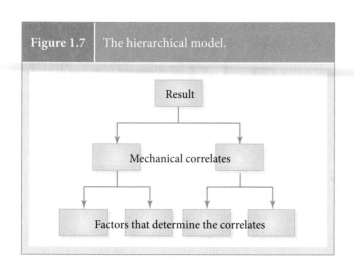

Figure 1.7 | The hierarchical model.

boxes on "Applied Research" will show you how these concepts are being used to solve everyday problems.

The second part of the book (Lessons 10–15) deals with the biological principles. You will begin with looking at the tissue level (muscle, tendon, bone, ligaments, cartilage) separately and then combine them into a single-joint system. You will then look at the specific joints in the lower extremities, trunk, and upper extremities. It is assumed that you have had at least some knowledge of anatomy, but looks at the topic in a slightly different way. For example, the quadriceps muscle group is not thought of as "knee extensors" but rather as producing a knee extensor torque. As you will see, the quadriceps group not only extends the knee but just as important, they also control knee flexion in many cases.

Lesson 16 deals with the principles of a multisegmented body, which you may have heard referred to as a *kinematic* or *kinetic chain*. Lesson 17 attempts to put the information from the previous 16 lessons together for a variety of activities of daily living and sport. It is hoped that at the end of this book you will be able to do the same thing for many activities not listed in this book.

Although some lessons and some sections within lessons may be self-contained, a majority of the sections and lessons build on the previous ones. Rather than putting all the review questions at the end of each lesson, you will see "competency checks" at the end of each major section. Do not skip over these. The major sections are blocks of information, and it is important that you understand the block before moving on to the next one. If you cannot complete the competency check, you should go back and review that block before moving to the next one. You will be glad that you did.

SUMMARY

In this lesson, you were introduced to the definition of biomechanics, the reasons for studying it, and the groups of principles that are derived from it. You were introduced to the importance of equations and hierarchical modeling, and then introduced to the topics of this book. It is time to get started with your first lesson: one-dimensional kinematics.

REVIEW QUESTIONS

Remember:

1. Define the following terms: biomechanics, kinematics, kinetics, mechanics.
2. List the four parts that are used for a symbol in this book. What is represented by each part?

Understand:

1. Why do you study biomechanics? How do you achieve these aims?
2. Describe the three sets of principles that are used in biomechanics.
3. Explain the difference between kinematics and kinetics.
4. Describe the rules for hierarchical modeling.

REFERENCES

1. Hatze H. The meaning of the term biomechanics. *Journal of Biomechanics.* 1974;7(2):189–190.

2. Lees A. Technique analysis in sports: a critical review. *Journal of Sports Sciences.* Oct 2002;20(10):813–828.

3. Latash ML. *Synergy.* Oxford: Oxford University Press, Inc; 2008.

4. Hay JG, Reid JG. *Anatomy, Mechanics, and Human Motion.* 2nd ed. Englewood Cliffs, NJ: Prentice Hall; 1988.

5. Bartlett R. *Sports Biomechanics: Reducing Injury and Improving Performance.* London: Spon Press; 1999.

PART I

The Whole Body Level

Lesson 2

Describing Motion: Linear Kinematics in One Dimension

© Maria Teijeiro/OJO Images/Getty Images

LEARNING OBJECTIVES

After finishing this lesson, you should be able to:

- Define the following terms: abscissa, absolute value, acceleration, average value, axis, body, cadence, direction, displacement, distance, frame of reference, gait, instantaneous value, kinematics, net value, ordinate, orientation, origin, point, position, relative speed, sense, scalar, slope, speed, step, stride, system, vector, and velocity.
- Explain the difference between speed and velocity.
- Write equations for the following concepts: distance, displacement, speed, velocity, and acceleration.
- Identify speed on a position–time curve.
- Identify velocity on a position–time curve.
- Identify acceleration on a velocity–time curve.
- Explain the difference between instantaneous and average kinematic measures.
- Describe situations in which velocity is more important than acceleration.
- Describe situations in which acceleration is more important than velocity.
- List the determinants of gait velocity.

The first key in unlocking the code to how we move in the world is to be able to describe the motion itself. This is the branch of mechanics called **kinematics**, which is the study of motion without consideration of the causes of that motion.

| Kinematics | The study of motion without considering what is causing the motion |

It involves both spatial and temporal characteristics of motion. In this lesson, we will begin by discussing the simplest case of motion: motion in a straight line going in one direction. The next section will examine motion in a more complex scenario—motion in two directions (but the same dimension). Throughout this lesson, examples from walking and running will be used to highlight these concepts, beginning with a familiar example from sport, then a test often used in physical education, and ending with issues involving gait and the elderly. You cannot adequately explain motion without first being able to describe it in detail, so it is very important that you master these fundamental ideas.

2.1 LINEAR KINEMATICS IN ONE DIRECTION

Section Question

Three men race the 100-meter sprint (**Figure 2.1**).[1] Runner A finishes first with a time of 9.83 seconds, followed by Runner B with a time of 9.93 seconds. It took Runner C 11.12 seconds to complete the race. Why did Runner A win the race? What would Runners B and C have to do to beat Runner A?

The logical answers are, "Runner A ran faster than Runners B and C" and "Runners B and C need to run faster." But what exactly does that mean? And is it very useful?

2.1.1 Preliminary Considerations: Representing Bodies of Interest and Establishing Reference Frames

Before you can analyze any movement, you must first ask two basic questions:

1. What is moving?
2. What is it moving in relation to?

11

Figure 2.1   Three runners are in a race. What advice can you give them to improve their performance?

© Maxisport/ShutterStock, Inc.

You first need a way to represent the thing you are interested in analyzing. In the case of the runners, you are currently interested in the whole persons. Later, you may also want to analyze just the hand or foot, or even an inanimate object like a ball or racquet. In these cases, you are interested in a single entity, which is a **body**, even if it is not a person's entire body. (Conversely, a **system** is more than one body.) How should you represent the thing of interest? To paraphrase Albert Einstein, "Keep things as simple as possible, but not simpler." Use the simplest representation that will answer your question adequately; more complex questions may require more complex representations. In the simplest case, you would represent the body as a **point** (a way to represent something without dimensions). More complicated representations of a body will be presented throughout the course of this book. But for now, representing the runners as points will suffice.

> **Body**   The object of analysis; it could be a whole person, a part of a person, or an inanimate object
>
> **System**   The object of analysis that is made up of two or more bodies
>
> **Point**   A way of representing a body that has no dimensions

Second, movement must always be described in relation to something. That something is called a **frame of reference**. For example, it does not make any sense to say that your school is located 5 miles away, unless you also state that it is 5 miles away *from a particular place* (such as your current location or your house). The finish line of the 100 m sprint is 100 m away *from* the starting line. You would

> **Frame of reference**   The perspective from which the movement is described

not tell somebody to go 5 blocks east, unless you had a mutually agreed-upon direction of east. So to set up your frame of reference you need to establish a location (called the **origin**) and **directions** from that origin.

> **Origin**   The place where the frame of reference begins
>
> **Direction**   A pointing toward something, determined by its orientation and sense

The origin is where your frame of reference begins. The origin can be anywhere, but if the laws described in this book are going to "work" without some complicated mathematics, the origin has to be fixed (that is, not moving). Usually, the origin is placed someplace physically meaningful, although sometimes it is placed where it will make the calculations easier. For example, when analyzing the 100-meter sprint, you would probably fix your frame of reference to Earth and put the origin at the starting line, although you could just as easily put the origin at the finish line. Technically, you could place it anywhere along the race course. It just would not be very meaningful.

After defining an origin, it is necessary to determine directions so that the frame of reference is complete. Directions are specified by **axes**, which pass through the origin and extend indefinitely on both sides of it. Directions have both an **orientation** and a **sense**. The orientation is specified in terms of particular reference lines (such as horizontal, vertical, north, south, east, or west), and the sense is specified by two points on that reference line. The concept of sense is best illustrated by an example. In **Figure 2.2**, a horizontal axis has two points, A and B. Going from point A to point B is one sense, and from B to A is the opposite sense. Because the sense

> **Axis**   A straight line running through the origin specifying a direction from the origin
>
> **Orientation**   A particular reference line
>
> **Sense**   Specified by two points; going from point *B* to *A* has the opposite sense (and opposite direction) of going from point *A* to *B* on the same line

Figure 2.2   The sense tells you that from *A* to *B* is the opposite direction from *B* to *A*.

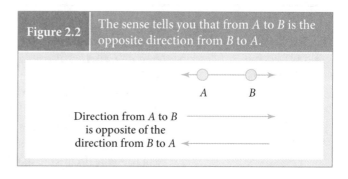

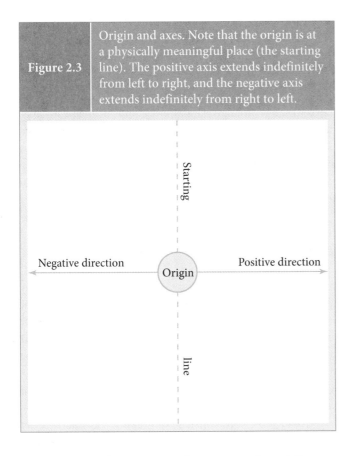

| Figure 2.3 | Origin and axes. Note that the origin is at a physically meaningful place (the starting line). The positive axis extends indefinitely from left to right, and the negative axis extends indefinitely from right to left. |

is opposite, the directions are also opposite. Pictorially, axes are defined by arrows, with the arrow pointing in the direction of the axis.

Now that you understand origins and axes, you can create your reference frame to analyze movement in one dimension. From the origin, one direction along an axis will be positive, whereas the other will be negative. For example, Figure 2.3 specifies an origin in the middle with a single axis running through it. Going from left to right is the positive direction, and going from right to left is the negative direction. Just as in identifying an origin, identifying an axis is somewhat arbitrary and usually defined by either convention or in some meaningful way. You could have just as easily stated that from right to left was positive and from left to right was negative, but because you read from left to right, it is usually more intuitive to make this the positive direction.

Important Point! To establish a frame of reference:
1. Locate an origin that is fixed and meaningful.
2. Define an axis, usually the horizontal and/or vertical along the lines of travel.
3. Specify a positive and a negative direction. The initial direction of travel and "up" is usually designated as positive.

Once you have a reference frame and a way to represent what it is that you wish to examine, you can begin your analysis.

COMPETENCY CHECK

Remember:
1. Define the following terms: body, direction, frame of reference, kinematics, orientation, origin, point, sense, and system.
2. What two things make up a frame of reference?
3. What does the positive or negative sign of an axis tell you?

Understand:
1. Why is it important to have a frame of reference?

Apply:
1. Pick an activity that only requires movement in one direction. Choose a reference frame. Why did you choose the reference frame you did?

2.1.2 Position

The next step is to determine a body's **position** (p; that is, its location) in the frame of reference. Position is the body's physical location in space. Because your current frame of reference consists of a single axis, mathematically the position is the location of that body on the axis. When describing a body's position, you have to state how far away it is located along the axis (magnitude) and on which side of the origin (direction). If body A is located 10 m to the right of the origin on the x-axis, it is $+10$ in the x direction. (Note that usually the "$+$" sign is omitted, and it is understood that if there is no "$-$" sign, it is a positive number.) If body B is located 10 m to the left of the origin on the x-axis, $^{x}p = -10$ m. Note that both are equidistant from the axis: 10 m. So unlike algebra, -10 is not less than $+10$, it just happens to be in the opposite direction.

> **Position** Location in a reference frame

Important Point!
1. A positive or negative sign establishes direction. Unlike algebra, a negative number is not less than a positive number.
2. If the sign is not specified, assume it is positive.

Oftentimes, you would like to look at a "picture" of a body's location, so you would construct a graph. You can

| Figure 2.4 | (a) A snapshot of the three runners is converted (b) to locations on an axis. |

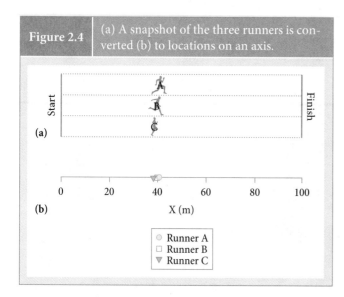

think of a graph as a map: a scaled-down version of a physical reality. Just as you would locate your school on a map, we can identify the position of a body on an axis of a graph. In the case of the runners, a particular snapshot in time can be converted into a graph (Figure 2.4).

2.1.2.1   Changing Position: Displacement and Distance

In biomechanics, you are usually interested in how things are *changing*. Kinesiology is the study of human movement, and movement implies change. Physically, change means that something is somehow different than how it was previously. Quantities that only have a magnitude are known as scalars; quantities that have both a magnitude and a direction are known as vectors.

> **Scalar**   A quantity that only has a magnitude
>
> **Vector**   A quantity that has both a magnitude and a spatial direction

Do not confuse a change in magnitude of a scalar with a spatial direction. For example, temperature is a scalar quantity. You can talk about the temperature going up (increasing) or down (decreasing); temperature can even be a negative number. But temperature does not have a spatial direction.

As mentioned in the preceding section, position is denoted by the symbol p. A prime ($'$) after the p denotes the position of the body at some other point in time (denote by a corresponding prime after the "t" symbol for time) because things change over time. A change in position is known as displacement (Δp). Mathematically, it is the difference in position

> **Displacement**   A change in position

(measured as a length) between two instances in time:

$$\Delta p = p' - p \qquad (2.1)$$

where the symbol Δ, delta, means "change in." So Equation 2.1 simply states the displacement (or change in position), Δp, is equal to the position at data point "1" minus the position at data point "0." The magnitude is simply how large the change is: A large number means a bigger change. Displacement is a measure of length, and so the units are in meters for the metric system. In the imperial system, the units are typically feet, yards, or miles.

Displacement is a vector quantity. This is not to be confused with distance (d), or the actual length of the route the body took to change its position, which is a scalar quantity. In the simplest case of a body moving in a straight line, the two are the same. From the start to finish of the sprint example, both the distance and displacement are 100 m. But if you ran a circuitous route, the distance could be over three times larger than the displacement (Figure 2.5). So if there is a large time difference between when you examine p and when you examine p', you may miss some valuable information.

> **Distance**   How far a body has traveled

If you wanted to construct a graph of how something is changing over time, one axis will no longer be adequate because you are now dealing with two dimensions (one being the direction of movement, the other being time). Your graph would have to have two axes to represent both of these dimensions (because you always need one axis per dimension on a graph). A typical 2-D graph will have a horizontal axis

| Figure 2.5 | If moving in one direction in a straight line, distance and displacement are equal. If the path of travel is not a straight line, the two values will be different. |

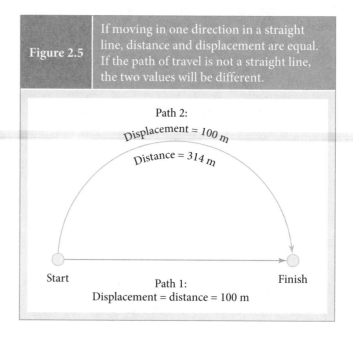

> **Abscissa**   The horizontal axis on a two-dimensional graph
>
> **Ordinate**   The vertical axis on a two-dimensional graph

(the **abscissa**) and a vertical axis (the **ordinate**). The abscissa is also referred to as the *x*-axis, and the ordinate is also known as the *y*-axis. By convention, time is always placed along the *x*-axis, or abscissa, and the variable that is changing with time along the *y*-axis, or ordinate. Do not be confused by the terminology: you could annotate the horizontal direction in physical space as the "*x*-" axis, but when you may plot it on the "*y*-" axis of a position versus time curve, as was done for the runners in **Figure 2.6**.

> **Important Point!**   Even if you are graphing a movement that is horizontal in space, it is placed on the vertical axis when plotted with respect to time.

Two things are worth noting on the graph in Figure 2.6. First, the change in position is not *directly* noted on the graph. The bigger the difference between the two positions would indicate a larger displacement (representing a greater change in physical position). Second, this change represents displacement and not distance—unless they happen to be equal. Essentially, the graph represents the positions of the runners at two snapshots in time: the beginning and end of the race. There is no information about the actual route (and thus distance) of each runner.

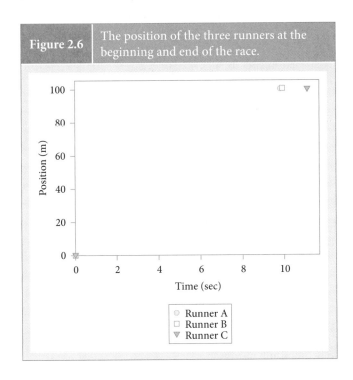

Figure 2.6	The position of the three runners at the beginning and end of the race.

2.1.3   Rates of Change

Simply knowing that something is changing is oftentimes not enough. Intuitively, you probably understand that something is different if two bodies change the same amount, but over a different time period. In the case of the sprinters, you know that each one covered a distance of 100 meters between the start and the end of the race but you are no closer to understanding why Runner A won the race. What is missing is how that change occurred *with respect to time*. This is called a **rate** (**Box 2.1**), and it gives you an indication of how some variable (such as position) is increasing or decreasing with time.

> **Rate**   How quickly a value is increasing or decreasing with time

2.1.3.1   Speed and Velocity

Speed and velocity are often used interchangeably in everyday language, and in fact, they can be used interchangeably if you are talking about bodies only moving in one direction. You

Box 2.1	Essential Math: Ratios and Rates

A **ratio** is simply one number divided by another number:

$$\text{ratio} = \frac{\text{one quantity}}{\text{another quantity}}$$

A **rate** is a ratio between the change in one quantity and the change in time:

$$\text{rate} = \frac{\Delta\text{one quantity}}{\Delta\text{time}}$$

The delta symbol (Δ) is shorthand for "change in." Think of the dividing line as "per," so we can think of a rate as a change in one quantity (position, velocity, force, work) per a change in a unit of time (seconds, minutes, hours). Rates are going to be very important in biomechanics. From algebra, you should be able to recognize that the rate will be larger if the change in the quantity is increased and/or the change in time is decreased.

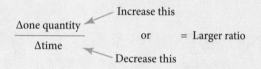

$$\frac{\Delta\text{one quantity}}{\Delta\text{time}} \quad \text{or} \quad = \text{Larger ratio}$$

Increase this — Decrease this

probably already have a notion about speed. So that would be a good place to start, and then you will learn about velocity.

Speed is how fast something is going. If you cover a greater distance in the same

Speed   How fast a body is moving

amount of time, or the same distance in a smaller amount of time, you have a greater speed. You are probably familiar with the concept every time you get into a car: the speedometer, or "speed meter," measures the speed of the car. What values does the car's speedometer give you? Miles per hour (or kilometers per hour). That gives you a clue that speed is a rate at which something is changing:

$$\text{speed} = \frac{\text{miles (kilometers)}}{\text{hour}} \tag{2.2}$$

But that is a very specific case. To make it useful in a greater number of situations, you need a more general form. Miles (kilometers) is a measure of distance covered, how far a thing traveled. Hour is a measure of how much time has elapsed (60 minutes). So, in the general form:

$$\text{speed} = \frac{\text{distance}}{\text{change in time}} = \frac{d}{\Delta t} \tag{2.3}$$

Speed is the rate of change of distance.

Speed is a scalar quantity. Suppose you were to create a reference frame where north on the freeway is positive and south is negative. Regardless if you were going north or south, your car's speedometer would only give you a magnitude (55 mph), not a direction (positive or negative).

Speed is a scalar quantity that is the time rate of change of the distance, another scalar quantity. In the last section, you learned that displacement was the vector change in position. **Velocity** is the vector quantity that is the time rate of change in position. If you substitute displacement for distance in Equation 2.3, you get

Velocity   How fast something is moving in a particular direction

$$\text{velocity}(v) = \frac{\Delta\text{position}}{\Delta\text{time}} = \frac{\Delta p}{\Delta t} = \frac{p' - p}{t' - t} \tag{2.4}$$

And if speed is how fast something is moving, velocity (being a vector) is how fast something is moving in a particular direction. The units are m/sec in the metric system and ft/sec in the imperial system.

In this relatively simple example of things moving in only one direction, displacement and distance are always the same, and thus velocity and speed will always have the same value. Biomechanicians are usually more concerned with vector quantities like displacement and velocity, and not distance and speed. As a scalar, speed is the magnitude of the velocity vector. If you are just comparing magnitudes (directions are not changing), it is acceptable to say "speed." Otherwise, you should get into the habit of using displacement and velocity when appropriate.

Returning to the 100-meter sprint example, you know the displacement (and distance traveled) was 100 meters (the length of the race). The times were recorded: 9.83, 9.93, and 11.12 seconds. Now calculate the velocity of each runner:

$$\text{Runner A} = \frac{\Delta p}{\Delta t} = \frac{100 \text{ meters}}{9.83 \text{ seconds}} = 10.17 \text{ m/sec}$$

$$\text{Runner B} = \frac{\Delta p}{\Delta t} = \frac{100 \text{ meters}}{9.93 \text{ seconds}} = 10.07 \text{ m/sec}$$

$$\text{Runner C} = \frac{\Delta p}{\Delta t} = \frac{100 \text{ meters}}{11.12 \text{ seconds}} = 8.99 \text{ m/sec}$$

It is no surprise that Runner A ran faster than Runner B, who ran faster than Runner C. This was the obvious answer from the beginning of the lesson. It does illustrate an important point about ratios, though: If you want to increase velocity, and the numerator (top number, displacement) is fixed, you have to decrease the denominator (bottom number, time).

Figure 2.7	Runners' position versus time. Note that velocity is the slope of the distance–time curve.

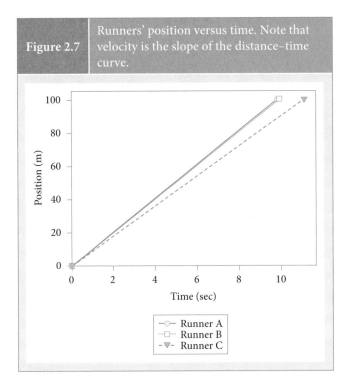

Box 2.2	Essential Math: Slopes

If you have ever walked up a hill, then you have an intuitive appreciation for what a slope is. In the case of the hill, the slope is the ratio between the change in vertical distance and the change in horizontal distance. If you plot the horizontal distance along the horizontal axis and the vertical distance along the vertical axis, you get a visual of what the slope looks like. Technically, the horizontal axis is known as the **abscissa**, and the vertical axes is known as the **ordinate**. It is helpful to think of them in those terms because you will not always be plotting the horizontal distance on the horizontal axis. Many times, you will want to know how something is changing over time, and by convention you always plot time on the abscissa (horizontal) and the variable of interest on the ordinate (vertical). In the race example, you are interested in how the (horizontal) distance is changing with time. In this case, time will be on the abscissa, and distance is on the ordinate.

Mathematically, the slope is

$$slope = \frac{\Delta ordinate}{\Delta abscissa}$$

Where, once again, the delta symbol (Δ) means "change in."

If the abscissa is kept constant and the ordinate is increased, there will be a larger number for the slope. This equates to a steeper slope. Walking up a hill with a steeper slope means that the change in vertical distance is increasing more than the change in horizontal distance. Walking on flat ground means that the vertical distance is not changing with horizontal distance—the slope is zero. Walking downhill, the vertical distance is decreasing while the horizontal distance is increasing. This is a negative slope. Positive, negative, and zero slopes mean the same things regardless of what you put on the abscissa or ordinate. You will look at slopes a lot in this book, so always keep in mind what they mean by visualizing the hill.

Recall that the equations give an output for a series of inputs. If the output was the time to finish the race, Equation 2.4 could be rearranged to put the time on the left-hand side and the other variables on the right:

$$\Delta t = \frac{\Delta p}{v}$$

(2.5)

Thus far, the concept of velocity was discussed physically and mathematically. A graph is a picture, and as the old saying goes, a picture is worth a thousand words. Graphically, you would "connect the dots" on Figure 2.6. The results are presented in **Figure 2.7**.

Notice that the slope of the position–time curve gives you the velocity: The steeper the slope, the greater the velocity (and in this case, speed; see Box 2.2). If you look at the slopes of the three runners' curves, you will note that Runner A has a steeper slope (and greater velocity) than Runner B, who has a steeper slope (and greater velocity) than Runner C. If a picture is worth a thousand words, then why is that not telling you much (certainly nothing you did not already know at this point)? It is because the resolution of your picture was too low to see all the details. For a higher resolution, you need more pixels. In this case, pixels are data points.

Slope The incline of a line on a graph from the horizontal axis

Terrain with positive, negative, and zero slopes.
© Stephan Scherhag/ShutterStock, Inc.

Average versus Instantaneous Velocity

All three representations (physical, mathematical, graphical) presented thus far tell you the same thing: Runner A covered the same distance (had a larger displacement) in a smaller amount of time (another way of saying had a larger displacement in the same amount of time), his velocity was greater than Runners B and C, and it had the steepest slope on the position–time curve.

But that does not provide very much information, and it certainly does not tell Runner B or C what they would have to do to beat Runner A in the future. Part of the problem may be that the values we calculated were over the duration of the entire race, or the average velocity. Average velocity assumes the velocity did not change throughout the race, but was this the case? If it took you 2 hours to get to grandma's house 100 miles

> **Average**   A number representing the value of a quantity if that quantity did not change (was constant) throughout the period of interest

away, then your average speed was 50 miles per hour. But you were not driving 50 miles per hour the whole time: you sped up, slowed down, stopped at a traffic light or two, and so on. Did the same thing happen to the runners?

To obtain more detailed information, you may wish to chop the race into small bits. If you could examine it so that the time changes were infinitesimally small (that is $t' - t$ is as close to zero as you can get), you could look at the instantaneous velocity—the veloc-

> **Instantaneous**   The value of a quantity at a particular moment in time

ity at a particular moment in time. (Incidentally, your car's speedometer measures instantaneous speed.) That would give you a clearer picture of not only the *outcome* of the race, but also what happened *during* the race—which could explain why the outcome was the way it was.

If timers were placed every 10 meters, then you would know how long it took to run each of the 10 10-meter segments. The data are recorded in Table 2.1. The table contains a lot of numbers, and it might be hard to figure out what is going on. A visual representation of this data might be helpful. Plotting the position of each runner as a function of time would look like Figure 2.8.

Physically, velocity is how fast an object is moving in a particular direction. Mathematically, it is the ratio of displacement (change in position) and change in time. Graphically, the velocity is the slope of position plotted with respect to time: the steeper the slope, the greater the velocity and the faster the speed (see Box 2.3 for more details). From

Table 2.1	The Time It Took Each Runner to Complete Each of the 10-Meter Segments of the 100-Meter Race		
	Time		
Distance	**Runner A**	**Runner B**	**Runner C**
Start	0	0	0
10	1.84	1.94	1.92
20	2.86	2.96	3.1
30	3.8	3.91	4.16
40	4.67	4.78	5.06
50	5.53	5.64	6.08
60	6.38	6.5	7.00
70	7.23	7.36	8.02
80	8.1	8.22	9.04
90	8.96	9.07	10.02
100	9.83	9.93	11.12

Figure 2.8, it appears as though the runners were very similar during the first two seconds (the slopes are practically on top of one another), but then they start to diverge after that. Calculating the velocity at every interval and plotting it as a function of time may give a better picture, as was done in Figure 2.9.

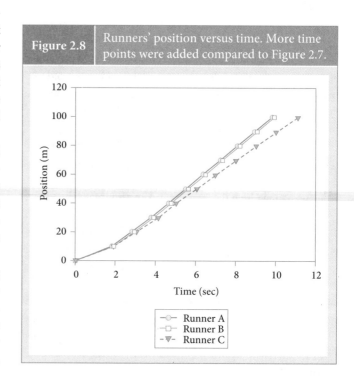

| Figure 2.8 | Runners' position versus time. More time points were added compared to Figure 2.7. |

Figure 2.9	Runners' velocity versus time. Note the Runner B actually had a slightly higher peak velocity than did Runner A, even though he lost the race.

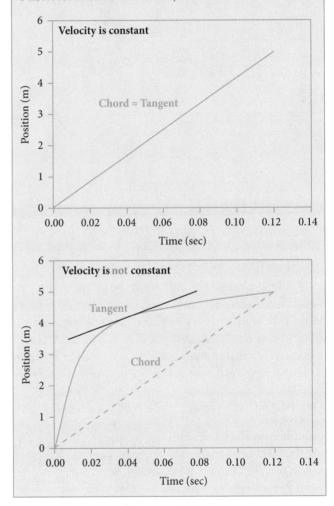

Box 2.3	Essential Math: Tangents and Chords

The slope of a curve compares the rates of change of two variables. In biomechanics, we are often interested in comparing the rate of change of one variable to the rate of change of time. This is often shortened to "the time rate of change of [insert whatever variable here], or even shorter to "the rate of change of [whatever variable]." For example, velocity, which is the time rate of change of position, is the slope of the position–time curve. If you are interested in the *average* velocity, you would examine the slope from the position at the first time period of interest to the position at the last time period of interest. If the velocity is constant, the position-time curve would actually be a straight line, and the slope would be identical everywhere, at every instant, on the "curve" (because it is constant). That rarely happens in human movement; usually the velocity is changing throughout the period of interest. In that case, the position–time curve truly is a curve. The *average* is still computed the same way: the ratio of the change in position to the change in time. Graphically, its slope is the **chord**, or straight line drawn from the start to the finish of the period of interest. If we want to know the *instantaneous* velocity, we would need to look at the slope of the **tangent** to the curve at that instant. A tangent is a straight line just touching the curve at a single point—the point being the instant you are interested in. If the velocity is constant, the tangent and chord will have

identical slopes. If the velocity is changing, at some points, the slope of the tangent will be larger than the slope of the chord (the instantaneous velocity is higher than average), and at other times the slope of the tangent will be smaller than the tangent of the chord (the instantaneous velocity is smaller than average). But at least at one point, the slopes of the tangent and chord will be the same (you cannot jump from 4 m/sec to 6 m/sec without going 5 m/sec for at least a brief instant).

2.1.3.2   Acceleration

Examining Figure 2.9, notice that Runner C did not attain the top velocity that the other runners did and could not hold his top velocity for very long. But there is a peculiar thing about the performance of Runner B: his top velocity actually exceeded that of Runner A! So why did he lose the race? This is an interesting question, and not readily apparent, unless the graph is examined very closely. It will be easier to see if Runner C is eliminated from the picture and just the first part of the race is examined (**Figure 2.10**).

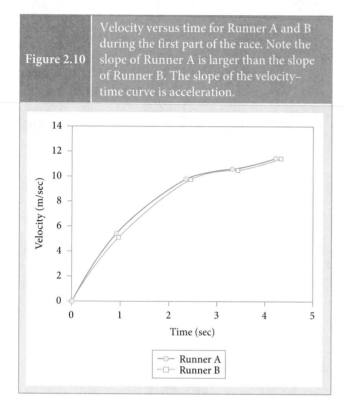

Figure 2.10 Velocity versus time for Runner A and B during the first part of the race. Note the slope of Runner A is larger than the slope of Runner B. The slope of the velocity–time curve is acceleration.

quickly velocity changes (either increases or decreases), or the rate of the change in velocity. A measure of a car's performance is how quickly it can go from "0 to 60," zero miles per hour to 60 miles per hour (a change of 60 mph), when you put your foot on the "accelerator." Mathematically, it is the ratio of the change in velocity per the change in time:

$$\text{acceleration} = \frac{\text{change in velocity}}{\text{change in time}} = \frac{\Delta v}{\Delta t}$$

$$= \frac{v' - v}{t' - t} = \frac{\text{meters per second}}{\text{second}}$$

(2.6)

And so it has units of miles per hour per second, or meters per second per second (m/sec^2, pronounced "meters per second squared"). So far, the discussion has been limited to the objects of interest (the runners) moving in one, single direction—which was identified as positive. In this circumstance, and this circumstance *only*, the sign (positive or negative) tells another useful piece of information: if the person is speeding up, the acceleration is positive. Conversely, if the person is slowing down, the acceleration is negative.

Important Point!   A positive acceleration means a body is speeding up *only if* there is a single direction of travel.

Looking closely, you will notice two things. First, although the two runners have reached the identical velocity during this time period, it took Runner B a little more time to get up to that velocity. Second, the **slope** of Runner A's curve is steeper than that of Runner B's. In other words, the **ratio** of the change in velocity to the change in time is different between the two runners.

Acceleration   How rapidly something is changing velocity (speeding up or slowing down)

This quality, the ratio of the change in velocity to the change in time, is **acceleration**. Physically, it is how

Just as there is a difference between the average and instantaneous velocity, there is a difference between the average and instantaneous acceleration. Compare the first four points on Figure 2.10 (the actual data are presented in Table 2.2). Calculating the average acceleration over this period, you will notice that there is a slight (2.6%) difference in acceleration between Runners A and B. Yet if you shrink the time intervals, distinct differences in the acceleration patterns emerge. Coming out of the blocks, Runner A had an 11.17% greater acceleration than Runner B, and this was crucial to

Table 2.2													
Data for the First Four Points of Runner A and Runner B													
	Runner A					**Runner B**				*a*		**Average *a***	
	t	*v*	Δt	Δv	*a*	*t*	*v*	Δt	Δv	*a*	% Difference	Runner A	Runner B	% Difference
0	0.00	0.00	0.00	0.00	0.00	0.00	0.00	0.00	0.00	0.00	0.00	0.00	0.00	0.00
1	0.92	5.43	0.92	5.43	5.90	0.97	5.15	0.97	5.15	5.31	11.17	5.90	5.31	11.17
2	2.35	9.80	1.43	4.37	3.06	2.45	9.80	1.48	4.65	3.14	−2.74	4.17	4.00	4.26
3	3.33	10.64	0.98	0.84	0.86	3.43	10.53	0.98	0.73	0.74	15.07	3.20	3.07	4.08
4	4.23	11.49	0.90	0.85	0.94	4.34	11.49	0.91	0.96	1.05	−10.47	2.72	2.65	2.60

his success. This important piece of information would have been lost had the intervals been too large. In fact, had you calculated the average acceleration of the two runners over the entire race, you would have found no difference between them! This finding again highlights the importance of a high number of data points: crucial information can be lost if the intervals are too large. Large numbers of data can be confusing in tabular form—just look how confusing Table 2.2 can be with only four points of data. Graphs are a great aid for this type of analysis.

Graphically, the acceleration is the slope of the velocity–time curve. Inspecting Figure 2.9 again, in several places along the curve Runner C appears to be slowing down. This can be verified by graphing the acceleration as a function of time (Figure 2.11).

Notice that in three places along the race Runner C "lost" speed, or decelerated. Comparing Runners A and B, we verify that Runner A "out accelerated" Runner B, which is why he won the race, even though Runner B had a greater instantaneous velocity. Runner A got too far ahead, and Runner B simply did not have time to catch up.

Average velocity will tell you who won the race (and average speed will tell you how long it will take you to drive to grandma's house). But it will not tell you **why** someone won the race. And you cannot use it as an excuse to get out of a speeding ticket ("But officer, my average speed was only 40 miles per hour!"). To figure out why someone won a race, you need to know the following: the top speed (instantaneous speed), the time it takes them to get to the top speed (acceleration), the duration they hold their top speed, and the difference between top speed and final speed.[2]

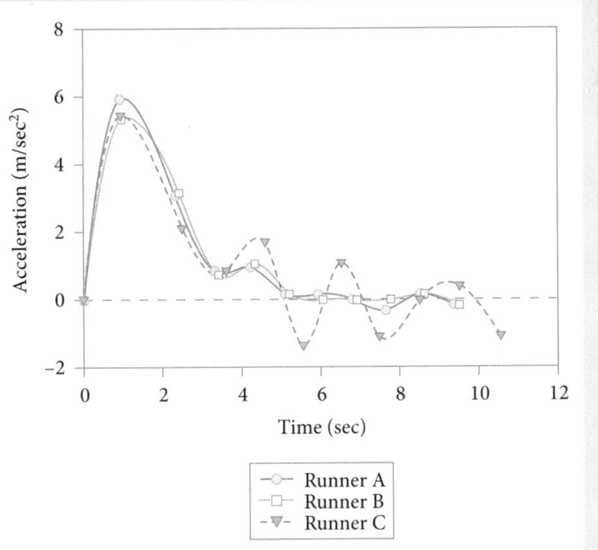

You are now armed with information that can assist Runners B and C. Runner B needs to work on his acceleration. Runner C needs to work on top speed and speed endurance.

2.1.3.3   Absolute versus Relative Velocity

So far, we have been discussing absolute motion, that is, motion of each runner relative to the (fixed) Earth. This is a very useful way to describe motion and is used often, but there are times when it may be useful to examine motion of one body moving relative to another (moving) body. Knowing how fast you are driving in your car is important to ensure that you can safely navigate the road and any fixed obstacles that may be on it (and avoid speeding tickets). But it is not enough to prevent an accident; you also need to know how fast you are going in relation to other cars. When discussing the velocity of one body moving in relation to another, it is called **relative velocity**, and the formula is

> **Relative Velocity** How fast one body is moving in relation to another body

$$v_{B/A} = v_B - v_A \qquad (2.7)$$

> **Important Point!**   Equation 2.7 is valid if both the velocity of A and the velocity of B were calculated in the same frame of reference.

Which is read as, "The velocity of B relative to A is equal to the velocity of B (relative to your fixed reference frame) minus the velocity of A (relative to your fixed reference frame)." This may sound a bit confusing, but think about it for a second. If the velocity of A is zero, then the velocity of B relative to A is simply the velocity of B in the fixed reference frame. If the velocity of B relative to A is zero, then the velocity of B and the velocity of A are equal.

Returning to the case of the two runners, A and B, Figure 2.12 is a graph of the velocity of B relative to A (using Equation 2.7). The negatives mean that A was running faster than B, whereas the positives mean that B was running faster than A. Figure 2.12 shows you what you already determined looking at the velocities and accelerations: A was running faster than B in the beginning part of the race, but B was running faster than A in the second part of the race. A got too far out in front of B, and B could not catch him. This again highlights the importance of acceleration when the

Figure 2.11 Runners' acceleration versus time. In this case, the negative acceleration means the runner is slowing down. Can you identify where Runner A and Runner C slow down?

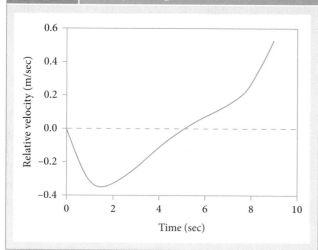

Figure 2.12 The velocity of Runner B relative to Runner A. Note that at the beginning of the race, Runner A is faster than Runner B (negative relative velocity), but Runner B is faster than Runner A (positive relative velocity) during the second half of the race.

movement times are short. Although it may appear that you did not gain any new information by examining the relative motion between the two runners, you will see how it is important in later lessons.

Section Question Answer

Several critical elements are involved in the race: peak speed, acceleration, length of time at peak speed, and the difference between peak speed and final speed. Runner A won the race because he had the best combination of these elements. Runner B needed to improve his acceleration, and Runner C needed to improve all but his acceleration. You would only know these things by examining the instantaneous velocities and accelerations of the entire race.

COMPETENCY CHECK

Remember:

1. Define the following: acceleration, average, instantaneous, rate, relative velocity, speed, slope, and velocity.

Understand:

1. Based on Equation 2.3, the time to complete a movement will decrease if the distance is ___decreased___ or the ___velocity___ is increased.

2. Calculate the velocity for the following:

p'	p''	Δp	Δt	v
0 m	10 m	10	1 sec	10
10 m	15 m	5	0.5 sec	10
0 m	15 m	15	.01 sec	1500
15 m	100 m	85	5 sec	17

3. Calculate the acceleration for the following:

v'	v''	Δv	Δt	a
1 m/sec	10 m/sec	9	1 sec	9 m/sec²
15 m/sec	10 m/sec	~5	0.5 sec	~10
0 m/sec	7 m/sec	7	.01 sec	700
150 m/sec	100 m/sec	~50	5 sec	~10

In each case, is the body speeding up or slowing down?

Apply:

1. Give an example of a movement or activity where speed or velocity is unimportant.
2. Give an example of a movement or activity where speed or velocity is important.
3. Describe a situation where velocity may be important but acceleration would be unimportant.

2.2   LINEAR KINEMATICS IN TWO DIRECTIONS

Section Question

The Shuttle Run (Figure 2.13) is one of the tests used to assess physical fitness by the President's Council on Physical Fitness and Sports, as well as other organizations. Two blocks are placed on a line 30 feet away from a starting line. On the command "go," the student runs to the first block, retrieves it, and returns to the starting line. After placing the block down, the student retrieves the second block in a similar manner. The score is the time it takes to complete the course. What does the Shuttle Run measure? Can we analyze it the same way we analyzed the 100-meter sprint?

At first glance, the 100-meter sprint and the Shuttle Run appear similar in that they both involve running as fast as you can, even though the Shuttle Run is much shorter (about 36.6 meters). The biggest difference, though, is that the sprint requires the runner to reach top speed as fast as he or she can and then maintain that speed through the finish line, where the Shuttle Run involves speeding up, slowing down, stopping, and changing direction. In the previous section, you

| Figure 2.13 | The Shuttle Run as described by the President's Council on Physical Fitness and Sports. |

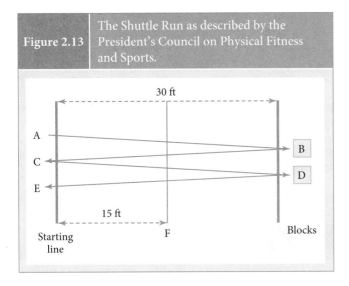

only dealt with changing magnitudes. You need to increase the sophistication of your analysis to deal with changing magnitudes *and* directions.

2.2.1 Displacement (Δp) and Distance (d)

Define your frame of reference with the starting line as the origin, and the direction from the starting line to the blocks as "positive." (Note that the lines should be superimposed on top of each other in Figure 2.13, but for purposes of the figure they are spread a little bit apart). If you wish to describe the length from the origin to where you pick up the first block (point "B"), you would say that is 30 feet. If you wish to describe the length from the origin to where you drop off the first block (point "C"), you see that you are back at the starting point (the origin), so the length, the displacement, is zero feet. But the actual distance you ran is 60 feet. The Shuttle Run illustrates that if you change directions, displacement and distance will not be the same.

> **Important Point!**   If a body starts and ends at the same spot, the displacement is always "zero," even though the round-trip distance could be great!

As a vector, displacement has both a magnitude and a direction. In one dimension (but two directions), you can extract three key pieces of information:

1. The axis
2. The magnitude of the change in position
3. The direction of the change in position

As long as you are working in one dimension, the axis will be self-evident because there is only one axis. As you start

working in 2-D and 3-D, this will become more important. The magnitude is simply how large the change is: A large number means a bigger change. Finally, there will either be a plus or minus sign in front of the number. (Note: If the direction is positive, the "+" sign is usually dropped and implied. Only the "−" sign is specified).

For example, let us say that the person is 15 feet away from the origin at time "1" (crossing line F), and arrives at point B at time "2." The displacement would be:

$$\Delta p = p'' - p'$$
$$\Delta p = 30\ \text{ft} - 15\ \text{ft} = 15\ \text{ft}$$

This means that the body displaced 15 feet in the positive direction. Now what happens if the positions are reversed (the person leaves Point B at time "2" and crosses Line F at time "3")?

$$\Delta p = p''' - p''$$
$$\Delta p = 15\ \text{ft} - 30\ \text{ft} = -15\ \text{ft}$$

This means the body displaced 15 feet in the negative direction. It is important to realize that, unlike in algebra, "−15" is not less than "+15." The magnitudes of the displacements are identical, 15. Case 2 just happens to be in the opposite direction of Case 1, which is what the negative sign implies. Now what happens if we have two displacements, with the runner going from A to F and then from F to B?

From A to F: $\Delta p = p'' - p'$
$$\Delta p = 15\ \text{ft} - 0\ \text{ft} = 15\ \text{ft}$$
From F to B: $\Delta p = p''' - p''$
$$\Delta p = 30\ \text{ft} - 15\ \text{ft} = 15\ \text{ft}$$
From A to B: $\Delta p = p''' - p'$
$$\Delta p = 30\ \text{ft} - 0\ \text{ft} = 30\ \text{ft}$$

> **Important Point!**   The plus or minus sign on a vector only indicates direction. It does not mean that positive values are larger than negative values.

Note that the displacement from t' to t''' is equal to the displacement from t' to t'' plus the displacement from t'' to t''', even though it is unnecessary to perform the first two calculations to get to the third. This is certainly not a problem *as long as the direction does not change*. But what if it does?

In this example, at t' the runner is 15 ft from the origin, at t'' the distance is 30 ft from the origin, and at t''' her position returns to where it was at t', 15 ft from the origin.

From A to F: $\Delta p = p'' - p'$

$$\Delta p = 30 \text{ ft} - 15 \text{ ft} = 15 \text{ ft}$$

From F to B: $\Delta p = p''' - p''$

$$\Delta p = 15 \text{ ft} - 30 \text{ ft} = -15 \text{ ft}$$

From A to B: $\Delta p = p''' - p'$

$$\Delta p = 15 \text{ ft} - 15 \text{ ft} = 0 \text{ ft}$$

Table 2.3	Position, Distance, and Time During the Shuttle Run			
Data point	Location	p(ft)	d(ft)	t(s)
0	A	0	0	0
1	B	30	30	2.5
2	C	0	60	5
3	D	30	90	7.5
4	E	0	120	10

What happened? The body originally displaced 15 ft in the positive direction, then displaced 15 ft in the negative direction, for a **net** displacement of 0 ft. This example illustrates two important points. First, if you look at the displacements over a large time interval (in this case, t' to t'''), then you may miss some important information. In this example, the body actually displaced (twice!) but if you only calculated the displacement from t' to t''', then you would come to the conclusion that the body did not move at all. Second, it illustrates the difference between displacement and distance. Distance is a scalar that tells us how much the body moved. A scalar has a magnitude, but no direction. Because direction is unimportant, you take the sum of the **absolute** values of the displacements. In the preceding example, the total displacement may have been 0 ft, but the distance the body traveled was 30 ft, the same as in the first example.

> **Net**   The total value after summing all the individual values

> **Absolute**   Magnitude of the value, regardless of the sign

If the body starts and ends in the same place, the displacement will always be zero, regardless of the distance it traveled. If a person runs around a quarter-mile track, the displacement would be zero, and the distance would be a quarter mile. These differences should be appreciated as you begin to learn more complex variables.

2.2.2   Velocity (v)

Now you can explore how changing direction changes the velocity vector. The velocity between several different points during the shuttle run was calculated from Table 2.3. The results are provided in Table 2.4. Note that when you go from Point A to Point B, the speed and velocity are identical: 12 ft/sec. What happens when you go from Point B to Point C? The magnitudes are the same (12 ft/sec), but the velocity has a negative sign in front of it. That negative sign indicates the direction, which in this case is negative (going from right to left across the page). Just like with displacement, "-12 ft/sec" is not less than "12 ft/sec," it is just in the opposite direction. More precisely, you should write this as "$^xv = -12$ ft/sec."

Table 2.4	Distance, Displacement, Speed, and Velocity During the Shuttle Run				
	Δt(s)	d(ft)	Δp(ft)	speed (ft/s)	v(ft/s)
From A to B:	2.5	30	30	12	12
From B to C:	2.5	30	-30	12	-12
From A to C:	5	60	0	12	0

> **Important Point!**   The sign of the velocity is always in the direction of the displacement.

As before, you can extract three pieces of information:

1. The axis (x)
2. The magnitude of the velocity (12 ft/sec)
3. The direction of the travel (negative, or right to left)

But what about the velocity between points A and C? What is the zero velocity all about? If the total change in position, or displacement, is zero (in other words, you started and ended in the same place), then the *total change* in velocity is also zero. That is why you need to be careful about the time period over which you average the movement. You get a better appreciation of the velocity throughout the movement by examining the velocity at each point in time: the instantaneous velocity. Examining lots of numbers can be tedious, and you really will not get a feel for what is going on. Graphing the data is a better alternative. The displacement and velocity, as functions of time, are plotted at the top and middle of Figure 2.14.

> **Important Point!**   Important information is lost if you look at the difference between the initial and final over a large period of time. Graphs give a visual representation of what is happening throughout the entire movement.

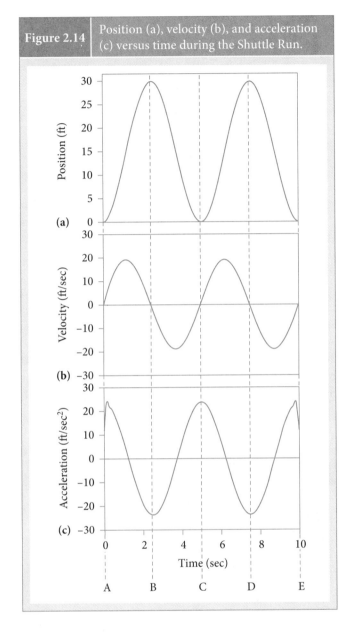

Figure 2.14 Position (a), velocity (b), and acceleration (c) versus time during the Shuttle Run.

the same: 12 ft/s. In fact, if you would calculate the velocity between the other data points (C to D and D to E), then you would come up with the same velocity at each interval, 12 ft/s. But this does not quite match the picture in Figure 2.14. At some points the velocity is zero, and at others the magnitude is much greater that 12 ft/s. If you were to *average* the velocity across all those data points it would be 12 ft/s, but that number represents only a few of the actual data points. In reality, the velocity is constantly changing throughout the run.

Important Point! Zero velocity on a curve indicates that either a body is not moving, or it changed direction.

A change from a positive direction to a negative one (and vice versa) will always require the velocity to be zero during the transition—however brief. Returning to the Shuttle Run example, it should become apparent that the "points" where the velocity is zero are A, B, C, and D. (Point "E" would probably not be zero if we are encouraging the person to "run through" the finish line, but it is set at zero for the purposes of this example.) Consistent with the reference frame, going from point A to point B and point C to D would be the positive directions, whereas going from point B to point C and point D to E would be in the negative directions. In between the points, you are speeding up and slowing down as your speed leaves and approaches zero.

2.2.3 Acceleration (*a*)

The speeding up and slowing down indicates that the velocity is changing. The rate at which velocity changes, or how quickly you are speeding up or slowing down in a particular direction, is acceleration. Acceleration can represent a change in the magnitude of velocity (speed), the direction of the velocity vector, or both.

Important Point! Acceleration represents a change in the magnitude of velocity, the direction of the velocity vector, or both.

Returning to the Shuttle Run example, calculate the acceleration between several different time points. The data are presented in Table 2.5, with the results in Table 2.6. You will notice a couple of peculiar things. First, going from A to F you have a positive acceleration, but going from F to B you have a negative acceleration—even though you are going in a positive direction. The same can be said for going from B to F to C (it is a negative direction, but you have a positive acceleration during the second half). Why is that? Second, going from A to B, B to C, and A to C, the accelerations are all zero. What does that mean?

Graphically, the velocity is the slope of the position–time curve: the steeper the slope, the greater the velocity. Check this by inspecting the top and middle of Figure 2.14. You should also notice that when the slope is negative, the velocity is also negative.

Important Point! A negative velocity does not mean a body is slowing down, only that it is moving with a certain speed in a negative direction.

Looking at the data in Table 2.4, you would be tempted to conclude that the velocity had not changed during the entire run. After all, at every point you examined the velocity was

Table 2.5	Position, Velocity, and Time During the Shuttle Run			
Data point	Location	p(ft)	v(ft/s)	t(s)
0	A	0	0	0
1	F	15	30	1.25
2	B	30	0	2.5
3	F	15	90	3.75
4	C	0	120	5

Table 2.6	Change in Time, Change in Velocity, and Acceleration During the Shuttle Run		
	Δt(s)	Δv(ft/s)	a(ft/s^2)
From A to F	1.25	18.9	15.1
From F to B	1.25	−18.9	−15.1
From A to B	2.5	0	0
From B to F	1.25	−18.9	−15.1
From F to C	1.25	18.9	15.1
From B to C	2.5	0	0
From A to C	5	0	0

Table 2.7	The Sign of the Acceleration Vector as a Function of the Sign of Direction and the Sign of the Velocity	
	Increasing Velocity (+)	Decreasing Velocity (−)
Positive Direction (+)	Positive Acceleration (+)	Negative Acceleration (−)
Negative Direction (−)	Negative Acceleration (−)	Positive Acceleration (+)

Notice that at each leg of the run, the acceleration is zero. That brings up a second principle regarding acceleration: Whenever the velocity starts at zero and ends at zero, the *average* acceleration must be zero. Remember that these values are averaged across the time periods and may represent few (if any) of the actual data points in the sample. The larger the time sample, the less information you will have about the acceleration. If you make the time periods very small, you can get a good approximation of the instantaneous acceleration at each point. Just like with velocity, looking at tabular data is only helpful if you have a small number of data points. With large amounts of data, it is more useful to look at a graph.

Let us tackle the first question. From A to F you are speeding up (increasing your velocity) in a positive direction: Positive change in velocity times positive direction equals positive acceleration. From F to B you are slowing down (decreasing your velocity) in a positive direction: Negative change in velocity times positive direction equals negative acceleration. A similar case can be made for going from B to F and F to C. Unlike velocity, which gives us both the magnitude and direction just by looking at the value (remember $^x v = -12$ ft/sec tells you that you are going 12 ft/sec in the $-x$ direction), you cannot determine the direction just by looking at the sign of the acceleration. (Remember that a negative acceleration means you were slowing down only in one direction.) Slowing down in the positive direction and speeding up in the negative direction both give −15.1 ft/sec^2. This principle is summarized in Table 2.7. To see how acceleration data is applied in the area of motor control, see Box 2.4.

Important Point!   A negative acceleration could indicate a body is speed up or slowing down. To indicate a body is slowing down, some people will say it is "decelerating." Likewise, they will reserve use for "acceleration" to indicate a body is speeding up, even though it is not technically correct.

Box 2.4	Applied Research: Biomechanics in Motor Control

When many people think of acceleration, they think of sporting activities. But in reality, accelerations are required any time you start and stop a movement. The simple act of reaching for a cup requires that you accelerate to start the movement (positive acceleration) and decelerate to stop the movement (negative acceleration). Going from positive to negative acceleration requires that the acceleration cross the zero line. In this investigation, researchers had subjects perform various reaching tasks of increasing complexity at both slow and fast speeds. They were able to show that if a movement was completely "preplanned," then the acceleration curves were smooth with one, single crossing of the zero line. Movements that used "online" feedback to make corrections to the movements had either: (a) multiple crossings of the zero line and/or (b) significant deviations from the smooth curve. Clearly, acceleration is an important variable for the central nervous system to consider when planning movement. Investigators study acceleration patterns to learn how movements are planned and executed in both healthy populations and persons with neurological impairments (stroke, Parkinson's disease, etc.).

Data from: van Donkelaar P, Franks IM. The effects of changing movement velocity and complexity on response preparation evidence from latency kinematic and EMG measures. *Experimental Brain Research* 1991; 83(3):618-632.

Important Point!   Acceleration will be zero whenever:

1. The velocity is zero
2. The velocity is changing from positive to negative
3. The velocity is not changing

Acceleration is the ratio of the change in velocity to the change in time. Graphically, ratios are represented by the slope of one value graphed as a function of the other. The velocity–time and acceleration–time curves of the Shuttle Run are represented in the middle and bottom of Figure 2.14, respectively. Notice that the acceleration is positive when the slope of the velocity–time curve is positive and negative when the slope is negative, as it should be (one leg of the run is displayed in Figure 2.15). You will also notice that the acceleration is zero every time the velocity reaches a peak, either positive or negative. The acceleration will be momentarily zero whenever you change from speeding up to slowing down (or vice versa).

Figure 2.15	Comparing velocity and acceleration profiles. When the velocity slope is positive, the acceleration is positive. When the velocity slope is negative, the acceleration is negative. At the peak velocity, the acceleration is zero. When the velocity starts and ends at the same point, there is no change in velocity, and the areas under the positive and negative acceleration curves must cancel.

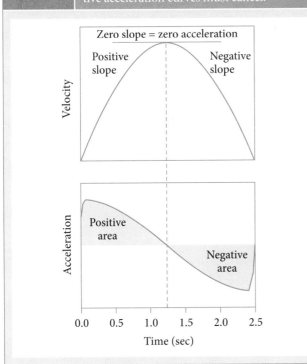

Whenever the change in acceleration is zero, the average acceleration must be zero. That does not mean that the acceleration is constant, or that the average positive acceleration is the same as the average negative acceleration. What it means is that the areas under the positive and negative acceleration curves need to cancel each other out. By visually inspecting the symmetry of the curve, it is apparent that this is the case. What happens when movements are not so symmetric?

Important Point!   The area under the acceleration–time curve represents a change in velocity. If the change in velocity is zero, the areas under the positive and negative acceleration curves must be the same.

Section Question Answer

The obvious differences between the 100 m sprint and the Shuttle Run are that the Shuttle Run has several changes in direction, and you do not run for more than ~9 m before you have to change direction. Because of the changes in directions, displacement and distance are not equivalent like they are in the 100 m sprint. This means that you must appreciate both scalars and vectors. Practically speaking, it takes about 50 to 60 m to reach peak speed in a sprint. Therefore, peak velocity is not a critical element in the Shuttle Run. Rather, it is a test of your capacity to accelerate and decelerate.

2.2.3.1   Asymmetric Acceleration Profiles

Section Question

In keeping with his new motto, "If you want to be fast, you have to train fast," Coach I. M. Strong began having his athletes perform all their squat exercises as rapidly as possible. Noting that they could not squat fast with a heavy load, Coach Strong decreased the weight so he could have them "moving really fast." Is this a good idea?

Take a closer look at the squat exercise by examining the path of the barbell (Figure 2.16). During a real squat, the

Figure 2.16	Squat.

barbell does not move perfectly linear, but for the purposes of this example, assume that it does. The person will start by standing erect, which will be called the top (T) position. She will then flex her hips, knees, and ankles until the bar reaches some predetermined depth (the bottom position, B). After a momentary pause, she will extend her hips, knees, and ankles to return back to the top position. There is also an intermediate point (I) halfway between the top and bottom positions.

The first thing you have to do is establish your frame of reference. In the previous running examples, the direction of movement was horizontal along the surface of Earth, and it was called the x-axis. The path of the barbell is in the vertical direction, so it is called the y-axis, and you should make it your reference axis. Locate the origin at the center of the barbell in the top position (1.37 m above the ground), and indicate "up" is the positive direction. Although you could have put the origin on the ground, it would make the math messier. Now you are ready to conduct your analysis.

First, see what happens when the exercise is performed at a relatively even cadence: 1 second on the way down, a 1-second pause at the bottom, and 1 second on the way up. The data for position, distance, and time are presented in Table 2.8, whereas the average velocity and acceleration data are presented in Table 2.9. Notice on the table that for velocity (v) and acceleration (a), a leading superscript "y" was added. It is a good habit to start including a leading superscript to denote the axis so that you know which direction you are talking about. (It will become even more important in the next section.) Compare Tables 2.6 and 2.9 and look for the following patterns: when you start and end in the same position, the displacement is zero and the velocity (and acceleration) is zero; and for any interval where the starting velocity is zero and the ending velocity is zero, the change in acceleration is also zero. Graphing the data (Figure 2.17), it is easy to see because the curves on the way down and on the way up are symmetric (maybe unnaturally so) that the areas

Table 2.9	Change in Time, Change in Position, Velocity, and Acceleration During the Squat			
	Δt(s)	$\Delta^y p$(m)	$^y v$(m/s)	$^y a$(m/s^2)
From T to I	0.5	−.225	.45	−.9
From I to B	0.5	−.225	−.45	.9
From T to B	1.0	−.45	.45	0
At the Bottom	1.0	0	0	0
From B to I	0.5	.225	.45	.9
From I to T	0.5	.225	−.45	−.9
From B to T	1.0	.45	.45	0
Complete Repetition	3.0	0	0	0

Figure 2.17	Position (top), velocity (middle), and acceleration (bottom) versus time during the squat. T = top, I = intermediate, B = bottom position.

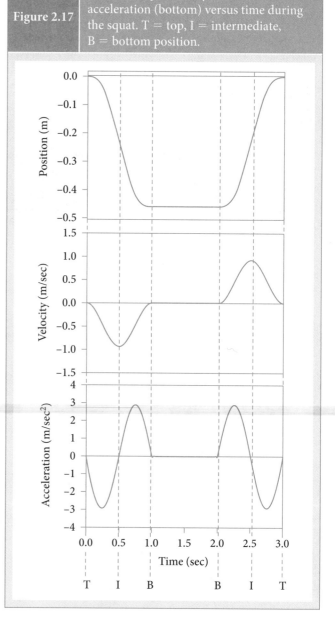

Table 2.8	Position, Distance, and Time During the Squat			
Data Point	Location	$^y p$(m)	d(m)	t(s)
0	Top	0	0	0
1	Intermediate	−.225	.225	0.5
2	Bottom	−.45	.45	1.0
3	Bottom	−.45	.45	2.0
4	Intermediate	−.225	.675	2.5
3	Top	0	.90	3.0

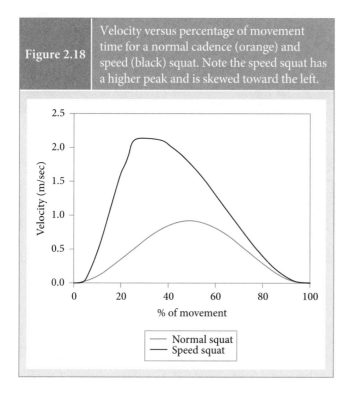

Figure 2.18 Velocity versus percentage of movement time for a normal cadence (orange) and speed (black) squat. Note the speed squat has a higher peak and is skewed toward the left.

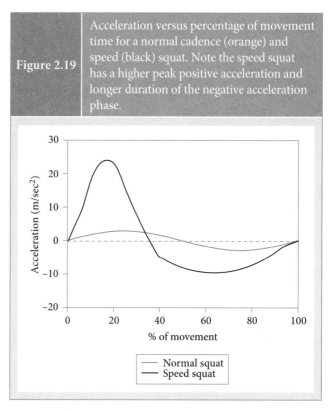

Figure 2.19 Acceleration versus percentage of movement time for a normal cadence (orange) and speed (black) squat. Note the speed squat has a higher peak positive acceleration and longer duration of the negative acceleration phase.

under the curves cancel. What happens when you examine Coach Strong's practice?

Let us limit the discussion to the "up" phase of the lift. If she were to move "as quickly as possible," the velocities would be very high for the first part of the movement. Because she is required to stop (have a zero velocity at the end of movement, otherwise she would leave the ground), the velocities toward the end of the movement would have to be comparatively low. Examining **Figure 2.18**, you will notice that this is the case: The velocities are skewed to the left.

How does such a movement affect the acceleration profile? Recall that if the movement begins with a zero velocity and ends with a zero velocity, then the average acceleration has to be zero. The positive and negative portions of the graph must cancel each other. The rapid increase in velocity in the beginning of the movement means correspondingly high accelerations. To have the area under the negative curve cancel the area under the positive curve, she must either: (a) have a very rapid deceleration at the end, or (b) spend a longer time with negative acceleration. Because it is somewhat difficult to rapidly stop at the end, it is more likely that she will spend less time accelerating and more time decelerating (**Figure 2.19**). This has been demonstrated experimentally (**Box 2.5**).

Alternate training methods should probably be employed. If the person were to jump up in the air, she would not be required to stop at the top of the lift (i.e., the velocity at the end

| **Box 2.5** | Applied Research: Speed Repetitions in Strength Training |

In an effort to get athletes to be more "explosive," some strength coaches have athletes perform repetitions of a weight exercise (such as the squat or bench press) as fast as possible. You cannot move it very fast if the weight is very heavy, so lighter weights are used during these "speed repetitions." In this investigation, subjects performed rapid repetitions of the bench press exercise with 45% of the heaviest weight they could lift. Remember, if you start with zero velocity and end with zero velocity, the total change in velocity is zero, and the areas under the positive and negative acceleration curves must cancel. With this movement, the acceleration phase was over 60% of the movement. That means the bar was decelerating for the other 40% of the movement! The investigators also had the subjects perform the same movement, but instead of stopping at the end, they threw the barbell. In contrast to speed repetitions, the acceleration phase was 96% of the movement and resulted in larger peak forces and velocities. They concluded that speed repetitions would not be an effective means of making an athlete more explosive.

Data from: Newton RU, Kraemer WJ, Hakkinen K, Humphries BJ, Murphy AJ. Kinematics, kinetics, and muscle activation during explosive upper body movements. *Journal of Applied Biomechanics* 1996;12(1):31–43.

Projectile An object in the air that is only subject to the force of gravity and wind resistance after it leaves the ground

would not have to be zero), and the acceleration phase would be prolonged. By becoming airborne, she would become a **projectile**, and gravity would slow her down.

Section Question Answer

The coach's idea to have his athletes move "fast" during this exercise is probably ill-advised because the areas under the positive and negative portions of the acceleration curve must cancel. To improve rapid movement, he would probably want to use an exercise that does not require the athlete to start and finish with zero velocity. Jumping would be one such exercise.

COMPETENCY CHECK

Remember:

1. Define the following: absolute value, acceleration, displacement, position, net value, vector, and velocity.

Understand:

1. Explain why you cannot tell the direction of travel from the acceleration vector.
2. Explain why the area under the acceleration and deceleration curves must be equal if the initial and final velocities are the same.
3. Calculate the velocity for the following:

p'	P''	Δp	Δt	v
0 m	10 m		1 sec	
15 m	5 m		0.05 sec	
125 m	15 m		10 sec	
150 m	100 m		0.5 sec	

For each case, what is the direction of travel?
4. Calculate the acceleration for the following:

v'	v''	Δv	Δt	a
20 m/sec	10 m/sec		1 sec	
15 m/sec	5 m/sec		0.5 sec	
0 m/sec	70 m/sec		0.01 sec	
150 m/sec	100 m/sec		5 sec	

In each case, is the body speeding up or slowing down?

Apply:

1. List activities where acceleration would be more important than maximum velocity.

2.3 GAIT

Section Question

Walking speed is an important ability that determines the functional independence of older adults (**Figure 2.20**). How can you improve walking speed? Is it the same for running?

Walking and running are forms of locomotion, referred to as **gait**. Average gait speed over a distance (over a time interval) is the average of speed of each **step** a person takes. A step is defined as the period from the initial contact of one foot to the initial contact of the other foot (**Figure 2.21**). Recall that speed is the ratio of distance per unit of time. If you know the length of each step (how "big" a step a person takes) and the **cadence** (or step

Gait Locomotion over land
Step The period from the initial contact of the one foot to the initial contact of the other foot
Cadence The number of steps taken in a given period of time

Figure 2.20	How can you improve the gait velocity of an older adult?

© piotrwek/ShutterStock, Inc.

Figure 2.21 | Step and stride length of the walking gait cycle.

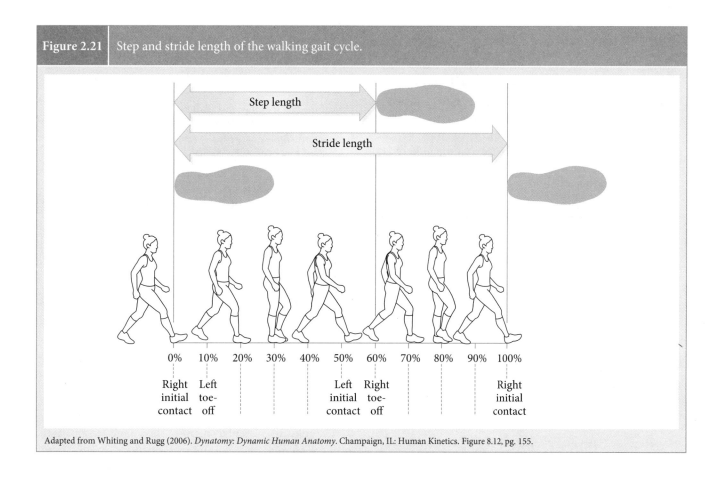

Adapted from Whiting and Rugg (2006). *Dynatomy: Dynamic Human Anatomy*. Champaign, IL: Human Kinetics. Figure 8.12, pg. 155.

rate; number of steps in a period of time), you can calculate gait speed as:

$$\text{Gait speed} = \text{step length} \times \text{cadence}$$

$$\text{Gait speed} = \frac{\text{length}}{\text{step}} \times \frac{\text{steps}}{\text{minute}} \tag{2.8}$$

$$\text{Gait speed} = \frac{\text{length}}{\text{minute}}$$

Remember, distance is a measure of length, so do not get thrown off by the preceding terminology. Sometimes, the word *frequency* will be used instead of *rate*. Both essentially mean the same thing. An alternate to Equation 2.6 is to calculate gait speed using stride length and stride rate (frequency). Two steps equal one **stride** (initial contact of one foot to the initial contact of the same foot; Figure 2.21). Either way will give you essentially the same answer (as long as the gait is symmetrical on both sides).

> **Stride**   The period from initial contact of one foot to the next initial contact of that same foot

Examining Equation 2.8, you will note that gait speed is the product of step length and step rate (Figure 2.22). Theoretically, if you double either one, you will double your gait speed. There are limits to how big a step you can take. So you cannot increase either one indefinitely. If you are working with an older adult who has impaired gait speed, you must first determine if the problem is one of step length or step rate (or both) because they are the only two variables in the equation. If the problem is one of step length,

Figure 2.22 | A model of the determinants of gait velocity.

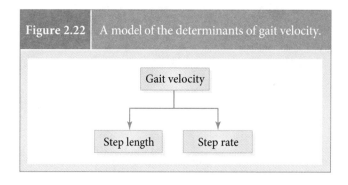

Range of Motion	The amount of movement available at a joint in a given direction

Range of Motion The amount of movement available at a joint in a given direction

Balance The ability to control the body over its base of support

Strength The ability to produce force

Power The ability to produce force while moving quickly

he or she may need to work on increasing either **range of motion** or **balance**. If the problem is due to a decrease in step rate, he or she may need to improve either **strength** or **power**. The principles outlined here are the same for walking and running gait.

Section Question Answer

Walking speed is determined by step length and step rate. Theoretically, improving either one will lead to an increase in walking speed. However, there is a limit to how big a step you can take. After that, further improvements in walking speed would come from increasing step rate. The same applies to running speed.

Box 2.6	Applied Research: Improving Gait Speed of the Elderly

In this investigation, researchers were interested in determining if changes in the walking speed of the elderly were due to aging or to the fact that they just preferred walking at a slower speed. They performed a full biomechanical analysis on older (65–84 years old) and younger (18–36 years old) adults walking at both their comfortable pace and a fast one. Changes that were different between the older and younger adults during both pace conditions were thought to be due to aging. These changes included decreased range of motion at the hip and strength of the calf. They concluded that these two areas should be targeted with exercise programs for the elderly.

Data from: Kerrigan DC, Todd MK, Della Croce U, Lipsitz LA, Collins JJ. Biomechanical gait alterations independent of speed in the healthy elderly: Evidence for specific limiting impairments. *Archives of Physical Medicine and Rehabilitation* 1998;79(3):317–322.

COMPETENCY CHECK

Remember:

1. Define the following: cadence, gait, step, and stride.

SUMMARY

In this lesson, the simplest motion (1-D) was described using the key concepts in Table 2.10. Knowing how position changes with time, you should be able to determine both velocity and acceleration. These are sometimes called

Table 2.10	Key Concepts

- Frames of Reference—origin, axes, and planes
- Scalars
- Vectors
- Distance
- Displacement
- Speed
- Velocity
- Acceleration
- Relative motion
- Vectors and scalars
- Absolute, net, relative, instantaneous, and average values

the spatiotemporal characteristics of movement because they relate to both space and time. You should be able to describe each of these concepts physically, mathematically, and graphically. Displacement is the change in position, or location in a reference frame. Velocity is the amount of displacement in a given amount of time, or how quickly a body is moving in a particular direction and is represented by the slope of the position–time curve. Acceleration is the change in velocity in a given amount of time: how quickly a body is speeding up, slowing down, or changing direction. It is the slope of the velocity–time curve. You should also be able to apply these concepts to human movement using your own examples. These same concepts will be used when the movement gets more complex, but the basic ideas are still the same.

REVIEW QUESTIONS

1. Define the following terms: abscissa, absolute value, acceleration, average value, axis, body, cadence, direction, displacement, distance, frame of reference, gait, instantaneous value, kinematics, net value, ordinate, orientation, origin, point, position, relative speed, sense, scalar, slope, speed, step, stride, system, vector, and velocity.
2. Write the equations for displacement, velocity, acceleration, and gait velocity.
3. Which has the greatest speed, 10 m/sec, or −20 m/sec? Why?

4. During a marathon, which kinematic variable is most important in determining the outcome of the race: average velocity or acceleration?

5. During a sprint, which kinematic variable is most important in determining the outcome of the race: velocity or acceleration?

6. List two ways you could improve gait velocity.

7. Pick any linear movement, and sketch a graph of position versus time, velocity versus time, and acceleration versus time.

REFERENCES

1. Wagner G. The 100-meter dash: theory and experiment. *Physics Teacher.* 1998;36(3):144.

2. Watkins J. *An Introduction to Biomechanics of Sport and Exercise.* Philadelphia: Churchill Livingstone; 2007.

Lesson 3

Describing Motion: Linear Kinematics in Two Dimensions

© Maria Teijeiro/OJO Images/Getty Images

After finishing this lesson, you should be able to:

- Define the following terms: apex, components, net value, parabola, plane, projectile, range, relative height, resultant, and trajectory.
- Given a resultant magnitude and direction, determine the components in the *x*- and *y*-directions.
- Given the components in the *x*- and *y*-directions, determine the resultant magnitude and direction.
- Write the equation for the range of a projectile that takes off and lands at the same elevation.
- List the determinants of a projectile's trajectory.
- Describe situations when a larger release angle is more advantageous.
- Describe situations when a smaller release angle is more advantageous.

Section Question

At kickoff, a kick returner (Figure 3.1) catches a football at his own goal line and returns it for a touchdown. He catches the ball at the south sideline and enters the opponent's goal line on the north sideline, running in a straight line. His average velocity is 8 yds/sec. It is recorded as a 100-yard touchdown return. Why did the run take him 14.18 seconds instead of the 12.5 seconds it would take to run 100 yards?

Intuitively you know that he ran more than 100 yards. How far did he run? To examine this situation, you must examine motion in two dimensions, requiring another set of tools.

Figure 3.1 A kick returner runs fom goal line to goal line at an average velocity of 8 yds/sec. Why did it take him 14.18 seconds and not 12.5 seconds?

© Mark Herreid/ShutterStock, Inc.

3.1 FRAME OF REFERENCE

When movement takes place purely along an axis, you can say that movement is taking place in one dimension. As you might have guessed, this rarely happens in the real world. So you need a way to express movement in two or three dimensions. You are aided in this by introducing the idea of **planes**.

> **Plane** A smooth flat space defined by two axes

The space defined by two perpendicular axes is called a plane. Because axes extend indefinitely, so does the plane.

34

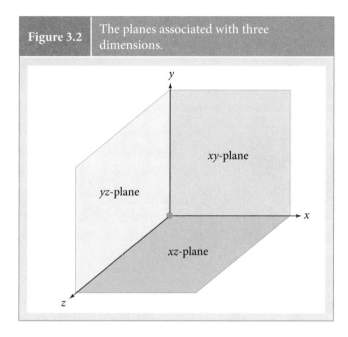

Figure 3.2 The planes associated with three dimensions.

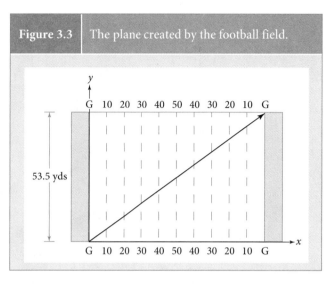

Figure 3.3 The plane created by the football field.

Any movement that takes place in that space is called planar motion. For example, if your desk represents a plane, moving your hand anywhere along the surface of the desk would be in the plane of the desk. If you lifted your hand off the desk, you are no longer moving in that plane.

We live in a three-dimensional world. Because each axis represents a dimension, and two perpendicular axes represent a plane, you can create three planes. If you have an x-, y-, and z-axis, you can create xy-, xz-, and yz-planes (Figure 3.2). Movement in each plane is just like moving your hand along your desk: If you move away from the plane you are no longer "in" the plane.

Returning the case of the kick returner, you should begin your analysis by establishing a frame of reference, starting with an origin. You could place the origin anywhere, but you will always want to pick an origin that is either physically meaningful or makes the analysis easier. In this case, placing the origin where the kick returner catches the ball satisfies both requirements. Like the 100-meter run in the

last lesson, you could put the axis in the direction of travel. However, the direction of travel is neither along the sidelines (east–west) nor along the yard line (north–south), but some of both. Expressing your axes in this way and explaining the returner's run in the plane made by these two axes is more appealing.

3.2 RESULTANTS AND COMPONENTS

During the kick return, if you express the axis made by the sidelines as the x-axis and the direction of the goal line (which is 90° to the sideline) as the y-axis, the movement will take place in the xy plane (Figure 3.3). The actual movement is called the **resultant** vector and is designated by the capital letter R. Any resultant can be thought of as the sum of two vectors (each coincident with the principal axes), which are known as its **components** (c). The relationship between the resultant and its components is reviewed in Box 3.1.

> **Resultant** A vector that is equivalent to the combined effect of two or more vectors
>
> **Components** Parts of a resultant vector, two or more vectors that are acting in different directions

Box 3.1   Essential Math: Trigonometry

A graph is, in many ways, just like a map.[1] Points on the map are determined by their coordinates. For example the origin of the reference system for my university's map is located on the southwest corner of the university. The horizontal axis extends eastward and is lettered from A to G. The vertical axis of the map, like most good maps, points northward and is numbered from 1 to 13. Any building can be determined by its two coordinates on the map. The building where my office is located is in F5, the library is D4, and so on.

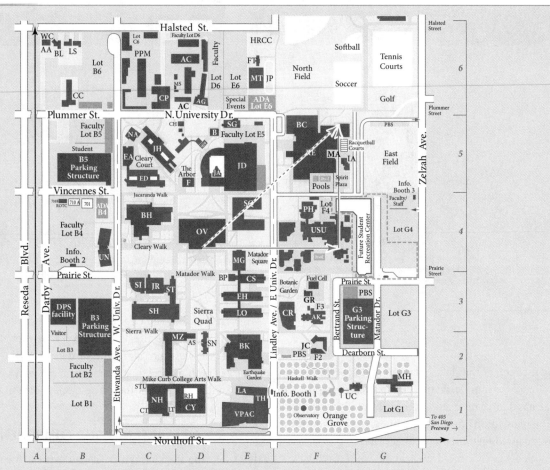

Map of California State Northridge (CSUN) campus.

Any point on a graph is also determined by its coordinates, which are always represented by the abscissa (horizontal) coordinate first and the ordinate (vertical) coordinate second. Now let us say I wanted to go from my office to the library, how would I get there? You could say that I had to go west one "unit" and south one "unit" (whatever unit the grid square represents, which in the case of my university map is a very nonuniform block). Conversely, if I wanted to go from the library back to my office, I would go east one block and north one block. But a lot of times I am in a hurry (I do not want to be late for class), and so I go directly from my office to the library, cutting through walkways, lawns, and even buildings if I have to. The distance I traveled if I stuck to the streets, are known as the components because they are along the axes of the graph. The actual distance I traveled as a "result" of my shortcuts in known as the resultant. The resultant is always a straight line between the two points. Sometimes we will know the components and want the resultant. More often, we will have the resultant and need to know something about the components. So it is important that you appreciate how to go back and forth between the two.

Any right triangle (that is, a triangle with one of the angles being 90°) will have some pretty unique properties that can be used

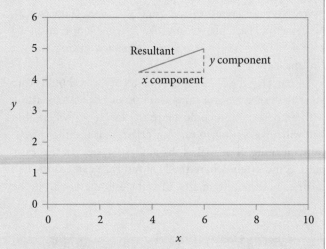

CSUN components.

to determine the components from the resultant and vice versa. For that reason, you always have coordinate systems that are at right angles to each other and always plot a coordinate versus time at right angles to each other. These properties are best explained by

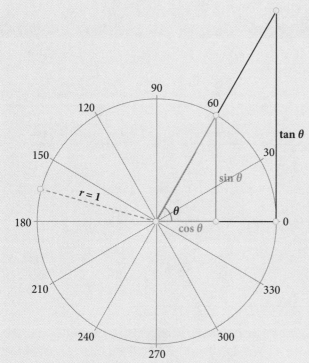

90
120
60
150
30
$r=1$
$\tan \theta$
180
θ
$\sin \theta$
$\cos \theta$
0
210
330
240
300
270

The unit circle and corresponding trig functions.

examining a right triangle within a circle. If the circle has a radius of one, the math is really easy, and it is called a unit circle.

Remember, the components are always along the horizontal (base of the triangle, the side adjacent to the angle of interest) and the vertical (height of the triangle, the opposite side of the angle of interest). The hypotenuse, or the side opposite the right angle, is the resultant. The resultant is obtained from the components by the following formula:

$$R^2 = a^2 + b^2$$
$$R = \sqrt{a^2 + b^2}$$

where R is the hypothenuse, and a and b are the two sides of the triangle. It should be apparent from the formula that R is always the largest side of a right triangle, so the resultant is always larger than the two components (or equal to one of the components if the other component is zero). The angle, θ, that the resultant makes with the adjacent side is

$$\theta = \tan^{-1}\left(\frac{\text{opposite side}}{\text{adjacent side}}\right)$$

What if you know the resultant but wish to find the components? The opposite side is simply the resultant times the sine of the angle the resultant makes with the adjacent:

$$\text{opposite} = R \times \sin\theta$$

Similarly, the adjacent side is the resultant times the cosine of the angle the resultant makes with the adjacent:

$$\text{adjacent} = R \times \cos\theta$$

In your first trigonometry class, you may have remembered the mnemonic Soh-Cah-Toa, which stood for:

- Sine is opposite over hypotenuse.
- Cosine is adjacent over hypotenuse.
- Tangent is opposite over adjacent.

So you end up with the following relations:

$$\sin\theta = \frac{\text{opposite}}{\text{hypotenuse}}$$

$$\cos\theta = \frac{\text{adjacent}}{\text{hypotenuse}}$$

$$\tan\theta = \frac{\text{opposite}}{\text{adjacent}}$$

1. Berlinski D. *A tour of the calculus.* New York: Random House, 1995.

You can use the information about resultants and components to know why the runner took longer to go from goal line to goal line. He actually ran 100 yards in the x direction and 53.5 yards in the y direction. These were the components of the movement, but not the resultant. The resultant (R), or total movement, is expressed as:

$$R = \sqrt{(^x c)^2 + (^y c)^2} \qquad (3.1)$$

And so in this example, the resultant distance run is determined in a similar fashion with the component distances:

$$^R d = \sqrt{(^x d)^2 + (^y d)^2}$$
$$^R d = \sqrt{(100 \text{ yds})^2 + (^y 53.5 \text{ yds})^2}$$
$$^R d = 113.4 \text{ yds}$$

which agrees with 14.18 seconds it took him to make the run. See **Figure 3.4**.

You should confirm for yourself that this is correct. Another way you could have expressed this was to say the 8 yds/sec was the resultant velocity. The component velocity in the x direction would then be found as follows:

$$\theta = \tan^{-1}\left(\frac{53.5}{100}\right)$$
$$\theta = 28.15°$$
$$^x v = {}^R v \times \cos\theta$$
$$^x v = 8 \text{ yds/sec} \times .88$$
$$^x v = 7.05 \text{ yds/sec}$$

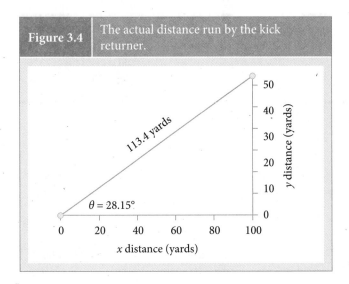

Figure 3.4 | The actual distance run by the kick returner.

Table 3.1 | Correction Factors

Multiply a resultant by its correction factor to determine the component along the x- or y-axis.

Angle		Correction Factor	
From the x-axis	From the y-axis	Along the x-axis	Along the y-axis
0	90	1	0
10	80	0.985	0.174
20	70	0.94	0.342
30	60	0.866	0.5
45	45	0.707	0.707
60	30	0.5	0.866
70	20	0.342	0.94
80	10	0.174	0.985
90	0	0	1

And you can verify once again that the time it took to travel 100 yards at 7.05 yds/sec was 14.18 seconds.

Whenever a vector is not parallel to an axis, you have to correct for the fact that the quantity (displacement, velocity, acceleration) is not going exactly in that direction. The cosine or sine of the angle acts as that correction factor.[1] The correction factor can never be greater than 1; it will always range from -1 to 1. A list of some of the common correction factors is presented as Table 3.1, and a graph of these correction factors from 0° to 360° is presented in Figure 3.5.

Important Point!   Whenever movement does not occur parallel to an axis in your reference frame, you must apply a correction factor to determine the component parallel to that axis.

In a general form, we can express the component of any vector (displacement, velocity, acceleration, etc.) as:

$$\text{component} = \text{resultant} \times \text{correction factor} \quad (3.2)$$

The correction factor will either be $\sin \theta$ or $\cos \theta$, depending on the relation of the component to the angle. Review Box 3.1.

A few things should become apparent from inspecting Table 3.1, Figure 3.5, and Equation 3.2:

- The maximum value along an axis occurs when the vector is parallel to that axis: The angle is 0, and the correction factor is 1.
- A larger angle will result in a smaller magnitude of the component parallel to that axis and a larger magnitude of the component perpendicular to that axis.
- The magnitudes of the components will be equal at 45°.

- The magnitude of the resultant is always larger than the two components, or equal to one of the components if the other component is zero.

Commit these relations to memory; they will help you to intuitively grasp 2-D motion.

As an example of how this works, say you are holding a box and you wish to accelerate that box 4 m/sec[2] at 45° from the horizontal (positive x) axis (Figure 3.6). What are the accelerations in x- and y-directions? First, you should already be aware of the fact the acceleration in the x- and y-directions

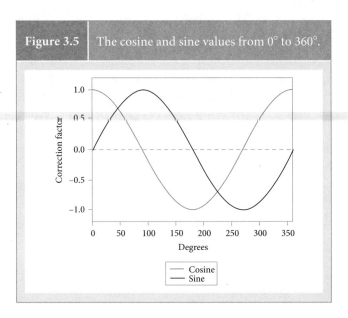

Figure 3.5 | The cosine and sine values from 0° to 360°.

| Figure 3.6 | The component accelerations for accelerating a box 4 m/sec² at 45° from the horizontal. |

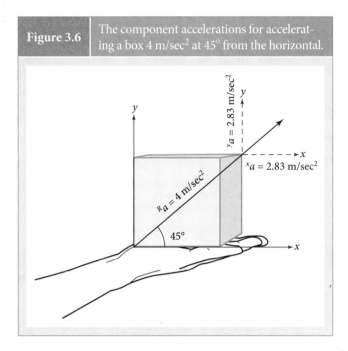

Understand:

1. For each of the following, state whether the horizontal or vertical component is larger:
   a. 4 m/sec at 35° from the positive x-axis
   b. 10 m/sec at 125° from the positive x-axis
   c. 150 m/sec² at 50° from the positive y-axis
   d. 25 m/sec² at 200° from the positive y-axis
2. For each of the preceding vectors, determine the correction factors for the x- and y-directions and calculate the components.
3. For each of the following, state whether the resultant is equal to 45°, less than 45°, or greater than 45° from the positive x-axis:

x component	y component
10 m/sec	50 m/sec
105 m/sec	90 m/sec
500 m/sec²	500 m/sec²
40 m/sec²	35 m/sec²

4. Calculate the resultant from each of the preceding components.

Apply:

1. What activity would occur mostly along a horizontal axis?
2. What activity would occur mostly along a vertical axis?
3. What activity would occur along a vector in-between the horizontal and vertical axes?

will be equal because the angle is 45°. Second, you should anticipate that the component values will be smaller than the resultant of 4 m/sec². To find your answer, you would use Equation 3.2 and the trig functions in Box 3.1:

$$^{x}a = {^{R}a} \times \text{correction factor}$$
$$^{x}a = {^{R}a} \times \cos(45°)$$
$$^{x}a = 4 \times .707 = 2.83 \text{ m/sec}^2$$
$$^{y}a = {^{R}a} \times \text{correction factor}$$
$$^{y}a = {^{R}a} \times \sin(45°)$$
$$^{y}a = 4 \times .707 = 2.83 \text{ m/sec}^2$$

And you would see that both of your anticipated values were correct.

3.3   NET VALUES

Section Question Answer

It took the kick returner longer to complete the run because he ran more than 100 yards. He actually ran a little over 113 yards because he ran at an angle to the sideline.

Section Question

A couple racing on a reality television show must cross a river to obtain their next clue, which is the location of the next pit stop (Figure 3.7). As the stronger swimmer, the woman decides to do it. Last year, she swam competitively and did the 100 m in 58.82 seconds. Standing on the south shore, she notices that the rushing current is moving at 1 m/sec, from west to east. Her husband approximates the width of the river to be 50 meters. Do you think if she enters the water directly across from the clue box that she will arrive at it on the north shore? If not, where should she enter the river to ensure she does not miss the clue box?

COMPETENCY CHECK

Remember:

1. Define: component, resultant, and plane.

As this example illustrates, sometimes velocities and accelerations (and forces) can come from more than one source. In such cases, the vectors should be added according to the

Figure 3.7	A woman needs to enter the water to swim across a river that has a moving current. Where should she enter the water so that she comes out in front of the orange box?

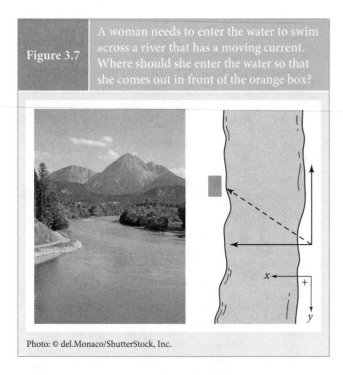

Photo: © del.Monaco/ShutterStock, Inc.

rules of vector addition to obtain the **net** value (such as velocity or acceleration). The net value will give you the overall effect, so it is very important.

Net	The total value after summing all of the individual values

Take the simple case of the woman swimming directly upstream. The origin is where she enters the water, and west is positive. The river runs east (at −1 m/sec), she is swimming westward (at 1.7 m/sec), and her net average velocity is 0.7 m/sec. You may have noticed this phenomenon if you ever flew across the United States. Going from California to New York is generally quicker than the flight in the other direction because the jet stream produces a tailwind (which is added to the plane's velocity) in the west–east direction and a headwind (which must be subtracted from the plane's velocity) in the east–west direction. Even though this is a simple example, the velocities were listed in the right column and then added together (Table 3.2).

Table 3.2	Determining the Net Velocity of a Swimmer Going Against the Current

	v (m/sec)
$v_{swimmer}$	1.7
v_{river}	−1
v_{net}	**0.7**

Return to the river. Even though the motions are not in the same direction, the principle is the same. One key to understanding this is to know that although the swimmer may be traveling at 1.70 m/s in the *x* direction, implicit in this statement is that she is traveling 0 m/sec in the *y* direction. Similarly, the current is traveling −1.0 m/sec in the *y* direction and 0 m/sec in the *x* direction. Using the rules of vector addition (reviewed in Box 3.2), you have to put the values in the appropriate columns and then simply add each row in the column (Table 3.3). Whenever you have more than one vector, it is also helpful to construct a vector diagram (see Box 3.3).

Box 3.2	Essential Math: The Rules of Vector Addition

The rules of adding in algebra are something you probably learned in grade school:

$$1 + 1 = 2$$
$$3 + (-5) = 3 - 5 = -2$$

Whenever you add two *scalars* together, we follow the same rules. Adding two *vectors* is a different story. Remember, vectors have both magnitude and direction. If you add the magnitudes of two vectors with different directions, you are essentially adding 3 apples to 5 oranges and saying you have 8 oranges. You cannot do it. So the trick is to make sure that you are only adding the magnitudes of vectors going in the same direction. You do this by first breaking the vector up into its components (with common directions) and then adding (or subtracting) the component parts. For example, let us say you wanted to add together two vectors, **A** and **B**, both with a magnitude of 100, and directions as shown:

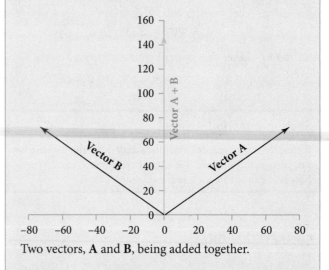

Two vectors, **A** and **B**, being added together.

First, construct a table with each component and each dimension. Then, using the trigonometric functions (Box 3.1), you would populate each cell of the table. Then you add each row in a column, and your answer is in the final row:

Table 1 Creating a Table to Add the Components of a Vector Together

	x	y
A	0.707	0.707
B	−0.707	0.707
A+B	**0**	**1.414**

Remember to put every component in its proper cell and only add down the columns so that you are only adding like components (x and x; y and y) together.

Graphically, vector addition uses the tip-to-tail method. Put the tail of one vector (vector **B**) next to the tip of the other vector (vector **A**). The resultant vector is drawn from the tail of the first vector (vector **A**) to the tip of the second vector (vector **B**).

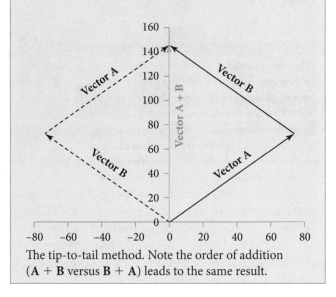

The tip-to-tail method. Note the order of addition (**A** + **B** versus **B** + **A**) leads to the same result.

Whenever there is more than one vector (displacement, velocity, acceleration, etc.) acting on a body, it is useful to construct a diagram to show all the vectors. For now, the body will be represented as a point. After annotating your frame of reference, you draw each vector in its correct orientation, as was done in the following figure.

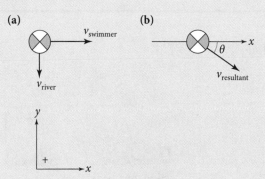

A vector diagram includes all the vectors acting on a body (a). The resultant can then be determined by inspection before you do your calculations (b).

This allows you to keep track of all the velocities, accelerations, and so on acting on a body. It is also helpful in giving you an idea of what the resultant would be. For example, if you know that two components are along the positive x- and positive y-axes, then the resultant must be somewhere in between the axes. If the value along the x-axis is larger than the value along the y-axis, then the resultant is less than 45° from the x-axis. If the value along the y-axis is larger than the value along the x-axis, then the resultant is greater than 45° from the x-axis. If the values are equal, then the resultant is 45° from each axis.

The net velocity is in two dimensions (x and y). Each motion is independent of the other; that is, movement in the x direction does not depend on movement in the y direction. Yet they occur simultaneously: They are related to each other

Table 3.3	Determining the Net Velocity in the x- and y-directions of a Swimmer Going Perpendicular to the Current	
	x_v (m/sec)	y_v (m/sec)
$v_{swimmer}$	1.7	0
v_{river}	0	−1
v_{net}	**1.7**	**−1**

through *time*. Let us see how that works. It would take the swimmer 29.41 seconds to swim 50 meters in the x direction. In 29.41 seconds, she would have traveled 29.41 meters east due to the river. Therefore, she should walk west 29.41 meters before entering the water if she wishes to land at the clue box.

Important Point! Components are independent of each other, but related to each other through time.

A warning here: Do not think that you need to know the resultant and angle! It is only the velocity in the x direction that determines how long it takes her to go 50 meters in the *x* direction. The only thing the *y* velocity affects is how far she goes in the *y* direction in the same amount of time.

Section Question Answer

If the woman entered the water directly across from the clue box, she would end up east of it due to the effect of the current. Therefore, she should start west of the box before entering the water.

COMPETENCY CHECK

Remember:

1. Define: net value.

Understand:

1. Determine the net acceleration when the following two accelerations are acting on a body:

a_1	a_2
150 m/sec^2 at 35° from the positive *x*-axis	200 m/sec^2 at 50° from the positive *x*-axis
50 m/sec^2 at 50° from the positive *y*-axis	35 m/sec^2 at 120° from the positive *y*-axis
25 m/sec^2 at 75° from the positive *x*-axis	10 m/sec^2 at 20° from the positive *y*-axis
10 m/sec^2 at 90° from the positive *y*-axis	15 m/sec^2 at 10° from the positive *x*-axis

3.4   PROJECTILE MOTION

Section Question

Tony is a skateboarder attempting a jump off a ramp (Figure 3.8). What determines how high he will go? What determines how far he will go?

Any body that is in the air is called a **projectile** and follows the same principles if the body happens to be a person or an object. The path that a projectile takes is called its **trajectory**. Wind has an effect on the trajectory of a body, but it will be ignored.

Let us start the discussion by conducting a thought experiment. What

> **Projectile**   An airborne body that is only subjected to gravity and wind resistance after it has left the ground
>
> **Trajectory**   The path of a projectile

Figure 3.8   What determines how high he will go? What determines how far he will go?

© shock/ShutterStock, Inc.

would happen if he tried this stunt in outer space? His trajectory would look something like **Figure 3.9**. That is right: He would go in a straight line, indefinitely. Although some of you probably wish you could do that, you know things here on Earth do not quite work that way. Why?

You were introduced to the concepts of displacement, velocity, and acceleration in that order. And learning them in that order makes a lot of sense: You cannot discuss the rate of change of velocity until you have an understanding of velocity. But in the world, cause and effect are reversed. You do not accelerate because you changed your velocity; you change your velocity because you have a net acceleration. So you should start to think of things this way. The correct hierarchical model is presented in **Figure 3.10**. Change in time is not included under the change in velocity because it would be redundant. So you can say that the final position of a body depends on its initial position, its initial velocity, the net acceleration, and the change in time. You will see these various elements come into play as you work through this section.

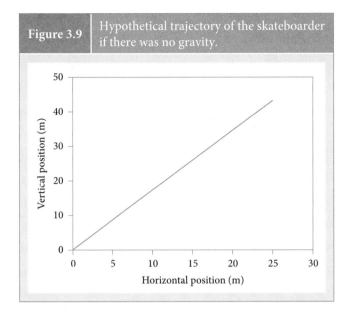

Figure 3.9 | Hypothetical trajectory of the skateboarder if there was no gravity.

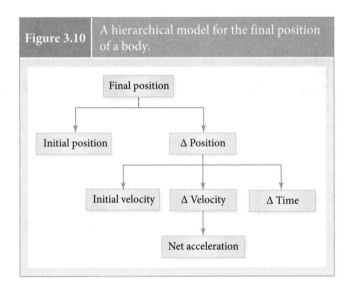

Figure 3.10 | A hierarchical model for the final position of a body.

Important Point!   A net acceleration causes a change in velocity, and a change in velocity causes a displacement.

   The last section discussed how you needed to add all the velocities acting on a body, and that the x and y velocities were related by time. On the surface of Earth, all bodies are subject to an acceleration due to the force of gravity. Suffice to say that any body on, or close to, the surface of Earth will be subject to a constant acceleration of 9.81 m/sec² or 32 ft/sec², which will be abbreviated as a_g. Because this acceleration is in the downward (negative direction), you must remember that a_g is equal to –9.81 m/sec² or –32 ft/sec². Unlike the velocity of the river, the velocity due to gravity is not constant. You see that:

$$a = \frac{\Delta v}{\Delta t} \tag{3.3}$$

Important Point!   It is important to keep the direction of the acceleration due to gravity straight. If you say that $a_g = -9.81$ m/sec², then a_g will be *positive* in the equations. If you say that $a_g = 9.81$ m/sec², then a_g will be *negative* in many of the equations. Either way will work as long as you do it consistently and remember: a_g always acts in the negative, vertical direction.

Replacing a with a_g and rearranging the preceding equation, you get:

$$\Delta^y v = a_g \times \Delta t \tag{3.4}$$

The y superscript is included to remind you that the velocity due to gravity is limited to the vertical (y) direction. Using a bit of complicated math, you can determine the change in position due to gravity as:

$$\Delta^y p = \frac{1}{2} a_g t^2 \tag{3.5}$$

You should be able to extract three important pieces of information from Equation 3.5. First, notice that gravity will alter the path of the projectile by $\Delta^y p$. Second, the magnitude of the alteration is not constant: As time increases, there will be a greater change in the trajectory of the projectile. Finally, this relation is not a linear one: Doubling the time will lead to a four-fold change in position because the time term in the equation is squared.

Important Point!   If a value is squared in an equation, then changes to that value will have a bigger effect on the outcome than changes that are not raised to a higher power.

   Let us see what that means. Suppose you ran off a 10-meter diving platform with a horizontal velocity ($^x v$) of 6 m/sec. Again, if you were in outer space, you would never hit the water. You would go in a straight line forever. But not only do you have a horizontal velocity, but you also have a vertical velocity because of the acceleration due to gravity. Therefore, your trajectory will be altered due to gravity. After establishing a frame of reference with the origin at the edge of the water right below the diving platform and with forward and up being positive, you can use the rules of vector addition

Table 3.4	Vertical and Horizontal Displacements as a Function of Time		
The jumper enters the water between time points 15 and 16.			
Point	Time	$\Delta^x p$	$\Delta^y p$
1	0	0	10.00
2	0.1	0.6	9.95
3	0.2	1.2	9.80
4	0.3	1.8	9.56
5	0.4	2.4	9.22
6	0.5	3	8.77
7	0.6	3.6	8.23
8	0.7	4.2	7.60
9	0.8	4.8	6.86
10	0.9	5.4	6.03
11	1	6	5.10
12	1.1	6.6	4.06
13	1.2	7.2	2.94
14	1.3	7.8	1.71
15	1.4	8.4	0.39
16	1.5	9	−1.04

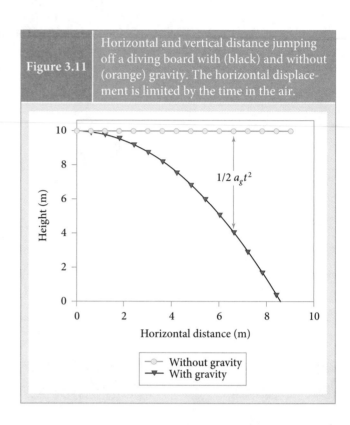

Figure 3.11 Horizontal and vertical distance jumping off a diving board with (black) and without (orange) gravity. The horizontal displacement is limited by the time in the air.

and Equation 3.5 to determine your trajectory. The results are presented in Table 3.4 and in Figure 3.11.

You can see that, as time progressed, you got further and further away from what your trajectory would have been had you not been subject to gravity. Also note that the horizontal distance was determined by how much time you spent in the air, and how much time you were in the air was related to how high you were off the ground (or in this case, water). The x and y terms are related by time. To illustrate this point, compare what would happen if you ran off the diving platform at 10 meters to what would happen if you ran off a 3-meter diving board (Figure 3.12). Even though your speed coming off the platform and board were identical, you went 8.57 meters from the 10 m platform but only 4.70 meters from the 3 m board. You traveled further from the platform because you started from a greater height and subsequently had more time in the air. Notice also that the platform was three times higher than the board, but you landed in water at a distance that was not even twice as far. That is because the time term in Equation 3.5 is squared—the longer you are in the air, the bigger impact it will have.

Now return to our skateboarder. Rather than going straight out like you did when you ran straight off the diving platform, he is leaving the ground at angle, θ, with some velocity, v. Suppose he leaves the ground at a 60° angle to the horizontal, and he has a resultant velocity of 5 m/sec.

The first thing you need to do is establish a frame of reference. Say the origin is where he leaves the ground, and up and forward are positive. Using the correction factors, you can calculate his horizontal velocity as 2.5 m/sec and his vertical velocity as 4.33 m/sec.

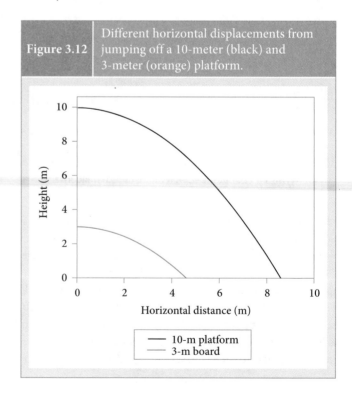

Figure 3.12 Different horizontal displacements from jumping off a 10-meter (black) and 3-meter (orange) platform.

Unlike the previous example, where you jumped off the diving platform, the skateboarder is leaving the ground with both a horizontal *and* vertical velocity, so his trajectory will not look like Figure 3.11. But he *is* subject to a velocity from the acceleration due to gravity, so his trajectory will not look like Figure 3.9, either. It will be something resembling a combination of the two, as in **Figure 3.13**.

A few things are worth noting in Figure 3.13. First, the shape of the trajectory is a **parabola**. The highest point on the trajectory is the **apex**, and the parabola is symmetric about its apex. The horizontal velocity is constant (2.5 m/sec), but the vertical velocity is not; it starts at the takeoff velocity

Parabola A type of plane curve

Apex The highest point of a trajectory

Figure 3.13 The motion of the skateboarder in the air (top) and the horizontal and vertical velocities (bottom).

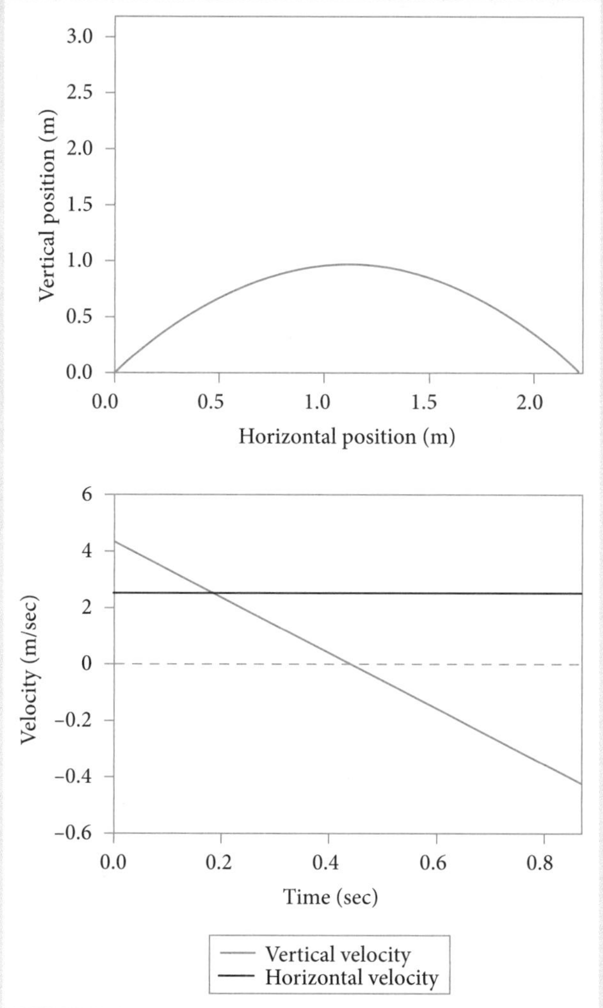

(4.33 m/sec) and decreases at a constant rate of $-9.81x^\Delta t$ m/sec. At the apex, the vertical velocity is zero. After that it becomes negative, increasing speed until he hits the ground. Recall that the velocity tells you the direction of the movement, so it makes sense that the downward direction should be negative.

Important Point! At the apex, the instantaneous vertical velocity is zero.

Now examine what determines how high he will go and how far he will go. Recall that velocity is the change in distance divided by the change in time. Rearranging Equation 3.3 and combining it with Equation 3.5, you get:

$$h = \frac{{}^yv^2}{a_g} \tag{3.6}$$

and use your correction factor:

$${}^yv = {}^Rv \times \sin\theta \tag{3.7}$$

So the height is determined by the velocity of takeoff and the angle of takeoff. If you think about it, it makes sense. A ball will go higher in the air if: (1) you toss it straight up (90° from the vertical, correction factor = 1) and (2) if you throw it faster.

Now think about how far he will go, which is also known as the **range**.

From Equation 3.8, you find that:

Range The horizontal displacement of a projectile

$$\Delta^xp = {}^xv \times \Delta t \tag{3.8}$$

So you know that the horizontal distance will depend on two things: the horizontal velocity and the time in the air. The horizontal velocity will also be dependent on two things: the velocity of takeoff and the angle of takeoff. What do you think time is dependent on?

First, it depends on the height of the takeoff, which is known as the **relative height**. Recall the example of jumping off the 10 m platform versus the 3 m

Relative height The difference between the vertical position at takeoff and the vertical position at landing

diving board: You went further because you were in the air longer, and you were in the air longer because you started from a greater relative height. The second thing that the flight time depends on is the vertical velocity of takeoff. Because you ran straight off of the diving board, you had no vertical velocity at takeoff, and were only subject to the vertical velocity due to gravity. The skateboarder has a vertical velocity at takeoff, which must be added to the velocity using the rules of vector addition. Combining the vertical velocity and horizontal

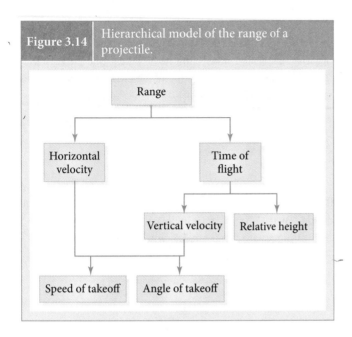

Figure 3.14 Hierarchical model of the range of a projectile.

Table 3.5	Angles That Produce the Same Range if the Object Takes Off and Lands With the Same Relative Height	
Angles		**Sin 2θ**
5	85	0.174
10	80	0.342
15	75	0.500
20	70	0.643
25	65	0.766
30	60	0.866
35	55	0.940
40	50	0.985
45		1.000

velocities, you will see that the trajectory of a projectile depends on three things (Figure 3.14):

1. The velocity at takeoff
2. The angle of takeoff
3. Relative height

Let us focus on the range. If the takeoff and landing occur at the same height (i.e., the relative height is zero), then the range of the projectile can be expressed in a succinct equation:

$$\text{Range} = \frac{{}^R v^2 \times \sin 2\theta}{a_g} \qquad (3.9)$$

There are two important things to note about Equation 3.9. First, the velocity term is squared. This means doubling the velocity will lead to a four-fold increase in distance. The velocity term obviously has a bigger impact than any other term in the equation. Second, notice the "sin 2θ" term. It has two important consequences.

Sin 2θ is the largest when the angle is 45°. If a projectile takes off and lands from the same height, and all other things are equal, the projectile's range will be greatest when the angle is 45°. If the object takes off from a greater height than where it lands, the angle that will produce the projectile's greatest range will be less than 45°. Conversely, if the projectile's takeoff is lower than its landing, the greatest range will occur with a takeoff angle greater than 45°.

Important Point!   Sin 2θ:
1. Is largest at 45°.
2. Except for 45°, two θ give identical sin 2θ.

Also, two angles (with the exception of 45°) will produce an equivalent value of sin 2θ. Examining Table 3.5, you will see that each angle is equidistant from 45°, and their sin 2θ values will always be less than 45° (which is 1). Even though two angles give the same range, the trajectories of projectiles using each angle will be different: Angles greater than 45° will have a steeper trajectory and less horizontal velocity than angles less than 45° (Figure 3.15). Consequently, angles less than 45° will arrive at the landing in less time, which may be

Figure 3.15 Two angles producing the same range. The steeper angle (black) results in a longer time in the air and a smaller horizontal velocity. Conversely, the smaller angle (orange) results in a shorter time in the air and a larger horizontal velocity.

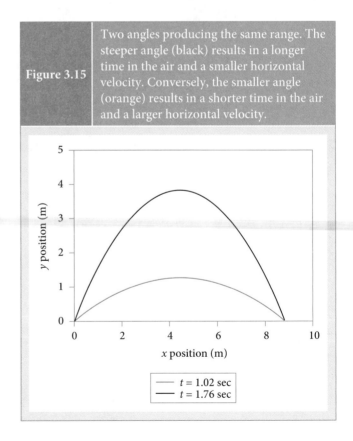

x position (m)

y position (m)

t = 1.02 sec
t = 1.76 sec

Figure 3.16   The dancer's time in the air can be manipulated by altering her takeoff angle.

© AYakovlev/ShutterStock, Inc.

beneficial for a quarterback passing to a receiver. However, the lower horizontal velocity of the higher angle means small errors will be better tolerated when attempting to have the projectile land in a specific spot (such as a basketball free throw or a golf drive).[2] In other words, when accuracy is more important than time, a steeper angle should be used.

These relations can also be manipulated by performance artists. If the dancer in **Figure 3.16** needs to land on a particular mark at a particular time, she can either jump at a larger angle to increase her flight time or at a smaller angle to decrease her flight time. This will affect her landing mechanics as well.

Section Question Answer

Four things determine the trajectory of a projectile, three of which you can control. You can control the velocity of takeoff, the angle of takeoff, and the relative height. You cannot control the acceleration due to gravity.

COMPETENCY CHECK

Remember:

1. Define the following: apex, parabola, projectile, range, relative height, and trajectory.
2. List the things that determine the range of a projectile. Which is more important? Why?

Understand:

1. Explain why a net acceleration causes a change in position instead of the other way around.

Apply:

1. List situations where you would want a flatter trajectory and situations where you would want a steeper trajectory.

SUMMARY

In this lesson, your ability to describe motion was expanded using the key concepts in **Table 3.6**. Motion was now described in two dimensions. You learned the key to 2-D motion was to analyze the movement as two separate but related 1-D motions. This requires the use of trigonometric functions that act as correction factors. This is extremely important when a body is not moving along a principal axis and/or is subjected to more than one acceleration or velocity. As you will see, this happens quite often in the real world. Thus far, all motions were considered equal. That is, a go-kart and a tractor trailer moving at the same velocity are considered equivalent. Is that truly the case? Before this can be answered you must now consider how to describe the second form of motion: angular motion.

Table 3.6	Key Concepts
• Plane	
• Correction factors using trigonometry	
• Projectile	
• Trajectory	
• Parabola	
• Range	

REVIEW QUESTIONS

1. Define the following terms: apex, components, net value, parabola, plane, projectile, range, relative height, resultant, and trajectory.

2. Which of the following would have a greater vertical velocity: a 10 m/sec resultant that is 10° from the horizontal, or a 10 m/sec resultant that is 30° from the horizontal?

3. True or False? If the horizontal component is larger than the vertical component, the resultant angle will be less than 45°.

4. List the three variables that determine projectile motion. Which one do you have the most control over? Which one has the biggest impact?

REFERENCES

1. Hay JG, Reid JG. *Anatomy, Mechanics, and Human Motion*. 2nd ed. Englewood Cliffs, NJ: Prentice Hall; 1988.

2. Hall SJ. *Basic Biomechanics*. 6th ed. New York: McGraw-Hill; 2012.

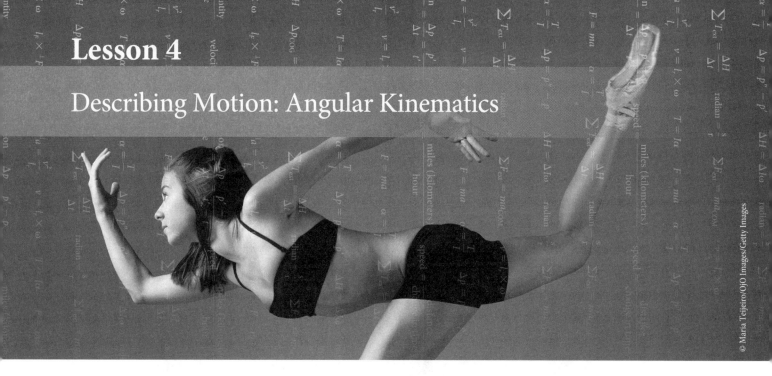

Lesson 4

Describing Motion: Angular Kinematics

LEARNING OBJECTIVES

After finishing this lesson, you should be able to:

- Define the following terms: rigid body, axis of rotation, angular position, angular displacement, angular velocity, and angular acceleration.
- Write equations for the following concepts: angular displacement, angular velocity, and acceleration.
- Convert angular velocity to linear velocity.
- Convert angular acceleration to tangential and centripetal acceleration.
- Identify angular velocity on an angular position–time curve.
- Identify angular acceleration on a angular velocity–time curve.
- Given angular displacement and time data, calculate angular velocity and angular acceleration.

4.1 ANGULAR KINEMATICS

Section Question

According to the Disability Statistics Center at the University of California, San Francisco,[1] 1.6 million Americans use wheelchairs (**Figure 4.1**). Understanding the motion of the wheel is essential in understanding the mobility of wheelchair users. Is the motion of the wheel the same as the motion of a body previously explored?

4.1.1 Rigid Bodies

Thus far, a body has been represented by a point, which was essentially dimensionless. That is, it had no length, width, or height. Now the representation of a body is going to change to a **rigid body** when discussing angular, or circular, motion. The

| Figure 4.1 | Is angular motion the same as linear motion? |

© gdvcom/ShutterStock, Inc.

rigid body will have length, but not height or width. By definition, a body is rigid if the distance between any two points on that body remain fixed. A baseball bat (**Figure 4.2**) would be a good example of a rigid

> **Rigid body** A body that maintains a constant shape

49

Figure 4.2	A baseball bat is a rigid body. A garden hose is not.

(left) © Roman Gorielov/ShutterStock, Inc., and (right) © Joe Belanger/ShutterStock, Inc.

body: the distance between the handle and barrel will not change (unless that bat is broken). A garden hose (Figure 4.2) would not represent a rigid body: the two ends could be brought next to each other by simply curling the hose.

The wheel, and more specifically its spokes, is an excellent example of a rigid body that undergoes angular motion. It will be used extensively throughout this section. You may be tempted to think that a wheelchair does not apply to what you want to do, but these concepts can be applied to the joints of the body, which undergo angular motion.

4.1.2   Frame of Reference and Axis of Rotation

Axis of rotation   A fixed line about which a body rotates

Angular motion always occurs in a plane about an axis, known as the **axis of rotation**. The axis is always perpendicular to the plane. Planes are defined in **Figure 4.3**. The axis of rotation for each plane will be parallel to one of the cardinal axes:

- the z-axis if the rotation is in the xy-plane
- the y-axis if the rotation is in the xz-plane
- the x-axis if the rotation is in the yz-plane

Figure 4.3	Rotations occur in a plane about an axis that is perpendicular to that plane.

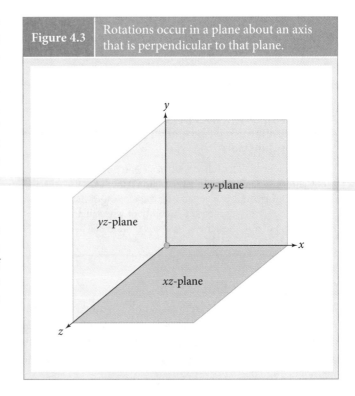

Figure 4.4	A rod rotating about an axis located at its leftmost edge, and rotating about a point in the center. All points on the body are rotating about the axis except the point at the axis of rotation.

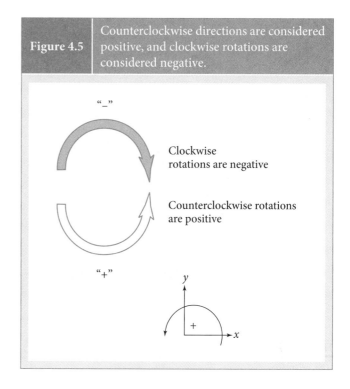

Figure 4.5	Counterclockwise directions are considered positive, and clockwise rotations are considered negative.

Clockwise rotations are negative

Counterclockwise rotations are positive

If the axis of rotation is located on the body, then no rotation is occurring at points along the axis. **Figure 4.4** shows you a rod rotating about two points. In the top figure, the rod is rotating about an axis located on its leftmost edge. In this case, a point on the center of the body is rotating. In the bottom figure, the rod is rotating about an axis located at its center. In this case, a point on the leftmost edge is rotating. In both cases, the rod is rotating in the xy plane, and the axis of rotation is parallel to the z-axis.

Now think of one of the spokes of a wheel. The spoke will rotate about a fixed line, which is the hub or axle. In mechanics, this fixed line is called the axis of rotation. Note that the axis of rotation may translate, or move, in space, but it is not rotating. The axis of rotation is always perpendicular (90°) to the direction of rotation. The wheel will rotate forward and backward, and the axis of rotation will be the axle, pointing side to side. With these basics in mind, you can begin to describe the angular motion.

As with linear motion, you need to first establish a frame of reference. With linear motion, you identified an origin and directional axes. With angular motion, you need to identify the axis of rotation and a reference axis. By convention, the positive axis of rotation will always be sticking out from the plane of the page. Rotation in the clockwise direction will always be negative, and rotation in the counterclockwise direction will be positive (**Figure 4.5**).

4.1.3 Angular Position (θ)

The first linear quantity examined was position. Now you wish to determine the **angular position**, or how far the body is rotated from the reference position. Linearly, position was denoted by a letter representing the axis or axes (x and y). The angular position is denoted by the Greek letter theta (θ). (All angular quantities are given Greek letters.) Angles can be measured as degrees (°), something that you are probably familiar with (there are 360° in a circle, perpendicular is 90°, an about-face is 180°, etc.). Another measure of angular position is radians (rads). Although degrees are probably more meaningful to you, radians will play an increasingly important part in biomechanics. If you are not familiar with them, check out **Box 4.1**.

> **Angular position** The orientation of a rigid body in reference to some axis

Reference frames and angular position are illustrated for the wheel by way of an example in **Figure 4.6**. The center of the circle represents the axle of the wheel, and the x-axis is our reference axis, representing the horizontal. The positive X direction is left to right, so that axis represents zero degrees. The thick lines represent the same spoke at different time points: A, B and C. The horizontal position, X, would be 0°. Positions A, B, and C would be 30°, 60°, and 205°, respectively.

Box 4.1	Essential Math: Radians

A radian is a ratio between an arc length (*s*) and the radius (*r*):

$$\text{radian} = \frac{s}{r}$$

The radius, as you recall, is the distance from the center to the perimeter of the circle. The arc length, *s*, is the length of a curve if you traced the curve with a piece of string, straightened it out, and then measured it. Why is this useful?

Every circle, regardless of its size, will have the same ratio between its circumference, *C*, and diameter, *D*:

$$\frac{C}{D} = \pi$$

This is a helpful measure because it is a ratio of an angular measurement (*C*), to a linear one (*D*). Because the diameter is equal to two times the radius (*r*), you can rewrite the equation as:

$$C = 2\pi r$$

But you also know that the diameter of any circle is equal to 360°. Substituting 360° for C, you get

$$360° = 2\pi r$$

And with a little algebraic manipulation:

$$r = \frac{360°}{2\pi}$$

$$r \approx 57.3°$$

So, to convert degrees to radians, you simply divide by 57.3. Radians, being a ratio of two lengths, have no units. So you can multiply an angular quantity by a linear quantity and get the units to match (in linear units). This is something you will see throughout the text.

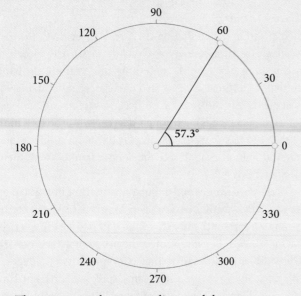

The conversion between radians and degrees.

Figure 4.6	The same spoke represented at different points in time (X = 0°; A = 30°; B = 60°; C = 205° or −155°).

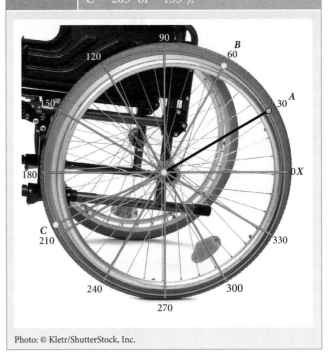

Photo: © Kletr/ShutterStock, Inc.

You should note that you could also represent the angular positions as −330°, −300°, and −155°, too, depending on how you got there (more on that in a minute).

4.1.4   Angular Displacement ($\Delta\theta$)

Recall that displacement was the change in position, represented in the following equation as:

$$\Delta p = p'' - p'$$

The angular equivalent, as you may have guessed, is **angular displacement**, which is the change in angular position. Equation 4.1 gives you:

> **Angular displacement**   The change in orientation of a rigid body in reference to some axis

$$\Delta\theta = \theta'' - \theta' \tag{4.1}$$

which tells you that the angular displacement is the difference in angular position between your two time periods of interest.

For example, say that spoke starts at the origin (X) at time point 0 and rotates to point B at time 1. The displacement would be:

$$\Delta\theta = \theta' - \theta$$
$$\Delta\theta = 30° - 0° = 30°$$

This means that the spoke rotated 30° in the positive direction. Now see what happens if the positions are reversed (the spoke rotates from Point A at time 1 back to X at time 2):

$$\Delta\theta = \theta'' - \theta'$$

$$\Delta\theta = 0° - 30° = -30°$$

This means the spoke rotated 30° in the negative direction. Just like linear displacement, "−30°" is not less than "+30°." The magnitudes of the displacements are identical, 30°. Case 2 just happens to be in the opposite direction of Case 1.

> **Important Point!**   Negative 30° is not less than positive 30°, it is just in the opposite direction.

Now look at what happens if you have two displacements, with the spoke rotating from X to A and then from A to B:

From X to A: $\Delta\theta = \theta' - \theta$

$$\Delta\theta = 30° - 0° = 30°$$

From A to B: $\Delta\theta = \theta'' - \theta'$

$$\Delta\theta = 60° - 30° = 30°$$

From X to A: $\Delta\theta = \theta'' - \theta$

$$\Delta\theta = 60° - 0° = 60°$$

Just like linear displacement, the angular displacement from t to t'' is equal to the angular displacement from t to t' plus the angular displacement from t' to t'', even though it is unnecessary to perform the first two calculations to get to the third. Again, this is not a problem as long as the direction did not change.

Now look at two different scenarios to prove the point. In both cases, the spoke starts at the origin (position X), at t' the spoke is at position A, at t'' it rotates to position B, and at t''' it rotates to position C. In the first scenario, it does this by displacements in the positive (clockwise direction):

From X to A: $\Delta\theta = \theta' - \theta$

$$\Delta\theta = 30° - 0° = 30°$$

From A to B: $\Delta\theta = \theta'' - \theta'$

$$\Delta\theta = 60° - 30° = 30°$$

From B to C: $\Delta\theta = \theta''' - \theta''$

$$\Delta\theta = 205° - 60° = 145°$$

From X to C: $\Delta\theta = \theta''' - \theta$

$$\Delta\theta = 205° - 0° = 205°$$

The second scenario starts off like the first: at t' the spoke is at position A, at t'' it rotates to position B, and at t''' it rotates to position C. However, at position B it changes direction and rotates clockwise to position C:

From X to A: $\Delta\theta = \theta' - \theta$

$$\Delta\theta = 30° - 0° = 30°$$

From A to B: $\Delta\theta = \theta'' - \theta'$

$$\Delta\theta = 60° - 30° = 30°$$

From B to C: $\Delta\theta = \theta''' - \theta''$

$$\Delta\theta = -155° - 60° = -205°$$

From X to C: $\Delta\theta = \theta''' - \theta$

$$\Delta\theta = -155° - 0° = -155°$$

Again, realize on a circle that 205° occupies the same position at −155° ($360° - 155° = 205°$). This example again illustrates how important it is to examine displacements over a small period of time. In the first case, everything works out the way it appears. In the second case, it would appear as though the spoke rotated just once: 155° in the negative direction (the net angular displacement). Yet the total angular distance was 275°. Even the angular distances are different in the two scenarios. In the first case, the angular distance is equal to the angular displacement, which is 205°. The second scenario the angular distance (275°) does not equal the angular displacement (−155°) or the angular distance of the first scenario (205°). And if both scenarios occurred over the same period of time, the spokes would have very different angular velocities.

> **Important Point!**   Important information is lost if the time intervals are too large.

4.1.5   Angular Velocity (ω)

Linear speed was defined as how fast a body was moving. Similarly, **angular speed** can be defined as how fast a body is rotating. Linear velocity was defined as how fast a body is moving in a particular direction and was mathematically expressed in the following equation:

> **Angular speed**   How fast a body is rotating

$$\text{velocity } (v) = \frac{\Delta \text{position}}{\Delta \text{time}} = \frac{\Delta p}{\Delta t} = \frac{p' - p}{t' - t}$$

The angular equivalent is **angular velocity**, denoted by the Greek letter omega, ω. Physically, it is how fast a body is rotating in a particular direction (in this case, positive or negative z-axis), or the rate at which the angular position changes, and is represented by Equation 4.2:

> **Angular velocity** How fast a body is rotating in a particular direction

$$\text{angular velocity } (\omega) = \frac{\Delta \text{angular position}}{\Delta \text{time}}$$

$$= \frac{\Delta\theta}{\Delta t} = \frac{\theta' - \theta}{t' - t} \qquad (4.2)$$

The direction, positive or negative, is about the axis of rotation (z-axis). The units are degrees (°/sec) or radians per second (rads/sec). Expressing angular velocity in °/sec is more appealing because you can visualize degrees better than radians. This is not a problem—most of the time. However, whenever you have to multiply the angular velocity by another quantity

(something explored later) you must make sure the angular velocity is in rads/sec before multiplying.

> **Important Point!**   Radians per second (rads/sec) and degrees per second (°/sec) are both measures of angular velocity.
> $$1 \text{ rad} = 57.3°$$

You can calculate the velocity between several different points (Table 4.1 for Scenario 1 and Table 4.2 for Scenario 2). The results are provided in Tables 4.3 and 4.4, respectively. In both scenarios, you ended up at an angular position of 205° in 3 seconds. By any standard, that is *extremely* slow, but it is useful in pointing out some things. First, the *average* velocities from A to C in Scenarios 1 and 2 are 68.33°/sec and −51.67°/sec. Just looking at that information, you would conclude that the wheel was rotating faster in Scenario 1. The angular speeds would tell a different story. As you might have

Table 4.1	Angular Position and Time for Scenario 1					
Data point	Location	$^z\theta(°)$	$\Delta^z\theta(°)$	$^z\theta(\text{rads})$	$\Delta^z\theta(\text{rads})$	$t(s)$
0	X	0	0	0	0	0
1	A	30	30	0.52	0.52	0.8
2	B	60	30	1.05	0.53	1.1
3	C	205	145	3.58	2.53	3.0

Table 4.2	Angular Position and Time for Scenario 2					
Data point	Location	$^z\theta(°)$	$\Delta^z\theta(°)$	$^z\theta(\text{rads})$	$\Delta^z\theta(\text{rads})$	$t(s)$
0	X	0	0	0	0	0
1	A	30	30	0.52	0.52	0.4
2	B	60	30	1.05	0.53	0.8
3	C	−155	−215	3.58	−3.75	3.0

Table 4.3	Change in Time, Change in Position, and Angular Velocity for Scenario 1				
	$\Delta t(s)$	$\Delta^z\theta(°)$	$\Delta^z\theta(\text{rads})$	$^z\omega(°/\text{sec})$	$^z\omega(\text{rad/sec})$
From X to A	0.8	30	0.52	37.5	0.65
From A to B	0.3	30	0.53	100	1.77
From B to C	1.9	145	2.53	76.3	1.33
From X to C	3	205	3.58	68.33	1.19

Table 4.4	Change in Time, Change in Position, and Angular Velocity for Scenario 2				
	Δt(s)	$\Delta^z\theta$(°)	$\Delta^z\theta$(rads)	$^z\omega$(°/sec)	$^z\omega$(rad/sec)
From X to A	0.4	30	0.52	75	1.30
From A to B	0.4	30	0.53	75	1.32
From B to C	2.2	−215	−3.75	−97.73	−1.70
From X to C	3	−155	−2.70	−51.67	−0.90

guessed, for Scenario 1 the angular speed and velocity are the same: 68.33°/sec. For Scenario 2, the average angular speed is 91.67°/sec! Because the angular speed is much higher than the angular velocity, you know the angular velocities must have been both positive and negative, which is verified by inspecting Table 4.4.

Tables 4.3 and 4.4 highlight the fact that different periods of interest have different average angular velocities. Hopefully, you are starting to get an appreciation for what is happening, but the perspective is still a bit limited. To get a true appreciation of the entire course of the movement, graph the angular position (top) and velocity (middle) as functions of time as in Figure 4.7.

Recall that graphically:

1. The velocity is the slope of the position–time curve.
2. The steeper the slope, the greater the velocity.
3. When the slope is negative, the velocity is also negative.
4. A change from a positive direction to a negative one (or vice versa) will always require the velocity to be zero during the transition.

The same principles apply to angular position and angular velocity. Scenario 1 started with zero angular velocity at 0° and end with zero velocity at 205°. So there will be a rise and fall of the angular velocity, but the angular velocity will always be positive (because the positive sign in angular velocity tells you that travel is in the positive direction). Scenario 2 is more complicated. Note that the first part of the graph looks like a steeper and compressed version of the entire graph of Scenario 1. That is because the wheel is starting with zero angular velocity at 0° and ending with zero angular velocity (however brief) at 60°. The second half of the graph looks similar to a mirror image of the entire graph of Scenario 1 as well. That is because the wheel is starting with zero angular velocity (however brief) at 60° and ending with zero angular velocity at −155°. The fact that there is a rapid reversal of direction means that there is going to be a brief moment when the angular velocity is zero. Rapid reversals in direction necessitate larger *peak* angular velocities: Notice

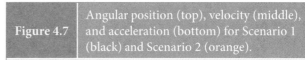

Figure 4.7   Angular position (top), velocity (middle), and acceleration (bottom) for Scenario 1 (black) and Scenario 2 (orange).

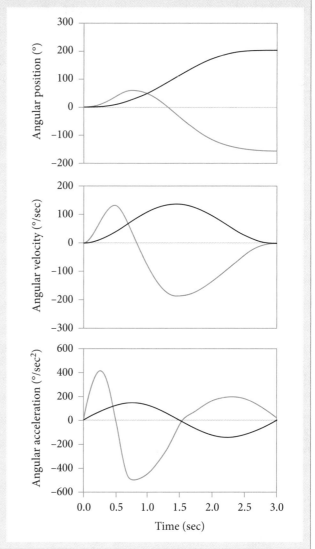

the peak negative angular velocity in Scenario 2 is almost twice that of the peak positive angular velocity in Scenario 1, even though the average angular velocities in Scenario 1 are larger than the average angular velocities in Scenario 2. It also means there is going to be larger angular accelerations.

> **Important Point!** Whenever there is a change in directions, the velocity will be zero at the instant the direction changes.

4.1.6 Angular Acceleration (α)

Analogous to linear velocity and acceleration, the time rate of change in angular velocity, or how quickly something is speeding up or slowing down its rotation in a particular direction, is called angular acceleration. It is denoted by the Greek lowercase α.

> Angular acceleration How quickly a body is speeding up or slowing down its rotation in a particular direction

Mathematically, linear acceleration was the rate at which linear velocity changes:

$$\text{acceleration } (\alpha) = \frac{\Delta \text{velocity}}{\Delta \text{time}} = \frac{\Delta v}{\Delta t} = \frac{v' - v}{t' - t}$$

Angular acceleration is the time rate of change of angular velocity. Remember, rates are always ratios. In this case, it is the ratio of the change in angular velocity to the change in time, or:

$$\text{angular acceleration } (\alpha) = \frac{\Delta \text{angular velocity}}{\Delta \text{time}}$$

$$= \frac{\Delta \omega}{\Delta t} = \frac{\omega' - \omega}{t' - t} \tag{4.3}$$

The units are degrees/sec^2 or rads/sec^2.

Let us calculate the acceleration between several different time points. The data are presented in Tables 4.5 to 4.8. As suspected, the angular accelerations are much larger in Scenario 2 because in the same amount of time (3 seconds) the spoke is covering a larger distance (70° more than Scenario 1) and switching directions. Both scenarios have positive and negative angular accelerations, but at different points. See what you can interpret from their graphs (bottom of Figure 4.7).

Once again, angular acceleration is the ratio of the change in velocity to the change in time. Graphically, ratios are represented by the slope of one value graphed as a function of the other. Comparing the middle and bottom graphs of Figure 4.7, angular acceleration is positive when the slope of the angular velocity–time curve is positive and negative when the slope is negative, as it should be. Just like in the previous examples

with linear acceleration, you will also notice that the angular acceleration is zero every time the angular velocity reaches a peak, either positive or negative, and the angular acceleration will be momentarily zero whenever there is a change from speeding up to slowing down (or vice versa).

> **Important Point!** The sign of the angular velocity provides information about the direction of travel; the sign of the angular acceleration does not.

Table 4.5	Time, Angular Position, and Angular Velocity for Scenario 1			
Data point	t(sec)	$^z\theta$(°)	$^z\omega$(°/sec)	$^z\omega$(rad/sec)
0	0	0	0	0.00
1	0.25	2.04	12.78	0.22
2	0.50	9.74	40.40	0.71
3	0.80	30.24	82.62	1.44
4	1.55	122.86	135.18	2.36
5	2.30	195.27	54.05	0.94
6	3.0	205.00	0	0.00

Table 4.6	Time, Angular Position, and Angular Velocity for Scenario 2			
Data point	t(sec)	$^z\theta$(°)	$^z\omega$(°/sec)	$^z\omega$(rad/sec)
0	0	0	0	0.00
1	0.25	9.57	75.88	1.32
2	0.50	39.30	131.01	2.29
3	0.80	60.00	14.59	0.25
4	1.55	−45.74	−184.68	−3.22
5	2.30	−143.78	−77.71	−1.36
6	3.0	−155	0	0.00

Table 4.7	Change in Time, Change in Angular Velocity, and Angular Acceleration for Scenario 1				
	Δt (sec)	$\Delta^z\omega$ (°/sec)	$\Delta^z\omega$ (rad/sec)	$^z\alpha$ (°/sec^2)	$^z\alpha$ (rad/sec^2)
From 0 to 1	0.25	12.78	0.22	51.12	0.89
From 1 to 2	0.25	27.62	0.48	110.48	1.93
From 2 to 3	0.30	42.22	0.74	140.73	2.46
From 3 to 4	0.75	52.56	0.92	70.08	1.22
From 4 to 5	0.75	−81.13	−1.42	−108.17	−1.89
From 5 to 6	0.70	−54.05	−0.94	−77.21	−1.35
From 6 to 7	3.0	0	0.00	0.00	0.00

Table 4.8	Change in Time, Change in Angular Velocity, and Angular Acceleration for Scenario 2				
	Δt(s)	$\Delta^z\omega$ (°/sec)	$\Delta^z\omega$ (rad/sec)	$^z\alpha$ (°/sec^2)	$^z\alpha$ (rad/sec^2)
From 0 to 1	0.25	75.88	1.32	303.52	5.30
From 1 to 2	0.25	55.13	0.96	220.52	3.85
From 2 to 3	0.30	−116.42	−2.03	−465.68	−8.13
From 3 to 4	0.75	−199.27	−3.48	−265.69	−4.64
From 4 to 5	0.75	106.97	1.87	142.63	2.49
From 5 to 6	0.70	77.71	1.36	103.61	1.81
From 6 to 7	3.0	0.00	0.00	0.00	0.00

The graphs of Scenario 1 should start to look familiar to you by now. The graphs of Scenario 2 may seem a bit strange to you, but it is not because they are graphs of angular data. Recall that at 60°, the wheel rapidly changed directions from positive to negative about the z-axis: there was no pause like there was in the previous examples. The spoke went from slowing down in the positive direction (negative acceleration) immediately to speeding up in the negative direction (also negative acceleration). Had there been a pause (when the acceleration would have been zero), there would have been a gap between the two acceleration periods (similar to the squat graphs). Because there was not, the two negative sections merged together like you see in Figure 4.7. Finally, toward the end of the movement in Scenario 2, the spoke is slowing down in the negative direction, which would produce a positive acceleration. Speeding up in a positive direction and slowing down in a negative direction both produce a positive acceleration; the same principles apply for both linear and angular acceleration.

The other principle you should recall from linear acceleration is that if an object starts and ends with the same velocity (in this case zero), then the change in acceleration is zero. If the change in acceleration is zero, the areas under the positive and negative acceleration–time curves must be equal. Although it might be hard to tell, this is the case in Figure 4.7.

Important Point!   If a rotating body does not change its angular velocity, the areas under the positive and negative acceleration–time curves must cancel.

Throughout this lesson, the angular measurement in both degrees and radians were included where possible. This was done just to show you that the difference between the two is a simple scaling factor: 1 radian equals 57.3°. Review Box 4.1 again if you are still unsure. Throughout the remainder of the

book, only one or the other will be used. When describing motion, degrees will be used because it is more familiar. Radians will be used when multiplying an angular quantity by some quantity (as will be done in Section 4.2).

4.1.7   Comparing Linear and Angular Kinematics

Most students have a difficult time grasping angular kinematics. The rotations may be a bit tricky, but it will be easier if you relate the angular measure to its linear cousin. Physically, position is a location—some distance from the origin. Angular position is the orientation (in degrees) of a rigid body in relation to a reference axis. Displacement is the change in position; angular displacement is the change in angular position, or orientation. Displacement, either linear or angular, is always *just* the change in position, whereas distance is how far the body traveled to get to that new position. Velocity is the amount of displacement in a given amount of time: how quickly something is moving in a particular direction. Angular velocity is the amount of angular displacement in a given amount of time, or how quickly a body is rotating in a particular direction. Finally, acceleration is the change in velocity in a given amount of time, or how quickly is a body speeding up, slowing down, or changing direction. Angular acceleration is the change in angular velocity in a given amount of time: how quickly is a rotating body speeding up, slowing down, or changing direction.

Important Point!   You should be able to describe each concept physically, mathematically, and graphically.

Mathematically, the linear and angular quantities are also very similar. Just remember to use the Greek lowercase letters for the angular quantities. The formulae are summarized in Table 4.9. Units for linear quantities are meters (or feet), meters per second, or meters per second per second (per second squared). Units for the angular quantities are degrees (or radians), degrees per second, or degrees per second squared.

Table 4.9	Comparing Linear and Angular Kinematic Variables	
	Linear	**Angular**
Displacement	$\Delta p = p' - p$	$\Delta\theta = \theta' - \theta$
Velocity	$v = \dfrac{\Delta p}{\Delta t}$	$\omega = \dfrac{\Delta\theta}{\Delta t}$
Acceleration	$a = \dfrac{\Delta v}{\Delta t}$	$\alpha = \dfrac{\Delta\omega}{\Delta t}$

A picture is worth a thousand words, so graphs are powerful visual aids. In addition, they allow you to examine the motion at every instant in time. Velocities and accelerations are ratios. Graphically, a ratio is slope: a change in one thing over a change in other. For kinematics, it is always over a change in time. Velocity is the change in position over a change in time, or the slope of the position–time curve. So angular velocity is change in angular position over the change in time, or the slope of the angular position–time curve. Acceleration is change in velocity over the change in time: the slope of the velocity–time curve. And (you guessed it) angular acceleration is the ratio of the change in angular velocity over the change in time. It is slope of the angular velocity–time curve.

Important Point!   Linear concepts are represented by Roman letters, whereas angular concepts are represented by Greek letters.

Section Question Answer

Angular motion is similar to linear motion in that there are positions, displacements, velocities, and accelerations. The relation between position, velocity, and acceleration are also the same. The main difference is that you are replacing a linear term with its angular equivalent.

COMPETENCY CHECK

Remember:

1. Define the following: rigid body, axis of rotation, angular position, angular displacement, angular velocity, and angular acceleration.
2. List the formulae for the following: angular displacement, angular velocity, and angular acceleration.

Understand:

1. Manipulate each equation so that the output you want is on the left-hand side, and the input is on the right-hand side. Explain how changing each of the inputs will either increase or decrease the output.
2. Calculate the change in angular position for the following:

$\theta(°)$	$\theta'(°)$	$\Delta\theta(°)$
0	100	
10	15	
0	150	
225	100	

3. Calculate the angular velocity for the following:

$\theta(°)$	$\theta''(°)$	$\Delta\theta(°)$	Δt	$\omega(°/sec)$
0	100		1 sec	
30	15		0.5 sec	
90	15		0.01 sec	
15	225		5 sec	

4. Calculate the angular acceleration for the following:

$\omega'(°/sec)$	$\omega''(°/sec)$	$\Delta\omega(°/sec)$	Δt	$\alpha(°/sec^2)$
1	10		1 sec	
15	10		0.5 sec	
0	70		0.01 sec	
225	100		5 sec	

Apply:

1. List activities for which angular motion would be important.

4.2   RELATING ANGULAR KINEMATICS TO LINEAR KINEMATICS

Section Question

Determining the angular velocity of the wheel is interesting, but what does that mean? For a given angular velocity, just how fast is the wheelchair actually going?

Angular motion is an important part of human movement. Motions at the individual joints of the human body are largely angular motions. Wheelchairs and bicycles (Figure 4.8), throwing things like the discus and softball (Figure 4.9), and swinging of bats and racquets (Figure 4.10) are all examples of activities that involve predominately angular motion. But in all these examples, angular motion is a means to an end. If you are interested in the outcome, then you want to know the linear speed of the wheelchair or bike, the linear speed of the end of the bat or racquet, or the linear distance of the discus or hammer.

4.2.1   The Relation between Linear and Angular Velocity

The linear velocity of any point on a rotating body is determined by two things: the angular velocity of the rigid body and the distance from the axis of rotation to the point on the body (Figure 4.11). Review Section 4.1.5 for a discussion on angular velocity. For the wheel, the distance is pretty simple: it is simply the distance from the wheel's axis of rotation to

Figure 4.8 | Wheelchairs' and bicycles' linear velocity is determined by angular velocity of the wheel and its radius.

(left) © Chris Mole/ShutterStock, Inc., and (right) © Diego Barbieri/ShutterStock, Inc.

Figure 4.9 | A projectile's trajectory is determined by its angle of release and velocity of release. The velocity of release of a softball or discus is determined by the angular velocity of the arm and length.

(left) © Glen Jones/ShutterStock, Inc., and (right) © Mark Herreid/ShutterStock, Inc.

| **Figure 4.10** | How an object responds to being struck is going to be determined by the mass and velocity of the striking implement. The linear velocity of the implement is determined by its angular velocity and length. |

(left) © Jeff Thrower/ShutterStock, Inc., and (right) © Neale Cousland/ShutterStock, Inc.

| **Figure 4.11** | A model of the determinant of linear velocity of a point on a rotating body. |

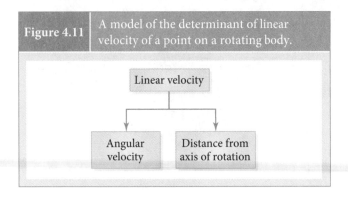

the tire (or the length of the radius, l_r). Mathematically, this relation is expressed as:

$$v = l_r \times \omega \qquad (4.4)$$

You can extract two key pieces of information from this equation. The first is intuitive: If you increase the angular velocity of the tires, then you will increase the linear velocity of the wheelchair. The second is that if you increase l_r,

or the distance from the axis of rotation to the tire (i.e., the length of the spoke), then you also increase the speed of the wheelchair.

> **Important Point!** Two things determine the linear velocity of a point on a rotating body:
>
> 1. The angular velocity of the body
> 2. The distance from the axis of rotation to the point

Let us look at an example. Suppose the length of the spokes (radius) of the chair is 0.3 meters and the angular velocity is 180 degrees/sec. (Remember to change the angular velocity from degrees/sec to radians/sec). The linear speed of the wheelchair would be

$$v = l_r \times \omega$$

$$v = 0.3 \text{ m} \times 3.14 \text{ rads/sec}$$

$$v = 0.94 \text{ m/sec}$$

Now what happens if you increase the radius of the wheel by 25%, from 0.3 meters to 0.375 meters? The new linear velocity would be:

$$v = l_r \times \omega$$

$$v = 0.375 \text{ m} \times 3.14 \text{ rads/sec}$$

$$v = 1.18 \text{ m/sec}$$

which is a 25% increase in the linear velocity of the wheelchair. Because the two terms, l_r and ω, are multiplied together, an increase in either one of the other will increase the linear velocity by the same amount. This is speaking in strictly mechanical terms. See Box 4.2.

Recall that velocity has both a magnitude *and* direction. What is the direction of the velocity vector in Equation 4.4? It is always perpendicular (or at right angles) to the rigid body, or tangent to the arc (or circle) that the point makes in space. See Figure 4.12. Using your trig functions, you can always determine the components of this vector (with the angle measured in radians):

$$^xv = {}^Rv \times -l_r \times \sin\theta \qquad (4.5)$$

$$^yv = {}^Rv \times l_r \times \cos\theta \qquad (4.6)$$

Determining the components for a wheel is normally not *that* critical because the vector is parallel to the ground when the wheel is in contact with it (Figure 4.13). It is more important when considering things like striking and throwing because the angle is important when the object is struck or released.

Important Point!   Sine and cosine are switched from their normal uses (cos for *x* and sine for *y*). This is because the tangential is 90° from the rotating body.

Box 4.2	Applied Research: The Velocities Produced Using Different Rear Wheel Diameters During Manual Wheelchair Propulsion

In this investigation, the average speed of propulsion over smooth concrete was compared for wheelchairs with rear wheel diameters of 64, 62, and 58 centimeters. Equation 4.4 would predict that the wheelchair with the largest diameter would produce the largest linear speed. This was not the case: The wheelchair with the diameter of 62 cm produced the largest speed, followed by the chairs with the 64 cm and then 58 cm diameter wheels. Why would a larger wheel not produce a larger linear speed? The answer will be explored later in this text.

Data from: Goswami A, Ganguli S, Chatterjee BB. Ergonomic analysis of wheelchair designs. *Clinical Biomechanics* 1986;1(3):135–139.

Figure 4.12	The resultant linear velocity (black) and its components (orange) from a rotating body.

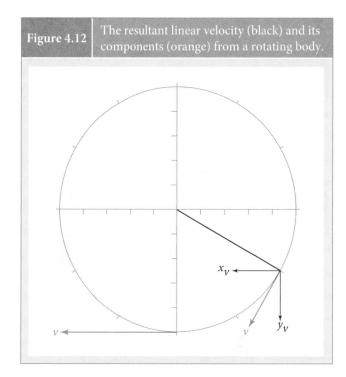

Figure 4.13	The linear velocity vector of the wheel is parallel to the ground during contact.

Photo: © Ljupco Smokovski/ShutterStock, Inc.

4.2.2   The Relation between Linear and Angular Acceleration

Remember that a vector has both a magnitude and a direction. If the velocity vector changes direction, even if the magnitude did not change, there would still be an acceleration. To see how this works, consider the hammer thrower in Figure 4.14.

Figure 4.14 | The hammer throw.

If you are not familiar with it, hammer throwing is an Olympic event in which the objective is to throw a heavy metal ball (7.257 kg for men and 4 kg for women) attached to a wire and handle (121.5 cm long for men and 119.5 cm long for women) as far as possible. You know from the lesson on projectile motion that the range of the hammer will be determined by the velocity at release, angle at release, and relative height. You also know that the velocity term is the most important (because it is squared). You now know that the linear velocity of the hammer will be determined by the angular velocity of the hammer and the length of the wire. If the length of the wire is fixed, linear velocity will be determined by the angular velocity of the hammer. Throwers usually complete several revolutions trying to get the angular velocity as high as possible for release. But they have to make sure they release it in the proper direction (toward the front of the circle and into the landing area).

For the purposes of this illustration, imagine that the hammer has attained its maximum velocity and is constant during the last revolution. Because the angular velocity is constant, the change in angular velocity ($\Delta\omega$) is zero, and the ratio of the change in angular velocity to the change in time (or angular acceleration) is also zero. If the angular velocity is constant, the magnitude of the linear velocity is, too. You know that the direction of the linear velocity vector is tangential to the circular path. That has to change because the hammer is making a circle with each revolution. So you have to be able to account for changes in the magnitude of the velocity vector and its direction.

There are two linear accelerations to consider for each angular velocity. One is tangential direction, just like the linear velocity. It represents the change in magnitude of the linear velocity. Its formula is

$$^{tan}a = l_r\alpha \qquad (4.7)$$

As you would expect if the angular acceleration is zero, the magnitude of the change in linear velocity is zero (if α is zero, ^{tan}a must also be zero). The other represents a change in direction of the linear velocity and is called the centripetal acceleration, $^{centripetal}a$. Its direction is always toward the center of the circle (axis of rotation). See **Figure 4.15**. The formula for the centripetal acceleration is

$$^{centripetal}a = \frac{v^2}{l_r} \qquad (4.8)$$

From Equation 4.4, you know that:

$$v = l_r \times \omega$$

So

$$v^2 = l_r^2\omega^2$$

| Figure 4.15 | Centripetal acceleration represents the change in the direction of the velocity vector. It always points toward the center of the circle. |

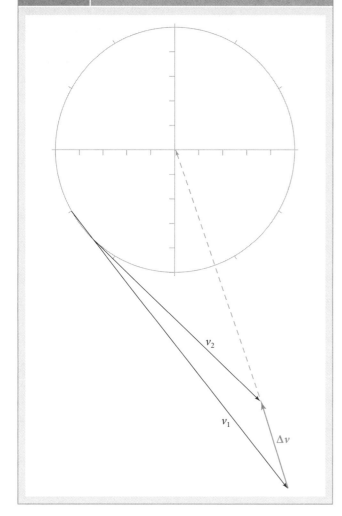

Substituting into Equation 4.8, you get an alternate equation for centripetal acceleration:

$$^{\text{centripetal}}a = l_r\omega^2 \tag{4.9}$$

Inspecting Equation 4.9 shows you that even if the angular velocity is constant, and the angular acceleration is zero, there will still be a centripetal acceleration. Although it may not appear so now, these equations will become important when discussing forces inside the body, at the various joints.

Important Point! There are two components to linear acceleration of a rotating body:

1. Tangential—which is due to the change in speed
2. Centripetal—which is due to the change in direction

SUMMARY

In this lesson, the concepts of one- and two-dimensional kinematics were expanded on to describe angular motion (Table 4.10). You should be able to describe each of these concepts physically, mathematically, and graphically. Angular position is the orientation (in degrees) of a rigid body in relation to a reference axis. Angular displacement is the change in angular position, or orientation. Angular velocity is the amount of angular displacement in a given amount of time, or how quickly a body is rotating in a particular direction and is represented by the slope of the angular position–time curve. Angular acceleration is the change in angular velocity in a given amount of time: how quickly is a rotating body speeding

Table 4.10	Key Concepts

- Rigid body
- Axis of rotation
- Angular position
- Angular distance
- Angular displacement
- Angular speed
- Angular velocity
- Angular acceleration

up, slowing down, or changing direction. It is the slope of the angular velocity–time curve. Can you apply these concepts to human movement using your own examples?

REVIEW QUESTIONS

1. Which has the greatest speed, 20 rad/sec, or −40 rad/sec? Why?

2. Would a point 10 cm away from an axis rotating at 10°/sec and a point 30 cm away from an axis rotating at 10°/sec have the same linear speed? Why or why not?

3. Pick any angular movement, and sketch a graph of position versus time, velocity versus time, and acceleration versus time.

4. List two ways you can increase the linear velocity of a point on a rotating body. Which factors can you control? Which one would have the biggest impact? Why?

REFERENCE

1. Kaye HS, Kang T, LaPlante MP. Mobility device use in the United States. Disability Statistics Report 14. 2000. Washington, DC, U.S. Department of Education, National Institute on Disability and Rehabilitation Research.

Lesson 5

Describing Motion: Inertia and Momentum

LEARNING OBJECTIVES

After finishing this lesson, you should be able to:

- Define the following terms: inertia, mass, momentum, center of mass, and moment of inertia.
- Explain how mass is related to weight.
- Describe how inertia is different for a stationary body, linearly moving body, and rotating body.
- Write the equations for linear and angular momentum.

Kinematics is the branch of mechanics that studies motion without worrying about the causes of that motion. Consider the question, Should all motions be treated as equal? This lesson will extend your study of kinematics to answer that question. The causes of motion are still not a concern, per se, although the lines are going to start to get blurred.

Section Question

My car can go from 0 to 60 mph in about 7.5 seconds (Figure 5.1). Elite sprinters can reach their top speed at around 5 to 6 seconds, or about 50 to 60 meters into the race.[1] Why can't my car go from 0 to 60 mph right now—instantaneously? Why does it take runners so long to get up to their top speed? Why do they not reach their top speed right out of the blocks?

| Figure 5.1 | Why does is take 7.5 seconds for my car to go from 0 to 60 mph? Why does it take 50 to 60 meters before a runner reaches top speed? |

(left) © Olaru Radian-Alexandru/ShutterStock, Inc., and (right) © Rob Bouwman/ShutterStock, Inc.

The answer to these questions is one word: **inertia**. Inertia is a body's resistance to change in velocity. Remember that velocity has

> **Inertia**   A resistance to change in motion, specifically a resistance to change in a body's velocity

both a magnitude and a direction. The greater a body's inertia, the harder it is to change its velocity (get it moving, stop it moving, speed it up, slow it down, or change its direction). The *concept* of inertia is the same regardless of what type of motion you are talking about: a body at rest, a body moving linearly, or a body rotating; it has a resistance to change (in either magnitude or direction) of velocity. How you *determine* its inertia, though, will be different in each of these three cases. Let us look at each one.

5.1   INERTIA FOR A BODY AT REST: MASS (*m*)

> **Section Question**
>
> Why is it harder to push a tractor-trailer than it is a "smart" car (**Figure 5.2**)?

You might be quick to say that the tractor trailer is heavier, or weighs more. But you would not be entirely accurate.

Weight is a force due to gravity, and it only acts in the downward, vertical direction—toward the center of Earth. You are not opposing gravity because you are pushing both vehicles in the horizontal direction, so their resistance to change in

> **Weight**   The force due to gravity; weight always acts in the downward, vertical direction and has a magnitude of 9.81 m/sec² times the body's mass
>
> **Mass**   The amount of matter in an object

velocity (inertia) is not due to weight (**Figure 5.3**). But you do know that the heavier truck is harder to push, so it is related to its weight. What determines an object's inertia when it is motionless is its **mass**.

Many physics and biomechanics textbooks define mass as "the quantity of matter." That definition is a bit dry. Look at it this way: everything around us (including us) is made up of "stuff." Mass, then, is the amount of "stuff" in an object. For our purposes, let us say that the "stuff" we are talking about is molecules. Larger molecules, a greater number of molecules, or a greater density of the molecules in a body will all result in an object having a larger mass (more stuff).

A very simple representation of the human body will have two components: fat mass, and fat-free mass. Let us look at how mass changes by examining two identical twins with the

Figure 5.2	Why is it harder to push a tractor-trailer than a smart car?

| Figure 5.3 | Weight only acts in the vertical, downward direction. |

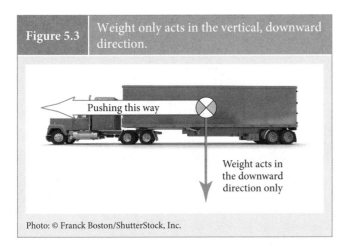

Pushing this way

Weight acts in the downward direction only

Photo: © Franck Boston/ShutterStock, Inc.

exact same body composition. If one of the twins begins a resistance exercise program, she can increase the size of her muscle fibers (cells), which would increase her mass. If the other twin did not exercise and ate a lot of high-fat foods, she would increase both the size and number of fat cells, also increasing her mass. If one of the twins increased the size of her muscle cells while the other increased the size of her fat cells by the same amount, the first twin would have a larger increase in mass because muscle is denser (more matter in the same amount of space) than fat.

How is mass related to weight? The two identical twins of identical body composition would both have identical mass because they are made up of the same amount of stuff. If one of the twins goes to the moon tomorrow while the other stays on Earth, they are both still made up of the same amount of stuff. Yet the twin on the moon would weigh less because the gravitational constant on the moon is less. Although weight is proportional to mass on the surface of Earth, it can change if the gravitational constant changes, and it only acts in the vertical, downward direction (Figure 5.3). Mass, the amount of stuff in an object, does not change in the short term (for most bodies) and is a measure of inertia in all directions. An implication of excess body mass on biomechanics is discussed in Box 5.1.

Important Point!   Weight is a vector, mass is a scalar. Mass does not have direction and can never be negative.

In most biomechanics contexts, the body being discussed is solid, and the mass would not change perceptibly during the time period of interest. If you were studying the motion of a rocket in physics, for example, the mass of the rocket would be changing as it was burning fuel. But for the purposes of this book, unless otherwise noted, always assume the mass does not change in the short term.

| Box 5.1 | Applied Research: Artificially Increasing Body Mass Decreases Sprint Times |

Increasing body mass can be a double-edged sword when it comes to human performance. In this investigation, 20 athletes sprinted in a normal condition and with vests that increased their mass by 10% and 20%. Increasing body mass by 10% increased their 10-meter sprint times by 7.5% and their 30-meter sprint times by 10.0%, whereas increasing their body mass by 20% increased their 10- and 30-meter sprint times by 9.3% and 11.7%, respectively. Yet, increasing muscle mass is a by-product of increasing strength. In almost all cases, increasing fat mass will have deleterious effects on performance. With increasing muscle mass, there is probably a point at which the positive effects of increasing force production are offset by the negative effects of increased body mass. This point is most likely different for every activity and may even be different for each individual.

Data from: Cronin J, Hansen K, Kawamori N, McNair P. Effects of weighted vests and sled towing on sprint kinematics. *Sports Biomechanics* 2008;7(2):160–172.

Section Question Answer

Your car cannot instantaneously go from 0 to 60 mph because of its mass. It is the same with the runners. Objects with a larger mass are harder to get moving (or up to speed).

COMPETENCY CHECK

Remember:

1. Define: inertia and mass.

Understand:

1. What is the difference between mass and weight?

Apply:

1. How might you decrease your inertia?

5.2 INERTIA FOR A BODY MOVING LINEARLY: LINEAR MOMENTUM (*L*)

Section Question

Do you have the same inertia running at your submaximal speed as you do when running at your maximum speed?

Because changes in body composition take time, you would not instantaneously change your mass. It has already been stated that mass is a measure of inertia, and that inertia is a resistance to change (in either magnitude or direction) of velocity. If mass were the only measure of inertia, it would be just as easy to speed up, slow down, or change direction if you were running at a slow jog (1 m/sec) or an all-out sprint (9 m/sec). In your experience, is that the case?

Important Point!   A change in velocity can be a change in magnitude and/or a change in direction.

Probably not: you have no doubt noticed that it is harder to slow down or change direction if you are moving very fast. Conversely, it is easier to increase your speed if you are already moving in that direction than if you are standing still. Mass does not tell the *whole* story of inertia if the body is moving; velocity also plays a part.

But mass does tell *part* of the story. Consider a tractor-trailer and a "smart" car. If both of them were moving at 20 mph, which one would be harder to stop or turn (change direction)? You are correct if you said the tractor-trailer. Why? Because it has more mass. So both mass *and* velocity are important when determining the inertia of a moving body.

This characteristic of a moving body is called its **momentum**. Most physics and biomechanics textbooks will echo Sir Isaac Newton and define momentum as "the quantity of motion." This definition is somewhat unsatisfactory. What exactly does "quantity of motion" mean? How much motion it has? In a way, yes. Physically, it is the resistance to change in velocity *of a moving body*. Mathematically, you would expect both mass and velocity to show up in the equation, and it does:

> **Momentum**   A resistance to change in velocity of a moving body

$$L = m \times v \qquad (5.1)$$

Linear momentum (L) is the product of mass and velocity, which are both "weighted" equally. In other words, doubling either the mass or the velocity will increase the linear momentum by a factor of two.

Look at a few examples in which momentum is changing during an activity. Consider three runners. Knowing the mass of each runner, and their top speed, you can calculate their change in momentum from the start of the race to their peak speed (Table 5.1). At the start, all three runners have the same momentum (because they all start with zero velocity). At this point, their inertia is determined solely by their mass. As they are running, their inertia is determined by their linear momentum: At the peak velocity of the run, Runner B has greater momentum than Runner A because he has more

Table 5.1	Inertial Characteristics of the Three Runners From the Start to Their Peak Velocity		
	Runner A	**Runner B**	**Runner C**
Mass (kg)	75	80	82
Start v (m/sec)	0	0	0
Start L (kg·m/sec)	0	0	0
Peak v (m/sec)	11.76	11.76	11.11
Peak L (kg·m/sec)	882.00	940.80	911.02
ΔL (kg·m/sec)	882.00	940.80	911.02

mass and they have the same velocity. Even though Runner C has more mass than Runner A or B, he has less momentum than Runner B because Runner B is running at a much higher velocity. Runner C does have greater momentum than Runner A even though he is slower because he has more mass.

In another example, a 62-kg female volleyball player jumps in the air to block a ball (Figure 5.4). On her way down, just before impact, she has a velocity of 2.8 m/sec. Just

Figure 5.4	As the volleyball player lands from her jump, her momentum will change from −173.69 kg · m/sec to 0 kg · m/sec.

Photo: © muzsy/ShutterStock, Inc.

| Figure 5.5 | (a, b) A 30° sidestep cutting maneuver in soccer. (c) The change in momentum associated with a 30° sidestep cutting maneuver. |

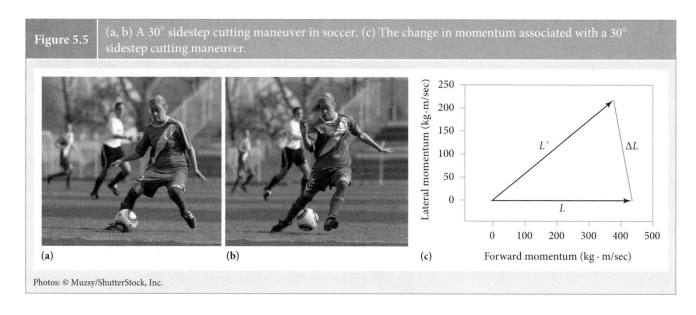

(a)   (b)   (c)

Photos: © Muzsy/ShutterStock, Inc.

before impact, then, her linear momentum is −173.69 kg·m/sec. (Remember, the negative sign means the negative direction, downward). After she lands, her momentum will go to zero, so her ΔL is the same amount but the opposite direction, +173.69 kg·m/sec.

> **Important Point!**   The negative sign in momentum indicates direction. The sign of momentum is always the same as the sign of velocity.

Finally, examine a 58-kg soccer player who is running 7.5 m/sec and then performs the cutting maneuver, changing her direction by 30° but maintaining her speed (Figure 5.5). You could think of this in one of two ways. First, the magnitude of her momentum does not change, only the direction. The second way is to examine the components of her momentum. Before the cut, she has 435 kg·m/sec in the x direction. After the cut, she has 376.72 kg·m/sec in the x direction and 217.5 kg·m/sec in the y direction. ΔL would then be −58.28 kg·m/sec in the x direction and 217.5 kg·m/sec in the y direction.

In each case, linear momentum is always proportional to, and in the direction of, velocity. But for momentum, the velocity is increased by a factor equal to the mass. Increasing either mass or velocity will increase linear momentum. Controlling linear momentum is an important consideration when examining both performance (Box 5.2) and injury (Box 5.3).

Inertia also helps explain how an athlete can elude a defender.[2] A defender in pursuit would be traveling at a high speed. If the athlete being pursued changes direction at the last possible moment, the defender could run right by him due to

his large momentum. This would only work if the defender is almost upon him; if he changes direction with too much distance between him and the defender, the defender will have time to react and adjust his momentum accordingly.

There is another, important point concerning linear momentum: if a body starts at zero velocity and ends at zero velocity, the total (or net) change in velocity is zero. If the total change in velocity is zero, the total change in momentum must also be zero. This does not mean that the momentum is not changing throughout the movement, only that the *net* change from the beginning to the end of the movement is zero.

| Box 5.2 | Applied Research: A Biomechanical Analysis of Failed Sit-to-Stand |

To stay functionally independent, older adults have to be able to complete several tasks, such as walking, negotiating stairs, and rising from a chair. The latter is also known as a sit-to-stand (STS). In this investigation, two types of "errors" were identified when elderly were unsuccessful in performing a STS. The first type of error was a "sitback," where the subjects were unable to generate sufficient momentum to stand up, and thus fell back into the chair. The second type of error was a "step," where the subjects were unable to control the direction of the momentum they did generate and had to take a step either backward or laterally to regain their balance. The causes of these errors require further investigation.

Data from: Riley PO, Krebs DE, Popat RA. Biomechanical analysis of failed sit-to-stand. *IEEE Transactions in Rehabilitative Engineering* 1997;5(4):353–359.

Box 5.3	Applied Research: Changes in Momentum May be Implicated in Knee Injuries

In this investigation, subjects ran at 3 m/sec under five different conditions: (1) straight ahead, (2) 30° sidestep cutting maneuver, (3) 30° crossover step maneuver, (4) 60° sidestep cutting maneuver, and (5) 60° sidestep cutting maneuver. Momentum decreased 13.4% and 50% in the x direction and increased 50% and 13.4% in the y direction for the 30° and 60° cuts, respectively. This led to loading on the knee that was two to six times larger than the straight-ahead condition. Faster running would lead to larger changes in momentum and even larger loading on the knee. These findings may help explain why these maneuvers are associated with a large number of noncontact ACL injuries.

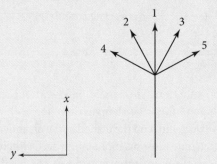

Running straight ahead (1), a side-step cut at 30° (2) and 60° (4), and a cross-over cut at 30°(3) and 60°(5).

Data from: Besier TF, Lloyd DG, Cochrane JL, Ackland TR. External loading of the knee joint during running and cutting maneuvers. *Medicine and Science in Sports and Exercise* 2001;33(7):1168–1175.

Important Point!   If a body starts with zero velocity and ends with zero velocity, the total change in momentum is zero.

To illustrate these two points, return to the example of the "up" phase of a very fast squat. The velocity started at zero, increased very rapidly, reached a peak, and then returned to zero. Suppose the athlete weighed 55 kg and squatted 1.5 times her body weight (82.5 kg). To be precise, you would have to account for the fact that her entire body is not moving (for example, her feet are stationary during the squat). But to keep the example simple, ignore that and say the mass

System	The bodies under consideration

of the **system** (body plus barbell) that is moving is 137.5 kg, and the velocity of the system's mass is the same as the velocity of the barbell in the earlier example. The graph is represented in **Figure 5.6**.

Figure 5.6	Momentum and velocity of the up phase of the squat. Note that momentum is always proportional to and in the direction of velocity.

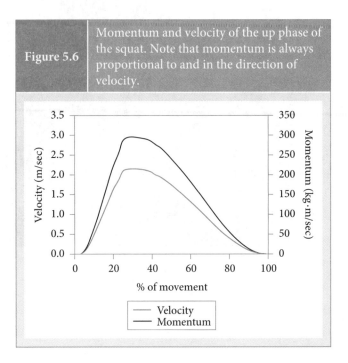

By examining the figure, you should be able to easily recognize the two points made earlier. First, at each instant in time, the momentum is proportional to, and in the direction of, the velocity. Second, the momentum started at zero and ended at zero, but varied in magnitude between these two times.

Section Question Answer

Even though you have the same mass, you do not have the same inertia with different running speeds. The faster you are going, the harder it is to stop or change direction. That is because you have more momentum with increased velocity.

COMPETENCY CHECK

Remember:

1. Define: momentum.
2. Write the equation for momentum.

Understand:

1. What happens to your momentum if you increase your mass? If you increase your velocity?
2. Will increasing mass or velocity have a more important effect on momentum? Why?
3. Who has the greater momentum: an 90-kg person running at 87 m/sec or a 80-kg person running at −8 m/sec?

5.3   INERTIA FOR ANGULAR MOTION

5.3.1   Moment of Inertia (*I*)

> **Section Question**
>
> Try this experiment (Figure 5.7): Pick up a baseball bat by the handle and swing it (Condition A). Now, slide your hands up higher on the handle (called "choking up"), and swing the bat with the same angular velocity (Condition B). Finally, turn the bat over and grab it by the barrel, and again swing it with the same angular velocity as you did the previous two times (Condition C). In which condition was it easiest to swing the bat? Which one was the hardest? Why?

Even if you did not perform this experiment but had some experience with bats, you would have easily determined that Condition C was the easiest to swing, followed by Condition B. Condition A was the hardest to swing. Why is that? After all, the mass of the bat did not change between conditions, and neither did the velocity. If how "easy" or "hard" it is to swing

| Figure 5.7 | Swinging a bat from the wrist: (a) at the handle; (b) choking up on the handle; and (c) at the barrel. |

something is a qualitative description of inertia, neither the mass nor the linear momentum adequately captures it. So what else do you need to know? What was different between the three conditions?

> **Qualitative**   Subjectively describing something using words rather than measuring it

Instead of the mass, it was how that mass was distributed relative to the axis of rotation. Depending on how you chose to swing the bat, the axis of rotation could have been your shoulder, elbow, or wrist. It really does not matter for the purposes of this discussion, so just say that it was your wrist. Most of the mass is concentrated in the barrel, with less mass near the handle. So in Condition C, most of the mass was located closer to the axis of rotation, making it easier to swing. Conversely, in Condition A, most of the mass was furthest away from the axis of rotation, making it more difficult to swing. As you choked up on the bat (Condition B), the barrel is getting closer to the axis of rotation, making it easier to swing.

Recall that every linear quantity has an angular analog. The angular equivalent to mass is something called the moment of inertia, whose symbol is the capital *I*. The moment of

> **Moment of inertia**   The angular equivalent of mass, gives an indication of how difficult it will be to rotate an object

inertia depends not only on how much mass something has, but also on how that mass is distributed in relation to the axis of rotation. It is not something that can be easily calculated because it is different for different axes (directions) and different locations of the axis of rotation. Because of this, it is usually determined experimentally. A very simple formula can be used to determine *I*, but first a new concept needs to be introduced, the center of mass (COM).

> **Center of mass**   A fictitious point where all the mass is considered to be concentrated

The COM is an abstract concept that you use to help visualize what is going on with a rigid body. It also makes it easier to perform calculations. You know that the mass of a rigid body is distributed throughout the body. But a rigid body behaves as if all that mass is concentrated into a single point, which is located at the COM. The COM is located at the geometrical center of the body if the body is symmetric with a uniform density (like a ruler). If the body has a uniform density but is not symmetric (like a baseball bat), the COM is located toward the larger area. Bodies with nonuniform density (like the human body) are harder to predict where the COM is located. The COM can even be located outside the body: the COM of a donut is located at the hole in the middle!

The formula for the moment of inertia for a point mass is

$$I_{\text{point mass}} = m\rho^2 \qquad (5.2)$$

where I is the moment of inertia, m is the mass, and ρ is the distance from the axis of rotation to the COM. You should be able to extract two very important pieces of information from this formula. First, increasing either the mass or the distance will increase I. Second, of the two, the distance (ρ) has a bigger effect because it is squared; doubling the mass will increase I by a factor of 2, but doubling ρ will increase I by a factor of 4.

Important Point!

1. I is determined by both m and ρ.
2. ρ has a bigger impact because it is squared.

So far, a body of interest has been considered as a point, a rigid body, and a point mass. Now you have to refine your thinking to consider a rigid body as a series of point masses. To determine I for a rigid body, you simply have to sum the moments of inertia of all the point masses:

$$I_{\text{rigid body}} = \Sigma m\rho^2 \qquad (5.3)$$

Let us begin with an easy example: consider a uniform rod that is 5 cm long and weighs 10 kg (**Figure 5.8**). If you

Table 5.2	The Moment of Inertia for a Uniform Rod, an Axis at Either End, and the Center of Mass

From Origin (A):				From COM (B):			
ρ	ρ^2	m	I	ρ	ρ^2	m	I
0.05	0.0025	2	0.005	0.2	0.04	2	0.08
0.15	0.0225	2	0.045	0.1	0.01	2	0.02
0.25	0.0625	2	0.125	0	0.0	2	0
0.35	0.1225	2	0.245	−0.1	0.01	2	0.02
0.45	0.2025	2	0.405	−0.2	0.04	2	0.08
		ΣI	0.825			ΣI	0.20

(mentally) break the rod into five segments, each weighing 2 kg, and place the center of mass in the middle of each section, you will get a chart like you see in **Table 5.2**. You should notice three things. First, the location the COM of the rod is at its geometric center. Second, even for this uniform, symmetric body, the moment of inertia changes depending on where the axis of rotation is located. Third, the moment of inertia is least when the axis of rotation is located at the COM and gets progressively larger the further away the axis is placed from the COM. The moment of inertia is over four times larger at the end of the rod compared to the COM.

Now examine a frustum—it is an important shape in biomechanics. Think of a frustum as a cone with the tip cut off. For comparative purposes, leave the segment parameters (length, mass) the same, but alter the distribution of the mass (**Figure 5.9**). The results for three different axes of rotation are presented in **Table 5.3**. What can you say about the differences between Tables 5.2 and 5.3? Notice that by changing the shape, the location of the COM of the object is changed, and I at every location is also changed.

Now imagine this object was a baseball bat, and axis A was at the handle, axis B was "choking up," and axis C was at the barrel. Do the changes in I match the difficulty in swinging the bat? Compare your thoughts with the experiment you conducted at the beginning of this section.

Most physics and engineering textbooks will give tables containing the moments of inertia for uniform bodies through their central axes, and using some mathematical procedures the moments of inertia at other locations and for other axes can be calculated. In almost all cases, the moment of inertia will take the general form:

$$I = \eta m\rho^2 \qquad (5.4)$$

where ρ is either the length or radius, and η is a "correction factor" that accounts for both the shape of the object

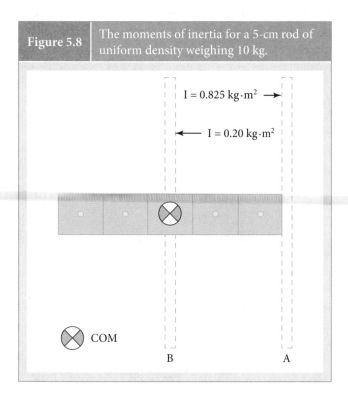

Figure 5.8	The moments of inertia for a 5-cm rod of uniform density weighing 10 kg.

$I = 0.825 \text{ kg·m}^2 \longrightarrow$

$\longleftarrow I = 0.20 \text{ kg·m}^2$

COM

B   A

| Figure 5.9 | The moments of inertia for a 5-cm frustum weighing 10 kg. Note that axis B is for comparative purposes with Figure 5.6 and does *not* represent the COM. |

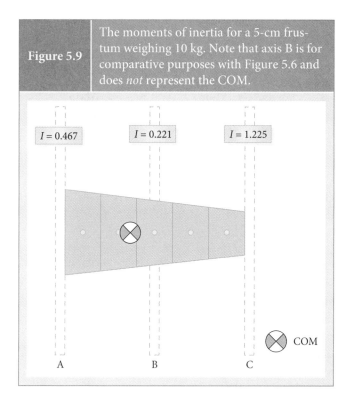

$I = 0.467$ $I = 0.221$ $I = 1.225$

⊗ COM

A   B   C

| Box 5.4 | Applied Research: Inertial Properties of Prostheses |

The inertial properties (mass and moment of inertia) are important when designing equipment that is swung, such as racquets and bats. Prosthetic limbs, particularly in the lower extremity, are also influenced by both of these characteristics, especially during the swing phase of gait. Common sense would suggest that the inertial properties of a prosthetic limb should match an intact, healthy limb. The gait characteristics of six ambulatory adults with a unilateral transtibial amputation under different prosthetic conditions were used to test this hypothesis. Investigators found that matching the prosthetic limb to the healthy limb actually decreased gait symmetry and resulted in a 6 to 7% increase in energy cost when walking at 1.34 m/sec. These findings suggest the development of artificial legs is more complicated than simply matching the inertial characteristics of the healthy leg.

Data from: Mattes SJ, Martin PE, Royer TD. Walking symmetry and energy cost in persons with unilateral transtibial amputations: Matching prosthetic and intact limb inertial properties. *Archives of Physical Medicine and Rehabilitation* 2000;81(5):561–568.

Section Question Answer

It is easier to rotate an object if more mass is located closer to the axis of rotation because, even though the mass is the same, the moment of inertia is less. So it is easier to swing a baseball bat from the center ("choking up") than from the barrel. And it is easier to swing a bat from the barrel than from the handle.

5.3.2 Angular Momentum (*H*)

Section Question

According to the equation $v = l_r \times \omega$, a wheel with a larger diameter should correspond to a faster linear velocity (velocity on the road) for the same angular velocity. Yet according to research, the wheel with the largest diameter did not produce the fastest linear speed for wheelchair users (Figure 5.10). Why?

and the location of the axis of rotation. For example, I for a rod rotating about its center is $1/12\ m\rho^2$, it is $1/3\ m\rho^2$ for a rod rotating about its end, and it is $m\rho^2$ for a ring rotating about a perpendicular axis through its center. Those details are not something that should be memorized. What is important is that you understand that η is not something that can be easily manipulated; it is going to be dictated by the shape of the object and the axis it is rotating about. The mass, m, can be considered when designing a piece of equipment or a prosthetic limb (see Box 5.4), but cannot be changed in the short term. The one variable that may be changed with different techniques is ρ, something that will be examined shortly.

Table 5.3	The Moment of Inertia for a Frustrum About an Axis at Both Ends and the Center of Mass							
	Location A			**Location B**			**Location C**	
m	ρ	ρ^2	I	ρ	ρ^2	I	ρ	ρ^2	I
3.84	0.05	0.0025	0.0096	0.2	0.04	0.1536	0.45	0.2025	0.7776
2.56	0.15	0.0225	0.0576	0.1	0.01	0.0256	0.35	0.1225	0.3136
1.70	0.25	0.0625	0.1062	0.0	0.0	0	0.25	0.0625	0.1062
1.14	0.35	0.1225	0.1396	−0.1	0.01	0.0114	0.15	0.0225	0.0256
0.76	0.45	0.2025	0.1539	−0.2	0.04	0.0304	0.05	0.0025	0.0019
		ΣI	0.467		ΣI	0.221		ΣI	1.225

| Figure 5.10 | A larger wheel does not necessarily produce a larger linear velocity. Why? |

© Daboost/ShutterStock, Inc.

Table 5.4	The Angular Momentum and Linear Velocity of a 62-cm and 64-cm Wheel	
See text for simplifying assumptions.		
	62 cm	**64 cm**
H	2.16 kg·m/sec	2.30 kg·m/sec
v	3.1 m/sec	3.2 m/sec

Every linear quantity has an angular analog. Linear momentum is the resistance to change in velocity of a (linearly) moving body. So angular momentum (H) is the resistance to change in angular velocity of a rotating body. The change in angular velocity can also be a change in magnitude or direction. Mathematically, linear momentum (Equation 5.1) is expressed as:

$$L = m \times v$$

If you substitute I for m and ω for v, you can express angular momentum as:

$$H = I \times \omega \tag{5.5}$$

In Equation 5.4, I was expressed as:

$$I = \eta m \rho^2$$

Substituting this expression in for I, Equation 5.5 can be expressed as:

$$H = \eta m \rho^2 \times \omega \tag{5.6}$$

Now compare data for the wheels from two chairs, one with a diameter of 64 centimeters and the other with a diameter of 62 centimeters. For simplicity's sake, assume that all other factors are equal (in reality, the larger wheel probably has a larger mass), and the wheel is rotating at 10 rads/sec. For purposes of illustration, a number of other simplifying assumptions will be made: The mass will be 2.25 kg, and the moment of inertia will be $m\rho^2$. Comparing the values for the two tires presented in Table 5.4, you will notice that the larger tire results in a 3.22% greater velocity, but also has a 6.48%

greater angular momentum. The increase in speed offered by the larger wheel is offset by the fact that it is also going to be harder to push. This finding illustrates an important point in biomechanics: whenever there is a **derived** variable (a variable that is formed by combining other variables), the largest value that can be obtained *mathematically* is not always possible to do *physically*. You will see many examples of this point throughout biomechanics.

> **Derived variable**   A variable that is formed by multiplying or dividing it by other variables

5.3.3   Angular Momentum for a System of Rigid Bodies

Thus far, the discussion has centered on a single rigid body, such as a wheel or a baseball bat. Those are good examples for you to learn the principles of center of mass, moment of inertia, and angular momentum. But they also only occur in a few circumstances. Your body is not a single rigid body, but can be thought of as a bunch of rigid bodies connected together, or a system of rigid bodies. Some interesting things happen with all three of these concepts when you apply them to a system of rigid bodies.

First, consider the center of mass. The center of mass for a rigid body is fixed. For a system of rigid bodies, it is not. To find the center of mass of the system, you must take a weighted average of the center of mass of each rigid body. For simplicity's sake, consider a system with two joints (J_1 and J_2) and two rigid body segments (S_1 and S_2) in **Figure 5.11**. Each segment has a uniform mass, so the center of mass of each segment is located at its geometric center. If the system is fully extended, then the COM of the system is located halfway between the COM of each segment, which happens to be at the location of J_2. If J_2 is flexed, then the COM of the system is still halfway between two COMs, but the two COMs are now closer together. Finally, if J_2 were flexed all the way, then the location of the COM of the system and the two bodies would be in the same place.

| Figure 5.11 | A 2-segment body (S_1 and S_2) with two joints (J_1 and J_2). The location of the center of mass of the system is the weighted average of the two segments and can change depending on the system posture. |

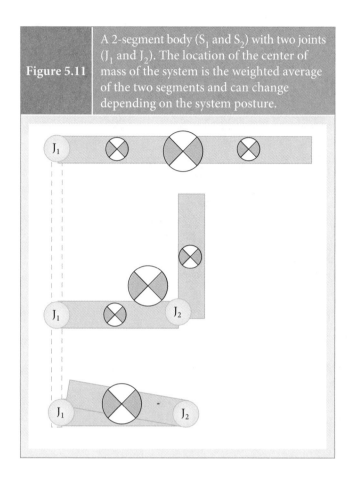

now closer to the axis of J_1. It is important to remember that this has a rather profound effect because ρ is squared in the calculation of I.

Important Point!   For a multibody system, because the location of the COM is not fixed, then neither is the moment of inertia.

Finally, consider the effects of changing the moment of inertia on the angular momentum and angular velocity. With linear momentum, you saw that momentum was always proportional to velocity because you cannot change your mass in the short term. But you can change your moment of inertia in the short term by changing your posture. This means two things for multibody systems. First, angular momentum is not proportional to mass. Second, you can change the angular velocity without changing the angular momentum. This is something you cannot do with linear momentum. Because mass is fixed, changing the linear velocity means that the linear momentum must also change.

Important Point!   You can change your angular velocity without changing your angular momentum by decreasing your moment of inertia.

This has a number of useful applications. If you flex your knee, then you will decrease the moment of inertia of the lower extremity. This will make it easier to swing that leg forward during running (**Figure 5.12**). Gymnasts, dancers, and figure skaters also take advantage of this concept. When turning or twisting in the air, they will pull the arms and legs in close to the trunk to decrease the moment of inertia, making it easier to turn or to increase their angular velocity without changing their angular momentum. When coming out of a twist or turn and they want to slow down their rotation, they can do so by extending their limbs and increasing their moment of inertia (see **Figure 5.13**).

The same goes for the COM of your body. During quiet standing, it is located at about the level of the S2 vertebra. (It is not located at the geometric center of your body because your lower body weighs more than your upper body. Remember: It is a weighted average of each segment.) If you were to raise both your hands over your head, you would raise the location of the COM because you raised the COM of each arm. Similarly, if you raised just your right hand, the COM of the body would go up and to the right. The COM of your body (or any multibody system) is not fixed; it can change.

Important Point!   For a rigid body, the location of the COM is fixed. For a multibody system it is not and will change, depending on the posture of the system.

Now think about what that does to the moment of inertia. In Figure 5.11, what would happen if you took the moment of inertia about J_1? In the fully extended position, the moment of inertia would be the largest because ρ is greatest. As J_2 flexes, ρ decreases because the COM of the system is

Section Question Answer

Although it is correct to say that a wheel with a larger diameter will produce a larger linear velocity for the same angular velocity, the larger wheel also has a larger moment of inertia, making it harder to turn or produce that linear velocity in the first place. The largest wheel does not necessarily create the greatest linear velocity if the person cannot turn it. So the best bet is to find the wheel that has the optimal interaction between v and I.

Figure 5.12 | The leg is easier to swing forward during running if it is bent. This decreases the moment of inertia.

(left) © Maxisport/ShutterStock, Inc., and (right) © Vidux/ShutterStock, Inc.

COMPETENCY CHECK

Remember:

1. Define the following: moment of inertia, center of mass, and angular momentum.
2. Write the equations for moment of inertia and angular momentum.

Understand:

1. Which has a larger moment of inertia: a 5-kg mass 10 cm away from the axis of rotation, or a 10-kg mass 5 cm away from the axis of rotation?
2. How can you change your moment of inertia without changing your mass?
3. Why do you flex your knee when swinging it forward while you run?

Apply:

1. If your physical education student was having a hard time swinging a bat or a racquet, what would you have them do to make it easier?
2. List activities where it is beneficial to increase your moment of inertia and activities where it is beneficial to decrease your moment of inertia.

5.4   COMPARING MEASURES OF INERTIA

As mentioned previously, inertia is a resistance to change in velocity. Mass is a measure of inertia, but only in a case where a body is not moving (which is called static). Mass is inadequate during dynamic movement be-

| Static | Not moving |
| Dynamic | Moving |

| Figure 5.13 | Bringing the arms and legs closer to the trunk decreases the moment of inertia and increases the angular velocity. Extending the arms and legs increases the moment of inertia and decreases angular velocity. |

© fstockfoto/ShutterStock, Inc.

| Table 5.5 | Comparing Linear and Angular Momentum |

	Momentum	equals	Mass	times	Velocity
Linear	L	$=$	m	$\times$	v
Angular	H	$=$	I	$\times$	ω

| Table 5.6 | Key Concepts |

- Inertia
- Mass
- Center of mass
- Moment of inertia
- Momentum—linear and angular

cause velocity also plays a role in determining a body's inertia. In dynamic situations, momentum is the measure of inertia and is the product of mass and velocity. For angular movement, the moment of inertia term replaces the mass term, and angular velocity replaces linear velocity. These relations are summarized in Table 5.5.

SUMMARY

In this lesson, the description of motion was further refined using the key concepts in Table 5.6. Inertia is a resistance to change in velocity. Mass is a measure of inertia, but only in static situations. In dynamic situations, momentum is the measure of inertia and is the product of mass and velocity. For angular movement, the moment of inertia term replaces the mass term, and angular velocity replaces linear velocity. You should be able to describe each of these concepts physically and mathematically. You should also be able to apply these concepts to human movement using your own examples. Armed now with an ability to describe motion, you can now focus your attention on the causes of motion.

REVIEW QUESTIONS

1. What is the difference between mass and weight?
2. Which person would have a harder time changing direction: a 70-kg man running at 11 m/sec, or an 80-kg man running at 10 m/sec? Why?
3. True or False? Two softball bats can have identical masses, but one could be harder to swing than the other.
4. Why would it be easier to swing my leg with my knee flexed than with my knee extended?
5. Examine the velocity graph of the shuttle run (Figure 2.14). Draw a graph of the momentum for an 88-pound girl. How are the momentum and velocity graphs similar? How are they different?

REFERENCES

1. Whiting WC, Rugg S. *Dynatomy: Dynamic Human Anatomy.* Champaign, IL: Human Kinetics, 2006.
2. Blazevich A. *Sports Biomechanics. The Basics: Optimising Human Performance.* London: A & C Black Publishers Ltd., 2007.
3. Selles RW, Korteland S, van Soest AJ, Bussmann JB, Stam HJ. Lower-leg inertial properties in transtibial amputees and control subjects and their influence on the swing phase during gait. *Archives of Physical Medicine and Rehabilitation* 2003;84(4):569-577.

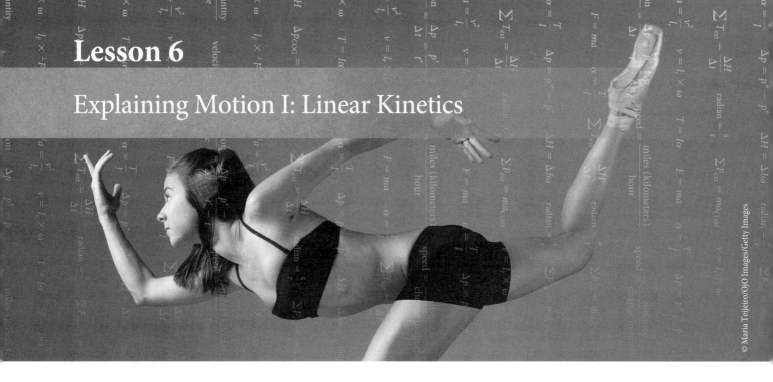

Lesson 6

Explaining Motion I: Linear Kinetics

LEARNING OBJECTIVES

After finishing this lesson, you should be able to:

- Define the following terms: force, impulse, rate of force development, weight, reaction force, net force, friction, fixed resistance, variable resistance, and accommodating resistance.
- List Newton's three laws of motion. Give an alternate to Newton's second law. Give examples of the three ways to change momentum.
- Manipulate the variables associated with Newton's second law to increase performance and decrease injury risk.
- List the types of forces usually encountered by the human body during movement.
- Combine several forces into a resultant force. Resolve a force into its components in the *x* and *y* directions.
- List the factors that affect friction. Distinguish between available and utilized coefficients of friction.
- List the types of forces that are used as resistance during exercise. Give examples of each.

You began your studies of biomechanics by describing motion using the concepts of time, position, displacement, velocity, and acceleration. By now, you should also appreciate that all motion is not equal (so to speak), and that to fully describe the motion of a body you need to discuss inertia and momentum because it will be harder to change the motion of a body with more momentum. But, you still do not know what is causing the change in motion. With this lesson, you will turn your attention to the causes of motion. The importance of this lesson cannot be overemphasized: If you are going to have any meaningful impact on changing someone's mechanics, you must know what causes the motion if you wish to

successfully intervene. You will begin with the very familiar Newton's laws of motion, followed by a discussion on a not-so-familiar, but very useful, take on Newton's second law. The lesson will conclude with some forces typically found during resistance exercise.

6.1 NEWTON'S FIRST LAW

When learning about momentum and changing momentum, an important point to remember is that, left alone, the momentum of a system will not change. To change momentum, you need a force. This is encapsulated in Newton's first law of motion:

> *A body will remain at rest or continue to move with a constant speed in a straight line unless acted upon by an outside force.*

In other words, momentum will not change without an outside force. The "outside" force needs some further clarification. Recall that a "system" is the thing that you are studying. Any forces that are inside the system (such as the forces that hold the body together) are called internal forces. Internal forces will not change the momentum of a system. Forces that are outside the system are external and will change the momentum of the system.

6.2 NEWTON'S SECOND LAW

This leads to the question: What is a force? It is interesting how hard it is for some people to articulate exactly what a force is. A **force** is simply a push or pull by one body on another. This definition

> **Force** A push or pull by one body on another

79

implies two bodies are always involved with a force, and those bodies are not part of the same system. It is important for you to remember that fact.

Although Newton's first law states that there needs to be an external force to change the momentum of a system, Newton's second law tells you the effect of that force. The law is usually stated as an equation:

$$F = ma \qquad (6.1)$$

Force is a vector, so it will have both a magnitude and a direction. Because mass is a scalar, the direction of force will be the same as the direction of the acceleration *caused by that force*. More specifically, the acceleration is caused by the force. Although you are probably very familiar with Equation 6.1, another way to conceptualize it may be to write Newton's second law as:

$$a = \frac{F}{m} \qquad (6.2)$$

This way shows you the universal law of cause and effect, but showing you the effect first. In other words, if you want this result (the left-hand side of the equation), then you need these causes (on the right-hand side):

$$result = causes$$

It should be clear looking at Equation 6.2 that if you desire a larger acceleration, you will either need to decrease the mass or increase the force. There are many benefits, with respect to both increasing performance and decreasing injury, by decreasing the mass. If you doubt the effect of mass on performance, try jumping as high as you can, and then put on a weighted vest and try jumping again. Will you be able to jump as high? The effect of mass on injury may not be as obvious. For a more detailed discussion on this topic, see **Box 6.1**.

The acceleration will always be in the direction of the force, but the acceleration is not always in the direction that the body is moving. To make matters more confusing, a negative acceleration could mean that the body was slowing down (if it was in the positive direction of movement) or speeding up (if it was in the negative direction of movement). Conceptually, to help keep physical interpretations clear, you learned that if a body was speeding up it was accelerating, and if it was slowing down it was decelerating.

A force will also not always be in the direction of movement, and a negative force is actually causing a body to speed up if it is in the negative direction. This can lead to similar confusion. To assist with physical interpretation of

Box 6.1	Applied Research: Obesity and Musculoskeletal Injury

These reviews examined the link between obesity and musculoskeletal injuries. Obesity has been associated with tendinopathies of the upper and lower extremities, plantar fasciitis, osteoarthritis of the hip and knee, fractures, and low back pain. Although the exact mechanisms have yet to be uncovered, obesity is found to increase loading and loading rate, as well as the overall magnitude and variability of motion. In addition, obesity is known to decrease endurance, shock attenuation, and neuromuscular control. These factors are likely to alter the mechanics associated with activities of daily living, work, and recreation, leading to an increased risk of musculoskeletal injury.

Data from: Wearing SC, Hennig EM, Byrne NM, Steele JR, Hills AP. Musculoskeletal disorders associated with obesity: a biomechanical perspective. *Obesity Reviews* 2006;7(3):239–250.

Wilder RP, Cicchetti M. Common injuries in athletes with obesity and diabetes. *Clinics in Sports Medicine* 2009;28(3):441–453.

the numbers, it is useful to think of a **propulsive force** as one that is causing the body to speed up and a **braking force** as one that is causing the body to slow down. If you are having trouble conceptualizing it, think of a propulsive force as stepping on the gas and a braking force as stepping on the brakes when driving a car.

Propulsive force	A force that is causing a body to speed up
Braking force	A force that is causing a body to slow down

> **Important Point!** A force will not always be in the direction of motion.

6.2.1 Weight

With the idea of force in mind, you can now return to the concept of weight. **Weight** is the force on a body due to gravity. Any two bodies, due to their masses, are attracted to each

Weight	The force on a body due to gravity

other. Your mass and the masses of the bodies in your immediate surroundings are not large enough to exert a very large force on one another. So this book will not move toward you if you move away from it. But Earth's mass is *huge* (5.9742×10^{24} kilograms). So you are going to be attracted to it by a rather large force if you try to move away from it. The magnitude of the weight:

$$Weight = F_g = ma_g \qquad (6.3)$$

where F is the force, m is the mass, and a_g is the acceleration due to gravity. Note that the subscript g after F tells us it is the force due to gravity. Although there may be minor fluctuations here and there, for the purposes of this lesson the acceleration due to gravity on or near Earth's surface is a constant 9.81 meters per second^2. If a body's mass is not going to change in the short term, then neither will its weight. The direction of weight is always going to be toward the center of Earth, or what is the vertical, downward direction. The sign of F_g will depend on the orientation of the gravity vector. Usually it is negative, but that is not always the case.

Important Point!   The acceleration due to gravity is not necessarily negative. It all depends on your frame of reference. Gravity is only negative when the upward vertical direction is positive.

6.2.2   Forces in Multiple Dimensions

Section Question

Why do you go down a hill faster when there is a steeper slope (**Figure 6.1**)?

Bodies do not only accelerate in one dimension. In fact, oftentimes they do not. If a force is acting in two dimensions, its effect will also be in two dimensions. To illustrate this point, consider the familiar example of going downhill. You could be going downhill on foot, bike, or skis, or (as in this example) snowboard. Intuitively, you know that the steeper the slope, the faster you will go downhill. But do you know why? You are probably guessing it has something to do with gravity.

Figure 6.1 | Why do you go downhill faster on a steeper slope?

© Ipatov/ShutterStock, Inc.

As you know by now, the first thing you have to do before analyzing a situation is pick a frame of reference. For a matter of convenience, pick a frame of reference where the x-axis is coincident with the slope of the hill, and consider the positive x-axis to be the direction that you predict the body will travel. The gravity vector is always oriented toward the downward vertical. Note that this vector does not coincide with any direction in your reference frame (second row of **Figure 6.2**). You know when this occurs you must examine the components of the vector in your frame of reference, as was done in the bottom row of Figure 6.2. To determine the components of the force due to gravity in your reference frame, you must apply correction factors:

$$^xF_g = ma_g \times \text{correction factor} \tag{6.4}$$

$$^yF_g = ma_g \times \text{correction factor} \tag{6.5}$$

Important Point!   Whenever a force vector is not coincident with an axis in your frame of reference, you must apply correction factors to examine the components in your reference frame.

Examining the angle of the hill, you should note that the angle between the g and y-axis is the angle the hill makes from the horizontal (the slope of the hill; θ_{hill}) and the angle from the x-axis is 90° minus θ_{hill}. The correction factors would be

$$^xF_g = ma_g \times \cos(90 - \theta_{hill}) \tag{6.6}$$

$$^yF_g = ma_g \times \sin(90 - \theta_{hill}) \tag{6.7}$$

You were originally interested in knowing why you go down a hill faster if there is a steeper slope. Equation 6.2 was the effect of force in a general form. More specifically, you would write:

$$^xa = \frac{^xF}{m} \tag{6.8}$$

and

$$^ya = \frac{^yF}{m} \tag{6.9}$$

To determine the acceleration due to gravity in the x direction, substitute Equation 6.6 into Equation 6.8:

$$^xa = \frac{^xF}{m} = \frac{ma_g \times \cos(90 - \theta_{hill})}{m} \tag{6.10}$$

Now, *without using a calculator*, can you guess the effect of the slope of the hill on the acceleration in the x direction? First, note that mass is included in both the numerator and

| **Figure 6.2** | A body going downhill. A steeper slope leads to a larger acceleration because a larger component of the weight is in the direction of travel. |

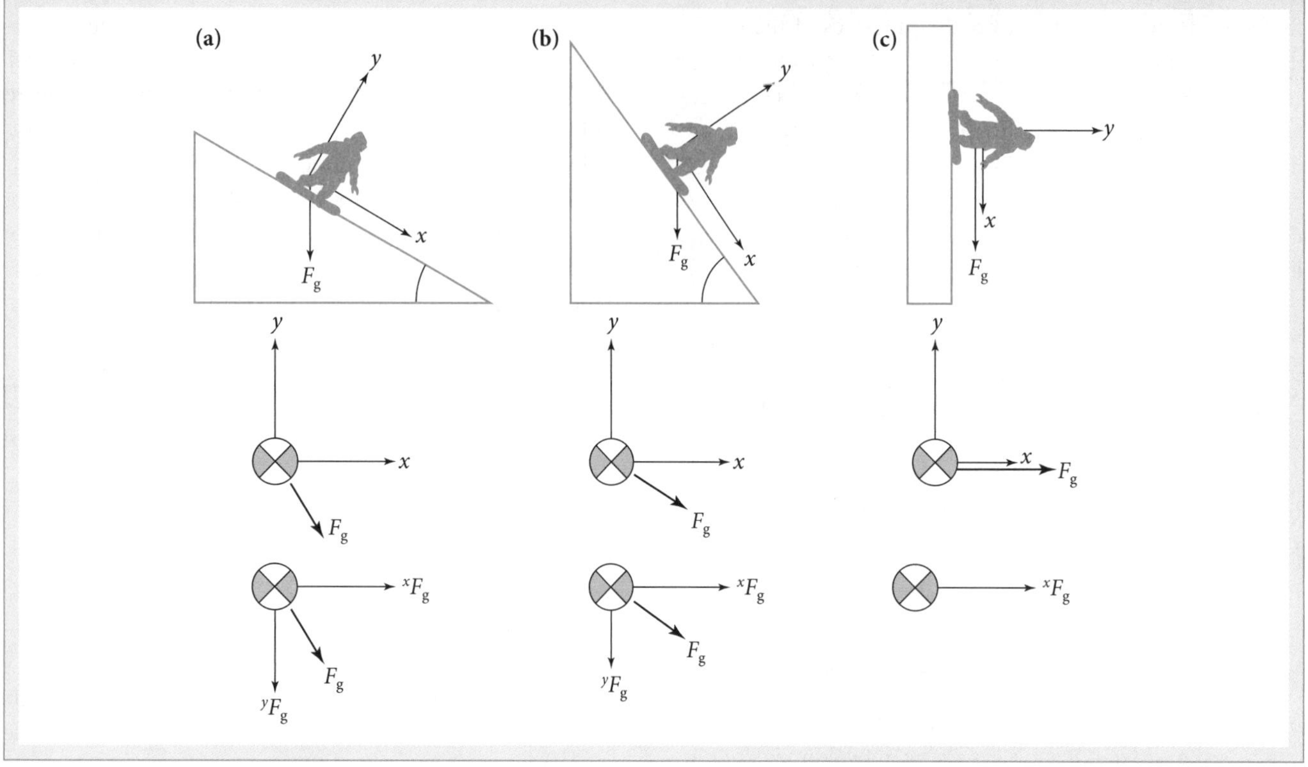

denominator of Equation 6.10, so they cancel. That means that the acceleration due to gravity will depend on g and the correction factor of $\cos(90 - \theta_{hill})$. As θ_{hill} gets larger, the number inside the parentheses gets smaller. As that number approaches zero, the cosine approaches 1. So at 90°, the number is zero, and the correction factor is 1. (Remember that the correction factor can never be larger than 1). The body is in free fall, and the acceleration is equal to g (Figure 6.2c). As the slope gets smaller, the correction factor is less, and so too is the acceleration due to gravity. If the slope is zero, the acceleration due to gravity in the x direction is also zero. Does this match your experience of going down a hill?

Whenever you resolve a vector into its components, you should consider all the components. What about the component ^yF_g? According to Equation 6.9, that component should cause the body to accelerate as well, either going through or leaving the ground. Why does it not accelerate? That question will be answered in the following sections.

Section Question Answer

You go down a steeper hill faster because the gravity vector is closer to the direction of travel.

COMPETENCY CHECK

Remember:

1. Define force, weight, propulsive force, and braking force.
2. State Newton's first and second laws.

Understand:

1. What will be the effect of a force on a body's acceleration if you double the force?
2. What will be the effect of a force on a body's acceleration if the body's mass is doubled?
3. What is the difference between mass and weight?
4. Resolve the following forces into their components:
   a. A 100 N force in the horizontal direction
   b. A 100 N force in the vertical direction
   c. A 100 N force 45° from the horizontal
   d. A 100 N force 60° from the vertical

Apply:

1. Are there ever situations when you may wish to apply a force in a direction that is slightly different from the angle of movement?

2. If you increase a body's mass, you must increase the force necessary to achieve the same acceleration. Does this mean that a person should be as light as possible? Why, or why not?

6.2.3   Effective (Net) Force

Section Question

While you are sitting there, you are not moving (Figure 6.3). But you have a force, your weight, acting on your body. So why are you not accelerating?

It is not enough to understand the concept of force. The concept of a **net** value has been used several times already. Think back to the woman swimming upstream: The river was running East at −1 meter per second, and she was swimming westward at 1.7 meters per second, so her net average velocity is 0.7 meters per second. It is the same thing with a net force—it is what you are left with after adding up all the forces acting on a body (using the rules of vector addition, of course).

The equation $F = ma$ is a little vague because it does not specify what force or what is accelerating. The force must be external to the system, and it tends to cause an acceleration of the center of mass (COM) of the system. Notice that the external force *tends* to cause an acceleration. It may not actually *cause* an acceleration because another force may counteract that effect. So you must account for all the forces acting on a body. The net force will determine if there is an

acceleration and in what direction. A more specific form of Equation 6.1 is

$$\sum F_{ext} = ma_{COM} \qquad (6.11)$$

Important Point!   A force does not cause an acceleration; only a net (effective) force does.

This form explicitly states that the vector sum of all external forces will cause an acceleration of the center of mass. The vector ΣF_{ext}, which is equivalent to ma_{COM}, is known as the "effective force." Again, if you wish to separate the causes from the effects, you may wish to rewrite Equations 6.8 and 6.9 as:

Effective force   The vector ma_{COM}, which is the effect of the net sum of all force vectors acting on a body

$$^x a_{COM} = \frac{\sum {}^x F_{ext}}{m} \qquad (6.12)$$

$$^y a_{COM} = \frac{\sum {}^y F_{ext}}{m} \qquad (6.13)$$

The equations open up the possibility of having more than one propulsive force, more than one braking force, or combinations of propulsive and braking forces (think stepping on the gas and hitting the brakes at the same time). It is the sum of those forces that determines the acceleration of the body's center of mass. To assist you in keeping track of all the forces acting on a body, you should draw a free-body diagram. See Box 6.2 for more information on constructing one.

Figure 6.3   While sitting quietly in a chair, your body is not moving. But your weight is acting on your body. Why are you not accelerating?

Box 6.2   Free-Body Diagrams

Free-body diagrams are a nice way to keep track of all the forces acting on a body, as well as the direction of those forces. They also give you an intuitive feel about the effect of those forces. To construct a free-body diagram, you:

1. Draw your system of interest isolated from the environment (which is why it is called a free-body) in the simplest representation possible. This usually means a stick figure or a simple outline shape.
2. Label the positive axes of your reference system.
3. Identify the center of mass and all other important points (such as contact points).
4. Draw and label all the external forces acting on the system.

You should note that you will not include internal forces, and generally you will have a contact force any time your system is touching something else.

In the simplest terms, we can represent the body as a point mass, which would be equivalent to the center of mass. Including two forces on a body, your free-body diagram would look similar to section (a) in the following figure. Looking at the free-body diagram, you should be able to tell that the effective force is going to be somewhere in between the negative x- and positive y-axes (b). If F_1 and F_2 were equal, then $F_{effective}$ would be 135° from the positive x-axis. $F_{effective}$ will be between 135° and 180° if F_1 is larger than F_2 and between 90° and 135° if F_2 is larger than F_1. The body will accelerate in the direction of the effective force. You can show this by substituting (ma) for the ($F_{effective}$). This is the result of all the forces acting on the body and is called a mass-acceleration diagram or a kinetic diagram.

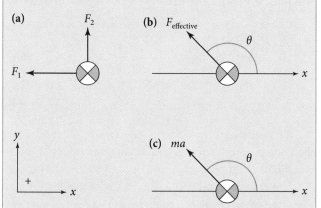

A free-body diagram with two forces (a), the effective force acting on the body (b), and the resulting acceleration mass acceleration diagram (c).

| Figure 6.4 | Holding a box statically requires the effective force to be zero. The force of the hand acting upward on the box must be equal and opposite to the force due to gravity acting downward. |

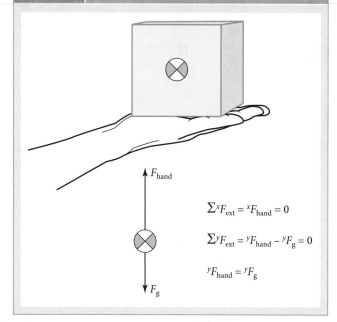

$$\sum {}^xF_{ext} = {}^xF_{hand} = 0$$

$$\sum {}^yF_{ext} = {}^yF_{hand} - {}^yF_g = 0$$

$${}^yF_{hand} = {}^yF_g$$

| Figure 6.5 | The effective force required to accelerate a 25 kilogram box 4 m/sec²: (a) horizontally, (b) upward vertically, (c) downward vertically, and (d) 45° from the horizontal. |

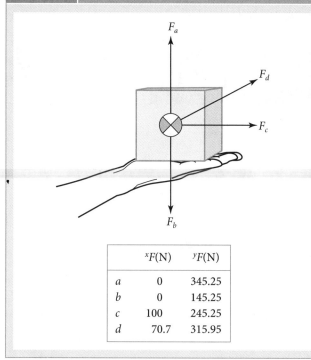

	^xF(N)	^yF(N)
a	0	345.25
b	0	145.25
c	100	245.25
d	70.7	315.95

Consider the forces required for you to hold a 25-kg box in the air (Figure 6.4). Assume that the only forces acting on the box are gravity and your hand, and gravity is the negative y direction. If the box is not moving, the acceleration must be zero. If the accelerations are zero, the net forces must be zero as well. In essence, your hand is cancelling out the force due to gravity. Because gravity is not acting in the x direction, your hand would not be applying a force in that direction, either. In the y direction, your hand would be applying a force on the box that is equal and opposite to that of gravity, which in this case would be 245.25 N.

Now consider a case where you would want to accelerate the box 4 meters per second², either (a) upward vertically, (b) downward vertically, (c) horizontally, or (d) 45° from the horizontal (Figure 6.5). What are the magnitudes of force that your hand would have to supply on the box in each case?

To produce the desired acceleration of 4 meters per second², you would need a net force of 100 N (25 kg × 4 m/sec²) in the desired direction, and gravity has a force of 245.25 N in the downward direction. In the case of the vertical upward direction, you would need to actually produce 345.25 N of force in the upward direction to net 100 N. In the downward direction, gravity is already producing 245.25 N, so you need to supply 145.25 N in the opposite direction to net 100 N. The numbers here are not as important as the realization that when accelerating against gravity, you need to produce more force than the weight of the object, and when accelerating with gravity, you would produce less force than the weight of the object. The horizontal direction is a bit more complicated: You would need to supply the desired 100 N in the horizontal direction (as you might have guessed), but you also need to supply 245.25 N force in the upward direction so that the box does not accelerate downward. In the case of 45° from the horizontal, the trick is to realize that accelerating in this direction is less than accelerating in either of the principal directions. The components of this force in the x and y directions are 70.7 N for each (because it was a 45° angle). Once you have figured this out, you need simply to add them like you did for the pure vertical and horizontal directions earlier.

Section Question Answer

If you are sitting in your chair, you are subject to the force of gravity. But if you are sitting in your chair, you are not accelerating, so the net force must be zero. If the net force is zero, there must be a force that is both equal in magnitude and opposite in direction to your weight—much like your hand supplied a force on the box when it was stationary. The free-body diagram is presented in Figure 6.6. What is the origin of this force? The answer is found in Newton's third law.

COMPETENCY CHECK

Remember:

1. Define an effective force. What is the direction of the effective force?

Understand:

1. In the example presented in Figure 6.5, are the forces of hand propulsive or braking forces?
2. If you were accelerating an object in the direction opposite of gravity, would the effective force be larger or smaller than the object's weight?

| Figure 6.6 | A free-body diagram of sitting in a chair. If your body is not accelerating, there must be an equal and opposite force to F_g. The origin of this force, F_1, will be discussed later in this lesson. |

$F_1 - F_g = 0$

3. If you were accelerating an object in the direction of gravity, would the effective force be larger or smaller than the object's weight?
4. If you were decelerating an object in the direction opposite of gravity, would the effective force be larger or smaller than the object's weight?
5. If you were decelerating an object in the direction of gravity, would the effective force be larger or smaller than the object's weight?

Apply:

1. Think of pushing something with both of your hands in several different directions. Can you apply the same effective force in each case? Why or why not?

6.3 CONTACT FORCES AND NEWTON'S THIRD LAW

Section Question

What forces cause you to move when you walk across the room (Figure 6.7)?

Newton's first law tells you that you must supply a net, external force to change the momentum of a system. Newton's second law tells you how effective a force (or system of forces) is in changing the momentum. Newton's third law clues you in as to the origin of the force(s) and states that for every force

| Figure 6.7 | Walking across a room requires forces. Where do they come from? |

© Radu Razvan /ShutterStock, Inc.

Let us tackle the vertical direction first. The effective force is determined by subtracting F_g from the y-component of the F_{GRF}:

$$\sum {}^{y}F_{ext} = {}^{y}F_{GRF} - F_g = m \, {}^{y}a_{COM} \qquad (6.14)$$

If ${}^{y}F_{GRF}$ is greater than F_g, then ΣF_{ext} will be positive, and the COM will accelerate upward. If F_g is greater than ${}^{y}F_{GRF}$, then ΣF_{ext} will be negative, and the COM will accelerate downward. If ${}^{y}F_{GRF}$ and F_g are equal, then ΣF_{ext} will be zero, and there will be no acceleration in the vertical direction. Because gravity does not act in the horizontal direction, the acceleration of the COM will be proportional to, and in the direction of, ${}^{x}F_{GRF}$:

$$\sum {}^{x}F_{ext} = {}^{x}F_{GRF} = m \, {}^{x}a_{COM} \qquad (6.15)$$

In all cases, F_{GRF} is equal in magnitude and in the opposite direction of the force you exerted on the ground. The ground does not produce a force (you do). But think of the ground as an integral link in the chain that causes your COM to accelerate. Via your muscles and gravity, you exert a force on the ground. The ground exerts an equal and opposite force back on you. The difference between the F_{GRF} and F_g determines the direction and magnitude of the acceleration of your body's COM. And because F_g cannot be changed or manipulated, performance is determined solely by the F_{GRF}.[1] Obviously, this means that understanding the F_{GRF} is a big part of understanding biomechanics.

Important Point!   The force and the reaction force are always acting on different bodies.

If a person is jumping straight up, the direction of the ground reaction force and the acceleration of the body's COM would be in the positive vertical direction. Upon landing, the F_{GRF} would also be in the positive vertical direction. In this case, the direction of the F_{GRF} would be the same for both takeoff and landing. In fact, the ${}^{y}F_{GRF}$ is always positive or zero. Can you explain why?

Important Point!   The vertical ground reaction force is always positive or zero.

The horizontal F_{GRF} does not follow this rule; it can be positive, negative, or zero. In the case of a long jump, the ${}^{x}F_{GRF}$ would be in the direction of the jump during takeoff and in the direction opposite of the jump during landing. While standing, there would be a ${}^{y}F_{GRF}$ equal to body weight, but the ${}^{x}F_{GRF}$ would be zero. During the stance phase of gait (Figure 6.8), the ${}^{x}F_{GRF}$ is initially negative at initial contact when the foot is in front of the COM. At the instant when the COM is directly over the foot, the ${}^{x}F_{GRF}$ is zero. During the

there is an equal and opposite reaction force. As was mentioned previously, two bodies are attracted to each other by the same force of gravity. Because of the large disparities in masses between you and Earth, you will have a relatively large acceleration, whereas Earth's will be imperceptible. Gravity is known as a force acting at a distance because you do not actually have to be touching the ground to have it affect you (think free fall). When two bodies are touching each other, they exert a **contact force** on one another. The forces are equal in magnitude, opposite in direction, and acting on different bodies. You exert a force, by your weight, on the ground. The ground will exert an equal and opposite **reaction force** on you. The harder you push on the ground, the harder the ground pushes on you (up to a point)—in the exact opposite direction. This reaction force is called the **ground reaction force** (F_{GRF}).

If you ignore air resistance (something you will do throughout this lesson), and if the person is not in contact with another object (such as an implement or another person) other than the ground, then the acceleration of that person is determined by F_g and F_{GRF}. To see how this works, remember that F_g only acts in the downward, vertical direction and that the effective force determines the acceleration of a body's COM.

> **Contact force**   The force created when two bodies are touching each other
>
> **Reaction force**   As a consequence of Newton's third law, for every force created by body A on body B, there is a force of equal magnitude and opposite in direction created by body B acting on body A
>
> **Ground reaction force**   The equal and opposite force the ground applies back on the person

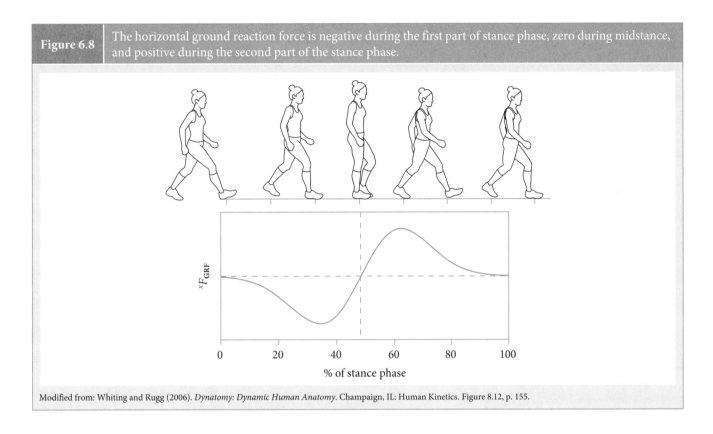

Figure 6.8 The horizontal ground reaction force is negative during the first part of stance phase, zero during midstance, and positive during the second part of the stance phase.

$^xF_{GRF}$

% of stance phase

Modified from: Whiting and Rugg (2006). *Dynatomy: Dynamic Human Anatomy*. Champaign, IL: Human Kinetics. Figure 8.12, p. 155.

second part of stance, when the COM is in front of the foot, the $^xF_{GRF}$ is positive. This same pattern holds for both walking and running gait. This means that the body's COM is decelerating during the first half of stance phase (when the foot is anterior to the COM) and accelerating during the second half of stance phase (when the COM is anterior to the foot).

6.3.1 Friction

No discussion of contact forces in general, or the F_{GRF} in particular, would be complete without mentioning friction. Contact forces, like any other forces, are vectors. And vectors can be resolved into their components. Typically, contact forces are resolved into components that are parallel and perpendicular to the contact area. In the last section, it was mentioned that the harder you push on the ground, the harder the ground pushes back on you—but only up to a point. Friction has a lot to do with that qualifier.

The force due to friction is always parallel to the contact area of two surfaces, and it opposes motion between the two objects in contact. Friction is caused by the irregularities in the contact surfaces between the two objects, which causes them to interlock to a certain degree. The general form of the equation for friction is

$$F_f = \mu\, ^nF \qquad (6.16)$$

Where F_f is the friction force, μ is the coefficient of friction, and ^nF is the normal force. The hierarchical model is presented in Figure 6.9. Remember that in the mathematical sense, "normal" means perpendicular. So the normal force is force perpendicular to the area in contact, and it can be thought of as the force holding the two surfaces together. The frictional force is proportional to the normal force: The larger the force holding the two objects together, the harder it is to

Figure 6.9 The hierarchical model for the force due to friction.

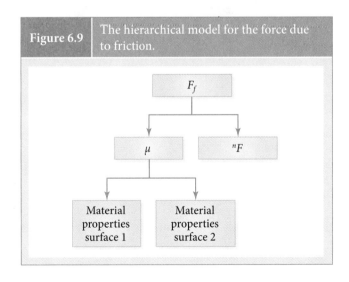

Figure 6.10	Holding a plastic cup. (a) Holding the cup on the table requires no friction, (b) holding the cup away from the surface requires a frictional force, and (c) the frictional force increases with the weight of the cup to prevent slipping.

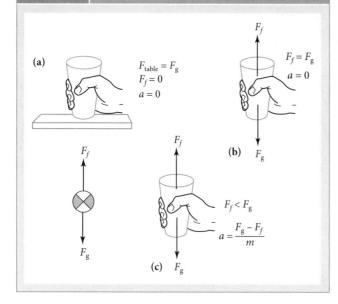

(a)

$F_{table} = F_g$
$F_f = 0$
$a = 0$

$F_f = F_g$
$a = 0$

(b) F_g

F_f

F_g

$F_f < F_g$

$a = \dfrac{F_g - F_f}{m}$

(c) F_g

reaction force of the table is equal and opposite to the force of gravity. Now pick up the cup and hold it still (Figure 6.10b). The reaction force of the table is no longer present, yet the cup is not moving relative to the hand. Your grip is producing a force, but the vector of that force is perpendicular to the surface of the cup (a normal force that, in this case, is in the horizontal direction). What prevents the cup from slipping is the friction between your hand and the cup. Now what happens if you pour your favorite beverage into the cup? The weight of the cup now increases, and so too must the force of friction. Otherwise, the cup will slip and start to accelerate (Figure 6.10c). Your immediate reaction to pouring something in the cup is to increase the force of your grip (^nF). Increasing ^nF increases F_f (Equation 6.16). Looking at the equation more closely, that is really the only thing you can do to increase F_f. The coefficient of friction is determined by the material properties of the two surfaces.

Friction often gets a bad rap. You hear people talking about trying to reduce friction. And it is true: Too much friction is a bad thing. But too little friction can also be a bad thing. Think about walking across a floor (**Figure 6.11**). Every time your foot makes contact with the ground, there are contact forces. The equal and opposite reaction forces determine the acceleration of your body's COM.

If there was not any friction between your shoes and the floor, you would be slipping every time you tried to push off for a step and every time your foot landed on the ground (think about walking across any icy surface with "slippery" shoes). In fact, you can calculate the amount of friction that you need to walk across a surface. If you substitute the $^xF_{GRF}$ for F_f and the $^yF_{GRF}$ for ^nF (see Figures 6.14 and 6.15)

slide the two surfaces over one another. The coefficient of friction depends on the characteristics of the two surfaces in contact and whether or not the objects are moving relative to one another.

To visualize these ideas, imagine holding a plastic cup (**Figure 6.10**). If you are holding the plastic cup on the table (Figure 6.10a), then there is no need for friction because the

Figure 6.11	The force of friction between the ground and shoe is necessary to prevent slipping and provide an effective force in the x direction.

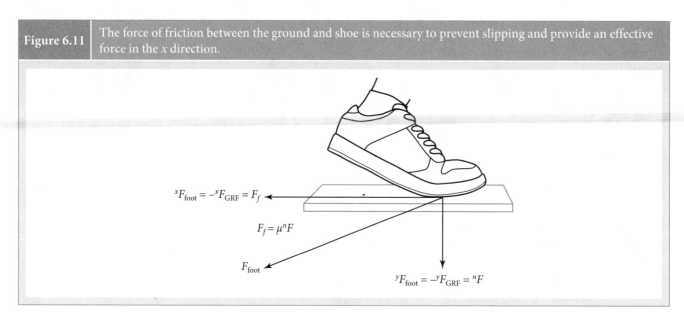

$^xF_{foot} = -^xF_{GRF} = F_f$

$F_f = \mu^nF$

F_{foot}

$^yF_{foot} = -^yF_{GRF} = {}^nF$

and rearrange Equation 6.16, you will see that:

$$\mu_{\text{utilized}} = \frac{^{x}F_{\text{GRF}}}{^{y}F_{\text{GRF}}} \qquad (6.17)$$

In other words, the utilized coefficient of friction is the ratio of the horizontal ground reaction force to the vertical ground reaction force. The larger the horizontal ground reaction force is in relation to the vertical ground reaction force, the more friction you need. If the utilized coefficient of friction is greater than the available coefficient of friction (usually determined experimentally for two unique surfaces), then a slip will occur (see Box 6.3). So you can reduce your utilized coefficient of friction by either reducing your horizontal ground reaction force, increasing your vertical ground reaction force, or a combination of the two.

But you do not want to maximize friction, either. Why do you have to replace the soles on your shoes (or the tread on your tires)? Friction causes a tiny bit of the material to be left on the ground. Tire skid marks are actually a part of the tire left on the road. Without friction, you could not walk across the floor, but because of it, your shoes will wear out. Other injuries can also occur if friction is too high because of your foot getting "stuck" to the ground (see Box 6.4).

Section Question Answer

The forces required to walk across the room are the ground reaction forces in the vertical and horizontal directions. Frictional forces also play a role in preventing the foot from slipping, particularly during initial contact and toe-off.

Box 6.3	Applied Research: Utilized Coefficient of Friction and Slips

In this investigation, subjects were asked to walk repeatedly down a 10-meter pathway with embedded force platforms. During one of the trials (unknown to the subjects), a lubricant was applied to the floor. Using a specialized device (known as a tribometer), investigators determined the available COF, and using Equation 6.15 they determined the utilized COF. Statistical techniques were used to predict when slips would occur. Results indicated a 1% probability of a slip occurring if the available COF exceeded the utilized COF by 0.077. There was a 50% probability of a slip occurring when the available COF was close (<0.006) to the utilized COF and a 99% probability of a slip occurring when the available COF was 0.090 less than the utilized COF. These values should be kept in mind when designing walkways.

Data from: Burnfield JM, Powers CM. Prediction of slips: an evaluation of utilized coefficient of friction and available slip resistance. *Ergonomics* 2006;49(10):982–995.

Box 6.4	Applied Research: High Coefficients of Friction and Knee Injury

Using a specialized device, researchers compared the resistance to twisting (similar to the coefficients of friction) of four different cleats and tracked the injury rates of football players wearing those cleats over three years. Those wearing the cleats with a higher resistance to twisting had an injury rate for the knee that was almost 3½ times greater than the other cleats.

A more recent study that examined the biomechanics of performing the side-step cutting maneuver found changes in biomechanics at the knee joint that put the knee at risk and help to explain these findings.

When it comes to the shoe-surface interface, too much friction can be just as injurious as too little friction.

Data from: Lambson RB, Barnhill BS, Higgins RW. Football cleat design and its effect on anterior cruciate ligament injuries—a three-year prospective study. *American Journal of Sports Medicine* 1996;24(2):155–159.

Dowling AV, Corazza S, Chaudhari AMW, Andriacchi TP. Shoe-surface friction influences movement strategies during a sidestep cutting task implications for anterior cruciate ligament injury risk. *American Journal of Sports Medicine* 2010;38(3):478–485.

COMPETENCY CHECK

Remember:

1. Define contact force, reaction force, ground reaction force, friction, and normal force.
2. State Newton's third law.

Understand:

1. What would be the direction of the ground reaction force if you were jumping 45° from the horizontal?
2. What would happen to the coefficient of friction if you replaced the plastic cup in Figure 6.10 with a glass? What does that do to the force of friction? How would you increase the force of friction in this case?
3. If you increase the surface area of your shoes, would you increase the force due to friction?
4. What happens to the utilized coefficient of friction if you decrease the angle of the force of the foot in Figure 6.11?

Apply:

1. Discuss a situation where you would want to increase the force of friction. How would you do it?

6.4   REVISITING NEWTON'S SECOND LAW

> **Section Question**
>
> Contrary to popular belief, Newton did not express his second law as $F = ma$. Many scientists feel the way he originally expressed it is more useful. How did he express it, and why is it more useful?

Newton's first law states that there must be a net, external force to change the momentum of the system. The second law determines the effect of that force. The third law states where the force comes from. But currently, the second law is stated in terms of acceleration (and coincidentally, not the way Newton wrote it[2]). There appears to be a link missing between the first and second laws. Remember that formula for acceleration is

$$a = \frac{\Delta v}{\Delta t}$$

If you substitute this equation for acceleration into Equation 6.11, you get

$$\sum F_{ext} = \frac{m\Delta v_{COM}}{\Delta t} \qquad (6.18)$$

Remember the formula for momentum:

$$L = m \times v$$

And the change in momentum:

$$\Delta L = \Delta(m \times v)$$

If the mass of the body does not change (an assumption you will continually make):

$$\Delta L = m(\Delta v)$$

The right-hand side of this equation looks just like the numerator in the right-hand side of Equation 6.18 (it is), so you see that:

$$\sum F_{ext} = \frac{\Delta L}{\Delta t} \qquad (6.19)$$

What does this tell you, exactly? It tells you that the net force is equal to the time rate of change in momentum of the system. You might be thinking that this equation is more complicated than the old $F = ma$, but it is much more useful. Let us see why. Multiply both sides by Δt:

$$\sum F_{ext} \times \Delta t = \Delta L \qquad (6.20)$$

Now you are really onto something: The force multiplied by the time is equal to the change in momentum. In other words, to be effective, a force has to be applied over a time (or through a distance). That quantity, $F \times \Delta t$, is so important it

has its own name: *impulse*. The impulse is equal to the change in momentum. You have learned how important momentum is, and now you know how you can change it: by applying an impulse, a force over a time. As with force, you can say that a propulsive impulse is increasing the momentum, whereas a braking impulse is decreasing momentum.

> **Impulse**   The product of average force and time that force is applied; it is equal to the change in momentum

> **Important Point!**   To be effective, a force has to be applied over time (or over a distance).

Equation 6.18 tells you that, to be effective, forces have to be applied over time (or over a distance). However, you cannot simply multiply any force value by time because a force is usually fluctuating over time. It is important to see what the force is doing over the entire time of interest. You could multiply the time by the *average* force, or graphically you could examine the area under the force–time curve. There are five important characteristics of the force–time curve:

- Peak force
- Rate of force development
- Rate of force fatigue
- Impulse
- Average force

Each of these will be discussed in turn and are illustrated in Figures 6.12 and 6.13.

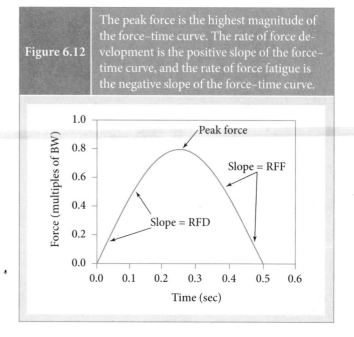

Figure 6.12   The peak force is the highest magnitude of the force–time curve. The rate of force development is the positive slope of the force–time curve, and the rate of force fatigue is the negative slope of the force–time curve.

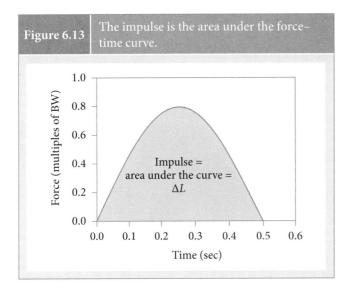

Figure 6.13 The impulse is the area under the force–time curve.

The magnitude is the amount of force. A larger force results in a larger acceleration. The peak magnitude is the largest amplitude of force during the period of interest. In Figure 6.12, the largest point on the curve is the peak force. This characteristic of a force is often measured, analyzed, and compared. But it is important to keep in mind that force is an instantaneous measure, and so the peak force is only the characteristic of one instant in time.

You could also be interested in how a force is changing with time. These characteristics are called rate of force development (RFD) and rate of force decline (to avoid confusion that could result from similar abbreviations, this characteristic will be referred to as rate of force fatigue, RFF). These values are the rates of change of force. A rate is a change in one variable over another. In this case, it is a change in force over a change in time:

$$\frac{\Delta F}{\Delta t} = \frac{F' - F}{t' - t} \qquad (6.21)$$

On a graph, it is the slope of the force–time curve. If the force is positive and the slope is positive (that is, force is increasing with time), then it is a rate of force *development*. If the force is positive and the slope is negative, then force is declining with time—it is *fatiguing*.

Impulse is the area under the force–time curve (Figure 6.13). Unlike force, impulse is a measure that captures the entire period of interest. Impulse tells you how much (and in what direction) the momentum will change over the period of interest.

Average force is what the magnitude of the force would be if the force were constant for the entire period of interest. The average force may represent few of those values, so some

would argue that it is not a very useful measure. But mathematically, the average force can also be calculated as:

$$F_{average} = \frac{\sum F \Delta t}{t' - t} = \frac{\Delta L}{\Delta t} \qquad (6.22)$$

This shows you another way of thinking about average force: It is the impulse divided by the time that force was applied. This is referred to as being normalized to time.

It should be clear that impulse is the key variable concerning momentum changes. In fact, the other variables listed earlier influence impulse, and therefore momentum. The hierarchical model is presented in **Figure 6.14**. Notice that the average force is determined by the peak force and how much fluctuation there is around the peak, or how much the force is changing over time. To see how these different variables can be manipulated, examine **Figures 6.15** to **6.17**. In Figure 6.15, impulse is increased by primarily increasing the peak force. In Figure 6.16, impulse is increased by primarily increasing the time. In Figure 6.17, impulse is increased by changing the overall "shape" of the curve by increasing the percentage of time the force is near the peak force. **Figure 6.18** is a side-by-side comparison of these different strategies. Notice how they all lead to the same changes in momentum.

To illustrate how this principle is applied, consider landing from a jump. There will be a decrease in momentum from whatever the impact velocity is to zero. Incidentally, you also know that this is going to require a braking impulse, and that braking impulse is produced by the vertical ground reaction force. Although that much is certain, lots of different strategies can be used to attain that impulse, but keep in mind that the same amount of braking impulse is necessary to come to

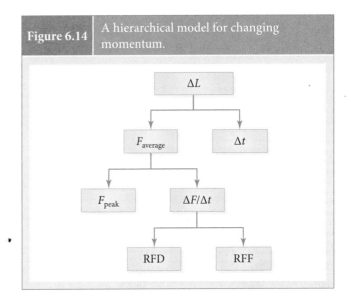

Figure 6.14 A hierarchical model for changing momentum.

Figure 6.15 | Increasing momentum by increasing the magnitude of the force.

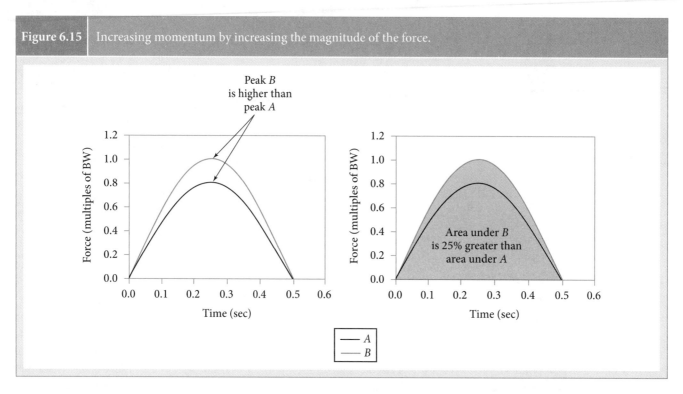

Figure 6.16 | Increasing momentum by increasing the duration of the force.

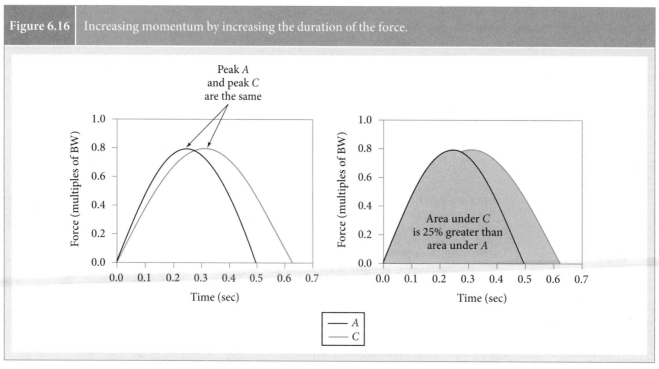

a complete stop, no matter which strategy is used. To contrast two extremes, consider a "hard" landing (where the person lands stiff with straight legs) versus a "soft" landing (where the person flexes the knees to land) in **Figure 6.19**. As the name "hard" implies, this type of landing will subject the body to a rather large force over a brief period of time. By flexing (bending) the knees, the person is actually increasing the time over which the braking force is applied. For the same change in momentum, increasing the time means that the forces that have to be applied are decreased.

Figure 6.17 | Increasing momentum by changing the overall shape of the force–time curve.

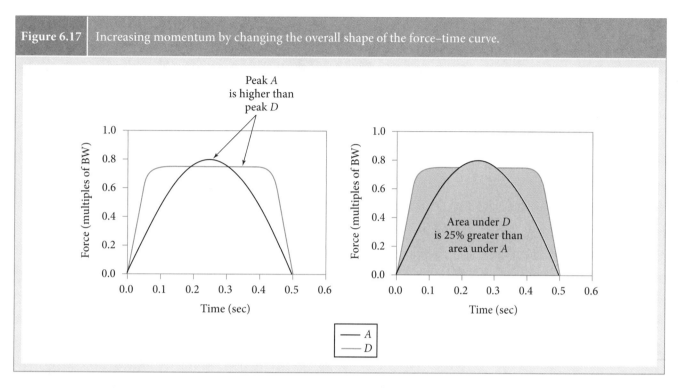

Figure 6.18 | A side-by-side comparison of the three strategies for increasing momentum.

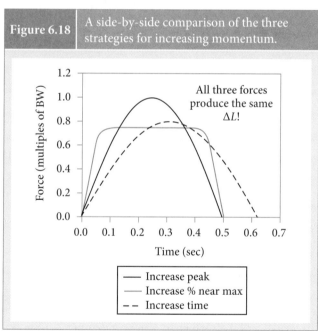

Figure 6.19 | A comparison of the ground reaction forces of a hard versus soft landing. The hard landing has a higher peak and shorter duration.

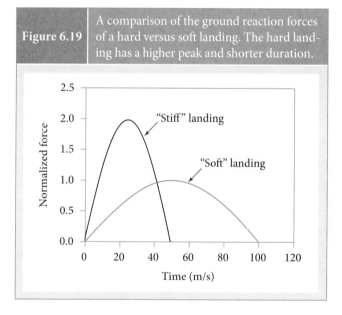

In the previous example, you were examining what happens when the body starts with a certain amount of momentum and the momentum decreases to zero. Now consider a movement where the initial and final momentum are both zero. This situation happens often; imagine reaching for a cup. The hand has an initial velocity of zero and has a final velocity of zero when it reaches the cup. You should be very

familiar with the general shape of the position, velocity, and acceleration curves for such a movement, as you have studied them extensively (**Figure 6.20**). Now look at the momentum and force/impulse graphs in **Figure 6.21**. In this case, the direction of travel of the hand is positive. Notice that in the first part of the movement, the momentum is increasing, so there is a propulsive impulse. Conversely, when the momentum is decreasing, there is a braking impulse. If the final *change* in momentum is zero, the propulsive and braking impulses

| Figure 6.20 | The (a) position, (b) velocity, and (c) acceleration graphs when the initial velocity is zero and the final velocity is zero. |

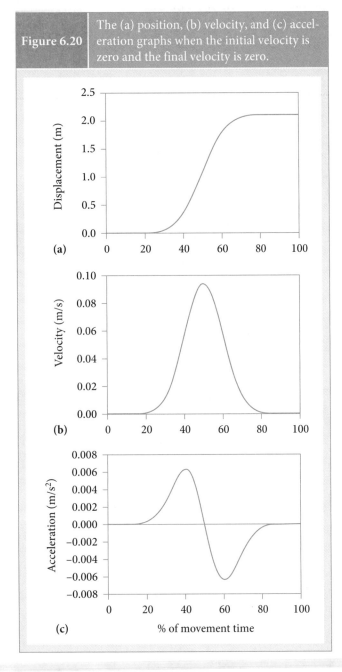

(a)

(b)

(c)   % of movement time

| Figure 6.21 | The momentum (a) and force (b) graphs for a movement that has an initial and final velocity of zero. |

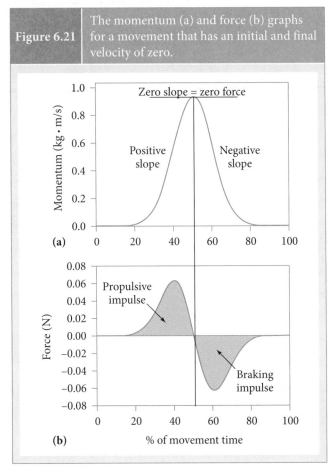

(a)

(b)   % of movement time

must be equal (same areas under the curve). In this case, the movement was very symmetric; the propulsive impulses and braking impulses each occurred for 50% of the movement time and are the same magnitude. What happens when they are not?

Important Point!   If a movement starts with zero velocity and ends with zero velocity, then the change in velocity is zero. If the change in velocity is zero, the change in momentum is zero. If the change in momentum is zero, the propulsive and braking impulses must be equal.

Consider the graphs in **Figure 6.22**. They present two alternate ways a movement could be performed. The top graph shows a large initial propulsive impulse followed by a long braking impulse of a small magnitude. The bottom graph shows an initial propulsive impulse followed by a period with no forces being applied. If there are no forces being applied, the momentum (and velocity!) will be constant. You may be familiar with the colloquial saying that "the momentum is doing the work." That is not exactly accurate, but for now realize that during that period the velocity is constant. Then there is a final braking impulse to bring the momentum back to zero. In both cases, the propulsive and braking impulses are equal, and the total change in momentum at the end is unchanged.

Finally, consider the case of the $^xF_{GRF}$ during gait (**Figure 6.23**). Assume the direction of travel is positive. Initially, there is a negative $^xF_{GRF}$ and a braking impulse. This occurs when the foot is in front of the body's COM. At the instant where the COM is directly over the foot, the $^xF_{GRF}$ is zero. When the foot is behind the COM, the $^xF_{GRF}$ is positive, and there is a propulsive impulse. This is always the case in both walking and running; there is initially a braking impulse followed by

| Figure 6.22 | Two alternate ways a movement could be performed. (a) A small braking force applied over a long duration; (b) a large braking force applied over a short duration. |

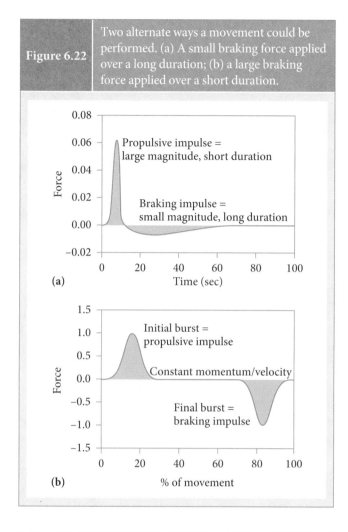

(a)

(b)

propulsive impulse would need to be larger than the braking impulse, and to slow down the braking impulse would have to be larger than the propulsive impulse.

Gait velocity is a product of step length and step rate. Increasing either one would improve gait velocity, but increasing step length would only work up to a point. Now you know why: The further the foot is in front of the body's COM, the larger the braking impulse. The larger the braking impulse, the more likely it is that you would lose velocity during that stance period.

Section Question Answer

Newton expressed his second law as a force creating a time rate of change of momentum. The impulse–momentum relation is derived from this version of the law. It is more useful because, to be effective, forces must be applied over time.

COMPETENCY CHECK

Remember:

1. Define: impulse, rate of force development, rate of force fatigue, peak force, and average force.
2. What two factors determine the change in momentum of a body?

Understand:

1. If the force is negative, what would a negative slope indicate?
2. If the force is positive, what would a negative slope indicate?
2. What will happen to the momentum of a body if the average force is increased?
3. For the same change in momentum, what happens to the average force if the time is decreased?

Apply:

1. List activities where it would be more beneficial to increase the average force rather than the time the force was applied.

| Figure 6.23 | The horizontal ground reaction force impulse during gait. |

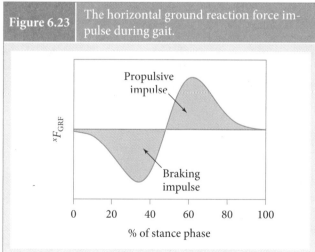

6.4.1 Using Newton's Laws to Improve Performance and Prevent Injury

Section Question

The egg (or water-balloon) toss game is summertime favorite. What strategy do you use to prevent the egg from breaking? Why does it work?

a propulsive impulse. In Figure 6.23, the propulsive impulse and braking impulse are equal. The areas under the curves are the same. So, in this case, the momentum (and thus velocity) is unchanged during this period of stance. To accelerate, the

Biomechanics is used to improve performance and reduce injury. The concepts outlined in this lesson and Equation 6.16 can be used to achieve these aims. When improving performance, maximization of the final velocity is often equated with performance. Rearranging Equation 6.19 to isolate this variable, you get

$$v'' = \frac{\sum F_{\text{average}} \Delta t}{m} + v' = \text{relative impulse} + v' \quad (6.23)$$

Note that four variables determine the final velocity: the average effective force, the time that force is applied, the mass, and the initial velocity. Notice that the numerator in Equation 6.21 is impulse. Impulse divided by mass is called *relative impulse*. So really, final velocity is determined by relative impulse and the initial velocity. These variables were used to create the hierarchical model in **Figure 6.24**. You can increase the final velocity

> **Relative impulse** The amount of impulse that can be generated relative to body mass

by increasing the average effective force, the time that force is applied, and/or the initial velocity. You can also increase the final velocity by decreasing the mass. In the short term, manipulating this variable is not really an option, but it does highlight the detrimental effect excess body fat can have on performance. Notice, in this case, the effective force is determined by the ground reaction force and the force due to gravity. If the movement is against gravity, mass imposes an additional penalty. The time the effective force is applied is an interesting variable. Theoretically, increasing the time will improve the performance, but oftentimes this is either not possible or not advised. For example, a "wind up" will allow more time to apply a force, but will also increase the time of the movement. In some activities, such as pitching, this would not be problematic. In other activities, such as boxing, it would be. Finally, note that increasing the initial velocity will increase the final velocity. If you try performing a standing long jump and running long jump, in which case would you jump further? With the running long jump, your body has an initial velocity prior to the last push-off. This is not the case in the standing long jump.

When trying to minimize injury risk, it is helpful to think in terms of trying to decrease the forces on the body. Using Equation 6.18 to minimize forces:

$$\sum F_{\text{average}} = \frac{m(v'' - v')}{\Delta t} \quad (6.24)$$

You will note that there are four variables that will decrease the need for an effective force: the mass, the initial and final velocities, and the time. The hierarchical model is presented in **Figure 6.25**. Like its effect on performance, decreasing mass will decrease the forces on the body. It is not a variable that can be manipulated in the short term, but it does highlight the fact that excess mass can also increase injury risk. Increasing the time over which the force is applied will also decrease the force. You saw that in the previous section.

This is the same strategy that is used in the egg toss game. Rather than abruptly changing the momentum when the egg hits your hand, you "give" with the egg to gradually change its

Figure 6.24	A hierarchical model for increasing the final velocity.

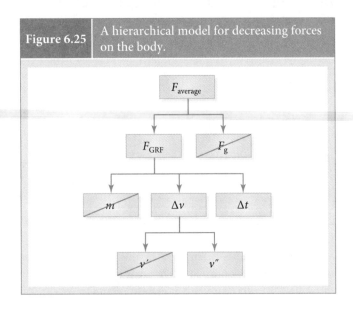

Figure 6.25	A hierarchical model for decreasing forces on the body.

momentum. By increasing the time over which the momentum is changing, the forces on the egg are less, and the shell does not break.

> ### Section Question Answer
>
> You prevent the egg (or water balloon) from breaking by slowly decreasing its velocity rather than stopping it abruptly by "giving" with it when it touches your hand. Because the time is increased for the same change in momentum, the average force is less. The same principle applies to the forces applied to your body.

COMPETENCY CHECK

Remember:

1. Define relative impulse.
2. List the factors that determine the final velocity of a body.
3. List the factors that determine the average force on a body.

Understand:

1. For each of the factors in Remember, Question 2, how will increasing the factor affect the final velocity? How will decreasing the factor affect the final velocity?
2. For each of the factors in Remember, Question 3, how will increasing the factor affect the average force? How will decreasing the factor affect the average force?

Apply:

1. Pick any activity, and show how Equation 6.19 can be used to improve performance or decrease injury risk.

6.5 TYPES OF LINEAR RESISTANCES USED IN EXERCISE

> ### Section Question
>
> Exercise against a resistance is popular as a recreational activity, to improve performance, and to rehabilitate after an injury. What are the differences between the various forms of resistance?

Thus far, you have examined the force due to gravity and contact forces in the form of the ground reaction force. But you will encounter other types of resistances to movement, particularly during exercise, including (1) constant (or fixed) resistances, (2) variable resistances, and (3) accommodating resistances.

6.5.1 Constant Resistance (Gravity)

You are already familiar with gravity. It is the only constant force experienced by the body. It has the form:

$$F_g = ma_g$$

6.5.2 Variables Resistance

Variable resistances actually change during the movement. As was mentioned previously, gravity is the only constant force. Some exercise machines are specifically designed to change the resistance they provide. Two variable forces that will be discussed here are inertial forces and elastic forces.

Inertial Force and d'Alembert's Principle

Inertial forces are a little difficult to understand, conceptually. It may be easier to explain it with an example. Imagine that you are moving a dumbbell in two cases: the first is in a direction that is across gravity, and the second is in a direction that is against gravity (Figure 6.26).

What forces are resisting the movement? In the first case, no forces are resisting movement, and it is a straightforward case of Newton's second law:

$$\sum F_{ext} = ma_{COM}$$

A larger mass or a larger acceleration will require a larger effective force. That can sometimes be a mouthful, both in speaking and in writing equations. For brevity's sake, you could say that a larger mass or a larger acceleration is creating a greater inertial force. To overcome the inertial force, you need a larger effective force. Mathematically, you arrive at this by first rearranging Newton's second law:

$$\sum F_{ext} - ma_{COM} = 0 \qquad (6.25)$$

The inertial force, $F_{inertial}$, is simply:

$$F_{inertial} - -ma_{COM} \qquad (6.26)$$

This means that:

$$\sum F_{ext} + F_{inertial} = 0 \qquad (6.27)$$

By creating this "fictitious" inertial force, you can easily solve problems that deal with motion as if they were static. It is called d'Alembert's principle, and it is really nothing more than a bookkeeping device so that you do not have to keep writing out Newton's second law (and you can just refer to the product of mass and acceleration as an inertial force), but it seems to either confuse a lot of people or be considered an unnecessary step (or two). Conceptually, it could aid you in keeping track of all the "forces" that are resisting movement (called resistive forces). For example,

Figure 6.26 | (a) Moving a dumbbell across gravity; (b) moving a dumbbell against gravity.

(a)

(b)

(a) © Lana K/ShutterStock, Inc. (b) © Philip Date/ShutterStock, Inc.

now consider that case of going against gravity. There are two forces resisting movement:

$$F_{\text{resistive}} = F_{\text{inertial}} + F_{\text{g}} \qquad (6.28)$$

The forces produced by the body must be equal to and in the opposite direction of the resistive force. The forces for this example are explicitly defined, using both Newton's second law and d'Alembert's principle, in Table 6.1. You will notice that d'Alembert's principle requires an extra couple of steps and may not seem like it is worth it. But things could get messy with Newton's second law if you have a lot of terms, and

sometimes it is easier (both conceptually and mathematically) to use the concept of inertial force. One word of caution: do not mix the methods. Do not include an inertial term as part of the sum of forces and then say this sum is equal to the mass times acceleration. You will have included the term twice!

Elastic Resistance

Elastic materials also provide variable resistance. A force is elastic if it follows Hooke's law:

$$F_{\text{resistive}} = -kl_{\text{stretch}} \qquad (6.29)$$

Table 6.1	A Comparison of Newton's Law and d'Almbert's Principle	
Newton's second Law		**d'Alembert's Principle**
$\Sigma F_{ext} = ma_{COM}$		$\Sigma F = 0$
$F_{body} - F_g = ma_{COM}$		$F_{body} + F_{resistive} = 0$
$F_{body} = ma_{COM} + F_g$		$F_{body} = -F_{resistive}$
$F_g = ma_g$		$F_{resistive} = F_{inertial} + F_{gravity}$
		$F_{inertial} = -ma_{COM}$
		$F_g = -ma_g$
$F_{body} = ma_{COM} + ma_g$		$F_{body} = -(-ma_{COM} - ma_g)$
	$F_{body} = m(a_{COM} + a_g)$	

Figure 6.27 Elastic tubing is an example of variable resistance. It provides greater resistance the more it is stretched.

© Kris Butler/ShutterStock, Inc.

where k is the stiffness and $l_{stretch}$ is the distance the material has been stretched. Elastic tubing, elastic bands, coils, and springs all make use of elastic resistances (**Figure 6.27**). The larger the k, the stiffer the material, and a larger force is required to stretch it. The negative sign indicates that the resistance will be in the direction opposite of that in which it is being stretched. The resistance force is not constant; it increases proportionally with the distance that the material is stretched beyond its resting length. The resistance is minimal at the beginning of the exercise and increases as the movement is performed. The resistance is related to the relative change in the original length of the material, and not the actual length that the material stretched:[3] Stretching a piece of tubing from its resting length to 0.25 meters would result in different forces than stretching that same piece of tubing from 0.25 meters to 0.5 meters. Even though the distance each material has been stretched is 0.25 meters, the resistances would be larger in the second case.

6.5.3   Accommodating Resistance Devices (Fluids)

Accommodating resistances provide resistance that is proportional to the lifter's effort: as the force is increased, so too is the resistance. Think of moving in water. It is quite a bit easier to move slowly in water than it is to move fast. Hydraulic and pneumatic devices (**Figure 6.28**) that use some sort of piston to push or pull a fluid (liquid for hydraulic and gas for pneumatic) through a cylinder have similar properties to movements that are performed under water. Fluid resistance follows the form:

$$F_{resistive} \propto \rho A v^2 \qquad (6.30)$$

This means that the resistance force is proportional to the product of: ρ, the fluid density, A, the surface area, and v^2, the square of the movement velocity. With certain exercise machines, the density of the fluid and the cross-sectional area

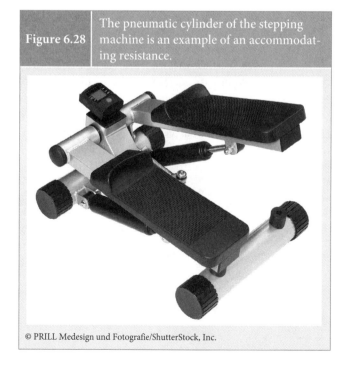

Figure 6.28 The pneumatic cylinder of the stepping machine is an example of an accommodating resistance.

© PRILL Medesign und Fotografie/ShutterStock, Inc.

of the piston are inherent in the design of the equipment. They cannot be altered by the exerciser. So the resistance that the machine provides to the exerciser increases with the velocity of the movement. Note that this is not a linear relationship; doubling the velocity will increase the resistance fourfold.

Section Question Answer

There are three types of resistance forces: constant, variable, and accommodating. A constant resistance does not change. The force of gravity is the only constant resistance force. Variable forces change throughout the range of motion and include inertial forces and elastic resistances. Accommodating resistances increase with the amount of force applied to them: The greater the force applied, the greater the resistance to movement.

COMPETENCY CHECK

Remember:

1. Define inertial force.
2. Describe d'Alembert's principle and Hooke's law.
3. What is the only truly constant resistance?

Understand:

1. How do you increase the force of a constant, variable, and accommodating resistance?

Apply:

1. Give examples of constant, variable, and accommodating resistances.

SUMMARY

In this lesson, you learned what causes linear motion (the key concepts are listed in Table 6.2). You began with a review of Newton's laws of motion, which you were probably familiar with, but they were presented in a slightly different way. To change the momentum of a system, you must apply an impulse from an effective force that is external to that system.

Table 6.2	Key Concepts
• Newton's laws of motion	
• Impulse–momentum relation	
• Friction	
• d'Alembert's principle	
• Constant, variable, and accommodating resistances	

The main two forces external to the body are gravity and contact forces, namely the ground reaction force. Because gravity is constant, the only force that can influence performance is the ground reaction force, which is a consequence of Newton's third law. Propulsive and braking impulse from the ground reaction force either increases or decreases momentum, respectively. The factors involved in impulse can be manipulated to improve performance and decrease injury risk. The importance of friction to the ground reaction force impulse was also demonstrated. Finally, various forms of resistance used in exercise were explored.

REVIEW QUESTIONS

1. Define the following terms: force, impulse, rate of force development, weight, reaction force, effective force, friction, fixed resistance, variable resistance, and accommodating resistance.
2. List Newton's three laws of motion.
3. Give an alternate to Newton's second law.
4. For each of the following forces, is the vertical or horizontal force greater?
   a. 400 N force 10° from the horizontal.
   b. 500 N force 80° from the horizontal.
   c. 1,000 N force 80° from the vertical.
5. What is the effective force of the following? Which way is the body accelerating? Is the GRF acting as a propulsive or braking force?
   a. $^yF_{GRF}$ 100 N and F_g of 500 N.
   b. $^yF_{GRF}$ 600 N and F_g of 500N.
   c. $^xF_{GRF}$ 100 N and F_g of 500 N.
   d. $^yF_{GRF}$ 1,000 N and F_g of 500 N.
6. Give examples of the three ways to change momentum.
7. List four ways to increase the final velocity.
8. List four ways to decrease the average force necessary to decrease momentum.
9. List the types of forces usually encountered by the human body during movement.
10. List the factors that affect friction.
11. List the types of forces that are used as resistance during exercise. Give examples of each.

REFERENCES

1. Zatsiorsky VM. *Kinetics of Human Motion.* Champaign, IL: Human Kinetics; 2002.

2. Frautschi SC, Olenick RP, Apostol TM, Goodstein DL. *The Mechanical Universe: Mechanics and Heat. Advanced Edition.* Cambridge: Cambridge University Press; 1986.

3. Simoneau GG, Bereda SM, Sobush DC, Starsky AJ. Biomechanics of elastic resistance in therapeutic exercise programs. *Journal of Orthopaedic & Sports Physical Therapy.* Jan 2001;31(1):16–24.

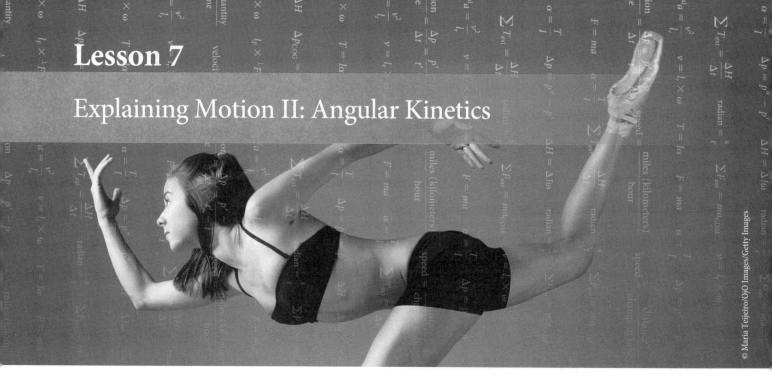

Lesson 7

Explaining Motion II: Angular Kinetics

© Maria Teijeiro/OJO Images/Getty Images

LEARNING OBJECTIVES

After finishing this lesson, you should be able to:

- Define the following terms: torque, moment of force, lever, lever arm, moment arm, mechanical advantage, propulsive torque, braking torque, effective torque, couple, angular impulse, and static equilibrium.

- List the angular versions of Newton's three laws of motion. Give an alternate to the second law.

- List the combined linear and angular effects of a force.

- Give examples of the three ways to change angular momentum.

- Manipulate the variables associated with the angular version of Newton's second law to increase performance and decrease risk of injury.

- List the ways torque is used as resistance during exercise. Give examples.

- List the conditions for static equilibrium. Give examples of when static equilibrium is important in human movement.

Angular motion is a big part of life. Wheels are a classic example of a rotating body. There are many other times when you would want to create angular motion: twisting a cap off a bottle or jar; swinging a racquet, club, or bat; and even opening a door. What may be even more important is that any movement your body makes is a combination of angular motions about your joints.

By now, you should appreciate momentum and its importance in determining the outcome of a movement. Recall that changing momentum requires an external influence: A body will not change its momentum on its own. Linearly, that external influence is a force or, more specifically, a mechanical impulse. It is now time to explore the angular equivalent of these concepts.

7.1 THE ANGULAR EQUIVALENT OF THE FIRST LAW

Section Question

Brittany is trying to change her flat tire, but the lug nut on the wheel is stuck (Figure 7.1). What can she do to improve her ability to rotate the nut so she can remove the wheel?

The essence of Newton's first law is that the linear momentum of a system will not change unless acted on by an outside (external) force. Likewise, you would say that the angular momentum of a system will not change unless acted on by something external to the system or body. Recall that, for biomechanical systems, linear momentum and velocity are proportional; you cannot change one without changing the other because the mass is fixed. For a single rigid body, the same relation holds true for angular momentum and angular velocity. However, for a system of connected rigid bodies, you can change the angular velocity without changing the angular momentum because you can change the moment of inertia by changing the posture. But consider for the moment how you would change the angular velocity if "I" was fixed. If you wanted to speed up or slow down the rotation of a body, how would you do it? To change

Figure 7.1	What can you do if a lug nut is stuck on a wheel?

© CCat82/ShutterStock, Inc.

Figure 7.2	A rod with two forces (F_1 and F_2) equal in magnitude, opposite in direction, and co-planar. The rod will not translate, but will rotate in the clockwise direction.

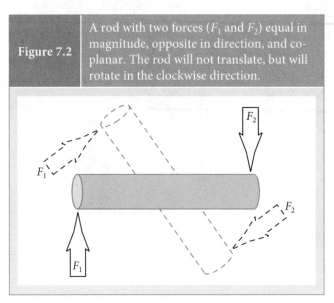

> **Torque**   The turning effect of a force
>
> **Moment of force** ("Moment" for short), is synonymous with torque; it is the turning effect of a force

7.1.1   Torque

What is torque? The idea of torque is one of the harder ones for students to grasp, but it need not be. Consider the rod pictured in **Figure 7.2**. Two forces are acting on the rod: F_1 and F_2. Both of these forces are equal in magnitude, opposite in direction, and in the same plane. From

$$\sum F_{ext} = ma_{COM}$$

it should be clear that the effective force is zero, and there will be no linear acceleration of the rod's center of mass because the forces cancel each other. Does that mean that the rod is not moving? No. Hopefully, you can see that the rod will be turning in a clockwise direction due to the application of these forces. That is what torque is: the turning effect of a force. Torque is sometimes called a moment of force, or "moment" for short. See **Box 7.1** for more details on the nomenclature associated with these terms.

the linear momentum, you would apply a force. To change angular momentum, you would apply an external **torque**, or **moment of force**.

Basic ideas about torque are exemplified by using a wrench to either tighten or loosen a nut on a bolt. To turn (rotate) the nut, you must create a torque. The torque comes from applying a force (your hand) to a rigid body (the wrench) some distance from an axis of rotation (the bolt). But not just any force will do. Consider four forces, each equal in magnitude, applied to the wrench in different directions as shown in **Figure 7.3**. From your everyday experience, you should be able to immediately recognize that the forces applied in 7.3(a) and 7.3(b) will not turn the nut. The forces in 7.3(c) and 7.3(d) will turn the nut, but in opposite directions.

What is the main difference between the forces in 7.3(c) and 7.3(d) compared to 7.3(a) and 7.3(b)? The forces in the latter two were applied parallel to the wrench, whereas the forces in the former were applied perpendicular to the wrench. Consider the formula for torque:

$$l_F \times {}^\perp F = T \tag{7.1}$$

Where T is the torque, l_F is the distance from the axis of rotation to the point of force application (the lever arm), and ${}^\perp F$ is the force that is perpendicular to the rigid body. This formula expresses the fact that the only forces (or components of forces) that create a torque are the ones that are perpendicular to the rigid body. If a force is applied at any angle other than perpendicular, the amount of torque it produces decreases; the further away from the perpendicular, the less amount of torque. Forces parallel to the body end up producing zero torque. To determine the magnitude of the torque, a force must be resolved into its components parallel and perpendicular to the rigid body using trigonometry (**Figure 7.4**). As

Box 7.1 | More on Moments

Part of the confusion around angular kinetic terms is that some authors use different words for the same thing, whereas others maintain that there is a difference between them. Consider again the forces in Figure 7.2. When two forces are equal in magnitude, opposite in direction, and in the same plane, they are called a couple. As you already determined, a couple will not cause a linear acceleration (because they cancel), but will potentially cause an angular acceleration. Some authors maintain that this moment of a couple is called torque, and that the turning effect of a force should be called a moment of force (usually shortened to just "moment"). Others use the terms interchangeably, as will be done in this book.

Another source of ambiguity lies in the terms *moment arm* and *lever arm*. Most authors used the terms interchangeably, but a few believe the lever arm is the distance from the axis of rotation (or pivot point) to the point of force application, and the moment arm is the perpendicular distance from the force vector to the axis of rotation. These two distances would be the same only if the force were acting perpendicular to the lever arm (see following figure). Because of the two different ways torque can be calculated, this distinction *will* be used throughout this book. The lever arm will be denoted by the symbol l_F, and the moment arm will be denoted by the symbol $^\perp d$.

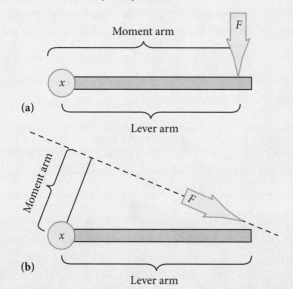

(a)

(b)

The differences between a lever arm and a moment arm. A lever arm is the distance along the lever from the axis of rotation to the point of force application. A lever arm is the perpendicular distance from the axis of rotation to the force vector. (a) The lever arm and moment arm are the same; (b) they are different because the force is applied at an angle to the lever.

Figure 7.3 Each of the forces have the same magnitude, but different directions. The force in (a) and (b) will not produce a torque. The force in (c) and (d) will produce torques in opposite directions.

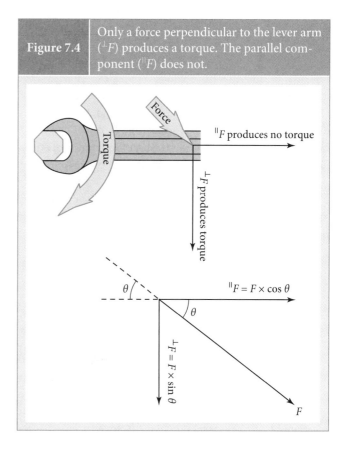

Figure 7.4 Only a force perpendicular to the lever arm ($^\perp F$) produces a torque. The parallel component ($^\parallel F$) does not.

Figure 7.5	The amount of force producing torque decreases in a nonlinear fashion as the angle moves away from the perpendicular.

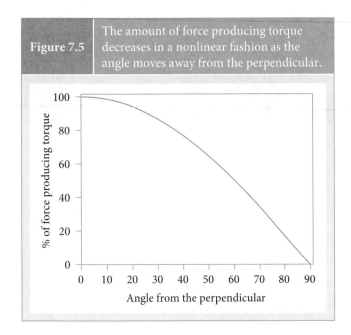

Figure 7.6	Torque is enhanced by increasing the length of the lever arm.

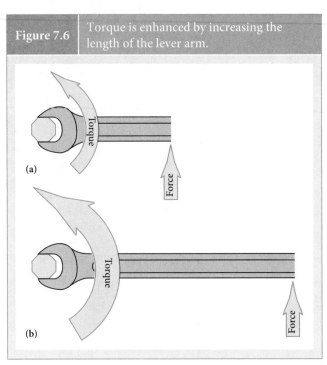

you can see from Figure 7.5, the decrease in the amount of force producing torque is not linear: A small change in the angle from perpendicular will have a small effect, but as the angle from the perpendicular gets larger, the decrease in force producing torque becomes more profound.

Another implication of Equation 7.1 is that the further away from the axis of rotation the force is applied (the longer the lever arm, l_F), the larger the torque it will produce. Again, you should note this from your everyday experience. If the nut is stuck, you would grab a wrench with a longer handle (Figure 7.6). Door handles are placed at the outer edge of the door, not next to the hinges, so that the force applied to the handle maximizes the torque turning the door. Hopefully, these everyday experiences help clarify the concept of torque. As you can see, if you wish to create a torque, three factors are involved: the magnitude of the force, the distance from the axis of rotation to the point of force application, and the angle between the force and the lever arm (Figure 7.7).

Important Point! You can increase the torque by increasing the magnitude of the force, increasing the length of the lever arm, or applying the force as close to the perpendicular of the lever arm as possible.

Oftentimes, biomechanicians speak of moment arms instead of lever arms (Box 7.1). Although conceptually these are different, the resultant torque produced by a force is equivalent no matter which way you think about it. The moment arm is the perpendicular distance from the force

vector to the axis of rotation. The important distinction is that the magnitude of the force remains unchanged, and the moment arm has changed when the force is applied at an angle different from the perpendicular. In such a case, Equation 7.1 becomes

$$F \times {}^{\perp}d = T \qquad (7.2)$$

where ${}^{\perp}d$ is the moment arm.

Important Point! If the force and the distance are not perpendicular to each other, you must correct either the force or the distance (but not both).

Figure 7.7	A hierarchical model of torque. Torque is determined by (1) the amount of force, (2) the distance from the axis of rotation to the point of force application, and (3) the angle between these two vectors.

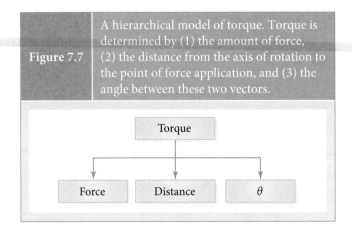

Returning to Figures 7.3(c) and 7.3(d), you no doubt correctly anticipated that the force applied in 7.3(c) rotated the nut clockwise (the negative direction, by convention), and the force in 7.3(d) rotated it counterclockwise (the positive direction). One of the right-hand rules (there are actually a few in mechanics!) is helpful in distinguishing the direction of torque for an applied force. Start with your hand on the axis of rotation, and point your fingers along the lever toward the applied force. Now curl your fingers in the direction that the force is applied. This is direction of the torque. Examine **Figure 7.8**, and notice that if you push up (a) or pull up (b) on the handle of the wrench, you have still produced a positive torque. Likewise, if you push down (c) or pull down (d) on the handle, you will produce a clockwise (negative) torque.

Because of the way most screws, nuts, and bolts are threaded, a clockwise torque will tighten the screw or nut, and a counterclockwise torque will loosen it. So it is important to know the direction of the force in addition to its magnitude when determining its effect.

Section Question Answer

If Brittany is looking to loosen the nut, she must produce more torque on it. She can do this by increasing the amount of force she produces, but that will be limited by her strength. In addition, she can ensure that all her force is producing torque by applying that force perpendicular to the wrench. But again, she is limited by the amount of force she can produce. Finally, she could increase the lever arm by using a longer wrench. Theoretically, with a long enough wrench, she could create as much torque as she needs. All these things assume that she is creating the torque in the correct direction.

COMPETENCY CHECK

Remember:

1. Define torque, moment of force, lever arm, and moment arm.
2. What is the angular equivalent of Newton's first law?

| Figure 7.8 | Pushing up (a) or pulling up (b) on the handle creates a clockwise torque in accordance with the right-hand rule. Pushing down (c) and pulling down (d) on the handle create a counterclockwise torque. |

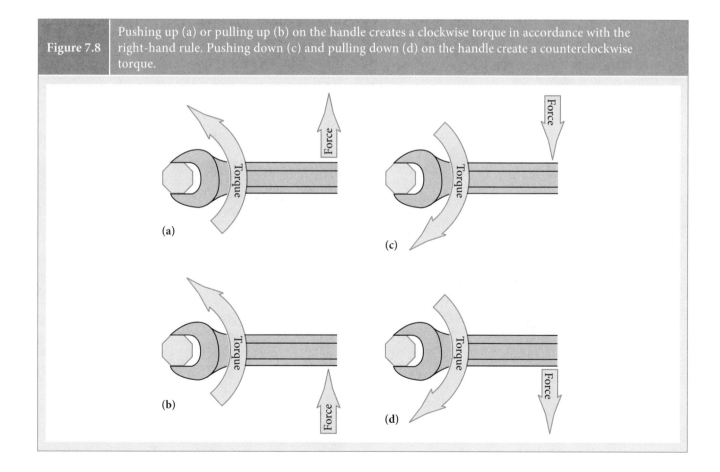

3. List the three factors that determine the torque on a body.

Understand:

1. What effect will doubling the force have on torque?
2. What effect will doubling the lever arm have on torque?
3. What effect will changing the angle of force application from perpendicular to 60° from the perpendicular have on torque?

Apply:

1. Can you always increase the lever arm? Why, or why not?
2. If not, what options do you have for increasing torque?
3. What is the typical direction when tightening objects (screws, caps, etc.)? What is the typical direction for loosening them?

7.2   THE ANGULAR EQUIVALENT OF THE SECOND LAW

Section Question

Diane was an avid bicyclist who, due to a recent accident, must now use a wheelchair (Figure 7.9). What can you say to instruct her on creating an acceleration of the wheel?

Now that you understand the concept of torque, you can begin to appreciate the effect of that torque. This is aided by reviewing Newton's second law:

$$F = ma$$

You just learned that torque (abbreviated with the Greek capital "T") is the angular analogue to force. The moment of inertia (I) is the angular analogue to mass, and the angular acceleration (α) is the analogue to linear acceleration. Substituting these variables into the equation

$$F = ma$$

gives you the angular equivalent to Newton's second law:

$$T = I\alpha \qquad (7.3)$$

Notice in Equation 7.1, torque was defined as the product of the lever arm and the force perpendicular to the lever arm. Equation 7.3 has torque as the product of the moment of inertia and the angular acceleration. Equation 7.1 (or 7.2) can be thought of as what is causing the torque and Equation 7.3 as the effect of the torque:

$$l_F \times {}^{\perp}F = T = I\alpha \qquad (7.4)$$

$$\text{cause} = \text{effect}$$

You may often wish to separate the effect from the causes. Equation 7.4 can be re-written to accomplish this for angular motion:

$$\alpha = \frac{T}{I} \qquad (7.5)$$

Consider the torque required to turn a wheel. A wheelchair with larger diameter wheels has the potential to create a faster linear velocity, but such wheels would have a larger moment of inertia and thus be harder to turn. In a self-propelled wheelchair, torque is supplied to the wheel via the force applied to the hand rim (Figure 7.10). Using

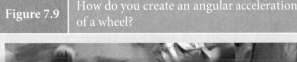

Figure 7.9 | How do you create an angular acceleration of a wheel?

© SVLuma/ShutterStock, Inc.

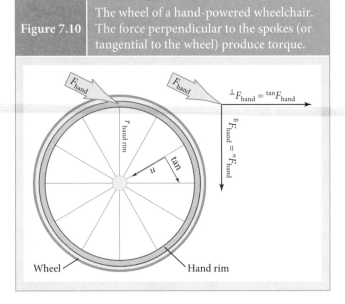

Figure 7.10 | The wheel of a hand-powered wheelchair. The force perpendicular to the spokes (or tangential to the wheel) produce torque.

Equation 7.5, you can determine the effect of the hand's force on rotating the tire:

$$\alpha_{\text{wheel}} = \frac{T_{\text{wheel}}}{I_{\text{wheel}}} = \frac{{}^{\perp}F_{\text{hand}} \times l_{\text{r-hand rim}}}{I_{\text{wheel}}} \quad (7.6)$$

If there is a coordinate system embedded in the wheel (Box 7.2), then the tangential force (${}^{\text{tan}}F$) is the same as the perpendicular component of the force (${}^{\perp}F$). Note that the hand rim is usually close to the same diameter as the wheel. This arrangement increases the lever arm of the hand's force, thus creating a more effective torque. If the hand rim was smaller, then the torque supplied for a given amount of force would be less. Also note that only the perpendicular force creates a torque; the force parallel to the spokes does not. Thus, hand position plays an important part in creating torque about the wheel. Unfortunately, the position of the hand in relation to the upper extremity does not always permit the most effective force applied to the hand rim (see Box 7.3). Finally, a larger moment of inertia of the wheel (I_{wheel}) will require a larger torque to overcome it and produce the same angular acceleration of the wheel.

The same idea applies not only to wheelchairs but also to bicycles. But instead of applying a torque to a hand rim, on a bicycle the force applied to the sprocket via the chain

---

In determining a frame of reference, you established an origin and directions from that origin. Mostly, you used x- and y-axes that were perpendicular to each other, and those axes were firmly attached to Earth. When the axis is fixed to Earth, it is called a **global reference frame**.

Sometimes, it is helpful to attach a reference frame to a body and have that reference frame travel with it. This is called a **local reference frame**. For example, consider three forces applied to the hand rim of a self-propelled wheelchair as in the following figure. For illustrative purposes, suppose the magnitude of each force is 100 N. In the global frame, the components of each force would be

Force	x-component	y-component
F_1	100	0
F_2	70.7	−70.7
F_3	0	−100

At this point, you may be tempted to think that each of these forces would produce a different torque on the wheel, but you would be wrong. All three forces have the same orientation with respect to the wheel (or the spokes). The component of force that is tangent to the wheel (or perpendicular to the spokes) will create a torque on the wheel, but that component is not fixed in the global reference frame. In such a case it would be helpful to think in terms of a local reference frame. Instead of calling it another x–y reference frame (which you could do), to avoid confusion it might be helpful to call it something else (and more descriptive) like n-tan reference frame, where the "n" stands for normal, or perpendicular, to the surface of the rim and "tan" stands for tangential to the surface of the rim. Hopefully, you will see that only the "tan" component contributes to the torque on the wheel because the "n" component has a moment arm of zero. With such a reference frame, you will note that the components of force are

Force	n component	tan component
F_1	0	100
F_2	0	100
F_3	0	100

showing you that each force produces that same torque on the wheel. You always want to pick a frame of reference that is physically meaningful and makes the math easier. In this case, the local n-tan coordinate system did both.

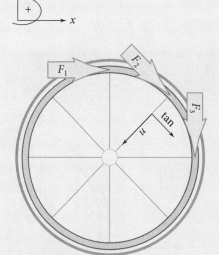

Three forces that are different in the x-y global frame are equivalent in the n-t local coordinate frame and produce the same effect on the wheel.

Box 7.3 | Applied Research: Propulsive Force in Self-Propelled Wheelchairs

Equation 7.6 would indicate that only the tangential force, ^{tan}F, is effective in applying a torque to the wheel, and thus moving a wheelchair. Yet even experienced wheelchair users have a substantial component of force in the noncontributing direction, ^nF. This investigation aimed to explain the discrepancy between what pure mechanics suggests someone should do and how they actually perform when propelling themselves in a wheelchair. Using mathematical modeling techniques, it was determined that the movement patterns that would produce the greatest mechanical torque on the wheel (largest ^{tan}F) required a large effort due to the arm posture during propulsion. In other words, the greatest mechanical effect also requires a huge biomechanical effort. Rather than trying to maximize the mechanical effect, wheelchair users attempt to strike a balance between mechanical effect and physical cost, at least during submaximal speeds. This study is just another example of the importance of "bio" in the study of biomechanics.

Data from: Rozendaal LA, Veeger HEJ, van der Woude LHV. The push force pattern in manual wheelchair propulsion as a balance between cost and effect. *J Biomech*. 2003;36:239–247.

Propulsive torque A torque that is increasing the speed of rotation

Braking torque A torque that is decreasing the speed of rotation

will cause the tire rotation to speed up (**Figure 7.11**). The created torque is thus a **propulsive torque**, analogous to a propulsive force. Applying a force to the rim via the brake pads will cause the rotation to slow down, creating an appropriately named **braking torque**. Because a negative torque can increase the angular velocity (in the clockwise direction) or decrease the angular velocity (in the counterclockwise direction), it is more appropriate to think in terms of propulsive and braking torques rather than positive and negative torques.

Important Point! A torque will not always be in the direction of motion.

7.2.1   Net (Effective) Torque

A force does not necessarily produce an acceleration. Rather, it is the net (effective) force that does. Similarly, a torque has the tendency to produce an angular acceleration, but it is the net, or effective, torque that determines the angular acceleration. Think of what happens if you pedal and apply the brakes at the same time, or consider a seesaw (**Figure 7.12**). If both F_1 and F_2 were the same magnitude and had the same lever arm ($l_{F\text{-}1} = l_{F\text{-}2}$), the seesaw would be balanced and not rotate. If F_2 was moved further away from the pivot point ($l'_{F\text{-}2}$), it would have a larger lever arm and produce more torque. The torques would be nonzero, and in this case, the seesaw would rotate clockwise. A more specific form of Equation 7.3 is

$$\sum T_{\text{ext-COM}} = I_{\text{COM}}\alpha \tag{7.7}$$

The value ΣT is known as the "effective torque." If you wanted to separate the effects from the causes, you would rewrite Equation 7.7 as:

Effective torque The vector $I_{\text{COM}}\alpha_{\text{COM}}$, which is the effect of the net sum of all torque vectors acting on a body

$$\alpha = \frac{\sum T_{\text{ext-COM}}}{I_{\text{COM}}} \tag{7.8}$$

Figure 7.11 | The force of the chain creates a propulsive torque on a bicycle.

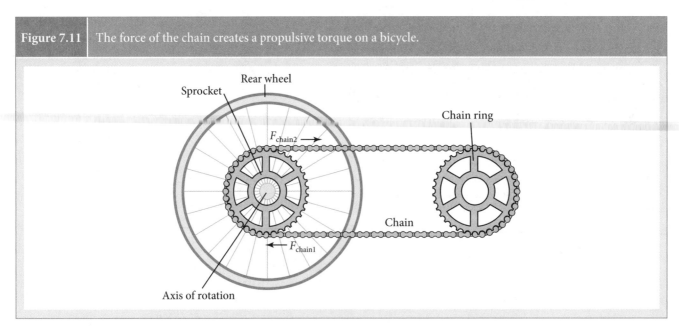

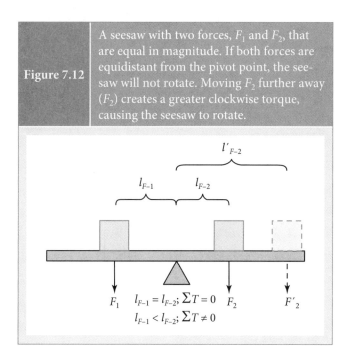

Figure 7.12 A seesaw with two forces, F_1 and F_2, that are equal in magnitude. If both forces are equidistant from the pivot point, the seesaw will not rotate. Moving F_2 further away (F_2') creates a greater clockwise torque, causing the seesaw to rotate.

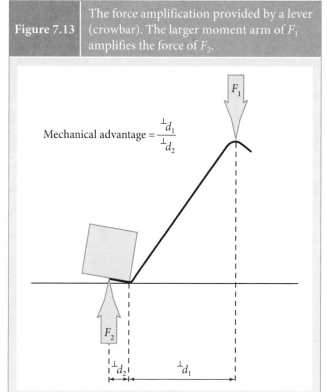

Figure 7.13 The force amplification provided by a lever (crowbar). The larger moment arm of F_1 amplifies the force of F_2.

Note that both the torques and the moment of inertia are determined relative to the system's center of mass, and that it is the torques external to the system that are of interest. If you were calculating the torques about any other point, you would have to also determine a new moment of inertia about that point.

7.2.2 Levers and Mechanical Advantage

A **lever** is a rigid body that is used in conjunction with a pivot point to multiply the force or speed applied to another body. Common examples of levers include a seesaw, a crowbar, a wheelbarrow, and a wrench. All levers have four components: a pivot point or fulcrum, rigid body, and two forces. Looking at a crowbar, you can see how a lever can act as a force multiplier (**Figure 7.13**). In a case where there is no rotation of the crowbar, the sum of the external torques must be zero. That is,

> **Lever** A rigid body that is used in conjunction with a pivot point to multiply the force or speed applied to another body

$$T_1 + T_2 = I\alpha$$
$$\alpha = 0$$
$$T_1 = -T_2$$

You know the causes of the torques are the two forces:

$$-F_1 \times {}^\perp d_1 = -(F_2 \times {}^\perp d_2) \qquad (7.9)$$

Rearranging Equation 7.9 gives you the result:

$$-F_1 \times \frac{{}^\perp d_1}{{}^\perp d_2} = -F_2 \qquad (7.10)$$

Notice the ratio ${}^\perp d_1 / {}^\perp d_2$ is the mechanical advantage of F_1. (Note that I used the moment arm, ${}^\perp d$, rather than the lever arm so that you can look at the whole force, and not just ${}^\perp F$). If the ratio of the lever arms is $2:1$, then the potential force on the other end of the lever is twice that of the original force.

A force further away will have a mechanical advantage over the other force because it will create a greater torque for the same amount of force. If both F_1 and F_2 have the same magnitude but F_2 is further away from the pivot point than F_1, the body will rotate clockwise because F_2 has a mechanical advantage (Figure 7.12).

Levers are often used to give you a mechanical advantage. A wheelbarrow (**Figure 7.14**) uses the mechanical advantage supplied by a lever: Because you grab the handles further from the pivot point than where load is, you have a mechanical advantage over the load and can lift a heavier load than you could just by picking it up.

An important point to keep in mind is that, with levers as with most things in life, you cannot get something for nothing. If a lever has a mechanical advantage, it means that you

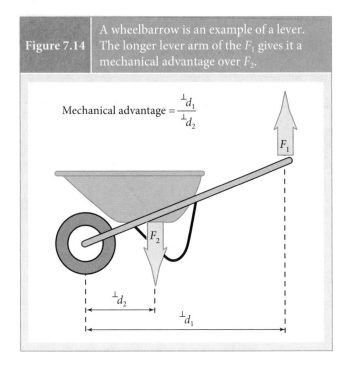

Figure 7.14 | A wheelbarrow is an example of a lever. The longer lever arm of the F_1 gives it a mechanical advantage over F_2.

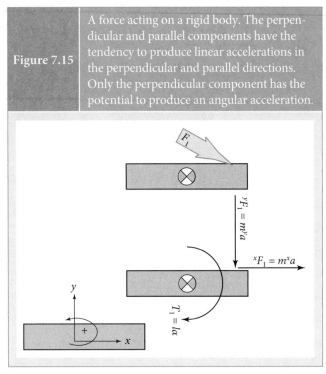

Figure 7.15 | A force acting on a rigid body. The perpendicular and parallel components have the tendency to produce linear accelerations in the perpendicular and parallel directions. Only the perpendicular component has the potential to produce an angular acceleration.

can get the same torque with less force. However, that force has to be applied over a greater distance. Assume that you are using a wrench that has a mechanical advantage of 4 : 1 over the nut (the actual numbers are not important here). Recall that if you wanted to determine the linear displacement, you would multiply the angular displacement (in radians) by the radius. Both the wrench and nut would have the same angular displacement, $\Delta\theta$. If the wrench's radius was four times larger than the nut's, it would have a linear displacement that was four times larger than the nut's as well. But because the force is so much less, applying it over a larger distance usually is not a problem.

7.2.3 The Combined Linear and Angular Effects of a System of Forces

Hopefully it is becoming apparent that rarely does a force have a single effect. A single force should be thought of in terms of its components. Each component then has the potential to produce both a linear and an angular acceleration (Figure 7.15). Note that it has the *potential* to cause an acceleration, not that it will. For example, the $^xF_1$ in Figure 7.15 does not produce an angular acceleration because it is parallel to the rigid body (moment arm is zero). In addition, the acceleration will be determined by the effective force and effective torque.

Examine the system of forces in Figure 7.16a. Assume that F_1 and F_2 are equal in magnitude, opposite in direction,

and coplanar. The net of these two forces is zero; they cancel each other's ability to produce a linear acceleration. Two forces that are equal, opposite, and coplanar are called a **couple**.

> **Couple** Two forces that are equal in magnitude, opposite in direction, and in the same plane; the effect of a couple is pure rotation with no translation

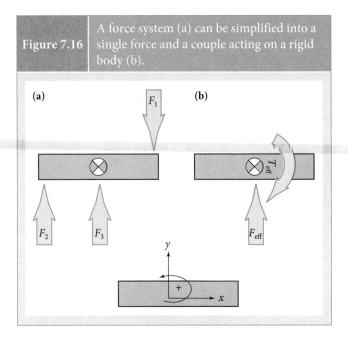

Figure 7.16 | A force system (a) can be simplified into a single force and a couple acting on a rigid body (b).

Couples produce no linear acceleration but pure rotation. Couples are also considered free vectors—that is, unlike forces, they can be applied anywhere on the rigid body. A good example of a couple would be if you had both hands on a steering wheel or bike handles. If you applied an equal force with each hand but in opposite directions, the wheel would turn without any linear forces. F_3 acts through the body's COM (the lever/moment arm is zero) and thus will not produce a rotation about the body's COM. It will produce a linear acceleration, but no angular acceleration. Only forces some distance from the pivot point will produce an angular acceleration.

When there are a number of forces acting on a body, the net effective force and the net effective torque determine the accelerations of the body. All systems of forces can be represented as a single force and a couple acting on a body, as in Figure 7.16b. Such a simplification allows you to appreciate the effect of all forces on a body.

Section Question Answer

Creating a change in angular velocity is caused by an angular acceleration, which in turn is caused by an effective torque. On a wheelchair, this torque is created by the perpendicular component of the force of the hand on the hand rim. Increasing the radius of the hand rim will create more torque for the same force of the hand. Increasing the radius of the wheel will make it harder to turn because the moment of inertia will be higher. These factors should all be considered when selecting and using a wheelchair. You can tell Diane that this is similar to a bicycle, with a few differences. First, the torque comes from the force of the chain on the sprocket. And unlike the hand, the force of the chain is always perpendicular to the sprocket.

COMPETENCY CHECK

Remember:

1. Define: propulsive torque, braking torque, effective force, and couple.
2. State the angular equivalent of Newton's second law.
3. What is the direction of the effective torque?

Understand:

1. What effect does the following have on the angular acceleration?
   a. Decreasing the force.
   b. Decreasing the moment of inertia.
   c. Decreasing the moment arm.

2. Identify the propulsive, braking, and effective torques of the following:
   a. A +30 Nm torque and a −40 NM torque while traveling in the clockwise direction.
   b. A +60 Nm torque and a −20 NM torque while traveling in the clockwise direction.
   c. A +30 Nm torque and a −40 NM torque while traveling in the counterclockwise direction.
3. What are the effects of the following?
   a. A 50 N force applied 20° from the long axis of the rigid body that is applied .02 m from the COM.
   b. A 100 N force applied 60° from the long axis of the rigid body that is applied at the COM.

Apply:

1. Give some examples of when creating a mechanical advantage may actually decrease performance.
2. Describe situations in which you can increase the angular acceleration without increasing the force.

7.3 THE ANGULAR EQUIVALENT OF THE THIRD LAW

Section Question

You are teaching your physical education class a unit on running and dribbling a soccer ball. Why do you teach them to rotate their arms in the opposite directions of their legs (Figure 7.17)?

Newton's third law states that for every force there is an equal and opposite reaction force. Similarly, for every torque exerted by one body on another, there is an equal and opposite torque exerted by the second body on the first. This may not seem intuitive, but several examples can illustrate this concept.

First, think about a diver, gymnast, long jumper, or any other athlete who is airborne and performs some variation of the pike maneuver (Figure 7.18). Generating a torque that creates a counterclockwise rotation of the lower body also creates a clockwise rotation of the upper body (Figure 7.19). Similarly, in the transverse plane, a torque creating a clockwise rotation of the upper body creates a counterclockwise rotation of the lower body (Figure 7.20). If the torques are equal and opposite, does that mean that both parts of the body rotate the same amount? Not necessarily. The angular equivalent of Newton's third law tells you:

$$T_{upper/lower} = -T_{lower/upper} \qquad (7.11)$$

| Figure 7.17 | Why are the arms and legs rotated in opposite directions when running? |

© Fotokostic/ShutterStock, Inc.

| Figure 7.18 | The pike maneuver performed in the air. |

(left) © Diego Barbieri/ShutterStock, Inc. (right) © Laura Stone/ShutterStock, Inc.

Figure 7.19 A clockwise torque on the upper body creates an equal torque in the counterclockwise direction on the lower body.

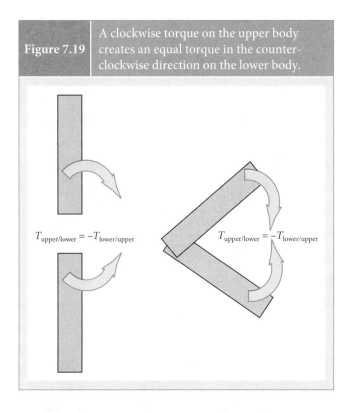

$$T_{upper/lower} = -T_{lower/upper}$$ $$T_{upper/lower} = -T_{lower/upper}$$

Figure 7.20 When twisting torque will rotate the upper and lower body in opposite directions.

© Aspen Photo/ShutterStock, Inc.

Substituting Equation 7.3 into Equation 7.11 produces

$$I_{lower} \times \alpha_{lower} = -(I_{upper} \times \alpha_{upper}) \qquad (7.12)$$

And finally rearranging the terms gives you:

$$\frac{I_{lower}}{I_{upper}} = \frac{-\alpha_{upper}}{\alpha_{lower}} \qquad (7.13)$$

The right-hand side of Equation 7.13 tells you that the angular accelerations would be in opposite directions. The magnitudes of the accelerations would be proportional to the magnitudes of the moments of inertia. The body with the smaller moment of inertia will have a larger angular acceleration. For example, say the ratio of I_{lower} to I_{upper} is 2:1. This is an unrealistic number, but it illustrates the point. This means that the ratio of α_{upper} to α_{lower} must be −2:1. In other words, the upper body would have twice the angular acceleration of the lower body.

Remember that, unlike mass, I is not fixed and can be manipulated during movement. If you flex your legs or extend your arms, you can decrease or increase I. This would cause a different ratio between I_{lower} and I_{upper} and thus different angular accelerations. These types of manipulations are often used in cases where a person (or animal) is airborne. See Box 7.4. But certainly you can move your upper body independent of your lower body, and vice versa. How so? Consider the pike example again (Figure 7.21). When standing on the ground

Box 7.4	Applied Research: Astronauts Maneuvering in Space

Have you ever wondered how astronauts maneuver in outer space? In this investigation, the mechanics were determined that would quantify how they could change their orientation without any reaction forces or employing dangerous equipment such as gas guns that could be subject to malfunction. Rotating in the sagittal plane about a medial–lateral axis can be completed by rotating the arms relative to the torso. Just like the case where you were pushed from behind, rotating the arms will cause the torso to rotate in the opposite direction. Rotating in transverse plane about a longitudinal axis can also be completed by rotating the arms relative to the torso. If the right leg starts anterior to the body and the left leg is posterior and they switch positions, the torso will rotate left. If the legs start in the opposite position (left leg in front), the torso will rotate right. Compare this to the rotations of the trunk induced by the legs during running. In a follow-up investigation, it was found that adding wrist weights will increase the amount of rotation of the trunk. Can you explain why?

Data from: Kane TR, Headrick MR, Yatteau JD. Experimental investigation of an astronaut maneuvering scheme. *J Biomech.* 1972;5:313–320.

Kane TR, Scher MP. Human self-rotation by means of limb movements. *J Biomech.* 1970;3:39–49.

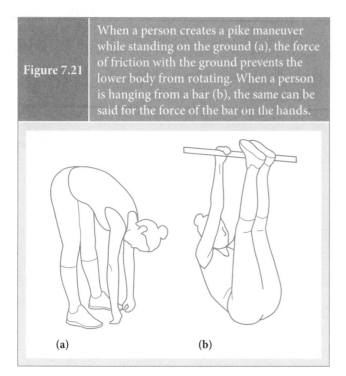

(a)   **(b)**

(a)

(b)

(top) © James M Phelps, Jr/ShutterStock, Inc. and (bottom) © Jose Gil/ShutterStock, Inc.

(Figure 7.21a), only your upper body will rotate in the anterior direction. Your legs will not move. Or if you are hanging from a bar (Figure 7.21b), your lower body will rotate anteriorly without your upper body doing so. Why? With the feet on the ground, the $^hF_{GRF}$ creates an additional torque that will oppose movement of the lower body. Another way to look at it is that when a body part is "attached" to the fixed earth, a massive I will oppose movement of the other part.

Many striking and throwing activities require rotation of the trunk (**Figure 7.22**). If the feet were not on the ground, rotation of the upper and lower body would be in opposite directions, lessening the amount of momentum that could be transferred within the body. This illustrates the need for a stable base when performing many activities.

There are other examples of a reaction torque within the body. Consider what would happen if you were standing on the edge of a pool and were pushed from behind. Your body starts to rotate in the anterior (or forward) direction, and you cannot take a step without stepping off the ledge and getting wet. Your initial reaction may be to rotate your arms as quickly as possible in the *anterior* direction. Notice that this is in the same direction as you are already rotating. Why is this effective? Because the anterior torque created by the trunk on the arms creates a posterior torque by the arms on the trunk. Because the moment of inertia of the trunk is much, much larger than the moment of inertia of the arms, the angular acceleration of the arms has to be very fast to produce a sufficient torque to rotate the trunk posteriorly and prevent you from getting wet. Incidentally, this is why a tightrope walker sometimes uses a long pole. The pole has a fairly large moment of inertia, and thus a smaller angular acceleration of the pole would produce a larger torque.

Finally, think about what happens when you walk or run. Typically, you are swinging your arms in the *opposite* direction that your legs are swinging: Your right arm is rotating posteriorly while your right leg is rotating anteriorly (relative to the body's center of mass). Why?[1] Viewed from above, the right leg moving anteriorly and the left leg moving posteriorly require a counterclockwise rotation of the pelvis. The counterclockwise rotation of the pelvis creates a clockwise rotation of the trunk. Having the trunk rotate while we are walking or running would be incredibly inefficient. Preventing a clockwise rotation of the trunk requires a counterclockwise torque. Requiring the spinal musculature alone to create that torque would be incredibly taxing. However, if we rotate our arms in the opposite direction, we can greatly reduce the demand on the trunk. Rotating your right arm posteriorly and your left arm anteriorly creates such a counterclockwise torque.

Section Question Answer

You will instruct your students to swing their arms in the opposite direction of their legs to reduce the torque placed on the spine due to the rotation of the legs. This decreases the load on the spinal musculature and helps them to run in a straight line.

COMPETENCY CHECK

Remember:

1. State the angular equivalent to Newton's third law.

Understand:

1. In which direction would you rotate your arms if you were trying to regain your balance after being pushed from behind? Why?

7.4   ANGULAR IMPULSE AND AN ALTERNATE VIEW OF THE SECOND LAW

Section Question

Billy hopes to break into competitive rodeo (Figure 7.23). During the tie-down roping event, he must throw a lasso around a calf. Why must he circle the rope several times over his head before throwing it?

An alternate form of Newton's second law establishes the idea of mechanical impulse. These ideas can be extended to

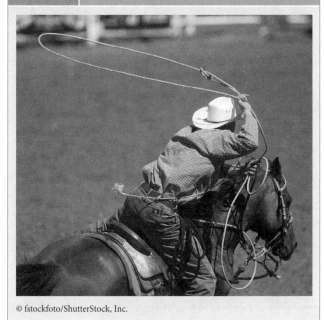

| Figure 7.23 | When using a lasso, why is it circled over the head several times before it is thrown? |

© fstockfoto/ShutterStock, Inc.

angular motion. Recall that angular momentum is the product of the moment of inertia and the angular velocity, and the change in angular momentum is such that:

$$\Delta H = \Delta I \omega \qquad (7.14)$$

Using a substitution procedure you can see that torque can then be stated as the rate of change of angular momentum:

$$\sum T_{ext} = \frac{\Delta H}{\Delta t} \qquad (7.15)$$

Rearranging the terms in Equation 7.15 shows you that angular impulse is the equal to the change of angular momentum:

$$\sum T_{ext} \times \Delta t = \Delta H \qquad (7.16)$$

Like linear impulse, angular impulse (and Equation 7.16) tells you that to be effective, a torque has to be applied over time (or over distance). Also like linear impulse, the angular impulse is torque multiplied by time only if the torque is constant. You could multiply the average torque by the time or examine the area under the torque–time curve. The important aspects of the torque–time curve include

- peak torque
- rate of torque development
- rate of torque fatigue
- angular impulse
- average torque

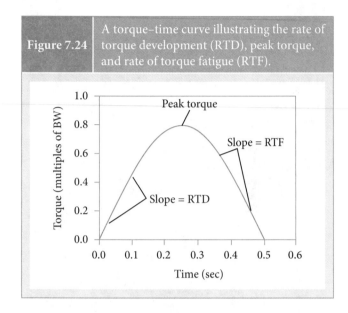

Figure 7.24 | A torque–time curve illustrating the rate of torque development (RTD), peak torque, and rate of torque fatigue (RTF).

The first three variables are presented in Figure 7.24. Note the similarity between this figure and Figure 6.12. They are identical, except for the fact that torque has replaced force on the ordinate axis. Angular impulse is the area under the torque–time curve (Figure 7.25). It is identical to Figure 6.13, except angular quantities have replaced linear ones.

7.4.1 Using the Angular Laws to Improve Performance and Reduce Injury Risk

Biomechanical analyses are used to assist you in improving performance or preventing injury. As with linear kinetics, you want to isolate the variable that you wish to maximize or

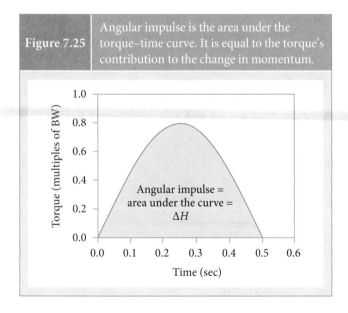

Figure 7.25 | Angular impulse is the area under the torque–time curve. It is equal to the torque's contribution to the change in momentum.

minimize. When looking to improve performance, you would want to maximize the final angular velocity:

$$\omega'' = \frac{\sum T_{average}\Delta t}{I} + \omega' \qquad (7.17)$$

Because there are four variables on the right-hand side of the equation, there are four possibilities to increase the final angular velocity: the average effective torque, the time the average effective torque is applied, the initial angular velocity, and the moment of inertia. If you increase the first three, angular velocity will increase. If you decrease the moment of inertia, you would also increase the final angular velocity. The average effective torque multiplied by the time that torque is applied is equivalent to the angular impulse. As with linear impulse, it is not always feasible to increase the time the torque is applied. Increasing the initial angular velocity will also increase the final angular velocity. Think of a discus or hammer thrower. They complete several circles before the final throw in an attempt to increase the initial velocity of the implement before the final torque is applied. Finally, the moment of inertia can potentially be decreased, but this option should be considered with caution. Neither mass nor the way it is distributed in a single rigid body can be changed in the short term. The length, l_r, of a multibody system can be changed in the short term by, for example, bending a joint. However, this same "l_r" is often used to determine the linear velocity of a point on a body ($v = l_r\omega$). Decreasing l_r, although it would decrease the moment of inertia, I, could also negatively affect the linear velocity that you are attempting to maximize with a maximum angular velocity.

Another goal of a particular movement pattern may be to decrease the average torque during the activity. This may be the case when you wish to minimize injury risk. As can be seen with Equation 7.18, three factors determine the average torque of a particular task:

$$\sum T_{average} = \frac{I_{COM}(\omega'' - \omega')}{\Delta t} \qquad (7.18)$$

Increasing the time will decrease the average torque necessary for the same angular impulse. As mentioned previously, this is not always possible given the circumstances, but it is a viable option in many situations. Minimizing the difference between the final and initial angular velocities will also minimize the required torque. Finally, the moment of inertia could be manipulated to decrease the torque requirements. For example, consider moving a box (either lifting it or rotating it side-to-side). Holding the box at arm's length would increase the torque requirements about the low back, whereas holding the box closer to the trunk would decrease the moment of inertia and thus the torque requirements.

Section Question Answer

When throwing an object, you know that the linear velocity is a key determinant in the projectile's trajectory. In many events, such as the lasso, the angular velocity (just prior to release) and the length of the radius determine the linear velocity. According to Equation 7.17, you would maximize the final angular velocity by maximizing the average torque and the time that torque is applied (assuming the moment of inertia cannot be changed). It should be emphasized that these quantities (torque and time) are only applicable during the final circling of the rope. In many ways, maximizing time and velocity are at odds with one another. One way this can be overcome is to start with a large initial velocity. This is where the repeated circling of the rope is critical. By circling the rope overhead several times, Billy will be maximizing the initial velocity before the final circle, which is the only one that matters for Equation 7.17.

COMPETENCY CHECK

Understand:

1. Which variables would you attempt to maximize if you wanted to maximize the final angular velocity?
2. Which variables would you attempt to minimize if you wanted to minimize average torque?

Apply:

1. Give examples of tasks where you want to maximize final angular velocity.
2. Give examples of tasks where you would want to minimize average torque.

7.5   APPLICATIONS OF ANGULAR KINETICS

7.5.1   Angular Resistances Used in Exercise

Various different resistance exercise machines are designed to provide a resistive torque that must be overcome by the body. Constantly subjecting the body to these torques will eventually increase the strength of the muscles that overcome them.

The simplest exercise machines use the concepts of levers (**Figure 7.26**). A resistance is either directly or via a cable attached to the end of a lever. In such cases, the resistance will vary with the range of motion. Alternately, the lever arm can be fixed to a shaft similar to wrench and provide a more constant resistance. In either case, the axis of rotation of the machine is aligned with the axis of rotation of

| Figure 7.26 | Many exercise machines are levers that provide a resistance to rotation of a lever arm. |

© CandyBox Images/ShutterStock, Inc.

the joint. A pad or handle is at the end of the lever. Applying force to the handle or pad creates a torque on the lever about the axis of rotation. As you know by now, the further the force is applied from the axis of rotation, the larger the mechanical advantage. However, the location of the pad or handle is usually determined by the anthropometrics of the individual, and it is not used to create a mechanical advantage so you can lift more weight. These machines are used to train a single joint rather than the multiple muscle groups that are trained with a bench press, squat exercise, or other multijoint exercise.

Single, fixed pulleys will change the direction of a force, but will not provide a mechanical (dis)advantage (**Figure 7.27**). A system of pulleys, or pulleys that move with the weight stack, may provide a mechanical advantage. The goal of a piece of exercise equipment is to provide an overload for the working muscles; this runs counter to the notion of designing an exercise machine to create a mechanical advantage. The benefit of a pulley machine is that

| Figure 7.27 | Pulleys will change the direction of force. In this case, the weight of the stack is changed to a horizontal force. However, pulleys do not necessarily provide a mechanical advantage. |

© Andresr/ShutterStock, Inc.

| Figure 7.28 | An isokinetic dynamometer will not allow the resistance arm to rotate at a greater velocity than what is programmed into the machine. |

the direction of force can be altered, so different movement patterns can be overloaded.

Another type of resistance equipment uses a flywheel. Flywheels are rotating disks, and the resistance torque is in the form:

$$T_{\text{fly wheel}} = ml_r^2\alpha \qquad (7.19)$$

Torque is provided to the flywheel via some sort of cord. The greater the torque provided to the flywheel, the faster its change in angular velocity. The greater the change in angular velocity, the greater the resistance will be. Therefore, the greater the torque supplied to the flywheel, the greater the resistance the flywheel will provide. Thus, flywheels provide an **accommodating resistance**—a resistance that changes with the amount of force (torque) provided to it.

A final type of machine, called an isokinetic dynamometer (**Figure 7.28**), prevents a lever arm from moving at a higher angular velocity than what was programmed into the machine. The harder one pushes against the lever, the greater the resistance torque provided by the machine so that a constant angular velocity is maintained. Therefore, isokinetic dynamometers are also classified as accommodating resistances.

> **Accommodating resistance** A resistance that increases with the amount of force or torque applied to it

7.5.2 Static Equilibrium

The following equations describe the motion of a body:

$$\sum F_{\text{ext}} = ma_{\text{COM}} = \frac{\Delta L}{\Delta t}$$

$$\sum T_{\text{ext-COM}} = I_{\text{COM}}\alpha = \frac{\Delta H}{\Delta t}$$

These two equations describe the combined effects forces and torques on a body. A net force or torque will produce an acceleration, or change the momentum, of the body. A special case, called **static equilibrium**, occurs when the combined forces and torques are balanced, creating a zero effective force and a zero effective torque:

> **Static equilibrium** A special case where both the linear and angular accelerations are zero, and thus the sum of the external forces and sum of the external torques are zero

$$a = 0 = \sum F_{\text{ext}} \qquad (7.20)$$

$$\alpha = 0 = \sum T_{\text{ext-COM}} \qquad (7.21)$$

Statics is an entire branch of engineering, and entire courses are devoted to the subject. Here, consider two cases requiring static equilibrium: standing balance and holding a glass.

Standing Balance

Understanding standing balance requires you to have a firm grasp of the following concepts: center of pressure (COP), center of gravity (COG), base of support (BOS), weight (F_g), and the ground reaction force (F_{GRF}).

Recall that all the ground reaction forces acting on the foot are concentrated at the center of pressure (COP). It should also be apparent that the ground reaction force is a contact force and must lie within the confines of each foot (if a person is standing on two feet). However, the resultant F_{GRF} and COP can lie outside the foot but must be within the perimeter of the base of support (BOS). The BOS is shown for various stances in **Figure 7.30**. In a simple stance with the weight equally distributed among both feet, the location

Figure 7.30 The base of support in a regular stance, tandem stance, and stagger stance.

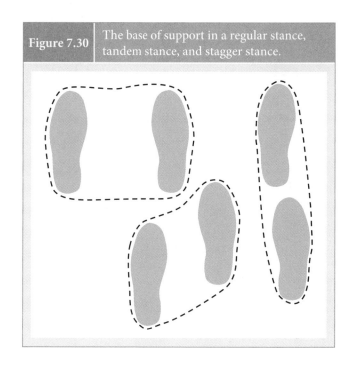

Figure 7.29 What factors are involved in standing balance?

© Flashon Studio/ShutterStock, Inc.

of the COP will be located at half the distance between your two feet. If you shift more weight onto your right foot, the COP will move to the right. Notice also that the BOS is very wide in the medial–lateral direction, but very narrow in the anterior–posterior direction. In a tandem stance, the BOS is longer in the anterior–posterior direction, but very narrow in the medial–lateral direction. The staggered stance seems to offer the best compromise between the width of the BOS in both the anterior–posterior and medial–lateral directions. Regardless of stance, the COP cannot be located outside the BOS.

This is important because the COP indicates the location of one force acting on the body, namely, the $^yF_{GRF}$. Another force acting on the body is your weight, the force due to gravity (F_g). As you now know, a person will be moving unless the effective force is zero. Thus:

$$\sum F_{ext} = ma_{COM}$$
$$^yF_{GRF} - F_g = ma_{COM}$$
$$a_{COM} = 0$$
$$^yF_{GRF} - F_g = 0$$
$$^yF_{GRF} = F_g$$

The $^yF_{GRF}$ must be equal to body weight. The other requirement for static equilibrium is that the effective torque must be zero. Because two forces are acting on the body, there are two potential sources of torque: F_{GRF} and F_g. To see how these

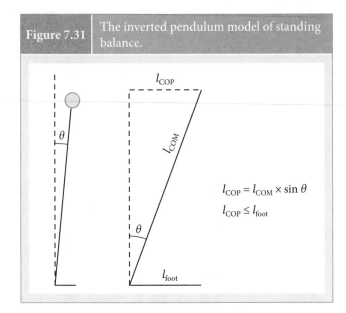

Figure 7.31 The inverted pendulum model of standing balance.

$$l_{COP} = l_{COM} \times \sin\theta$$
$$l_{COP} \leq l_{foot}$$

two forces interact, consider a simple model with a massless foot connected by a hinge to a mass representing the COM on the end of a massless rod. This model resembles an inverted pendulum, as is presented in **Figure 7.31**.

Assume the pivot point to be the hinge. Now consider the effect of gravity. If the COG is located directly above the hinge, then the moment arm of the COG is zero, and there is no torque about the hinge. If the COG is displaced by an angle, θ, then the moment arm increases by a distance that is equal to the length of the rod multiplied by the correction factor ($\sin\theta$). This will create a torque that is equal to:

$$T_g = F_g \times l_{COM} \times -\sin\theta \quad (7.22)$$

Note that both the negative sign before the sign of θ and the sign of θ are important. If θ is in the anterior (clockwise) direction, the torque due to gravity will be clockwise; if θ is posterior, the torque will be counterclockwise. Assume for the moment that the torque due to gravity is in the clockwise (anterior), negative direction. To satisfy the conditions for static equilibrium, torque from the F_{GRF} must be equal and opposite—in the positive, counterclockwise direction. Knowing that F_{GRF} and F_g are equal in magnitude but opposite in directions, equal torques means that their moment arms must be equal. In other words, the COP must be directly under the COG to maintain static equilibrium.

The fact that static equilibrium is only maintained if the COP is directly under the COG has a number of interesting implications. First, there is a limit to the amount of lean in any direction. This is called the limit of sway, and

it produces an elliptical cone describing how far one can lean in any direction. Because the COP must be within the confines of the BOS and the COP must be under the COG, static balance requires that the COG must be over the BOS. Increasing the BOS will potentially increase the size of this cone. Second, the displacement of the COG is defined by the equation:

$$\Delta p_{COG} = l_{COM} \times -\sin\theta \quad (7.23)$$

Rearranging these terms, you see that the angle, θ, is equal to:

$$\theta = -\sin^{-1}\left(\frac{\Delta p_{COG}}{l_{COM}}\right) \quad (7.24)$$

Because the distance from the pivot point to the COG must be equidistant from the pivot point to the COP, Equation 7.24 becomes

$$\theta = -\sin^{-1}\left(\frac{\Delta p_{COP}}{l_{COM}}\right) \quad (7.25)$$

The value of the arc sine must be between 0 and 1. The closer it is to zero, the smaller the value of θ. The closer it is to 1, the closer θ approaches 90° (an impossible angle of sway). Thus, the larger the value inside the parentheses, the larger the angle of sway may be. This can be accomplished in one of two ways: increasing Δp_{COP} or decreasing l_{COM}, in other words, by either increasing the BOS (spreading the feet) or lowering the COM (bending the knees).

The difference between the limit of sway and the actual sway would be the "safety margin" of standing balance. The closer the actual sway is to the limit, the greater the chance that the actual sway would surpass the limit of sway. If this occurs, a person would either have to take a step or risk falling. Although normal, healthy adults appear to have a large safety margin, persons with neurological conditions, such as Parkinson's disease described in **Box 7.5**, do not seem to enjoy such a large safety margin and are more at risk for falling.

Section Question Answer

To increase her balance, May needs to either increase her base of support or decrease her center of gravity.

Holding an Object

Section Question

Following a stroke, Janice is having trouble holding a glass in her hand (**Figure 7.32**). What factors are involved with this task?

Box 7.5	Applied Research: Parkinson's Disease and the Limits of Sway

Parkinson's disease (PD) is a progressive neurological impairment with no known cause that leads to degenerative changes in the basal ganglia (gray matter embedded within the white matter of the brain), with losses or decrease in dopamine. PD leads to tremor, muscle rigidity, slowness of movement, and decreased balance. Although there is no known cure for PD, levodopa (L-dopa) is a drug that is used to combat some of its functional consequences. However, L-dopa has periods where it is effective in controlling symptoms ("on") and times when it is not ("off"). This investigation looked to compare the actual amount of sway with the limit of sway in three groups: healthy controls, patients with PD "on" L-dopa, and patients with PD "off" L-dopa. It was hypothesized that a larger actual amount of sway compared to the limit of sway would lead to a decreased safety margin and potentially an increased risk for falling. It was found that, compared to healthy controls, patients with PD had both an increased sway and a decrease in the limit of sway. When "on" L-dopa, patients had both an increase in the limit of sway and a decrease in sway (increasing the safety margin), but still not to the level of healthy controls. In addition to the beneficial effects of L-dopa, this study demonstrated that patients with PD are at a greater risk of falls, particularly during the "off" phase.

Data from: Menant JC, Latt MD, Menz HB, Fung VS, Lord SR. Postural sway approaches center of mass stability limits in Parkinson's disease. *Mov Disord.* 2011;26:637–643.

Figure 7.32	What factors are involved in holding a glass or cup?

© Kasco Sandor/ShutterStock, Inc.

As mechanically complex as the simple task of maintaining quiet balance is grasping and holding an object such as a cup or glass with the four fingers and thumb using a prismatic grip is even more complex, because at least 11 forces and 10 torques have to be controlled (Figure 7.33). The force generated by each finger and the thumb can be separated into components that are normal (or perpendicular) to the surface, ^nF, and tangential (or parallel) to the surface ($^{\tan}F$). Thus, five fingers produce 10 forces. In addition, the weight of the object, F_g, is another force. Assuming that the torques are determined about the COM, then F_g does not contribute to the sum of the torques, but the force provided by the fingers and thumb will. Static equilibrium means[2]

$$^nF_1 + {}^nF_2 + {}^nF_3 + {}^nF_4 + {}^nF_5 = \sum {}^nF \quad (7.26)$$

$$^{\tan}F_1 + {}^{\tan}F_2 + {}^{\tan}F_3 + {}^{\tan}F_4 + {}^{\tan}F_5 + F_g = \sum {}^{\tan}F \quad (7.27)$$

$$^nF_1\,{}^nl_1 + {}^nF_2\,{}^nl_2 + {}^nF_3\,{}^nl_3 + {}^nF_4\,{}^nl_4 + {}^nF_5\,{}^nl_5$$
$$+ {}^{\tan}F_1\,{}^{\tan}l_1 + {}^{\tan}F_2\,{}^{\tan}l_2 + {}^{\tan}F_3\,{}^{\tan}l_3$$
$$+ {}^{\tan}F_4\,{}^{\tan}l_4 + {}^{\tan}F_5\,{}^{\tan}l_5 = \sum T \quad (7.28)$$

This is quite a large number of variables to keep track of! A simplified model considers all the fingers involved as one "virtual finger,"[3] which will be referred to as "finger." In the normal direction, the force of the thumb has to be equal to the force of the finger:

$$^nF_{\text{thumb}} = -{}^nF_{\text{finger}} \quad (7.29)$$

Otherwise, the object will move left or right. The forces in the tangential direction have to be equal to the object's weight:

$$^{\tan}F_{\text{thumb}} + {}^{\tan}F_{\text{finger}} = -F_g \quad (7.30)$$

or the object will slip. The torque produced by the four forces must cancel to prevent the object from tilting:

$$^nF_{\text{thumb}}\,{}^nl_{\text{thumb}} + {}^nF_{\text{finger}}\,{}^nl_{\text{finger}} + {}^{\tan}F_{\text{thumb}}\,{}^{\tan}l_{\text{thumb}}$$
$$+ {}^{\tan}F_{\text{finger}}\,{}^{\tan}l_{\text{finger}} = 0 \quad (7.31)$$

This creates a complex situation where precise control of a number of forces and torques is required. The nervous system seems to be able to do this with ease under normal

| **Figure 7.33** | The forces and moment arms for the thumb and fingers holding a glass. |

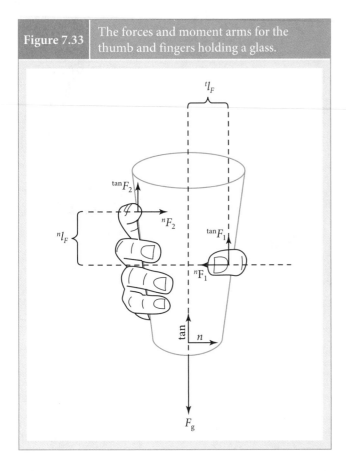

| **Box 7.6** | Applied Research: The Effects of Aging on Grasping |

It is generally well known that aging results in a decline in the maximum force that a muscle or group of muscles can produce. It is also well known that many everyday tasks requiring dexterity of the hand (eating with a fork, drinking from a glass, writing with a pen or pencil) require precise control of both forces and torques created by the fingers and thumb. What is not well understood is how aging affects the ability to control these forces and torques. In this experiment, both elderly and young subjects pressed on force transducers that were affixed to a handle with the goal of either producing maximal force or torque or precisely controlling submaximal force or torque of the digits. The investigators determined the forces and torques created by the fingers and thumb as described by Equations 7.26 to 7.28. Not only did the elderly have impaired force and torque production, but they also demonstrated a decreased ability to accurately control submaximal force and torque production. It was found that elderly subjects produced higher forces and moments necessary for the task, increasing the "safety margin" necessary to prevent an object from slipping out of the hand. This was interpreted as a compensation for the inability to control variations of force and torque accurately.

Data from: Shim JK, Lay BS, Zatsiorsky VM, Latash ML. Age-related changes in finger coordination in static prehension tasks. *J Appl Physiol.* 2004;97:213–224.

circumstances. However, as we age, this seemingly simple task becomes more difficult. See **Box 7.6** for details.

Section Question Answer

Each finger and the thumb holding a glass produce forces and torques. In addition, the weight of the glass is another force acting on it. Assuming that all the fingers and thumb are holding the glass, 11 forces and 10 torques must be precisely controlled to hold the glass. It should come as no surprise that neurological impairment can make this seemingly simple task quite difficult.

COMPETENCY CHECK

Remember:

1. Define: accommodating resistance and static equilibrium.
2. List the equations for static equilibrium.

Understand:

1. Why are many exercise machines considered to be variable resistance devices?
2. Why would a pulley not provide a mechanical advantage?
3. Why must the center of gravity remain within the base of support during standing?

Apply:

1. List several activities that require static equilibrium.

SUMMARY

In this lesson, you learned about the causes of angular motion using the key concepts in **Table 7.1**. The modifications of Newton's three laws of motion to angular movement were discussed, as were a view of the second law that explained angular impulse and its effect on angular momentum. Static equilibrium, where the linear and angular accelerations are zero and thus the effective force and effective torque are zero, was examined in the case of standing balance and holding an object.

Table 7.1	Key Concepts

- Torque (Moment of force)
- Couple
- Lever
- Lever arm
- Moment arm
- Angular impulse
- Static equilibrium
- Derivative

REVIEW QUESTIONS

1. Define the following terms: torque, moment of force, propulsive torque, braking torque, effective torque, couple, accommodating resistance, and static equilibrium.

2. Explain the concept of mechanical advantage. Give three examples of how you would use it in everyday life.

3. State the angular equivalents of Newton's three laws. Describe how you would apply each of them.

4. List three ways of increasing angular impulse.

5. Examine several pieces of exercise equipment in your gym. Describe the levers being used and how they overload the muscles.

6. Describe the conditions for static equilibrium. Explain how they are used to analyze standing balance and holding an object steady.

REFERENCES

1. Blazevich A. *Sports Biomechanics. The Basics: Optimising Human Performance.* London: A & C Black Publishers Ltd., 2007.

2. Zatsiorsky VM, Gao F, Latash ML. Prehension synergies: effects of object geometry and prescribed torques. *Exp Brain Res.* 2003;148:77–87.

3. Zatsiorsky VM, Latash ML. Prehension synergies. *Exerc Sport Sci Rev.* 2004;32:75–80.

Lesson 8

Work–Energy

© Maria Teijeiro/OjO Images/Getty Images

LEARNING OBJECTIVES

After finishing this lesson, you should be able to:

- Define the following terms: energy, kinetic energy, potential energy, gravitational potential energy, strain potential energy, work, mechanical energy expenditure, efficiency, economy, and power.
- State the conservation of energy and the first law of thermodynamics.
- Give examples of linear kinetic energy, rotational kinetic energy, and gravitational potential energy.
- Determine the amount of work done by a force or torque.
- Manipulate the variables associated with the center of mass equation to increase performance and decrease injury.
- Explain why walking is more efficient than running.
- Explain how mechanical energy expenditure and power are important in human movement.

Investigating the causes of motion using the concepts of force (torque), impulse, and momentum is very useful, has solved a variety of problems, and provides a great deal of information about what is going on with a system. However, it does not tell the whole story. Some subtle things are going on that would be missed if you only used Newton's methods to analyze movement. Techniques of analysis, requiring new concepts, were introduced to uncover these subtleties. In this lesson, you will learn about the concepts of energy, work, efficiency, and power and see how they aid your understanding of human movement.

8.1 ENERGY

Section Question

True or False: You can jump up into the air because the sun is shining (**Figure 8.1**).

To begin, you first need to understand the concept of energy. The way energy is used by scientists differs from the way it is used in everyday language. To make matters worse, most physics and mechanics books jump right in to talking about energy without ever really defining it, or they employ a rather circular definition. Even the Nobel prize–winning physicist Richard Feynman stated that he did not have a really good definition for it[1] (and if he did not, who does?). One of the better definitions is: Energy is a state of matter that makes things change, or has the potential to make things change.[2] Note that energy is a *state* that characterizes a body or system. For a body or system to change, there needs to be energy. If there is no energy, the body or the system will not change.

> **Energy** The state of matter that makes things change, or has the potential to make things change

There are many different types of energy (for different types of changes): nuclear, chemical, electromagnetic, acoustic, and mechanical (**Figure 8.2**). The one that you are most interested in for biomechanics is mechanical energy (in physiology, you are probably more interested in chemical energy) because mechanical energy is needed to change the mechanics of a body or system, and that change could be in position,

124

Figure 8.1 | True or False: You can jump up into the air because the sun is shining.

Kinetic energy   The energy that a body has due to its motion

Potential energy   The energy a body has that has the potential to change something, but it is not currently changing anything

it has to have energy. There were two types of motion: linear (or translational) and angular (or rotational). It should come as no surprise that there are two types of kinetic energy: linear kinetic energy (E_{LK}) and angular kinetic energy (E_{AK}). If the object is moving, it has velocity, so that is going to be a factor in quantifying the amount of kinetic energy. And you may have guessed that objects that have more mass are going to have more energy. Both terms show up in the equation:

$$E_{LK} = \frac{1}{2}mv^2 \tag{8.1}$$

And if you substitute in the angular equivalents of each factor you have

$$E_{AK} = \frac{1}{2}I\omega^2 \tag{8.2}$$

It is important to realize that the velocity term in the equation is squared. That means that it is more important than mass. If you double the mass, you will double the kinetic energy (for the same velocity). But if you double the velocity (for the same mass), you will quadruple the amount of kinetic energy.

velocity, or shape. There are two types of mechanical energy: **kinetic energy** and **potential energy**.

Kinetic energy is the energy that a body has because it is moving. If it is moving, it is changing (position) and if it is changing,

Important Point!   Because the velocity term is squared, doubling the velocity will increase the kinetic energy by a factor of four.

As the name implies, potential energy is the potential to change. Unlike kinetic energy, the body is not currently

Figure 8.2 | Different types of energy.

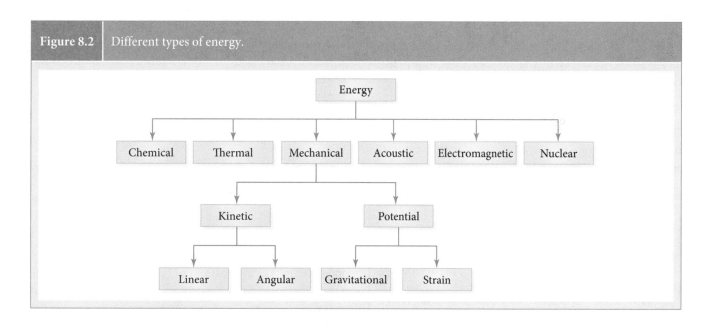

> **Gravitational potential energy**   The potential energy that a body has due to its position
>
> **Strain potential energy**   The energy a body has due to its deformation

changing—but it may. There are two types of potential energy: gravitational potential energy and strain potential energy. Gravitational potential energy is the potential an object has to change due to its position relative to the earth. Think about a rock lying on the ground. Currently, it has no potential to change its position: It is just lying there. If you pick the rock up and hold it at shoulder height, it now has potential energy: If you let go of the rock, it will fall to the ground. The potential is there. All you have to do is release it. As you can imagine, the mass of the rock, its height above the ground, and the acceleration due to gravity all influence this potential:

$$E_{GP} = ma_g h \qquad (8.3)$$

Technically, height can be measured from any convenient place that makes the math easier. It does not have to be measured from the ground. In this lesson, it will (almost) always be measured from the ground because it is pretty meaningful for your types of analyses.

The other type of potential energy is strain potential energy. To visualize strain energy, think of a rubber band. On its own, the rubber band is not going to do anything. It has no energy, so nothing happens. If you stretch the rubber band, and let it go: snap! It does something. Once it is stretched, it has the potential to do something. All you have to do is release it. Strain energy will not be covered in detail in this lesson. For this lesson, you will concentrate on kinetic and gravitational potential energy.

When describing the state of a system, you simply add up all the energies in that system. So for any system, the mechanical energy would be equal to:

$$\text{Energy}_{\text{system}} = E_{LK} + E_{AK} + E_{GP} + E_{SP} \qquad (8.4)$$

Of course, for a complete description of the system, you would need to consider other forms of energy as well. But in many situations, you will limit your analysis to the mechanical energy of a particular system.

Section Question Answer

In a weird sort of way, you can jump up into the air because the sun is shining: The radiant energy of the sun provides for the chemical energy in food; the chemical energy in food is stored in body; the chemical energy in the body is used for muscular contractions; the energy from the muscular contractions is used to jump.

COMPETENCY CHECK

Remember:

1. Define the following terms: energy, potential energy, kinetic energy, gravitational potential energy, and strain potential energy.

Understand:

1. Which of the following has more energy?
   a. A 100 kilogram mass moving at 50 meters per second.
   b. A 50 kilogram mass moving at 100 meters per second.
2. Which of the following has more energy?
   a. A 100 kilogram mass located 1 meter above the ground.
   b. A 50 kilogram mass located 5 meter above the ground.

Apply:

1. What different types of energy are involved when someone is jumping in the air?

8.2   WORK

Section Question

You are working with Anne on a comprehensive rehabilitation program. When performing an exercise such as bench press, how much work is she doing (**Figure 8.3**)? How much mechanical energy is she expending?

So far, you have learned about two of the three great conservation laws in physics: the conservation of linear momentum and the conservation of angular momentum. Essentially, these laws state that the momentum of a system will not change unless an external force (or torque) is applied to the system. Can you guess what the conservation of energy states?

Energy can neither be created nor destroyed; it can be transformed from one state to another, or transferred into or out of the system, but the total amount of energy never changes.

As a consequence, the total amount of energy in an isolated system is constant. Think for a moment about what those statements mean. Compare it to the conservation of momentum. The momentum of the system will not change unless there is an external impulse. Similarly, the amount of energy in the system will not change unless energy is either added to it, or taken away from it.

Figure 8.3When performing repetitions that begin and end in the same position, how much work is being done? How much mechanical energy is being expended?

Say you jumped in the air. If you were a point mass, you can state that E_{AK} and E_{SP} are zero and:

$$E_{system} = E_{LK} + E_{GP} \qquad (8.5)$$

While you are in the air, energy will not be entering or leaving the system defined by you and the earth. Therefore, the change in energy, ΔE, will be zero. This means that:

$$0 = \Delta E_{LK} + \Delta E_{GP} \qquad (8.6)$$

If you expand those two terms, you will get:

$$0 = (E_{LK2} - E_{LK1}) + (E_{GP2} - E_{GP1}) \qquad (8.7)$$

And rearranging the equation gives you:

$$E_{LK1} + E_{GP1} = E_{LK2} + E_{GP2} \qquad (8.8)$$

If you consider the height of takeoff as your reference (something you will modify a bit later), you will have zero E_{GP} at takeoff and a certain amount of E_{LK} determined by your takeoff velocity. That is the total amount of energy in the system. While you are in the air, as the distance from the ground is increasing, your velocity is decreasing. The E_{LK} is transforming to E_{GP}, but the sum of the E_{GP} and E_{LK} must be the same as it was at takeoff (Figure 8.4). At the apex of the jump, your velocity (and E_{LK}) will be momentarily zero, but your E_{GP} will be maximized. This is a simple way to determine jump height. If $E_{GP\,takeoff}$ and $E_{LK\,apex}$ are both zero, you know that $E_{LK\,takeoff}$ equals the $E_{GP\,apex}$, and:

$$\frac{1}{2}mv^2 = ma_g h \qquad (8.9)$$

Rearranging the terms show you that jump height is a function of velocity and gravity:

$$h = \frac{v^2}{2a_g} \qquad (8.10)$$

And because you cannot do anything about gravity, jump height is determined by takeoff velocity. And because the velocity term is squared, it will have a profound effect: Doubling the takeoff velocity will quadruple the jump height.

During the descent, the E_{GP} at the apex is transformed back into E_{LK}. The E_{LK} at landing is the same as the E_{LK} at takeoff. Where did the initial E_{LK} at takeoff come from? Where did the energy go after you landed? The energy was either transformed or transferred.

Figure 8.4   When a projectile is in the air, the linear kinetic energy in the vertical direction is transformed into gravitational potential energy during the ascent. During the descent, the gravitational potential energy is transformed into linear kinetic energy in the vertical direction.

$E_{LK} = 0$ $E_{GP} = E$

$E_{LK} = \frac{1}{4}E$ $E_{GP} = \frac{3}{4}E$

$E_{LK} = \frac{1}{2}E$ $E_{GP} = \frac{1}{2}E$

$E_{LK} = \frac{3}{4}E$ $E_{GP} = \frac{1}{4}E$

$E_{LK} = E$ $E_{GP} = 0$

E = Total energy

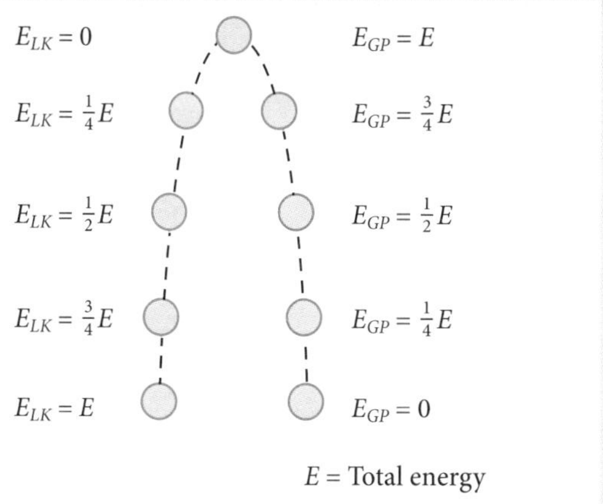

The process of changing the amount of energy in the system is called **work**. Work is defined as a change in energy. In other words, work is changing (either increasing or decreasing) the amount of energy in the system. So whereas energy defines the *state* of a system, work is a *process* of moving energy into or out of that system. This distinction is sometimes confusing because both energy and work have the same units (Joules, or J, which is a Newton-meter. Joules are used instead of Nm because those units are reserved for torque). Although performing work is not the only form of changing the energy in a system, it is the main one of interest in biomechanics.

> **Work** The process of changing the amount of energy in a system

Say you wanted to increase the potential energy of a system, such as a barbell and Earth. How would you do it? You would have to supply an average force (equal to the weight of the barbell, ma_g) over a distance (h) raising the barbell from position A to position B in **Figure 8.5**. This would increase the potential energy in the barbell–Earth system (by the amount $ma_g h$). Notice that the amount of time it took to raise the barbell was irrelevant in determining the energy. All that mattered was that you raised it by a certain amount (Δh). This helps you understand the formula for work:

$$W = F \times \Delta p \tag{8.11}$$

Work is equal to the product of force and displacement. Note that this formula only holds when force is constant; otherwise, you must determine the area under the force–position curve (**Figure 8.6**). (This is just like how you determined the change in momentum as the area under the force–time curve). Also note that the path of the barbell was irrelevant; all that mattered was that the barbell was raised a particular height (Δh). A force that is applied in a direction other than

Figure 8.6 | The amount of work performed is the area under the force–position curve.

(a)

(b)

the displacement does no work, so the proper formula for work is

$$W = F \times \Delta p \times \cos\theta \tag{8.12}$$

where θ is the angle between the force and displacement vectors. This concept is shown in **Figure 8.7**, and the hierarchical model for work is presented in **Figure 8.8**. It is important to remember that if the displacement is zero, the work is zero regardless of how large the force is. Just like you saw with impulse–momentum, where in order for a force to be effective it had to be applied over time, work–energy concepts tells you that the force has to be applied over a distance to be effective in changing energy.

> **Important Point!** Work only occurs if there is a displacement.

Work is considered a scalar; it has a magnitude but no direction. However, the plus or minus sign (+/−) in front of the work number is very important. If the force and the

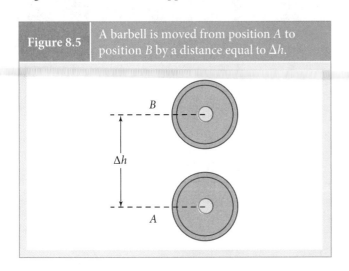

Figure 8.5 | A barbell is moved from position A to position B by a distance equal to Δh.

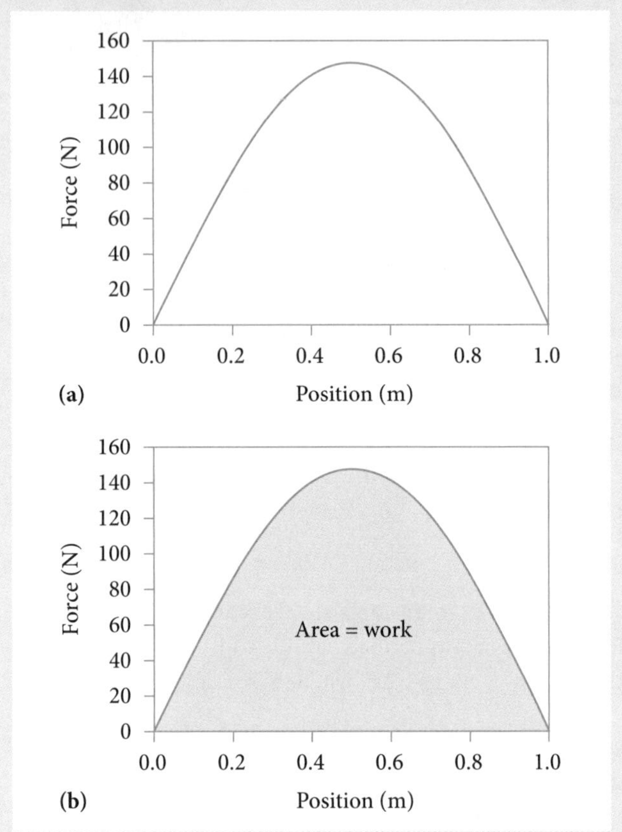

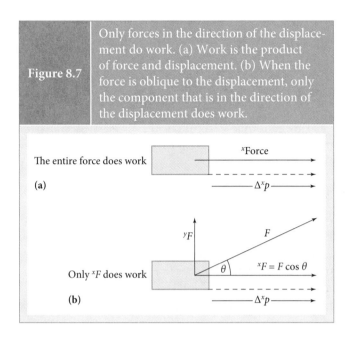

Figure 8.7 Only forces in the direction of the displacement do work. (a) Work is the product of force and displacement. (b) When the force is oblique to the displacement, only the component that is in the direction of the displacement does work.

The entire force does work

*Force

(a) Δ*p

ʸF   F

Only *F does work   *F = F cos θ

(b) Δ*p

displacement are in the same direction, the work will be positive. When work is positive, the energy of the system increases. If the work and displacement are in opposite directions, the work is negative. When work is negative, the energy of the system decreases. The energy is transferred to objects outside the system. Note that the actual direction of the force and displacement does not matter, only the relative direction between the two.

Important Point! Positive work means that the energy is entering the system; negative work means energy is leaving the system.

When raising the barbell from point A to point B in Figure 8.5, the energy in the system increased, so positive work was performed *on* the barbell–Earth system *by* the person lifting it. If the barbell was lowered from point B to point

Figure 8.8 A hierarchical model of work. Work is determined by (1) the amount of force, (2) the displacement, and (3) the angle between these two vectors.

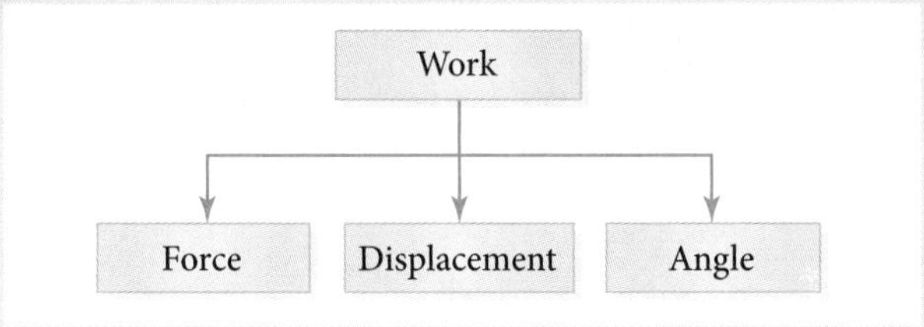

Work → Force, Displacement, Angle

A, the amount of energy in the system decreased. Negative work was performed. This is sometimes stated as the barbell–Earth system performed work *on* the person. This phrase can be confusing, so you would probably be better off saying that the person performed negative work because he or she is supplying the force.

Even this simple example illustrates that you should not just say "work." You should specify which force is doing the work. This is even more important when more than one force (source of energy) is acting on a system, like the tug-of-war example in Figure 8.9.[3] In this case there are two forces and a displacement. Each force is doing work. F_1 is producing a larger force and doing more work. In this case, it happens to be positive work. However, F_2 is producing a force and doing negative work.

There are two ways you can look at this situation. First, because work is a scalar, you could simply sum the various sources of energy ($W_1 + W_2$) and determine the net work ($W_{net} = 10$ J). Second, you could determine the resultant force ($F_{effective}$; 10 N) and multiply that by the displacement (1 m) to get the net work ($W_{net} = 10$ J). Either way, you get the same result (10 J). Three work quantities are involved here: W_1, W_2, and W_{net}. This is why it is so important to specify what is doing the work.

What about gravity? It is a force; does it not do work? Why was it not included in the preceding barbell example? That largely depends on how you define your system. Earlier,

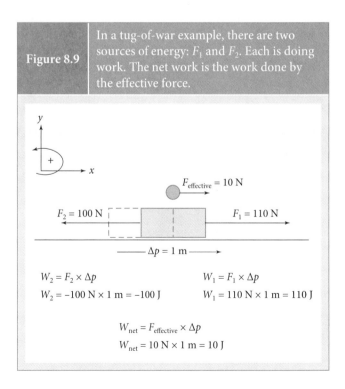

Figure 8.9 In a tug-of-war example, there are two sources of energy: F_1 and F_2. Each is doing work. The net work is the work done by the effective force.

$F_{effective} = 10$ N
$F_2 = 100$ N $F_1 = 110$ N
Δp = 1 m

$W_2 = F_2 \times \Delta p$
$W_2 = -100$ N × 1 m = -100 J

$W_1 = F_1 \times \Delta p$
$W_1 = 110$ N × 1 m = 110 J

$W_{net} = F_{effective} \times \Delta p$
$W_{net} = 10$ N × 1 m = 10 J

it was carefully stated that the barbell–Earth was a system. Gravity is internal to the system. The person lifting the barbell is outside this system and does work on it, increasing the energy of the system by separating the barbell from Earth (kind of like pulling a rubber band):

$$W = \Delta E$$

$$W_{person} = \Delta E_{GP}$$

$$W_{person} = ma_g h$$

If the system includes only the object under consideration (in this case, the barbell), gravity is external to that system and does work on that body:

$$W = \Delta E$$

$$W_{person} + W_g = 0$$

$$W_{person} - ma_g h = 0$$

$$W_{person} = ma_g h$$

Either way you include it is fine (the work performed by the person is the same); just do not make the classic mistake of including gravity as both performing work on the system and a source of potential energy inside the system. It is either one or the other, depending on your perspective.

Important Point!   Gravity can be placed on the work side or the energy side of the equation, but not both.

Recall that the potential energy increased irrespective of the path taken by the barbell. If, instead of pushing the barbell straight up, you curled the barbell by flexing a single joint, you would still increase the potential energy of the barbell–Earth system by the amount $ma_g h$ (Figure 8.10). However, instead of a force doing the work, a torque would be doing the work.

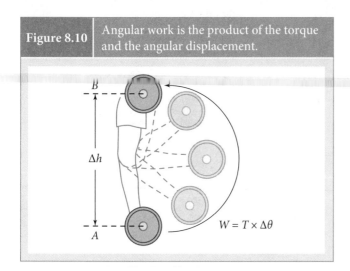

Figure 8.10   Angular work is the product of the torque and the angular displacement.

And, in this example, the linear displacement corresponded to an angular displacement. The work determined by the torque would thus be

$$W = T \times \Delta\theta \tag{8.13}$$

Again, it should be emphasized that Equation 8.13 is only valid if the torque is constant. If not, the area under the torque–angular displacement curve needs to be determined (Figure 8.11).

8.2.1   Mechanical Energy Expenditure

Hopefully, it is clear that if work has been done, then there is a change in energy, and that energy is in the form of either kinetic energy or gravitational potential energy. If the mass of the system does not change (an assumption we constantly make with the human body), then that means either the velocity or the height above the ground had to have changed. If the

Figure 8.11   Angular work is the product of the torque and angular displacement only when the torque is constant. Otherwise, it is the area under the torque–angular position curve. Compare to Figure 8.6.

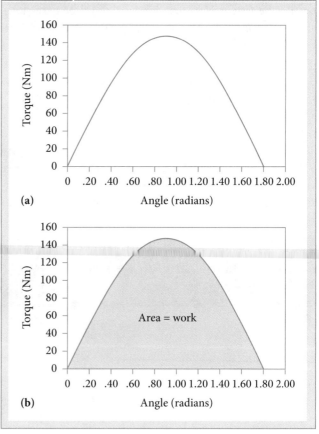

velocity did not change and the height above the ground did not change, then the net overall work was zero. This mechanical definition of work makes for some paradoxical observations when analyzing human movement.

Consider the case where you lift the barbell from point A to point B in Figure 8.5, and then back from point B down to point A. This is, of course, the normal way of performing a repetition. How much mechanical work was performed? The answer is zero: neither the kinetic nor potential energy changed, so no work was done. But certainly physiological effort was expended during the repetition. The problem is that the positive work during the "up" phase was cancelled by the negative work during the "down" phase. This is a case where a purely mechanical concept does not translate well to a biological system.

To fix this problem, a new concept is needed. Instead of simply summing the work values, you could sum the absolute values of work (see **Box 8.1**). This term is called **mechanical energy expenditure** (MEE).[4] To see the difference, examine the equation for total work:

> **Mechanical energy expenditure** The amount of mechanical energy necessary for a prescribed motion

$$W_{total} = W_{positive} + W_{negative} \qquad (8.14)$$

and mechanical energy expenditure:

$$MEE = W_{positive} + |W_{negative}| \qquad (8.15)$$

In the case where a 45 Newton barbell was lifted a distance of 1 meter, the work would be 45 Joules. If that same barbell was lowered by 1 meter to the original starting position, the work would be −45 Joules. The total work would be 0 Joules because there was no change in energy. However, the mechanical energy expended during that time frame would be 90 Joules. Notice that the first term (work) is in strict accordance with the mechanical definition of work, whereas the second (MEE) is the amount of energy spent by the biological system and is more in line with your everyday experiences. Certainly, lifting and lowering weights "cost" something.

However, you should be aware of some of the assumptions associated with MEE. First, it assumes that the "cost" of doing positive and negative work is the same. That is, the requirements for lowering the weight are the same as raising it. This assumption is challenged in **Box 8.2**. Second, it assumes that none of the energy expended in lowering the weight can be stored and reused for lifting it. You have probably surmised that this assumption is not 100% accurate, either. If you were to perform two jumps, one where you crouched down and waited before jumping up (a squat jump)

The absolute value is how far a number is from zero. Think of it as a distance (having only a magnitude), not a displacement (also having a direction). So an absolute value is never negative. The absolute value is denoted by placing the value inside vertical bars. So the absolute value of work, W, is |W|. Technically, to calculate the absolute value of some variable, you would first square it and then take the square root of the squared value. For example:

$$|-3| = \sqrt{-3^2} = \sqrt{9} = 3$$

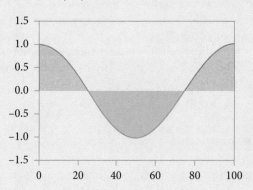

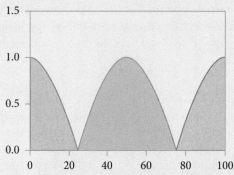

The negative area (in orange) becomes positive when absolute values are calculated.

An alternate way of finding the absolute value on computer spreadsheet would be to use an "if, then" command to multiply numbers greater than or equal to zero by 1, and numbers less than zero by −1. This procedure will also give you all positive numbers.

Graphically, to determine the absolute value, you would flip any negative values over the zero line. This is illustrated in the above figure. Notice that the absolute values also affect the area under the curve. In the top graph, the positive areas under the curve (in gray) are equal to the negative area under the curve (in orange). The total area under the curve is zero. This would be analogous to work. In the bottom graph, all the areas under the curve are positive, and their sum would be a positive value. This would be analogous to mechanical energy expenditure.

Box 8.2	Applied Research: The Cost of Negative Work

The "cost" of any activity is usually determined by controlling for and measuring the gases (O_2 and CO_2) entering into and leaving the body. Work can be determined from measuring the forces/weights and displacements. In this experiment, investigators examined the relationship between metabolic cost and mechanical work associated with repetitively raising or lowering four different loads. Using regression analysis, they found that the metabolic cost of lowering a load was approximately 30 to 50% that of raising it. Although there were many limitations to this study, it was one of the few that attempted to quantify the differences in cost between positive and negative work. Clearly, negative work "costs" something, but it does not cost nearly as much as positive work. The exact percentage probably varies and may be dependent on the activity and the person performing the task.

Data from: de Looze MP, Toussaint HM, Commissaris DACM, Jans MP, Sargeant AJ. Relationships between energy-expenditure and positive and negative mechanical work in repetitive lifting and lowering. *J Appl Physiol.* 1994;77:420-426.

and one where you quickly bounced down and immediately jumped up (a countermovement jump), which one would result in the higher jump? The countermovement jump would be higher. Why? Because some of the energy absorbed on the way down can be stored in the tendons and recouped on the way up.

Important Point!   Mechanical energy expenditure assumes that both positive and negative work have the same metabolic cost and that none of the energy during the negative portion of a movement can be recouped and used during the positive portion.

Another paradoxical aspect of applying the concept of work to biological systems involves cases where there is no movement (statics). If you were to place this book on the table, the table would not be doing any work on the book. If you were to pick up the book and hold it at arm's length, you would not be doing any work on the book while you were holding it still. Yet after a while, it would take considerable effort to continue holding the book. Although MEE is a useful concept in dealing with negative work, there is no concept for dealing with energy expenditure in biological systems during static conditions. It is probably best to analyze these situations using impulse.

Important Point!   Work should not be confused with physiological effort.

Section Question Answer

If you have a patient that is performing exercises in an up-and-down fashion (such as a bench press or a squat), then the total change in position is zero. If the total change in position is zero, then the total work is zero. Because you know it takes a lot of effort to perform these repetitions, this answer is not very appealing. The mechanical energy expenditure (ignoring the energy needed to move the limbs) is equal to two times the weight times the distance moved in one direction. The actual physiological cost of performing the exercise is probably somewhere between these two extremes.

COMPETENCY CHECK

Remember:

1. Define the following terms: work, positive work, negative work, and mechanical energy expenditure.
2. List the assumptions associated with mechanical energy expenditure.

Understand:

1. How much work is required to do the following? Is the energy entering or leaving the system?
   a. Raise a 10 kilogram mass 0.5 meters.
   b. Lower a 5 kilogram mass 0.1 meters.
   c. Raise a 15 kilogram mass by 0.5 meters, and then lower it back to its original starting position.
2. How much mechanical energy expenditure is required to do the following?
   a. Raise a 10 kilogram mass 0.5 meters.
   b. Lower a 5 kilogram mass 0.1 meters.
   c. Raise a 15 kilogram mass by 0.5 meters, and then lower it back to its original starting position.

Apply:

1. List several activities where:
   a. Work would be an appropriate measure for analysis.
   b. Mechanical energy expenditure would be an appropriate measure for analysis.
   c. Neither work nor MEE would provide useful insights.

8.3   LOCOMOTOR WORK, THE CENTER OF MASS EQUATION, AND THE FIRST LAW OF THERMODYNAMICS

Section Question

You are working with Max, a basketball player, to improve his jump height (Figure 8.12). Before you can begin to improve his ability to jump, you must first ask yourself: What determines how high he can jump?

Figure 8.12	What determines how high he can jump?

©dotshock/ShutterStock, Inc.

If the body was a point mass, and all the forces acted on that point mass, you would see that:

$$\sum F_{ext} \times \Delta p_{COM} = \Delta\left(\frac{1}{2}mv_{COM}^2\right) \qquad (8.16)$$

Note that in this equation there is only one kind of energy and only one input into the system. You could also think of this as if the center of mass was fixed to a rope, and this quantity was the energy expended to pull it.[5] The work required to change the velocity of the COM is often called "external" work, although some authors prefer the term "locomotor" work.[5]

> **External (locomotor) work**
> Energy required to change the motion or location of the center of mass

Equation 8.16 is contrasted with the first law of thermodynamics, which states:

$$W = \Delta E - Q \qquad (8.17)$$

where W is the work done on the system, ΔE is the change in energy, and Q is the energy that is lost as heat to the

environment. Notice that the COM equation considers a very narrow range of possibilities: The system of interest is considered a point mass, the forces are acting on that point mass, and velocity of the COM is changing as a result. The first law of thermodynamics is more complete and yields additional information.[6] To illustrate these differences, consider a person performing a vertical jump. Say you wished to determine how high the person could jump. As you already learned, when in the air, energy in the system is conserved, and

$$0 = \Delta(E_{LK} + E_{GP}) = \Delta E_{system}$$

At takeoff, the person has a certain amount of potential energy and a certain amount of kinetic energy. Because energy is conserved in the air, the kinetic energy will be transformed into kinetic energy, and:

$$\frac{1}{2}mv^2 = ma_g h$$

So the change in height is determined by the kinetic energy at takeoff. Where does the kinetic energy at takeoff come from? Using the COM equation, you would note that the kinetic energy at takeoff is determined by the locomotor work:

$$({}^yF_{GRF} - F_g) \times \Delta^y p_{COM} = \Delta\left(\frac{1}{2}mv_{COM}^2\right) \qquad (8.18)$$

Do not get thrown off by the $-F_g$ in the equation. You can look at it in a few different ways. First, you can say that it is the net external force (and not the ground reaction force) that determines the locomotor work. Second, you could recognize the F_g is equivalent to ma_g, and $\Delta^y p_{COM}$ is the same as $ma_g h$, the work done by gravity. The net work is the work done by the ground reaction force minus the work done by gravity. By knowing the F_{GRF} and using a point mass model, you can determine the velocity of the COM at takeoff and, subsequently, the height of the COM at the apex of the jump. You have lost any information about *time* using work–energy methods, but see how easy it was to calculate the jump height.

If you rearrange the terms in Equation 8.18, you get

$${}^yF_{GRF} \times \Delta^y p_{COM} = \Delta\left(\frac{1}{2}mv_{COM}^2\right) + ma_g h \qquad (8.19)$$

This equation is also useful. The terms on the right side of the equation are the kinetic energy at takeoff and the height of the center of mass (or potential energy) at take-off (**Figure 8.13**). Collectively, they are called the effective energy at takeoff.[7] The left-hand side of the equation is the locomotor work required to change the effective energy. The right side of the equation is a more complete determination of the jump height because it includes the potential energy

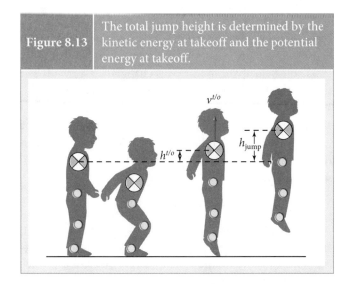

Figure 8.13 | The total jump height is determined by the kinetic energy at takeoff and the potential energy at takeoff.

at takeoff (**Figure 8.14**). In other words, to jump high, you want to extend your legs as much as possible before takeoff so that you: (1) increase $^y\Delta p$, and (2) increase the potential energy at takeoff.

Using the COM equation only gives you information about the velocity (and potentially the position) of the COM. Using the first law of thermodynamics for the same problem provides more information:

$$-ma_g h = \Delta\left(\frac{1}{2}mv^2_{COM}\right) + \Delta E_{K\text{-internal}} + \Delta E_{thermal} + \Delta E_{chemical}$$

$$(8.20)$$

A word about what the terms are on the right-hand side of the equation. $E_{K\text{-internal}}$ is the energy required to move the segments of the body relative to the COM. If the body is

Figure 8.14 | A hierarchical model of vertical jumping.

```
                    Jump height
                         │
                         ▼
                    E_effective
                     ┌────┴────┐
                     ▼         ▼
                v_takeoff   h_takeoff
                     │         │
                     ▼         ▼
                  ʸF_GRF    Δʸp_COM
```

modeled as a point mass, then this term would be zero. $E_{thermal}$ is the energy that gets lost as heat. $E_{chemical}$ is the energy required to create the muscle contractions in the first place. Ultimately, this is the energy that comes from food. It is the "cost" of performing the movement.

Now consider the landing. At the apex of the height, you know that the potential energy is equal to $ma_g h$, and the kinetic energy is zero. You also know that in the descent energy is conserved and at the instant of contact the potential energy is transformed into kinetic energy. As the person lands, that kinetic energy is then absorbed by the body. Again using the COM equation, you see that:

$$(^yF_{GRF} - F_g) \times \Delta^y p_{COM} = \Delta\left(\frac{1}{2}mv^2_{COM}\right) \quad (8.21)$$

Assuming that the mass of the person did not change while they were in the air, you find that negative work by the ground reaction force is necessary to change the kinetic energy of the person to zero. Knowing that work is the average force multiplied by the displacement that the force is applied, you can see that the greater the distance, the smaller the magnitude of the average force has to be. If a person lands "stiff," without bending the legs during landing (**Figure 8.15**), the displacement of the COM will be less, and consequently the average forces will be higher than if the person landed with a "soft" landing (bending the knees; Figure 8.15).

The COM equation does not provide you with the same information as the first law of thermodynamics. Can you tell where the energy went? Remember, the first law of thermodynamics

Figure 8.15 | During landing, the kinetic energy must be absorbed by the body. A stiffer landing results in a smaller displacement (Δp_1) and consequently higher forces. A softer landing results in a larger displacement (Δp_2) and less forces.

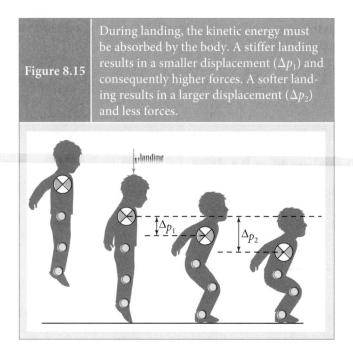

states that energy cannot be created or destroyed, so it has to go somewhere. Where did it go? The energy left the skeletal system and was ultimately dissipated as heat.

Section Question Answer

Jump height is determined by the difference between the magnitudes of the ground reaction force in the vertical direction and body weight (the effective force) and the displacement of the body's COM while on the ground. If Max would increase one or both of these variables, he would increase his jump height.

COMPETENCY CHECK

Remember:

1. State the first law of thermodynamics.
2. List the equations for the center of mass equation and the first law of thermodynamics.

Understand:

1. What information is provided by the first law of thermodynamics that is not provided by the center of mass equation?

8.4 EFFICIENCY AND ECONOMY

Section Question

Bob has recently suffered from mild stroke. He is complaining that he just does not seem to have the energy like he used to when he is walking (Figure 8.16). Is there anything you can do to help?

The first law of thermodynamics states that you cannot get something from nothing. In essence, you must perform work to change the amount of energy in a system. The second law of thermodynamics states that you cannot even break even.[8] In other words, the amount of energy you can get out of a system is always less than the energy that you put into it. Rearranging the terms in Equation 8.20 to put the inputs on one side and the outputs on the other:

$$\Delta E_{\text{chemical}} = \Delta \left(\frac{1}{2} m v_{\text{COM}}^2 \right) + m a_g h + \Delta E_{\text{K-internal}} + \Delta E_{\text{thermal}}$$

$$(8.22)$$

E_{chemical} is the energy that comes from the three energy systems inside the body: ATP-PC, anaerobic glycolysis, and aerobic glycolysis. The first two terms on the right side of the equations

Figure 8.16 What determines how much energy is expended during gait?

© Dwight Smith/ShutterStock, Inc.

are the outputs in terms of the center of mass. These are the quantities that you are probably most interested in, and they determine things such as jump height or velocity of the center of mass. The $E_{\text{K-internal}}$ is part of the cost of doing business and may not lead to any productive external work. Consider moving your right arm up while simultaneously moving your left arm down. There is a metabolic cost of doing so, but because the location of the COM has not changed, there is no change in energy of the COM. Similarly, using the COM equation always underestimates the amount of work being done because it does not account for changes in $E_{K\text{-internal}}$. But even if it did, one more term still needs to be accounted for: the thermal energy, $E_{thermal}$. This is the work required to overcome friction and other passive forces in the body, and it has been referred to as "hidden" work.[5] This work ends up as heat, and it is one of the reasons why your core body temperature increases during physical activity. It is what people mean when they say they must "warm up" prior to exercise or activity.

The amount of energy that is lost to these hidden sources is quite high. It is estimated that only ~20 to 30% of the chemical energy can be converted to mechanical energy by a muscle fiber.[9] This may sound bad, but it is higher than the chemical energy that can be converted to mechanical energy by your car.

Oftentimes, you will want to compare the amount of energy that you put into a system to the amount of work (or more correctly, the amount of mechanical energy expended) that system can do with that energy. This is known as **efficiency**. The basic idea is the same with your car's fuel efficiency: How many miles can you go (output) with a gallon of gas (input)? Once again, you will look at ratios:

> **Efficiency**   The amount of mechanical energy that can be expended with a given amount of energy

$$\text{Efficiency} = \frac{\text{output}}{\text{input}} = \frac{\text{mechanical energy expended}}{\text{metabolic energy}} \quad (8.23)$$

Typically, though, the ratio is expressed as a percentage:

$$\%\text{Efficiency} = \frac{\text{Mechanical Energy Expended}}{\text{Metabolic Energy}} \times 100 \quad (8.24)$$

So you are saying the same thing if you say that a system has an efficiency of 0.23 or 23%. The other 77% of the energy is lost as heat.

Important Point!   No activity is 100% efficient.

> **Economy**   The amount of energy required to perform a certain amount of work (MEE)

Related to the concept of efficiency is **economy**.[10] Efficiency is usually thought of as how much work (MEE) can be performed for a given amount of energy. Economy, on the other hand, refers to how much energy it takes to perform a given amount of work (MEE).[11] A lower amount of energy for the same amount of work means the movement is more

economical. The two terms only differ in perspective. If the activity under consideration requires x amount of MEE, then you would probably be looking at minimizing the metabolic cost, so you would be concerned with economy of movement. If you have x amount of energy and you want to maximize the amount of MEE you can do, then you would be concerned with efficiency.

8.4.1   The Energetics of Gait

With a projectile (when a person was in the air), you have already seen energy exchanges in action. Potential energy was transformed into kinetic energy (and vice versa) without any additional external work being performed on the system. This idea can be extended to a pendulum, where a mass is suspended from a pivot point on some sort of string. Unless work is performed on the mass, it will just sit there. But if you raise the mass a certain height (i.e., perform work on it), that potential energy can be converted to kinetic energy (**Figure 8.17**). Provided that no energy is lost to friction at the pivot point, the pendulum would continue to swing back and forth forever. This is because the energy is constantly being interchanged between the two forms: When the E_{GP} is at its greatest, E_{LK} is zero, and when E_{LK} is at its greatest, the E_{GP} is zero. When this occurs, it is said that the two forms of energy are 180° out of phase. See **Box 8.3** for more details.

Walking and running gaits provide interesting contrasts between energy exchanges and provide insights into why running requires so much more energy than does walking. You can use some simple models to help you really understand what is happening with both forms of locomotion.

Walking has often been described as acting like an *inverted* pendulum, with the COM located on top of a rigid strut (**Figure 8.18**). With this model, the strut makes contact with the ground at some angle, θ, during initial contact. The

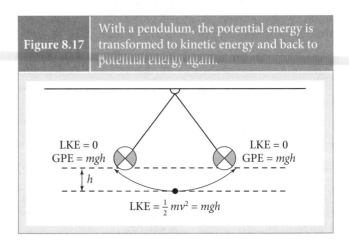

Figure 8.17   With a pendulum, the potential energy is transformed to kinetic energy and back to potential energy again.

| Box 8.3 | Essential Math: Phase Difference |

The terminology concerning phase difference can be a bit confusing because for any cyclic motion, a full cycle (or motion) represents 360°, even if the body is oscillating at something less than 360° (like the pendulum example in the text, or a mass bobbing linearly on the end of a spring). Conceptually, the phase difference represents the relative timing (expressed as degrees) of key events (such as a maximum or minimum) that have the same frequency. If the maxima (or minima) of two curves occur at the same point in time, the phase difference is zero, and the curves are said to be **in phase**. If they occur at different points in the cycle, they are **out of phase**. For example (top part of figure), if the one curve reaches its minimum at the halfway point in its cycle, and the other curve reaches its minimum three-quarters through its cycle, the minima are off by a quarter of a phase. Because a cycle is 360°, the two curves are out of phase by 90°.

Mathematically, if the times of the key events are known, the phase difference, φ, can be calculated as:

$$\varphi = \frac{t_2 - t_1}{t_{total}} \times 360°$$

where t_2 and t_1 represent the time of the key events, and t_{total} is the overall time. For example, in the bottom part of the figure, the total time is 100 seconds. The maxima for the two curves occur at 100 seconds and 50 seconds. The total time is also 100 seconds. Thus:

$$\varphi = \frac{100 - 50}{100} \times 360° = 180°$$

The two curves are 180° out of phase.

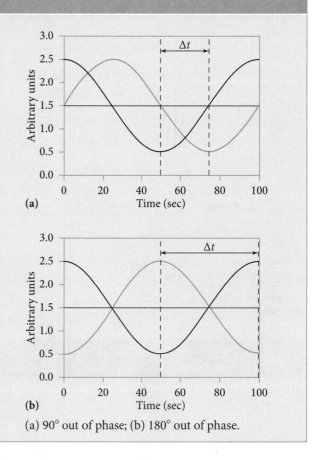

(a) 90° out of phase; (b) 180° out of phase.

| Figure 8.18 | During walking, the body acts like an inverted pendulum, transforming potential energy to kinetic energy and back to potential energy again. |

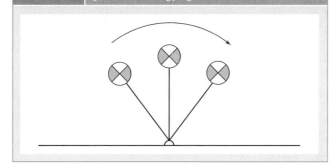

| Figure 8.19 | During running, the body acts like a mass spring. Because the kinetic and potential energies are in-phase, energy cannot be transformed between the two types of energy. |

COM then vaults over strut, reaching its apex in height at midstance.

Running, on the other hand, is modeled as a mass on top of a spring (Figure 8.19). Like walking, the spring initially makes contact with the ground at initial contact. In contrast

to running, the spring then compresses during stance and is at its lowest height at midstance.

Examine the graphs in Figures 8.20 and 8.21. If you look at the energetics of the COM, you would note that the E_{LK} is maximized at takeoff and is minimized at midstance.

Figure 8.20 The potential and kinetic energies during walking.

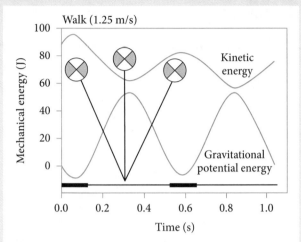

Data from: Farley CT, Ferris DP. Biomechanics of walking and running: center of mass movements to muscle action. *Exercise and Sport Sciences Reviews, Volume 28, 1998.* 1998;26:253–285. Figure 5, pg. 259.

Figure 8.21 The potential and kinetic energies during running.

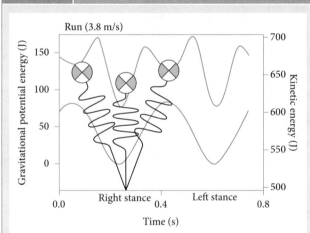

Data from: Farley CT, Ferris DP. Biomechanics of walking and running: center of mass movements to muscle action. *Exercise and Sport Sciences Reviews, Volume 28, 1998.* 1998;26:253–285. Figure 10.6, pg. 260.

In walking, the E_{GP} is maximized at midstance; during running, E_{GP} is minimized at midstance. These fluctuations are the defining differences between walking and running gaits.[12]

To understand the energetic exchanges, you must examine the phases between E_{LK} and E_{GP}. Begin by reviewing Box 8.3. When you return to Figures 8.20 and 8.21, you will notice that during walking the E_{LK} and E_{GP} are out of phase by about 180°.

When this happens, energy is being transformed between the two types of energy. Consider the projectile in Figure 8.4. At takeoff, the E_{LK} is at its maximum, and the E_{GP} is zero. Notice that, once in the air, the E_{LK} is decreasing by the same amount the E_{GP} is increasing. At the apex, the E_{GP} is at its maximum and the E_{LK} is zero. During the downward motion, the E_{GP} is converted back to E_{LK} until the return, where the E_{GP} is once again zero and the E_{LK} is the same at takeoff.

Now consider the E_{LK} and E_{GP} of gait. Do you see the similarities between the energy exchanges for walking gait and the projectile? E_{LK} and E_{GP} are not completely 180° of phase during walking, but they are pretty close (199° ± 9°).[13] And walking is not 100% efficient (nothing is). It is similar to the efficiency of muscle (~20%). But it is estimated that approximately 60% of the mechanical energy can be recovered during normal walking, and the efficiency would decrease by a factor of two if this was not the case.[13]

During running, the E_{LK} is approximately in phase with E_{GP}; both are at a maximum during takeoff and at a minimum at midstance. This means that the E_{GP} cannot be transformed to E_{LK} and vice versa. Thus, the energy costs associated with running are much higher, and the efficiency is much lower (~2–11%).[14]

Section Question Answer

Many patients with a neurological impairment (such as stroke or cerebral palsy) like Bob use more energy when they walk. If you rearranged the terms in Equation 8.23, you would see that the energy used is determined by the amount of work they perform and the efficiency of their movement. Interventions would focus on either: (1) decreasing the mechanical energy expended (such as decreasing the vertical displacement of the center of mass or the change in energy of the segments relative to the center of mass), or (2) improving the efficiency (the exchange between the kinetic and potential energies).

COMPETENCY CHECK

Remember:

1. Define: efficiency and economy.

Understand:

1. What is the difference between efficiency and economy?

Apply:

1. List activities for which efficiency would be very important.
2. List activities for which efficiency would not be important.

8.5 POWER

Section Question

Wayne is a collegiate football player hoping to go into the professional league. During the scouting combine, he is tested on a maximum-height vertical jump (Figure 8.22). What does this measure? Is it testing the same thing as a squat?

During a vertical jump, you leave the ground when your legs are fully extended and your feet are no longer touching the ground. There is a limited amount of time from the initiation of the jump until you are airborne. Therefore, it is not only important that you perform a large amount of work; you must also do that work fairly quickly.

Power is the time rate of doing work. You know that a rate is a ratio, and in this case the ratio is the change in work to the change in time:

Power The time rate of doing work; alternatively, how quickly energy is entering or leaving the system, or how much force can be produced while moving quickly

$$P = \frac{W}{\Delta t} \qquad (8.25)$$

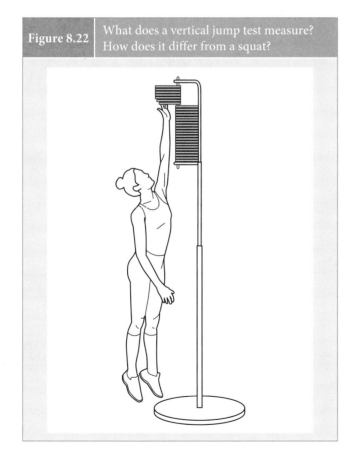

| **Figure 8.22** | What does a vertical jump test measure? How does it differ from a squat? |

The unit for power would be Joules per second, or Watts. Again, the power in Equation 8.25 is an average power; it approaches the instantaneous power as the time interval becomes infinitesimal.

Because work is a change in energy, you can also think of power as being the rate at which energy is entering or exiting the system:

$$P = \frac{\Delta E}{\Delta t} \qquad (8.26)$$

An alternate way of looking at power is

$$P = \frac{F \times \Delta p}{\Delta t} = F \times \frac{\Delta p}{\Delta t} = F \times v \qquad (8.27)$$

The angular equivalent would be determined by making the appropriate substitutions:

$$P = \frac{T \times \Delta \theta}{\Delta t} = T \times \frac{\Delta \theta}{\Delta t} = T \times \omega \qquad (8.28)$$

In simple terms, this means that power is the ability to produce force (or torque) while moving quickly. Power is an important variable whenever the movement speeds are high.

Many activities are defined by the velocity at some instant: the velocity of takeoff, the velocity at impact, and so on. The velocity term shows up in both the impulse–momentum and center of mass energy equations. If you were to create a hierarchical model for this activity, you could choose to use either equation to guide your model building. But with impulse–momentum, you come to another paradox:[15] To increase momentum, you have to apply a force over a greater period of time. Yet when moving fast, you want to minimize the time. Power is helpful here. Instead of maximizing the time, you wish to minimize the time while maximizing the distance over which the force is applied. This is important for many human movements.

Section Question Answer

The vertical jump is a test of lower extremity power. You can appreciate this by understanding that, with the jump, you have to have a large change in energy in a very short period of time. At first glance, the squat may seem to be measuring the same thing as the vertical jump. The movement patterns are very similar, and with the squat you are performing a large amount of work (the weight lifted multiplied by the upward displacement). However, the squat does not have the element of time associated with it as does the jump. In fact, you may notice that movement speeds are very slow when lifting maximal weights. So the squat can be thought of as a test of how much work can be performed, or how much force can be generated by the lower extremities. However, this is fundamentally a different movement quality than power.

Table 8.1	Key Concepts

- Energy
- Work
- Mechanical energy expenditure
- Efficiency
- Power

SUMMARY

In this lesson, you learned about an alternative to Newton's laws for analyzing human movement. This method involved the concepts of work, energy, and power (Table 8.1). Because of some issues with using these concepts with biological systems, mechanical energy expenditure was introduced. The first law of thermodynamics was compared to the center of mass equation, and efficiency and economy were introduced. Impulse–momentum and work–energy methods provide complementary information and a more complete analysis of movement for several different tasks.

REVIEW QUESTIONS

1. Define the following terms: energy, kinetic energy, potential energy, gravitational potential energy, strain potential energy, work, mechanical energy expenditure, efficiency, economy, and power.
2. State the conservation of energy and the first law of thermodynamics.
3. What is meant by the term *negative* work or power?
4. What information is provided by using work–energy methods that is not provided by using impulse–momentum methods?
5. What information is provided by using impulse–momentum methods that is not provided by using work–energy methods?
6. Which requires more work?
   a. Increasing the velocity of a 10 kilogram object from 5 meters per second to 10 meters per second.
   b. Decreasing the velocity of a 5 kilogram object from 10 meters per second to 5 meters per second.
   c. Lifting a 10 kilogram object from the ground to 2 meters above the ground.
   d. Holding a 100 kilogram object in place 1 meter above the ground.

7. Which requires greater power?
   a. Increasing the velocity of a 10 kilogram object from 5 meters per second to 10 meters per second in 1 second.
   b. Decreasing the velocity of a 5 kilogram object from 10 meters per second to 5 meters per second in 2 seconds.
   c. Lifting a 10 kilogram object from the ground to 2 meter above the ground in 0.5 second.
   d. Holding a 100 kilogram object in place 1 meter above the ground for 10 seconds.
8. Describe movements where you would primarily use work–energy methods to analyze them.
9. Describe movements where you would primarily use impulse–momentum methods to analyze them.
10. Describe movements where efficiency is important.
11. Describe movements where efficiency is not important.
12. Describe movements where power is important.
13. Describe movements where power is not important.

REFERENCES

1. Feynman RP. *Six Easy Pieces.* New York: Basic Books, 1995.
2. Watson D. Energy Definition. 2011. Internet Communication.
3. Zatsiorsky VM. *Kinetics of Human Motion.* Champaign, IL: Human Kinetics, 2002.
4. Aleshinsky SY. An energy-sources and fractions approach to the mechanical energy-expenditure problem. 1. Basic concepts, description of the model, analysis of a one-link system movement. *J Biomech.* 1986;19:287–293.
5. Zatsiorsky VM, Gregor RJ. Mechanical power and work in human movement. In: Sparrow WA, editor. *Energetics of Human Activity.* Champaign, IL: Human Kinetics, 2000: 195–227.
6. Sherwood BA. Pseudowork and real work. *American Journal of Physics.* 1983;51:597–602.
7. Bobbert MF, van Ingen Schenau GJ. Coordination in vertical jumping. *J Biomech.* 1988;21:249–262.
8. Park RL. *Voodoo Science: The road from foolishness to fraud.* New York: Oxford University Press, 2000.
9. He ZH, Bottinelli R, Pellegrino MA, Ferenczi MA, Reggiani C. ATP consumption and efficiency of human single muscle fibers with different myosin isoform composition. *Biophys J.* 2000;79:945–961.
10. Enoka RM. *Neuromechanical basis of human movement.* 4th ed. Champaign, IL: Human Kinetics, 2008.
11. Whiting WC, Rugg S. *Dynatomy: Dynamic Human Anatomy.* Champaign, IL: Human Kinetics, 2006.
12. Farley CT, Ferris DP. Biomechanics of walking and running: Center of mass movements to muscle action. *Exercise and Sport Sciences Reviews, Volume 28, 1998.* 1998;26:253–285.
13. Ortega JD, Farley CT. Minimizing center of mass vertical movement increases metabolic cost in walking. *J Appl Physiol.* 2005;99:2099–2107.
14. Kyrolainen H, Belli A, Komi PV. Biomechanical factors affecting running economy. *Med Sci Sports Exerc.* 2001;33:1330–1337.
15. Bartlett R. *Introduction to Sports Biomechanics: Analysing Human Movement Patterns.* 2nd ed. Abingdon: Routledge, 2007.

Lesson 9

Collisions, Impacts, and the Conservation Laws

LEARNING OBJECTIVES

After finishing this lesson, you should be able to:

- State the conservation of momentum and the conservation of energy laws.
- Define the following terms: elastic collision, inelastic collision, coefficient of restitution, and effective mass.
- State the difference between an elastic and inelastic collision.
- Determine the transfer of momentum between two colliding bodies in both one and two dimensions.
- Explain why inelastic collisions can be so damaging.
- Explain how manipulating the coefficient of restitution of the objects used in a sport can give someone an advantage.
- Explain the importance of effective mass during impacts.

INTRODUCTION

You should be familiar with both the conservation of momentum and the conservation of energy. In this lesson, you will consider what happens during impacts—when two objects collide with one another. This is actually a very common occurrence in human movement. Examples include striking, such as when a bat or racquet hits a ball, or when one American football player hits another. But every time your foot hits the ground it is an impact. So even if you are not very interested in the aforementioned sports, you should have an idea about impacts.

In this lesson, you will learn all the basic ideas about collisions and resulting impacts by looking at a two point-mass system that is traveling along a single axis. Once you understand them, you can apply them to more complicated scenarios, such as when the collision takes place in two dimensions or when the bodies are not point-masses. Impacts and collisions are really just special cases where the sum of the external forces is zero, and thus momentum is conserved. Knowing this fact allows you to make several important observations without needing any detailed knowledge about the forces that are involved. You will see how this type of analysis is useful in many different scenarios.

9.1 SIMPLE COLLISIONS OF POINT-MASSES

> **Section Question**
>
> Assuming that you actually hit the ball (**Figure 9.1**), would it go further if the ball was pitched at 60 miles per hour or 80 miles per hour? Why?

9.1.1 Defining Your System: Internal and External Forces

To help you understand the basic ideas behind collisions and impacts, you will begin by studying a two point-mass system. Remember how important it is to define your system before you begin to analyze any problem from a biomechanical perspective. To illustrate this point, examine two bodies that collide with each other by defining the system in two different ways.

> **Important Point!** The first step in performing any type of biomechanical analysis is to define your system.

© Aspen Photo/ShutterStock, Inc.

Figure 9.1   Assuming you actually hit the ball, would it go further if the ball was pitched at 60 mph or 80 mph?

In the first case, let us define two systems: one for body *A* and one for body *B* (Figure 9.2). To distinguish them, you should draw dashed circles to represent the boundaries of each system. Before impact, body *A* is at rest, and body *B* is moving with some sort of linear momentum, L_A. Because body *B* is at rest, you know its momentum must be zero. At impact, body *A* contacts body *B*. You also know that whenever two bodies are in contact, there are equal and opposite forces acting on the two bodies. In this case, a force from body *A* is acting on body *B*, and a force from body *B* is acting on body

A. After impact, would you expect the two momenta would be the same? Would you expect body *A* to be moving at the same speed and in the same direction as before it collided with body *B*? Probably not. How about body *B*? Unless its mass is huge compared to body *A*, you would expect it to move. In fact, it probably would move regardless, but it just might be imperceptibly small. There was an external force acting on each body, and the result of that force was a change in momentum. These observations are in keeping with Newton's laws of motion. Also note that because there were external forces and changes in velocities, there was also work performed on each system.

> **Important Point!**   Effective forces external to a system change the momentum and perform work on that system.

In the second case, imagine that body *A* and body *B* are part of the same system (Figure 9.3). Remember that Newton's laws apply to *external* forces, not *internal* ones. Internal forces do not change the momentum of a body and perform no work (do not confuse internal forces with internal work). If you were to define a system as your leg, for example, there would be equal and opposite forces between your femur and tibia. Those forces would be equal in magnitude and opposite in direction, and therefore they would cancel. The forces between your femur and tibia would not change the momentum of your leg as a whole. The same is true with bodies *A* and *B*. Even though they may start relatively far apart, they are part of the same system, and any forces between them are considered to be internal forces. If

Figure 9.2   Defining your system determines whether forces are internal or external to that system. In this case, bodies *A* and *B* are two separate systems. The reaction forces at impact (*A* on *B* and *B* on *A*) are external to the system.

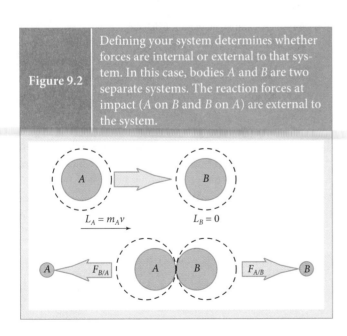

Figure 9.3   When bodies *A* and *B* are part of the same system, the reaction forces between them are internal to the system and do not affect a change in momentum of that system.

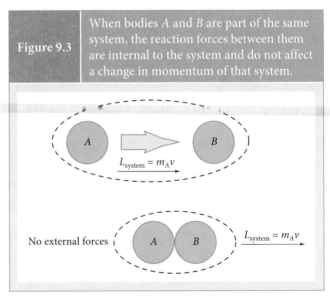

there are no external forces acting on the system, then the change in momentum must be zero:

$$\sum F_{ext} = \frac{\Delta L}{\Delta t} = 0 \qquad (9.1)$$

As you will see, that does not mean that the momentum of each body does not change—only that all changes of momentum must stay within the system. In other words, the increase in momentum of one body must be compensated for with a decrease in momentum of another body by the same exact amount. If the momentum of the system is not changing with time, it is conserved. Similarly, if there are no external forces doing work on the system, you can also say that the total energy within the system is conserved. Using the conservation of momentum and energy concepts, you can obtain a lot of information about what happens during impacts and collisions without needing detailed information about the forces involved. Pretty cool, huh?

Important Point!   Forces internal to a system do not change the momentum of or perform work on that system.

9.1.2   Conservation of Momentum

After identifying your system, you need to identify your frame of reference. In this section, the frame of reference is going to be one dimension, so there is no need to specify any axes. The other thing you need to do is list your assumptions. For the collisions that you will be looking at, assume that there are no frictional (or other external) forces, and the time of impact is extremely small. Because the change in momentum of the system is zero in each case, you know:

$$\Delta L = L' - L = 0 \qquad (9.2)$$

where the prime sign will indicate the time immediately after impact, and the momentum without the prime sign is the momentum immediately before impact. You know that momentum is the product of mass and velocity:

$$L = mv \qquad (9.3)$$

It is also important to note that the momentum of any system is the sum of the momentum of each body in a system:

$$L = \sum (m_i \times v_i) \qquad (9.4)$$

where the symbol i represents the number of bodies in the system. If there are two bodies (a and b) in your system, then:

$$L = m_a v_a + m_b v_b \qquad (9.5)$$

If you substitute Equation 9.5 into Equation 9.2, you get:

$$\Delta L = (m_a' v_a' + m_b' v_b') - (m_a v_a + m_b v_b) = 0 \qquad (9.6)$$

Important Point!   The conservation of momentum law states that if there is no external, effective force, then the momentum of a system will not change.

Equation 9.6 is a very general equation for the conservation of momentum. In biomechanics, you will often make the assumption that the masses of objects do not change during the period of your analysis. So you can remove the primes from the masses:

$$\Delta L = (m_a v_a' + m_b v_b') - (m_a v_a + m_b v_b) = 0 \qquad (9.7)$$

By manipulating Equation 9.7, you can find out some pretty interesting things about what happened immediately before or after an impact. Let us look at a few examples.

To begin, consider the case of two pennies with equal masses in Figure 9.4a (you can try this experiment yourself). Before impact, penny b is motionless and has an initial momentum of 0. Penny a has some momentum, $m_a v_a$, as it moves toward penny b. At impact, penny a comes to a complete stop ($L_a = 0$). Meanwhile, penny b speeds off in the original direction that penny a was traveling. Because linear momentum was conserved, you know that penny b now has the same momentum as penny a. And because they have the same mass, you know that the final velocity of penny b is equal to the initial velocity of penny a.

In this case, penny b bounced off penny a without any deformation. In these cases the collision is considered to be **elastic**. Another example of an elastic collision would be if two pennies were going in the same direction, but at different speeds (Figure 9.4b).

> **Elastic collision**   A collision where two objects bounce off each other without any deformation or loss of heat

In this case, imagine that penny a is traveling at twice the speed of penny b and in the same direction. Because penny a is behind penny b, they will obviously collide. Momentum is transferred from penny a to penny b, and penny b speeds off at the higher velocity penny a originally had, while penny a slows down to the original speed of penny b.

Important Point!   If two objects with the same mass collide head-on in a perfectly elastic collision, the momentum of each body will transfer to the other.

For the final example, consider two pennies traveling in opposite directions at the same speed (Figure 9.4c). What happens when these two pennies collide? If it was a direct, head-on elastic collision, the two pennies would bounce off

Figure 9.4	Two pennies colliding in a perfectly elastic collision. (a) The momentum of penny *A* is transferred to the motionless penny *B*, and the momentum of penny *A* becomes zero. (b) The momentum of the faster penny *A* is transferred to the slower-moving penny *B* while penny *A* assumes the momentum of the slower-moving penny *B*. (c) pennies *A* and *B* collide and go off in opposite directions.

(a)

Before: $L_a = mv$ $L_b = 0$

After: $L_a = 0$ $L_b = mv$

- -

(b)

Before: $L_a = m2v$ $L_b = mv$

After: $L_a = mv$ $L_b = m2v$

- -

(c)

Before: $L_a = mv$ $L_b = -mv$

After: $L_a = -mv$ $L_b = mv$

Figure 9.5	Two American football players collide head-on in a perfectly inelastic collision. What is the final velocity of the two players combined?

Before: L_a L_b

After: L_{ab}???

Photo: © Alexey Stiop/ShutterStock, Inc.

each other. They would return in the opposite directions from which they came with the same speed.

Pennies and billiard balls are good examples of objects that exhibit elastic collisions. But even these are not perfectly elastic. That is an ideal that is never reached. On the opposite end of the spectrum are two objects that stick together after they collide. Such collisions are called inelastic. An example of an inelastic collision would be two American football players colliding.

> **Inelastic collision**   A collision in which two objects stick together after they collide

Important Point!   A perfectly elastic collision is an ideal that is never quite reached in the real world.

Think about what would happen if two football players were running at each other and made a head-on, inelastic collision (**Figure 9.5**). With inelastic collisions, both objects have the same velocity after impact. Equation 9.7 becomes

$$(m_a + m_b)v'_{ab} - (m_a v_a + m_b v_b) = 0 \qquad (9.8)$$

where v'_{ab} is the velocity of the combined two bodies after impact. In cases like these, you often want to know what happens. Does player *A* keep making forward progress, or does player *B* drive him back? Rearranging Equation 9.8 helps you obtain the answer:

$$v'_{ab} = \frac{m_a v_a + m_b v_b}{m_a + m_b} \qquad (9.9)$$

Easy examples are when the masses of the two players are equal. You should have an intuitive feel that if player *A* is moving faster than player *B* before impact, player *A* continues to advance. Similarly, if player *B* is moving faster than player *A* before impact, player *B* will drive player *A* back. If they are moving at the same speeds, then both will stop dead in their tracks on impact. You do not even need to plug in the numbers to know that. What if their masses are not equal?

If player *A* weighs 5 more kilograms than player *B*, player *B* would have to run 5 meters per second faster than player *A* (because both mass and velocity are weighted equally in the momentum equation). This is a lot easier said than done.[1] If player *A* has a mass of 130 kilograms and can run the 40-yard

Table 9.1	Theoretical Velocities That a Player of Different Masses Would Need to Achieve a Linear Momentum of 792.48 kg·m/sec	
Weight		**40-yard-dash time**
120		5.54
110		5.08
100		4.62
90		4.15*
80		3.69*

* Below current NFL record

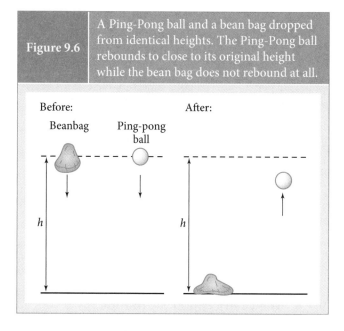

Figure 9.6	A Ping-Pong ball and a bean bag dropped from identical heights. The Ping-Pong ball rebounds to close to its original height while the bean bag does not rebound at all.

dash in 6 seconds, he would hypothetically have an average momentum of 792.48 kilograms times meters per second. Table 9.1 shows the corresponding 40-yard-dash times that player *B* would need to have to generate the same momentum, depending on his mass. What is remarkable here is that the current National Football League Scouting Combine record for the 40-yard dash is 4.24 seconds.

Increasing the momentum of a football player is not an easy answer. It is not as simple as just getting a smaller man to run faster than a bigger man. A slower player would do well to increase his speed, but a player who is already near his top speed will not be able to increase it very much. A player who is smaller should potentially add more mass, but adding more mass may make him slower. There needs to be a delicate balance between the player's mass and running speed.

9.1.3 Conservation of Energy

To begin understanding the role of energy conservation in impacts, think of two objects (a Ping-Pong ball and a bean bag) dropped from a height onto the ground (Figure 9.6). The Ping-Pong ball bounced back up into the air, but never quite reached the same height at which it was dropped. The bean bag, on the other hand, hit the ground with a thud. It did not bounce at all. Why?

The Ping-Pong ball was an example of an elastic collision. As it impacted the ground, it deformed, increasing the strain potential energy in the ball. The ball regained its shape, and the strain potential energy was converted back into kinetic energy, which in turn was converted into potential energy as the ball increased its height. Remember that no process is 100% efficient. Some energy will be lost as heat. So the ball will never return to its original height.

The bean bag was a different story. The bean bag deformed, but never returned back to its original shape after the impact. All the energy was absorbed by the bean bag on impact. The bean bag was an example of an inelastic collision.

Energy concepts can help refine the definitions of elastic and inelastic collisions. In a perfectly elastic collision, all the kinetic energy is preserved. The kinetic energy after impact is equal to the kinetic energy before impact. In a perfectly inelastic collision, kinetic energy is not conserved; the kinetic energy after impact is not the same as the kinetic energy before impact.

Important Point! In a perfectly elastic collision, kinetic energy is conserved. In a perfectly inelastic collision, kinetic energy is not conserved. In both types of collisions, momentum and the total energy of the system are conserved.

When kinetic energy is conserved during perfectly elastic collisions, you have another set of useful equations:

$$\Delta E_{LK} = \left(\frac{1}{2}mv_a'^2 + \frac{1}{2}mv_b'^2\right) - \left(\frac{1}{2}mv_a^2 + \frac{1}{2}mv_b^2\right) = 0 \quad (9.10)$$

If you substitute Equation 9.10 into Equation 9.8 and do some tricky algebraic manipulations,[2] you will get the following result:

$$v_a - v_b = v_b' - v_a' \quad (9.11)$$

This equation has an interesting interpretation. It says that during head-on, perfectly elastic collisions, the relative speed of the two bodies is the same after the collision as it was before. And notice that the mass terms do not appear in Equation 9.11. This means that the relative speed is independent of the mass. You already saw this for two objects colliding with the same mass: Momentum was transferred between them, and the slower object had the velocity of the faster object (and vice versa) after the impact. If a heavier, faster object collided with a smaller object, the smaller object would speed up, and the faster one would slow down. What would happen if a faster, lighter object collided with a heavier one? The heavier one would speed up, and the lighter one would bounce off it, moving back in the opposite direction so that the relative velocities between them remained constant.

Remember: In any case, the total energy is conserved. Energy cannot be created or destroyed, only moved from one place to another or transformed from one type to another. The kinetic energy "lost" during an impact is converted to nonkinetic forms (such as heat, sound, and the breaking of materials). These nonkinetic forms of energy may even cause injuries. The amount of energy that is converted to nonkinetic forms is explored a little later in this lesson.

9.1.4 The Coefficient of Restitution

The discussion thus far has been about perfectly elastic and perfectly inelastic collisions. And after this section, you will return to looking at these cases. Here, you will briefly examine a case where these ideal conditions are not met. It follows naturally once you get a grasp on the conservation of momentum and the conservation of kinetic energy.

Remember, in all cases momentum must be conserved, and the total energy must be conserved. But the total kinetic energy does not have to be conserved. In a perfectly elastic collision, 100% of the kinetic energy is conserved, whereas in a perfectly inelastic collision 0% of the kinetic energy is conserved. Most collisions lie somewhere along a continuum of these two extremes. The value assigned to the "elasticity" of a collision is called the **coefficient of restitution**, and it is represented by the letter e.

> **Coefficient of restitution**
> The measure of elasticity of a collision between two objects

To determine the coefficient of restitution, take the ratio of the relative velocities before and after impact:

$$e = \frac{v_b' - v_a'}{v_a - v_b} \qquad (9.12)$$

In a perfectly elastic collision, the value of e would be one. This follows from Equation 9.11, where the relative velocities before and after impact would be equal. But perfectly elastic collisions never happen; the ball never returns to its original height from which it was dropped. In the case of perfectly inelastic collisions (e.g., the bean bag), e would be zero. So you see that e will always have a value between zero (perfectly inelastic) and one (perfectly elastic).

Lots of factors determine the value of e. Dropping a tennis ball on a hard court would give a different bounce than dropping it on a clay court or dropping it on grass. So it may be best to think of the coefficient of restitution like the coefficient of friction—it is unique to the two materials in question. Nonetheless, in an attempt to standardize equipment, many governing bodies have put limits on the coefficient of restitution for certain sporting implements. You can imagine what an unfair advantage someone may have if they manipulate the coefficients of restitution for either the ball or implement in such sports as baseball, softball, golf, and tennis.

9.1.5 Combining Information from the Two Conservation Laws

Returning to the two football players example in Figure 9.5, say that player A has a mass of 100 kilograms and a velocity of 7 meters per second. Player B has a mass of 90 kilograms and a velocity of -7.5 meters per second. After impact, the players "stick" together, so there is a perfectly inelastic collision. What is their resulting velocity? How much energy was converted to nonkinetic forms?

The results are listed in Table 9.2. Player A has a greater initial momentum than player B. You should automatically conclude that player A will be able to continue to advance, albeit much more slowly, because he has more momentum. To determine the final velocity, you would use Equation 9.9:

$$v_{ab}' = \frac{m_a v_a + m_b v_b}{m_a + m_b} = \frac{700 - 675}{100 + 90} = \frac{25}{190} = 0.13 \text{ m/sec}$$

If this were a goal-line stand, and no other players were involved, he would probably score.

Constructing tables like Table 9.2 is a good way to keep track of the mass and the momentum of the system. In both cases they are additive. And remember that the momentum of the system is the same after impact as it is before impact. Do *not* make the mistake of thinking that velocity can also be added. Notice how it is blank for the system in Table 9.2. It

Table 9.2	Mass, Velocity, Momentum, and Linear Kinetic Energy of the Two-Football-Player System		
	Player A	**Player B**	**System**
m (kg)	100	90	190
v (m/sec)	7	-7.5	—
L (kg·m/sec)	700	-675	25
E_{LK} (kJ)	2.45	2.53	4.98

does not work that way. Even if the two football players had the same mass, the final velocity after impact is not simply the net velocity of the two players. The mass values in the numerator and denominator are never equal in Equation 9.9.

Once you have determined the final velocity of the system, you can then use this information to determine the change in kinetic energy of the system:

Before impact:

$$E_{LKa} + E_{LKb} = \frac{1}{2}mv_a^2 + \frac{1}{2}mv_b^2 = \frac{1}{2}(100)(7)^2$$
$$+ \frac{1}{2}(90)(-7.5)^2 = 4981.25 \text{ J}$$

After impact:

$$E_{LKab} = \frac{1}{2}mv_{ab}^2 = \frac{1}{2}(190)(0.13)^2 = 1.6 \text{ J}$$

Total energy converted to other forms:

$$\Delta E_{LK} + \Delta E_{\text{other}} = 0$$
$$\Delta E_{\text{other}} = -\Delta E_{LK} = -(1.6 - 4981.25) = 4979.65 \text{ J}$$

As you can see, almost all the kinetic energy from each player was converted to nonkinetic forms. You can begin to appreciate why some football players say that playing the game is like being in a car wreck every week.

Section Question Answer

It may be surprising to you that, all other things being equal, the ball hit off an 80 miles per hour fastball would go further than the one hit off a 60 miles per hour fastball. The answer is determined using Equation 9.11. You can ignore the coefficient of restitution if you assume that it will be the same in both cases. Rewrite Equation 9.11 as $v_{\text{bat}} - v_{\text{ball}} + v_{\text{bat}}' = v_{\text{ball}}'$, and realize that the ball is originally traveling in the negative direction. A larger negative value of v_{ball} will make for a larger positive value of v_{ball}', which is exactly what you want. Of course, you still have to hit the ball, and one could argue that it is harder to hit an 80 miles per hour ball than a 60 miles per hour one. But all other things being equal, the ball hit off an 80 miles per hour pitch will go further than the ball hit off a 60 miles per hour pitch.

COMPETENCY CHECK

Remember:

1. Define inelastic collision, elastic collision, and coefficient of restitution.
2. State the conservation of momentum and conservation of energy laws.

Understand:

1. Explain the difference between an elastic and inelastic collision.
2. Explain why inelastic collisions are potentially more damaging than elastic collisions.
3. Determine the resultant velocity for both a perfectly elastic and perfectly inelastic collision between bodies *A* and *B*:

Body A		Body B	
Mass	Velocity	Mass	Velocity
100	10	90	−10
80	5	80	−8
10	−8	20	10
50	6	80	−5
90	−5	80	5

Apply:

1. Check with governing bodies of various sports such as baseball, softball, tennis, table tennis, and golf. Do they have regulations for the coefficients of restitution for the equipment used in that sport?

9.2   MORE COMPLICATED COLLISIONS OF POINT-MASSES

Section Question

You have been asked to provide assistance for a golfer who wants to increase the distance of her drives (Figure 9.7). How can you use your knowledge of collisions to help?

The previous section highlighted the important concepts related to collisions. In this section, you will extend these concepts to impacts that occur in two dimensions. The ideas are the same; you just conserve momentum in two dimensions instead of one.

9.2.1   Conservation of Momentum

Remember, when you are conducting your analyses, you first need to define your system and your frame of reference. Earlier, you saw how it was convenient to establish the bodies involved in a collision as being part of the same system. Remember to choose a reference frame that either makes the most physical sense or makes the math easier. Regardless of

Figure 9.7	How can you use your knowledge of collisions to improve the distance of a golf drive?

© iofoto/ShutterStock, Inc.

Figure 9.8	Two American football players collide at an angle while one is running along the sideline. What is the outcome of the play?

© Nicholas Moore/ShutterStock, Inc.

the chosen reference frame, momentum must be conserved in each dimension:

$$\Delta^x L = (m_a{}^x v_a' + m_b{}^x v_b') - (m_a{}^x v_a + m_b{}^x v_b) = 0 \quad (9.13)$$

$$\Delta^y L = (m_a{}^y v_a' + m_b{}^y v_b') - (m_a{}^y v_a + m_b{}^y v_b) = 0 \quad (9.14)$$

Important Point!   During collisions, momentum is conserved in every plane.

To make life simpler, it is always best to establish the positive x direction as the initial direction of travel of one of the bodies. For example, return to the two American football players described earlier. Instead of making a goal-line stand, player A is running down the sideline. Instead of hitting him head-on, player B is going to hit him from an angle. This is shown in Figure 9.8. And instead of sticking to him, we will assume a perfectly elastic collision where player B is just trying to run into player A and knock him out-of-bounds. Keep the masses and the initial velocities the same. How does changing the angle of impact change the result?

This example is complicated by the fact that the players do not have the same mass. If they did, then you know when player B hit player A, player B's momentum is transferred to player A, and player B's momentum goes to zero. Because player B has less mass than player A, player B will actually bounce back a little bit after the impact. Ignore the mathematics for now, and see if you can intuitively grasp what would happen to player A.

You already saw that if player B hit player A head-on from the front (impact angle of 180°), then he would slow down player A—but player A would keep moving forward. I am sure that you could also guess that if player B hit player A from behind (impact angle of 0°), player A would speed up in the forward direction. This might cause him to lose his balance and fall, but it is probably not a strategy that should be counted on. In both cases, player B is either adding or subtracting momentum to player A in his direction of travel (along the x-axis).

Player A currently has no momentum in the y direction. If player B hits him precisely at 90°, he will transfer a large percentage of his momentum in that direction to player A. Momentum is a vector, and you know how to add vectors:

Figure 9.9	An impact angle from 1° to 89° is in quadrant 1, whereas an impact angle for 91° to 179° is in quadrant 2.

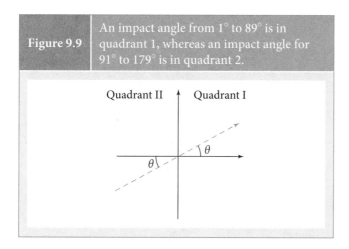

Figure 9.10	(a) Two automobiles collide at an angle to one another, and (b) the resulting vector of the SUV/car system after impact.

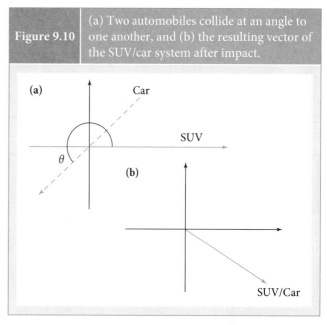

break them apart into their components, add the components together, and then determine a resultant after all the necessary additions. After being hit precisely at 90°, player A's new momentum (and velocity) vector will be right around 45° from the x-axis. If he is running along the sideline, there is probably a very good chance that he would run out-of-bounds.

If the impact angle is anywhere from 1° to 89° (which we will call quadrant 1; Figure 9.9), the momentum of player A will increase in both the positive x and y directions. The resultant momentum of player A will be less than 45° from the x-axis. Conversely, if the impact angle is anywhere from 91° to 179° (which we will call quadrant 2), player A's momentum will increase in the positive y and decrease in the x direction. However, the resultant angle will still be less than 45°.

9.2.2   Conservation of Energy

Recall that kinetic energy is a scalar, so the same ideas from the previous section will apply here. There is not kinetic energy in different planes! And remember, during perfectly elastic collisions, kinetic energy is conserved. During perfectly inelastic collisions, kinetic energy is not conserved. In all cases, the total energy and momentum are conserved.

> **Important Point!**   Energy is a scalar. Although there is an equation for momentum in each plane, there is only one energy equation.

9.2.3   Combining Information from the Two Conservation Laws

In the previous sections, you examined the collisions that occur between two players in American football. In this section, you will leave the area of sports and examine violent impacts that are an all-too-familiar occurrence: automobile accidents. Consider a car that has a mass of 1500 kilograms and

an SUV that has a mass of 2600 kilograms. The SUV is driving on an icy road at 13 meters per second and the car ploughs into it driving at 20 meters per second. For this example, we will say the SUV is traveling along the positive x-axis, and the car is traveling 200° from the positive x-axis (Figure 9.10a).

First, can you guess in which direction the combined SUV/car will travel after the impact? Examining Table 9.3, you should note that the SUV had a larger initial momentum in the positive x direction than the car had in the negative x direction. This tells you that the combined SUV/car will continue to travel in the positive x direction. In the y direction, the SUV had an initial momentum of zero, but the car had an initial momentum in the negative y direction. This should tell you that the combined SUV/car will also travel in the negative y direction. The combined direction of travel should be between 270° and 359° from the positive x-axis. If you did the math, you would find that the SUV/car had a final velocity of 2.85 meters per second at about 299° from the positive x-axis (Figure 9.10b).

Table 9.3	Mass, Velocity, Momentum, and Linear Kinetic Energy of the Two-Automobile System		
	SUV	**Car**	**System**
m (kg)	2600	1500	4100
v (m/sec)	13	20	
Theta (°)	0	200	
^xL (kg·m/sec)	33800	−28190.8	5609.2
^yL (kg·m/sec)	0	−10260.6	−10260.6
E_{LK} (kJ)	219.7	300	519.7

Now think about the change in kinetic energy. The SUV originally had an initial kinetic energy of 219.7 kilojoules. The car had an initial kinetic energy of 300 kilojoules. Combined, their initial kinetic energy was 519.7 kilojoules. After the crash, their combined kinetic energy was only 16.7 kilojoules. Over 500 kilojoules (97% of the initial kinetic energy) was converted to nonkinetic forms. There should be little wonder why auto crashes are so damaging to both the vehicles and their drivers.

Section Question Answer

Driving a golf ball can be thought of as a collision. Assuming no spin to the ball, the hierarchical model is presented in **Figure 9.11**. To begin, remember that the range of a projectile (in this case, the golf ball) is due to its velocity of takeoff, angle of takeoff, and the acceleration due to gravity. The acceleration due to gravity is a factor that you cannot control. The angle of takeoff and velocity of takeoff of the ball are determined by the mass of the ball, the linear momentum of the clubhead before and after impact, and the coefficient of restitution (assuming that the momentum of the ball before impact is zero). If the ball takes off and lands at the same elevation, the angle to takeoff should be 45°. To achieve this angle of takeoff, the velocity of the ball in x and y directions after impact must be the same. This means that the clubhead must strike the ball at such an angle to impart equal momenta to the ball in both directions.

COMPETENCY CHECK

Remember:

1. How many equations are there for an impact that occurs in two dimensions?

Understand:

1. Determine the resultant velocity for both a perfectly elastic and perfectly inelastic collision between bodies A and B:

Body A		Body B		Impact
Mass	Velocity	Mass	Velocity	Angle (°)
100	10	90	−10	45
80	5	80	−8	90
10	−8	20	10	135
50	6	80	−5	225
90	−5	80	5	270

Apply:

1. Create a hierarchical model for the collision between an implement and a ball in a sport.

9.3 EFFECTIVE MASS

Section Question

Manny and Floyd (**Figure 9.12**) are two boxers that have the same body mass and can deliver punches with the same velocity. Yet Manny "hits harder" than Floyd. Why is that?

The discussion up to this point has been limited to a system that consisted of two bodies that were represented as point-masses. A point-mass has mass, but no dimensions. By extension, anytime a point-mass is involved in a collision all

| **Figure 9.11** | A hierarchical model for the range of a golf ball. |

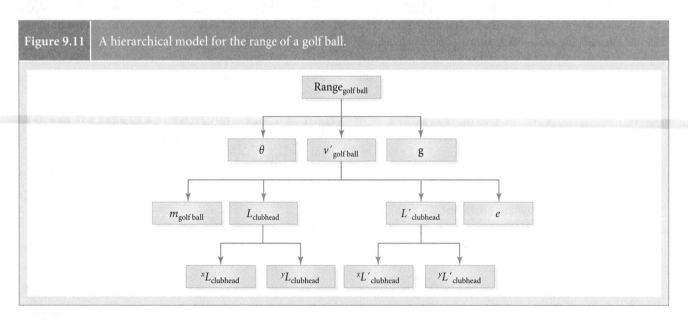

Figure 9.12	Why is it that two boxers can have the same body mass and deliver punches with the same velocity, yet one boxer hits harder than another?

© Diaframma/ShutterStock, Inc.

Figure 9.13	Three identical bars impacting the ground. With the first two bars, the entire mass is involved with the impact. With the third bar, only a portion of the mass is involved in the impact.

its mass is as well. But real objects are not point-masses. And not all of a body's mass is necessarily involved in a collision.

Think of three bars with identical masses and dimensions making contact with the ground (Figure 9.13).[3] The first bar makes impact with the ground in an upright position. The second bar makes impact with the ground in a flat position. The third bar makes impact with the ground at an angle. When the first two bars make contact with the ground, all their mass is involved with the collision. When the third bar makes contact with the ground, only a portion of its mass is involved with the impact; the rest is still in motion as it falls toward the ground. That portion of the mass that is involved with the collision is called the effective mass. It is the mass of an imaginary rigid body that replaces the body, and it has the same speed as the body part involved in the collision.[4]

> **Effective mass** The portion of a body's mass that is involved with a collision

This same idea is employed when the highway patrol uses the PIT maneuver to stop a car in a high-speed chase (Figure 9.14). As the police approach the speeding car, they bump into the back end of the car rather than hitting it head-on. This causes the car to spin out and eventually come to a stop. The PIT maneuver is a lot less damaging than hitting the car head-on because only a portion of the vehicle's masses is involved in the collision.

Now that you have the idea of effective mass, you can extend it to human movement. What happens when your foot makes contact with the ground? In essence, that is another collision. Thus far, when you have studied impacts with the ground (such as during running or landing from a jump), you have looked at it from the whole-body level, and you have examined what happened with the body's COM. Alternately, you could

Figure 9.14	The PIT maneuver is not as damaging as the head-on collision because the entire masses of the vehicles are not involved in the collision.

look at it from the foot's perspective and what happens during the foot–ground collision.[5] In such cases, you would be interested in the velocity of the foot and the effective mass of foot.

Just like the rod, the effective mass of the foot is not just the mass of the foot, and it is not necessarily the mass of the entire person. That is because the COM of each segment may be moving relative to the portion of the body that is involved in the collision—in this case, the foot. If you were to land "stiff," there is a harder impact with the ground because more of your mass stopped abruptly with the impact. The effective mass was higher. If you were to land "soft," the impact forces would not be as high because more of your mass is still moving after the impact, just like the rod that landed at angle. The effective mass was lower.

Equation 9.3, and all the equations that followed it, are fine for dealing with point-masses. When dealing with multisegmented bodies, they should really be more explicit with the mass and velocity terms. When dealing at the whole-body level, the momentum equation would be

$$L = mv_{COM} \qquad (9.15)$$

where it is understood that m is the mass of the entire body and v is the velocity of the center of mass. When you are dealing with a multisegmented body and you want to know the momentum of the segment involved in a collision, then Equation 9.3 becomes

$$L = m_{effective}v_{segment} \qquad (9.16)$$

Subsequent equations (Equations 9.4–9.14) should make similar distinctions when dealing with bodies that are anything other than point-masses.

In many activities, you may be striking an object: baseball, softball, tennis, golf, boxing, or martial arts. When striking an object, you actually want to create a large impact. Performance is generally improved if you transfer a large momentum to the object you are striking. So how would you transfer a larger momentum to the object? The beginning of the hierarchical model is presented in **Figure 9.15**. Momentum transfer depends on the effective mass and the velocity of the segment or implement that is actually striking the object. The velocity component needs no further elaboration. The effective mass can be increased in one of two ways.

First, the mass of the segment or implement can be increased. In some cases, this is not possible. A martial artist cannot appreciably increase the mass of his fist, although wearing a boxing glove could have this effect. In other cases, it comes with trade-offs. Remember that momentum is the product of mass and velocity, and the two affect each other. In general, as the mass of the object increases, it is going to be harder to increase its velocity. So these two competing factors

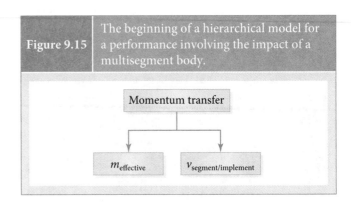

| **Figure 9.15** | The beginning of a hierarchical model for a performance involving the impact of a multisegment body. |

must be balanced. The optimal mass/velocity combination will be different for every activity and for every individual participating in that activity.

Second, the effective mass can be increased by linking the mass of the segment/implement to the masses of other segments in the body. This is done by a proper positioning of the segments and the appropriate muscle activation to stiffen the extremity. This is similar to stiffening the legs during a "hard" landing.

Effective mass is an important concept in any physical activity that involves impacts, either with a part of your body or with an implement. It can help explain why helmet-to-helmet contacts in football are so dangerous, particularly to the defender (see **Box 9.1**). It also explains the answer to

| **Box 9.1** | Applied Research: Effective Mass and Head Injuries in American Football |

Head injuries in American football are a serious problem, particularly those resulting from helmet-to-helmet contact. In many cases, the injury to the offensive player receiving the impact is greater than the injury sustained by the striking, defensive player. In this investigation, the researchers provide an explanation for why this is the case. Reconstructing actual, recorded game-time head injuries using instrumented dummies in the laboratory, they found that the striking player aligned their head, neck, and torso (called spearing), increasing the effective mass of the striking player to 1.67 times that of the player being hit. In a follow-up investigation, they compared these impacts to punches to the head delivered by Olympic-caliber boxers. They found these impacts did not transfer as much linear momentum as the football head strikes due to the lower effective mass of the fist.

Data from: Viano DC, Pellman EJ. Concussion in professional football: biomechanics of the striking player—Part 8. *Neurosurgery.* Feb 2005;56(2):266–278.

Viano DC, Casson IR, Pellman EJ, Bir CA, Zhang LY, Boitano MA. Concussion in professional football: comparison with boxing head impacts—Part 10. *Neurosurgery.* Dec 2005;57(6):1154–1170.

the question at the beginning of this section: Why is it that one boxer can hit harder than another, even if they have the same mass and can deliver a punch with the same velocity?

Section Question Answer

Just because two boxers have the same mass and punch with the same velocity, it does not mean that they can deliver the same momentum to their intended target. Research[4,6] has shown that highly skilled fighters do indeed impart more momentum to the objects they are striking than lesser-skilled fighters with the same mass and punching velocity. The investigators determined that this must be due to the more highly skilled practitioners developing a greater effective mass by proper positioning of the arm and effectively stiffening the appropriate joints. Manny can hit harder than Floyd because he has a greater effective mass.

SUMMARY

In this section, you learned about collisions, the resulting impact and transfer of momentum. The key concepts are listed in Table 9.4. Using the conservation of momentum law, you saw how momentum was transferred from one body to another during elastic collisions. You also saw how damaging inelastic collisions can be because a large percentage of the initial kinetic energy of the two bodies is changed into non-kinetic forms. Collisions in two dimensions follow the same principles as in one dimension, but momentum is conserved in both dimensions. Finally, the importance of effective mass in both injury and performance was highlighted. The topic of effective mass plays an important role with multijoint systems.

REVIEW QUESTIONS

Remember:

1. State the conservation of energy and the conservation of momentum.
2. Define the following terms: elastic collision, inelastic collision, coefficient of restitution, and effective mass.

Apply:

1. Explain the differences between an elastic and inelastic collision.
2. Explain why collisions can be so potentially damaging.

Understand:

1. Create a hierarchical model involving collisions. Explain how each variable can be increased or decreased to improve performance to decrease injury potential.

REFERENCES

1. Hay JG, Reid JG. *Anatomy, Mechanics, and Human Motion.* 2nd ed. Englewood Cliffs, NJ: Prentice Hall; 1988.

2. Giancoli DC. *Physics.* 4th ed. Englewood Cliffs, NJ: Prentice Hall; 1995.

3. Lieberman DE. Biomechanics of foot strikes and applications to running barefoot or in minimal footwear. 2010; http://barefootrunning.fas.harvard.edu/4BiomechanicsofFootStrike.html. Accessed February 20, 2012.

4. Neto OP, Magini M, Saba MMF. The role of effective mass and hand speed in the performance of kung fu athletes compared with nonpractitioners. *Journal of Applied Biomechanics.* May 2007;23(2):139–148.

5. Lieberman DE, Venkadesan M, Werbel WA, et al. Foot strike patterns and collision forces in habitually barefoot versus shod runners. *Nature.* Jan 28 2010;463(7280):531–U149.

6. Smith PK, Hamill J. The effect of punching glove type and skill level on momentum-transfer. *Journal of Human Movement Studies.* 1986;12(3):153–161.

Table 9.4	Key Concepts
• Conservation of momentum
• Conservation of energy
• Elastic collisions
• Inelastic collisions
• Effective mass

PART II

Tissue Level

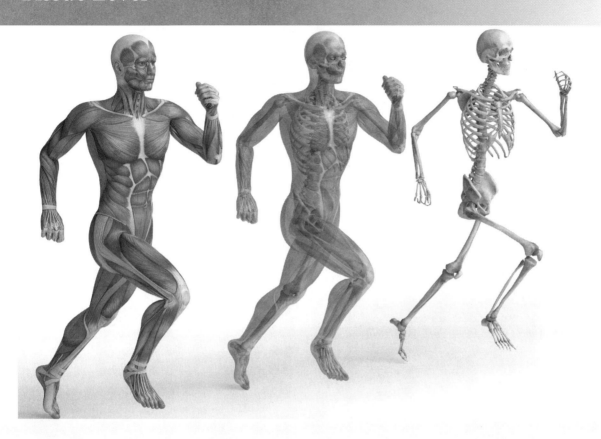

Lesson 10

Mechanics of the Human Frame

© Maria Teijeiro/OjO Images/Getty Images

LEARNING OBJECTIVES

After finishing this lesson, you should be able to:

- Define the following terms: load, compression, tension, shear, bending, neutral axis, area moment of inertia, torsion, polar moment of inertia, strength, deformation, ductile, brittle, elastic deformation, plastic deformation, yield point, stiffness, compliance, strain potential energy, toughness, stress, strain, elastic modulus, strain energy density, stress-relaxation, creep, hysteresis, strain–rate dependency, boundary lubrication, fluid film lubrication, squeeze film lubrication, hydrodynamic lubrication, elastohydrodynamic lubrication, wear, interfacial wear, fatigue wear, adhesion wear, abrasion wear, material failure, failure tolerance, margin of safety, acute injury, chronic injury, bone mineral content, bone mineral density, anisotropic, fracture, osteopenia, osteoporosis, osteoarthritis, and sprain.
- List the factors that determine the effect of a load on a body.
- Describe the different types of loading, and give an example of each.
- Write the equations for stiffness, compliance, stress, strain, and Young's modulus.
- Sketch a load-deformation curve, and label the following: toe region, elastic region, yield point, plastic region, ultimate strength, stiffness, and strain energy.
- Sketch a stress-strain curve, and label the following: toe region, elastic region, yield point, plastic region, ultimate strength, elastic modulus, and strain energy density.
- Explain the time-dependent properties of viscoelastic materials, and describe how they can be applied.
- Explain how each of the three fluid film lubrication modes reduce friction.
- Explain how material failure occurs.
- Use a hierarchical model to give concrete examples of how to reduce the risk of injury.

- List the common injuries that occur to bones, tendons, and ligaments.
- Describe the effects of disuse, aging, and exercise on the mechanical properties of bones, tendons, and ligaments.

INTRODUCTION

This lesson is going to be a little bit different than others. In other lessons, you were introduced to the fundamental concepts used in mechanics (statics and dynamics) to study the behavior of *rigid* bodies *in* the world. Notice the emphasis on the word *rigid*, and the behavior *in* the world. This lesson is going to be a little different in that you will be throwing the rigid body assumption out the window. Here, you will be studying the *internal* responses of a body to forces that are imposed on it.

> **Important Point!** In most branches of mechanics, you are interested in the external effects that forces and moments produce on a body. With the mechanics of materials, you are interested in the internal effects of forces and moments.

To study these effects, you will have to become familiar with another branch of mechanics: the mechanics of materials. Engineers use the concepts found in the mechanics of materials when designing structures like bridges or deciding which walls can be knocked down as part of remodeling project. Essentially, they are trying to determine if the frame can handle the loads on it.

Biomechanicians use the concepts of mechanics of materials for the same reason, but for them the frame is the skeletal system. Similar to a bridge or a house, you want to know if the frame can handle the loads because if it does not, an injury will occur. Unlike a bridge or a house, the skeletal frame has to be able to move. For this reason, you need to explore not only the structures that give the body its shape (bones), but also the structures that allow the body to move (the cartilage and ligaments of the joint).

In this lesson, you will learn some of the basics of the branch of mechanics known as the mechanics of materials. You will then examine some unique properties of a class of materials known as viscoelastic because almost all tissues of the body are viscoelastic to some degree. By the way, because the material that makes up the human frame (bone, cartilage, and ligaments) is connective tissue, biomechanicians will sometimes call this area tissue mechanics instead of material mechanics. After that, you will spend some time looking at what causes materials to wear down and break. Because the aforementioned topics are a different branch of mechanics, they come with their own, specialized vocabulary. So a lot of new concepts will be presented in this lesson. After you have a handle on them, you will see how these concepts apply to injuries in a general way before turning your attention to the specific tissues that make up the human frame: bone, ligaments, and cartilage. So let us get started with the basic ideas.

> **Important Point!** Tissue mechanics is the mechanics of materials of human connective tissue (bones, ligaments, cartilage, and tendons).

10.1   BASIC MECHANICS OF MATERIALS

> **Section Question**
>
> Why can someone (or eggs, **Figure 10.1**) lay on bed of nails and come away unharmed?

In the mechanics of materials, an externally applied force is often referred to as a **load**. You will generally see that distinction made in this text.

> **Load**   An externally applied force

When discussing movement mechanics, I will use the term *force*, and when discussing material mechanics, I will use the term *load*. How an object responds to a load is determined by seven factors:[1]

- Magnitude
- Location
- Direction

Figure 10.1

© Alta Oosthuizen/ShutterStock, Inc.

- Duration
- Frequency
- Variability
- Rate

These factors will be discussed in detail throughout the lesson.

10.1.1   Types of Loading

There are several different types of loading to which a body can be subjected (**Figure 10.2**).[2] Most of these distinctions are made based on the location and direction of the external forces applied to a body. Axial loads act along a line, which is the longitudinal axis of a body. With a compressive load (or **compression**; **Figure 10.3a**), two forces are directed toward each other, squeezing the body together. With a tensile load (or **tension**; **Figure 10.3b**), two forces are oppositely directed, pulling a body apart. You

> **Compression**   A load that squeezes the parts of a body together
>
> **Tension**   A load that pulls the parts of a body apart

can think of them as being linear types of loading. The response that a body has to these types of loads is proportional to its cross-sectional area (see **Box 10.1**).

Figure 10.2	The types of loads.

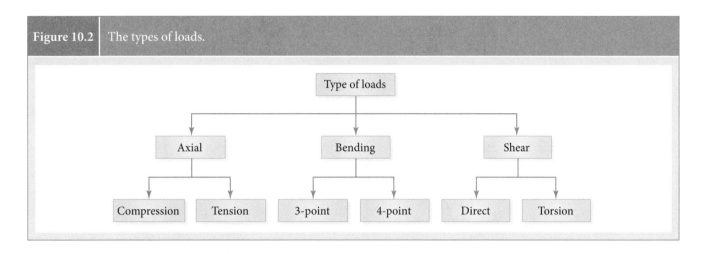

Figure 10.3	Uniaxial loading. (a) Compression; (b) tension.

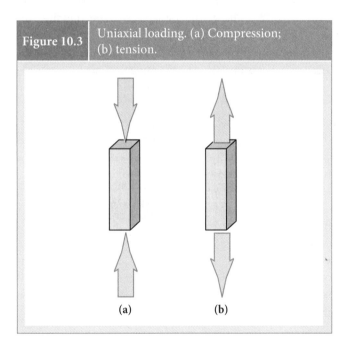

(a)   (b)

Bending	A load applied perpendicular to the longitudinal axis of a body, causing it to curve

The second type of load is bending (Figure 10.4). A bending load, as the name implies, occurs when a load is applied perpendicular to the longitudinal axis of a body, causing the body to curve about some axis. There are two types of bending with which you should be familiar. Three-point bending has three forces applied to the body. These forces create two bending moments about the point of force application of the middle force (Figure 10.5a). Four-point bending has four forces applied to a body: two "outer" forces and two "inner" forces (Figure 10.5b). Of course, you could also have bending in more than one direction (think about a fishing pole with a fish fighting

Box 10.1	Essential Math: Volume, Mass, and Cross-Sectional Area

The volume of an object is how much space it occupies. The mass is how much matter is located within that space. The volume of an object such as rectangular cuboid is

$$\text{Volume} = \text{length} \times \text{width} \times \text{height}$$

For cylinders, width and height are replaced with π and the radius squared:

$$\text{Volume} = \text{length} \times \pi \times r^2$$

The cross-sectional area depends on which cross section you are looking at. The rectangular cuboid has three different cross-sectional areas:

Cross Section 1: length × width
Cross Section 2: length × height
Cross Section 3: width × height

So it is important to identify the cross sections to which the loads are being applied.

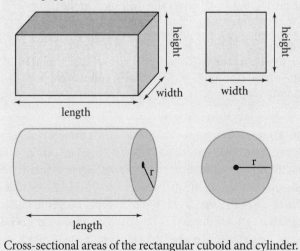

Cross-sectional areas of the rectangular cuboid and cylinder.

Figure 10.4	A bending load.

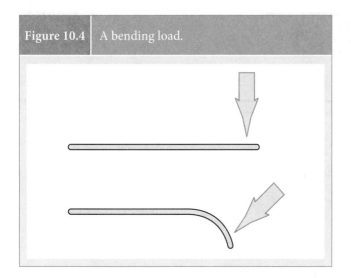

Figure 10.5	Types of bending. (a) 3-point bending; (b) 4-point bending.

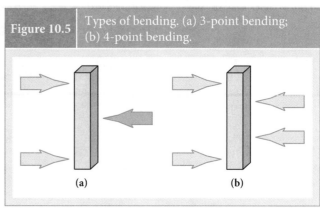

(a)   (b)

at the end of the line), but those types of bending add more complexity to the problem and will be left for more advanced treatments on the subject.

The effect that bending has on a body is determined by three factors. The first two are the cross-sectional area and distribution of the material around the neutral axis. This is similar to mass moment or inertia. In fact, it is called the **area moment of inertia**, and it is a measure of the body's resistance to bending (similar to the way the mass moment of inertia is a resistance to change in angular momentum). A larger area moment of inertia means that a body has a greater resistance to bending. You know that the turning effect of a force (torque) depends on both the mass moment of inertia and the distance that the force is applied from the axis of rotation. Similarly, the bending effect of a force depends on the area moment of inertia and the length of the body.

> **Area moment of inertia**   A measure of a body's resistance to bending

> **Important Point!**   A body's response to bending is determined by the cross-sectional area, the distribution of the material around the neutral axis, and the length of the body.

To see an example of this, examine the three scenarios in **Figure 10.6**. In Figure 10.6a, a 2′ × 4′ × 6′ is laid down so that the board is 2′ high, 4′ wide, and 6′ long. In Figure 10.6b, the same board is reoriented so that it is 4′ wide, 2′ wide, and 6′ long. In both cases, a 100 N weight is placed in the middle of the board. Even though the board did not change, the bending in Figure 10.6a is greater than the bending in Figure 10.6b. This is due to the area moment of inertia. In Figure 10.6c, the board is identical to Figure 10.6a except doubled in length (from 6′ to 12′). The weight is still placed in the middle of the board (in this case, at the 6′ mark). The board in 10.6c will bend more than the board in 10.6a, even though they have the same material properties and the same thickness.

The effects of a bending load are not the same throughout the entire body. When an object is bent, you will notice a concavity on one side and a convexity on the other. Interestingly, there is also a region that experiences no loading. This is called the **neutral axis**, and in a symmetrical body it is located in the geometric centroid of the body. The material on the convex side experiences tensile loading, whereas the material on the concave side experiences compressive loading. Increasing the distance from the neutral axis increases the amount of compressive and tensile loading on the material. See **Figure 10.7**.

> **Neutral axis**   The line along which there is neither compressive nor tensile loading on a body during loading

> **Important Point!**   During bending, one side of the body experiences tensile loading, and the other side experiences compressive loading.

A **shear** load (**Figure 10.8**) is similar to a compressive load in that the two forces are directed toward each other, but the forces are not along the same line. Rather, they are directed parallel to the surface of a structure. (The term *shear* is also used in conjunction with loading on a joint. In that case, shear is causing the two surfaces to move in opposite, parallel directions). This causes one part of the body to move parallel past another part. Shear forces are responsible for cutting paper.

> **Shear**   A load that causes one part of a body to move parallel past another part

Figure 10.6   The effects of cross-sectional area and length on the bending moment of a board. Figures (a) and (b) have the same dimensions, yet (a) bends more than (b) because of the different cross-sections. Figures (a) and (c) have the same cross-sections, but (c) is longer than (a) so it bends more.

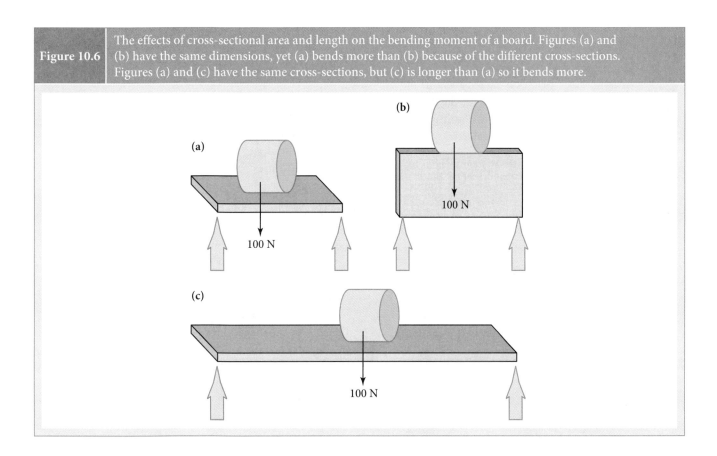

Figure 10.7   During bending, one side experiences tensile loading, and the other side experiences compressive loading. The magnitude of the load increases with the distance from the neutral axis.

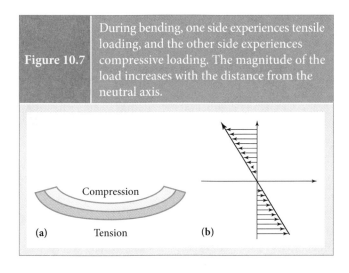

Figure 10.8   Shear loading.

Figure 10.9   Torsional loading.

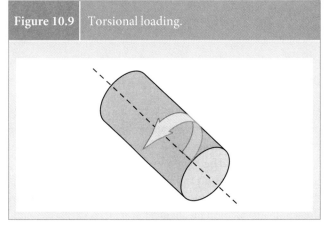

Torsion   The type of loading that exists when there is a twist around the neutral axis

The last type of loading is **torsion** (**Figure 10.9**). You can also think of it as a rotational type of loading, like bending. While bending curves around an axis, torsion is a twisting around an axis. Similar to bending, the effect of a torsional load depends on the distribution of material around the neutral axis. For twisting, it is called a **polar moment of inertia**. During torsion, the material

Polar moment of inertia   The resistance to torsional loading

experiences shear. There is no shearing of material located at the neutral axis; shear loading increases with the distance away from the neutral axis.

> **Important Point!**   Shear forces increase as the distance from the neutral axis increases.

The types of loading presented in Figure 10.2 are rather simple. You may have gotten the impression that these types of loading occur in isolation. Do not make that mistake! Loading can occur in more than one direction, be around more than one axis, and involve more than one type. The loading presented here was just to give you an idea of the various ways an object could be loaded. The combinations are limitless.

COMPETENCY CHECK

Remember:

1. Define the following terms: load, compression, tension, shear, bending, neutral axis, area moment of inertia, torsion, and polar moment of inertia.
2. List the factors that determine how an object responds to a load.

Understand:

1. Describe the differences between the five types of loading.
2. Describe the difference between three-point and four-point bending.

Apply:

1. Give examples of the different types of loading.

10.1.2   Mechanical and Material Properties

How a body responds to one of the loads described earlier depends on both its mechanical and material properties. The mechanical properties tell you how the body as a whole responds to a load. The material properties tell you how the material that makes up the body responds to loading. This distinction will make more sense after you have a handle on the whole idea behind mechanical properties.

Mechanical Properties

Mechanical properties give you an indication of how a body will respond to a load. Typical things you would want to know about include a body's:

- Strength
- Deformation
- Stiffness
- Toughness

Each of these properties, and their relation to each other, will be discussed next.

Load and Strength   As mentioned previously, the load is the external force applied to a body. The ultimate **strength** of the material is the largest load that a body can withstand before failure (something that will be discussed more

Strength   The amount of loading an object can withstand before failure

in the next section). Of course, what you consider to be the ultimate strength of the material depends on how you define failure. Failure can be the point when the material first starts to break apart, or it can be its complete rupture. A material that can withstand a large load is said to be strong, whereas one that can withstand a small load is considered weak.

Deformation   When subjected to a load, a material will deform. A **deformation** is change in dimension. If you are dealing with tensile

Deformation   A change in dimensions of a body

or compressive forces, this deformation is determined as a change in length. Think about stretching a rubber band: As you stretch it, you change its length. A similar thing happens when you squeeze a water bottle—you have changed the shape of the water bottle. Objects that cannot undergo very large deformations are called **brittle** (think of glass), whereas objects that can undergo large deformations are called **ductile** (think of gum or Silly Putty).

Deformations can be elastic or plastic. With an **elastic deformation**, the

Brittle   A characterization of an object that can only undergo very small deformations
Ductile   A characterization of an object that can undergo very large deformations
Elastic deformation   A deformation in which the object returns to its original dimensions after the deformation

body will always return to its original shape. Again, think back to the rubber band, which experiences mostly elastic deformations. The rubber band keeps returning to its

Plastic deformation   A deformation in which the object does not return to its original dimensions after the deformation

original shape. With a **plastic deformation**, the object has been "stretched out of shape" and will not ever return to its original dimensions. Think of the little plastic thingy (the scientific term) that holds a six-pack (of soda) together. Have you ever tried to put a can back in the plastic thingy after you have taken it out? Chances are you were not very successful because the thingy underwent a "plastic" deformation.

When will an object undergo a plastic versus an elastic deformation? If an object is deformed too much, there is ac-

Yield point   The amount of deformation that marks the transition from elastic to plastic deformations, and deformation beyond this point results in a permanent deformation

tually a microtearing of the material. The point where it is deformed too much is called the **yield point**: Any deformations beyond the yield point result in permanent (plastic) deformations.

Stiffness and Compliance   Loads and deformations are related to one another—a load causes a deformation. This

Stiffness   The ratio of the change in load to the change in deformation

relation is characterized by either stiffness or compliance. **Stiffness** is the amount of load needed for a one-unit change in deformation, or the ratio to the change in load to the change in deformation:

$$\text{Stiffness} = \frac{\Delta \text{Load}}{\Delta \text{Deformation}} \qquad (10.1)$$

Compliance   The ratio of the change in deformation to the change in load; it is the opposite of stiffness

Compliance is the opposite of stiffness. It is the amount of deformation for a one-unit change in load, or the ratio of the change in deformation to the change in load:

$$\text{Compliance} = \frac{\Delta \text{Deformation}}{\Delta \text{Load}} \qquad (10.2)$$

A body that is stiffer (or less compliant) will require a greater amount of load for the same amount of deformation. Conversely, a body that is less stiff (or more compliant) will undergo greater amounts of deformation for the same load.

Toughness   Thus far, you have been learning about material mechanics by discussing forces. As you have learned in earlier lessons, there are really two ways to explain motion: force–

impulse and work–energy. Similarly, in material mechanics you can also look at things from an energy perspective. When a body is subjected to a load, it deforms. As it deforms, it absorbs energy. This is another type of potential energy called **strain energy**. It is a potential energy because of the possibility that the energy absorbed by the body can be reused (and you will learn about what happens when it is not a little bit later). The amount of energy that can be absorbed by a body before failure is a measure of **toughness**.

Strain potential energy   The amount of energy absorbed by a body as a result of deformation
Toughness   The amount of energy that can be absorbed by a body before failure

Load-Deformation Curves   Load, deformation, stiffness, and toughness are all related. These relations can be best understood by examining a load-deformation curve, as in **Figure 10.10**, where the load is plotted as a function of the deformation. Although load-deformation curves are unique to each type of material, the load-deformation curve of the biological tissues discussed in the second part of this lesson share some pretty common traits, so it is worth while to spend some time with it.

First, notice at low loads that there is a very nonuniform deformation response. This region is known as the toe region. It is probably due to slack being taken up in the materials that make up the biological tissues.

After the toe region, there is very linear response between load and deformation—up to a point. That point is the yield point mentioned earlier. To the left of the yield point is the elastic region, and to the right of it is the plastic region. Again,

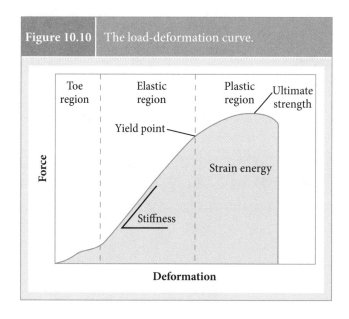

Figure 10.10   The load-deformation curve.

this is the point where tiny microfailures in the material begin, and any deformations beyond that region are permanent. As a greater and greater load is applied, more and more of the material starts to fail, until the body finally breaks apart. At that point, the load drops immediately to zero because the body is no longer whole and subjected to the load.

Typically, within the elastic region you will characterize the stiffness of the body. It is the slope of the force-deformation curve in this region. Because it is a fairly linear response, you can characterize it by one number. In the plastic region, you would need to look at the tangent of the curve at a particular point. Do not confuse strength and stiffness; they are not related. A material can be strong and stiff, weak and stiff, strong and compliant, or weak and compliant. Do not assume stiffer is stronger.

The area under the force-deformation curve is the energy absorbed by the material, its toughness. Note the similarity in motion mechanics: Work, a change in energy, was the area under the force-displacement curve. Here, a change in strain potential energy is the area under the force-deformation curve. Again, do not automatically assume that a stiffer or stronger object can absorb more energy. Because both factors (force and deformation) are equally important, it is possible for a weaker, more compliant object to absorb as much energy as a stronger, stiffer one.

Material Properties

Although the mechanical properties tell you about how a body will respond to a particular force, it does not give you much of an idea as to why the body is behaving that way. For that answer, you may have to look at the body's material properties. There is almost a one-for-one correspondence between mechanical and material properties (**Table 10.1**).

Stress   Have you ever wondered why someone can lay unharmed on a bed of nails? It has nothing to do with any mystical powers. They simply picked the right number of nails. Had they tried to lie on only four nails, they may have experienced a very different outcome. Why?

The answer lies (pun intended) not only in how much force there was, but also in how that force was distributed. This

Table 10.1	Correspondence Between Mechanical and Material Properties
Mechanical property	**Material property**
Force	Stress
Deformation	Strain
Stiffness	Elastic modulus
Strain energy	Strain energy density

quantity is known as **stress** or pressure. Technically, engineers will tell you that there is a distinction and that pressure deals with external forces and stresses deal with internal forces. But they both have the same units (N/m^2) and are essentially dealing with the same thing, so you need not worry about getting too hung up on technicalities.

> **Stress**   The way a force is distributed within a body

Stress is the force per unit area:

$$\sigma = \frac{\text{Force}}{\text{Area}} \qquad (10.3)$$

Here, the area is the cross-sectional area, or contact area.

If a large force is distributed over a small area, then the amount of force each piece of the area is subjected to is less. When lying on a bed of nails, each nail is only taking up a very small amount of the total force. As the number of nails decreases, the amount of force per nail will increase. Once the force is large enough, it will pierce the skin, but until that time you are safe. So you can see why stress may be a better indicator than force when understanding how a body is responding to a load.

Strain   Earlier you learned that a deformation was a change in dimension. For our purposes, stick with the dimension of length. Now, what if I told you that a body stretched (increased in length) by 2 millimeters? Is that a lot or a little? The answer is: You do not know because you do not know how long the body was to begin with. It is a lot if the body we are interested in is 2 millimeters, but it could be fairly insignificant if the body is 2 meters. So the change in length, relative to its original length, is important. This quantity is called strain. Note that some authors will call a deformation a strain, and strain as it is defined here as relative strain. In this lesson, strain will be a material property defined by the equation:

> **Strain**   The change in dimension normalized to the original dimension

$$\varepsilon = \frac{\Delta l}{l_0} \qquad (10.4)$$

Elastic Modulus   If you want to know the material property analogous to stiffness, you would determine the ratio of stress to strain:

$$Y = \frac{\sigma}{\varepsilon} \qquad (10.5)$$

This value is referred to as the **elastic modulus**, or **Young's modulus**.

> **Elastic modulus**   The ratio of stress to strain
>
> **Young's modulus**   The ratio of stress to strain

Strain Energy Density Strain energy was the amount of energy that the body could absorb as it was deformed. **Strain energy density** is the amount of energy that can be absorbed by a normalized piece of tissue.

> **Strain energy density**
> Relative amount of energy stored by the material

Stress–Strain Curves Similar to the load-deformation curve, you can also construct a stress–strain curve (**Figure 10.11**). The corresponding points are very similar. Instead of the ultimate strength, you have the ultimate stress. Instead of deformation to failure, there is strain to failure. The slope of the stress–strain curve is elastic modulus, and the area under the curve is the strain energy density.

Two bodies could have very similar load-deformation curves, but very different stress–strain curves. Likewise, two bodies could have very different load-deformation curves, but similar stress–strain curves. This is because the mechanical properties are determined by the material properties, the quantity of that material, and how that material is distributed within a body.

An example of this idea is presented in **Table 10.2**. Notice that both bodies, A and B, have the same mechanical properties (force, deformation, and stiffness). However, the two bodies have different cross-sectional areas and initial lengths. Therefore, they have different stresses, strains, and elastic moduli.

> **Important Point!** How a body responds to a load depends on the amount of material, how the material is arranged or distributed, and its material properties.

Table 10.2	Two Bodies, A and B, Have the Same Mechanical Properties, However, They Have Different Material Properties	
	A	**B**
ΔF (N)	100	100
Δl (mm)	10	10
k (N/mm)	10	10
CSA (mm²)	50	25
l (mm)	2	4
σ (N/mm²)	2	4
ε	5	2.5
Y	0.4	1.6

Section Question Answer

If you look at a person lying on a bed of nails, you should recognize that they are in a state of static equilibrium. Although the net force is zero, two forces are actually acting on him: his weight and the reaction force from the nail bed. Think about his posture. Whether he is standing, side-lying, or lying supine, the force from the bed of nails is always the same. It is equal to his body weight. What makes lying supine so appealing is that the surface area has increased. The larger area means that loading is spread. This will decrease the stress on the areas directly over the nails. Decreasing the nails that are supporting the weight will increase the stress over those areas. Lying on the bed of nails is safer than standing on them. Standing on one foot is potentially even more dangerous. Jumping up and down is still worse because the reaction forces are now larger than his body weight. Hopefully you can begin to see how these different factors affect the way the body responds to a given load.

COMPETENCY CHECK

Remember:

1. Define the following terms: strength, deformation, ductile, brittle, elastic deformation, plastic deformation, yield point, stiffness, compliance, strain potential energy, toughness, stress, strain, elastic modulus, and strain energy density.
2. Write the equations for stiffness, compliance, stress, strain, and elastic modulus.

Understand:

1. Describe the differences between an elastic deformation and a plastic deformation.

Figure 10.11	The stress–strain curve.

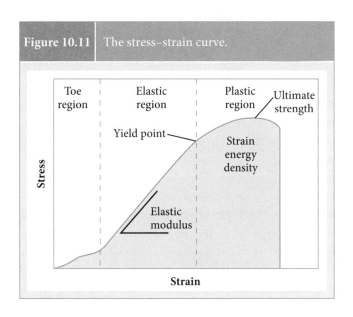

2. Sketch a load-deformation curve, and label the following: toe region, elastic region, yield point, plastic region, ultimate strength, stiffness, and strain energy.
3. Describe the differences between the mechanical properties and material properties of a body.
4. Sketch a stress–strain curve, and label the following: toe region, elastic region, yield point, plastic region, ultimate strength, elastic modulus, and strain energy density.

Apply:

1. Give examples of situations where there is high mechanical stress and low mechanical stress.

10.2 PROPERTIES OF VISCOELASTIC MATERIALS

Section Question

Cindy is a client of yours that needs to increase the flexibility (extensibility) of her muscles (Figure 10.12). How can you use your knowledge of viscoelastic materials to instruct her on proper stretching techniques?

Most, if not all, human tissues exhibit viscoelastic properties. These properties arise due to the large fluid (water) content

within the tissues. Four viscoelastic properties should be of interest to you:

- Stress relaxation
- Creep
- Hysteresis
- Strain rate dependency

Take a closer look at each of these properties, using stretching as an example of each.

> **Important Point!**   Biological tissues exhibit both elastic and viscous properties.

10.2.1   Stress Relaxation

Think about what happens if you stretch a muscle to a certain length, and then hold it. (Better yet, try it!) At first, there could be quite a bit of discomfort because the stress is quite high. But as you continue to hold the stretch, the discomfort decreases. Not discounting the fact that there could be a very large neurological component to this experience (you may just become desensitized to the stretch), it does highlight the phenomenon known as **stress-relaxation**.

> **Stress-relaxation**   A decrease in stress when the strain is held constant for a given period of time

Whenever the strain is held constant, there is a decrease in the stress of the tissue (Figure 10.13). This decrease in stress occurs over time and does not decrease indefinitely to zero. But there is a significant drop in the stress values.

Figure 10.12

© Blend Images/ShutterStock, Inc.

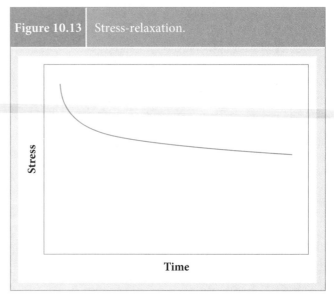

Figure 10.13 | Stress-relaxation.

Stress

Time

10.2.2 Creep

The opposite side of the coin for stress-relaxation is creep.

> Creep An increase in strain when the stress is held constant for a period of time

With creep, if you hold the stress constant, you will see a gradual increase in strain (Figure 10.14). Think about (or try) the stretching example again. Assume that your discomfort level is an indicator of stress, even though this may be a bad assumption. As you initially go into your stretch, there is a certain level of discomfort. As you just saw, that discomfort level will quickly go down. So for the same discomfort level, you can go further into the stretch with time. That is creep.

10.2.3 Hysteresis

Stress-relaxation and creep were two time-dependent properties that occurred when either the deformation or load were held constant. Hysteresis

> Hysteresis A loss of energy to heat when unloading after loading

is observed when the tissue is cyclically loaded and unloaded. During a loading cycle, a tissue is stiffer than when it is unloaded (Figure 10.15). Looking at Figure 10.15, you will also note that the area under the unloading curve is less than the area under the loading curve. Realizing that the area under the curve represents energy, you will quickly note that area in between the two curves represents the energy that is "lost." Do you know where it went? The energy represented by the area between the curves was dissipated as heat. That is energy that cannot be recouped, showing you once again that human movement cannot be 100% efficient.

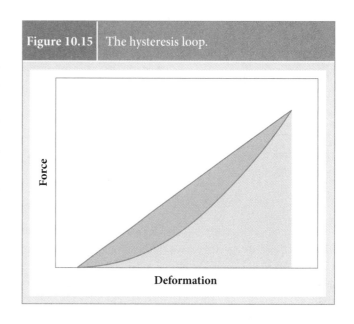

Figure 10.15 | The hysteresis loop.

At first glance, you are probably thinking that hysteresis is a bad thing. This is not necessarily so. If you were looking to improve your client's flexibility by stretching the muscle, you would note that as she is coming out of the stretch, the muscle is less stiff then when she went into it. Performing another repetition of stretching during this phase will mean that the muscle is not as stiff as it was going into the last stretch, and she can go further, or the same distance with less discomfort, then she could on the previous repetition (Figure 10.16). Try it for yourself and see how much further you can go on subsequent repetitions of a stretch.

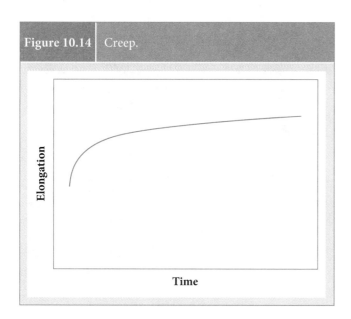

Figure 10.14 | Creep.

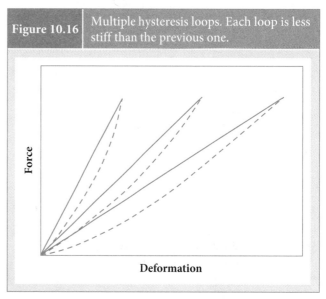

Figure 10.16 Multiple hysteresis loops. Each loop is less stiff than the previous one.

10.2.4 Strain–Rate Dependency

The final viscoelastic property is **strain–rate dependency**.

> **Strain–rate dependency**
> When mechanical properties are dependent on the rate of change of strain

You can probably figure out what that means by breaking the term apart. It means that the mechanical properties of a viscoelastic material are dependent on the rate of strain. The rate of strain is how quickly the strain is changing over time:

$$\dot{\varepsilon} = \frac{\Delta \varepsilon}{\Delta t} \qquad (10.6)$$

In general, the strength, stiffness, and toughness of the viscoelastic material increases with increasing strain rate (Figure 10.17). This is generally considered a good thing, as these are properties that you want to see in most of your tissues. However, when it comes to stretching, it could be a drawback. While stretching, you want your muscle–tendon complexes to be less stiff, not more. So moving more slowly would be more beneficial if you were looking to stretch.

> **Section Question Answer**
>
> The muscles (actually, the muscle–tendon complex) is a viscoelastic material and has the properties of stress-relaxation, creep, hysteresis, and strain–rate dependency. Each of these factors should be kept in mind when trying to stretch Cindy because you want to maximize the elongation of the muscle while minimizing the stress. You would instruct Cindy that the discomfort that she feels will soon decrease (stress-relaxation), and she should be able to stretch further after holding the stretch for few seconds (creep). She should also perform several repetitions of the stretch because her muscles will be less stiff with each repetition (hysteresis). Finally, you want her to stretch slowly to decrease the stiffness (strain rate dependency).

COMPETENCY CHECK

Remember:

1. Define the following terms: stress-relaxation, creep, hysteresis, and strain–rate dependency.

Understand:

1. Explain the time-dependent properties of viscoelastic materials.

Apply:

1. Show how the time-dependent properties of viscoelastic materials are applied to stretching muscle–tendon complexes.

| Figure 10.17 | Strain–rate dependency. |

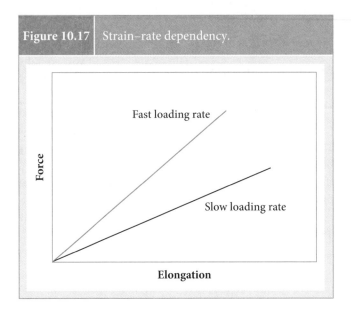

10.3 GENERAL MECHANICS OF INJURY

> **Section Question**
>
> The new football coach came to you for advice about how he can keep his team free from injury (Figure 10.18). What can you tell him?

10.3.1 Lubrication and Wear

Although the human frame shares some similar traits to the frame of your house, it also has movable joints. Because moving joints involve two surfaces in contact moving relative to one another, they could potentially be subjected to high frictional forces, which must be prevented if the joints

| Figure 10.18 | What advice can you give the new coach to decrease the risk of injury to his players? |

© Arthur Eugene Preston/ShutterStock, Inc.

are going to move freely without significant wear. Frictional forces are often decreased through lubrication.

> **Important Point!**   Lubrication prevents or reduces wear that occurs from two bodies rubbing on one another.

Lubrication

With lubrication, a fluid is both absorbed by the articulating surfaces and placed between them. There are several mechanisms by which lubrication can reduce friction (**Figure 10.19**). When the fluid is absorbed by the articulating surfaces, it is called **boundary lubrication** because it creates a layer of fluid that prevents direct surface-to-surface contact (contact is between the layers of fluid). The fluid may also be protected by decreasing any heat that may develop on the surfaces. With **fluid film lubrication**, some sort of movement between the contact surfaces increases the amount of fluid, thus providing greater separation between them. Greater load is supported by the pressure of the fluid. This can occur in one of three ways. With **squeeze film lubrication**, articulating surfaces move closer together in the direction that is perpendicular to their joint surfaces (joint compression). This creates pressure on the fluid. With **hydrodynamic lubrication**, the joint surfaces are moving tangentially to each other (which is part of normal joint motion). As the surfaces move, they create a wave of fluid that lifts the front half surface—creating greater separation. This type of lubrication also moves the fluid around the articulating surfaces.

> **Boundary lubrication**   The lubricating fluid prevents direct surface-to-surface contact
>
> **Fluid film lubrication**   Movement increases the amount of fluid between articulating surfaces, thus increasing their separation

> **Squeeze film lubrication**   The amount of separation between articulating surfaces is increased by the fluid when the two surfaces are compressed together

> **Hydrodynamic lubrication**   The amount of separation between articulating surfaces is increased by the fluid when a wave of fluid is created by the two surfaces moving tangentially to one another
>
> **Elastohydrodynamic lubrication**   The amount of surface area is increased by the pressure of the fluid

Elastohydrodynamic lubrication is similar to hydrodynamic lubrication. But in addition to lifting surfaces and moving fluid around the articulating surfaces, pressure of the fluid also creates a deformation of the articulating surfaces. This deformation increases the surface areas, which in turn decreases the stress on them.

Wear

Despite lubrication, articulating surfaces can still wear. **Wear** occurs when the surface material is deformed and removed by frictional forces. You should be aware of two types of wear (**Figure 10.20**). The first type is **interfacial**, which happens when two surfaces come in contact with each other. The other type is **fatigue wear**, which occurs with the accumulation of microdamage.

> **Wear**   Surface material is deformed and removed by frictional forces
>
> **Interfacial wear**   Wear that occurs when two surfaces come in direct contact
>
> **Fatigue wear**   Wear that is the result of microdamage

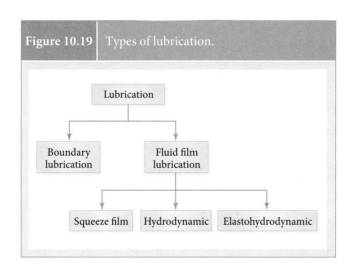

Figure 10.19 | Types of lubrication.

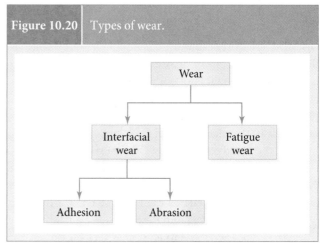

Figure 10.20 | Types of wear.

> **Adhesion wear**   Interfacial wear that occurs as a result of surfaces sticking to one another and then tearing apart
>
> **Abrasion wear**   Interfacial wear that occurs as a result of a hard surface scraping a softer one

In addition, interfacial wear can occur when either two surfaces come in contact and then stick. When they get unstuck, the surfaces tear. This is called **adhesion**. Interfacial wear can also occur when a harder surface scrapes a softer one. This is called **abrasion**.

10.3.2   Fatigue, Cracking, and Failure

Articulating surfaces at joints are not the only damage that can be sustained by the frame. Any body can get damaged, the result of which is an injury. As mentioned previously, excessive stresses/strains can cause any material to break apart, first at the microscopic level, and then at the macroscopic level. When this happens, **failure** occurs. Note that some authors

> **Material failure**   A breaking apart of the material

will say that failure occurs when a body completely breaks apart, but here it will mean any breaking apart of the tissue. And do not confuse material failure with task failure.

The easiest way to think about failure is that maximum stress exceeds the ultimate stress of the material. Another way to think about it is that strain exceeded the maximum strain of the material. There is some debate over if the stress or strain ultimately causes the failure. You will see one such case in the muscle–tendon complex. In most cases, when dealing with loading in a single direction, it is usually adequate to think about it in one of these two ways.

Motion mechanics includes two fundamental ways of analyzing movement: impulse–momentum and work–energy. Work–energy methods give you insights that are not possible with impulse–momentum methods. Similarly, you can use energy methods to examine material failure.

Remember that the strain energy is the amount of energy absorbed by a material during loading, and it is represented as the area under the force-deformation curve. Remember also that the fundamental laws of thermodynamics tell you that energy cannot be created or destroyed, only moved from one place to another or transformed from one type to another. Also, nothing is 100% efficient: Not all the energy that gets absorbed by a body is reused on unloading. Some of it is lost as heat.

With these ideas in mind, ask yourself the question: What happens if the magnitude of the strain energy exceeds its capacity? Certainly energy can leave the body as heat. But what if energy is entering into the body more quickly than it is leaving? The energy has to go somewhere—it cannot just

disappear. And the answer is: The energy can form a crack in the material. It takes energy to form a new surface (form a crack) and propagate that crack (make the crack grow). Energy methods are very helpful when there is a combination of loading types and directions.

10.3.3   Models of Injury

You are now at the point where you can appreciate how an injury occurs. Here, you are going to look at three different models of injury,[3] which will be based on a simple assumption of uniaxial loading and failure occurring as a result of the maximum stress (demand) exceeding the capacity of the material. In each case, you will first create a hypothetical **failure tolerance**, which will be related to how much stress the tissue can handle before failure (capacity). Second, you will determine

> **Failure tolerance**   The stress level above which failure will occur

the actual stress to which the body is subjected (demand). If the stress is above failure tolerance, injury will occur. The severity of injury will depend in part on how much the actual stress exceeds failure tolerance. If the actual stress is below the injury tolerance, the difference will be known as the **margin of safety**. See Figure 10.21a. The failure tolerance is not a static line; over time, failure tolerance can lower with tissue fatigue.

> **Margin of safety**   The difference between the failure tolerance and actual stress applied to a body

> **Important Point!**   Injury occurs when the stress level exceeds the failure tolerance.

In the first model of injury, one single external load creates enough stress to the tissue that it exceeds the failure tolerance, and an injury occurs (Figure 10.21b). In this case, the injury is considered to be **acute** because it happens immediately as the direct

> **Acute injury**   An injury that happens immediately

consequence of the applied load. An example of an injury that occurs with this model is a fracture that occurs when you drop a large weight on your foot.

In the second model (Figure 10.22a), the external load does not provide sufficient stress for an injury to occur (i.e., there is a margin of safety). However, the load is repeatedly applied to the tissue. As you will see in the next section, if the load is applied at the right frequency (not too small and not too high) biological responses will occur in the tissue that will improve its ability to handle stress; that is, it will increase the failure tolerance. If the frequency of the loading is too high, the tissue will fatigue, and the failure tolerance will begin to

Figure 10.21	(a) If the failure tolerance exceeds the stress on the tissue, there is a margin of safety, and no injury will occur. (b) If the stress on the tissue exceeds the failure tolerance, an injury will occur.

(a)

Stress

Margin of safety

— Demand
— Capacity

Time

(b)

Stress

— Demand
— Capacity

Time

Figure 10.22	Chronic models of injury with (a) multiple loadings over time; and (b) a single load held for a long period of time.

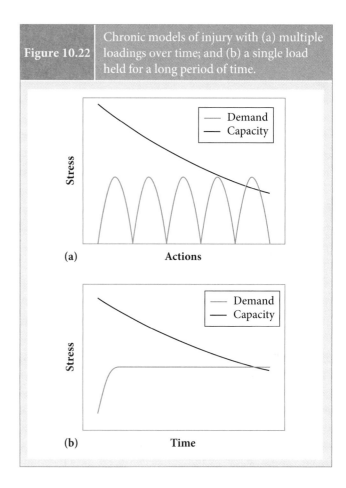

(a)

Stress

— Demand
— Capacity

Actions

(b)

Stress

— Demand
— Capacity

Time

decline, as will the margin of safety. Eventually, the fatigue will be so great that a load that was once below the failure tolerance now exceeds it. A stress fracture is a type of injury that occurs under this scenario.

The third model (Figure 10.22b) is very similar to the second, in that an initial load does not apply enough stress to damage tissue. There is also a margin of safety. Whereas there was a repeated load applied in the second model, a constant load is applied in this one. Once again the tissue fatigues, decreasing the failure tolerance and margin of safety. And once again, the initial load that was once below the failure tolerance exceeds it with time. Pressure sores in the diabetic foot or bedsores would be an example of this type of injury. Because the injuries in the second and third model occur over time, they are referred to as chronic injuries.

Chronic injury An injury that develops over time

These models also provide insight on how injuries can be prevented. Can you figure out what they are? The first model suggests that injury occurs when the stress exceeds the failure tolerance. So the two ways you can increase the margin of safety are to decrease the stress or increase the failure tolerance. Remember the equation of stress:

$$\sigma = \frac{\text{Force}}{\text{Area}}$$

As you can see, there are also two ways that you can decrease the stress: decrease the load or increase the area. You can decrease the load (force) by increasing the time over which the force is applied (remember impulse-momentum?):

$$F = \frac{\Delta L}{\Delta t}$$

or the distance over which the force acts (remember work-energy?):

$$F = \frac{\Delta E}{\Delta p}$$

You can really think about increasing area in two ways as well: increase either the contact area over which the force is applied or the cross-sectional area of the tissue taking the strain. The first is often used in the manufacturing of safety equipment, like helmets. The second is done through exercise, which is

Figure 10.23 | A hierarchical model for decreasing injury risk.

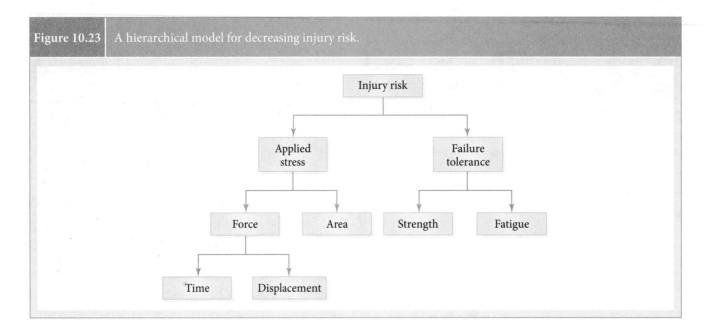

also how you improve failure tolerance, and will be discussed further in the next section.

The second and third models would also include decreasing the stress or increasing the margin of safety as a way of decreasing injury risk. They also suggest that decreasing the rate of fatigue (also accomplished through exercise) would be an effective strategy. In addition, either decreasing the number of exposures or increasing the interval between exposures to the load (second model) or decreasing the duration of the exposure to the load (third model) would also be effective strategies for decreasing the risk of injury. The hierarchical model for decreasing injury risk is presented in Figure 10.23.

Section Question Answer

Injuries occur when the stresses exceed the failure tolerance of the tissue. So, in general, to decrease the risk of injury, you can either increase the failure tolerance or decrease the stress imposed on the tissue. Increasing the failure tolerance is done through a proper exercise program that enhances both the strength and endurance of the tissue. Decreasing the stress on the tissue occurs through proper protective equipment, correct technique, and monitoring the training program to ensure adequate rest between the applied stresses.

COMPETENCY CHECK

Remember:

1. Define the following terms: boundary lubrication, fluid film lubrication, squeeze film lubrication, hydrodynamic lubrication, elastohydrodynamic lubrication, wear, interfacial wear, fatigue wear, adhesion wear, abrasion wear, material failure, failure tolerance, margin of safety, acute injury, and chronic injury.

Understand:

1. Explain the difference between boundary lubrication and fluid film lubrication.
2. Explain how each of the three fluid film lubrication modes reduces friction.
3. Explain how material failure occurs.
4. Create a hierarchical model for decreasing injury risk.

Apply:

1. Use your hierarchical model to give concrete examples of how you can reduce injury risk.

10.4 BIOMECHANICS OF THE HUMAN FRAME: BONE, CARTILAGE, AND LIGAMENTS

Section Question

Your friend Janice sprained her ankle while hiking (Figure 10.24), your brother Tony broke his leg during a soccer game (Figure 10.25), and you just found out your Uncle Joe has osteoarthritis (Figure 10.26). How did these injuries occur? Did they all happen the same way?

Figure 10.24

© Warren Goldswain/ShutterStock, Inc.

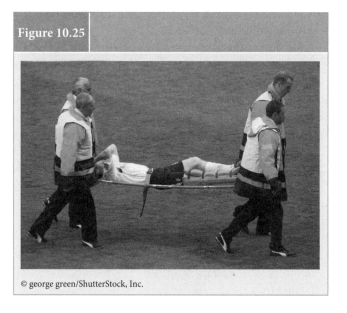

Figure 10.25

© george green/ShutterStock, Inc.

Figure 10.26

© OneSmallSquare/ShutterStock, Inc.

10.4.1   Bone

Function and Structure

Bone is one of the hardest structures in the body (dentin and enamel in the teeth are harder).[4] Bone has several functions, including[5]

- Providing a rigid framework that gives the body its shape
- Functioning as a set of rigid links and mechanical levers
- Providing an attachment site for the muscle–tendon complex
- Protecting the central nervous system and various internal organs
- Storing minerals
- Producing red blood cells

In biomechanics, you are predominately interested in the first four functions. The last two, although important, will not be a focus of this section. Here, you will just focus on how bone responds to loading.

Macroscopically, there are two types of bone (**Figure 10.27**). The outer layer, or shell, is called cortical (or compact) bone. Under the shell is trabecular (also called spongy or cancellous) bone. Cortical bone always surrounds trabecular bone—you will never see the trabecular bone without the

| Figure 10.27 | Two types of bone, cortical and trabecular. |

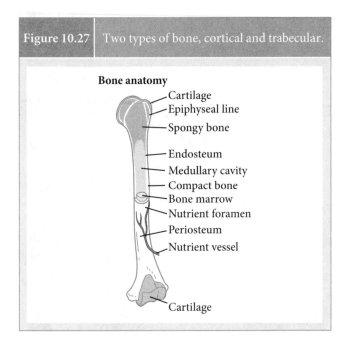

Bone anatomy

- Cartilage
- Epiphyseal line
- Spongy bone
- Endosteum
- Medullary cavity
- Compact bone
- Bone marrow
- Nutrient foramen
- Periosteum
- Nutrient vessel
- Cartilage

hard, compact shell, although the relative amounts of each type differ between bones. The two types of bone are made of the same material; the differences are in how those materials are put together. Compact bone is solid and denser. Trabecular bone appears to be more randomly organized. The body plates, or trabeculae (where the name comes from), can be thought of as small beams, rods, or struts. They are laid down to provide support against the stresses that are imposed on the bone.

Mechanics

The mineral (such as calcium) level of bone is thought to affect its mechanical properties. Like most other things, there appears to be an optimum level of mineralization in bone: too little, and the bone becomes weak; too much, and the bone becomes brittle.[6] To determine the level of mineralization in bone, two measures are commonly used: **bone mineral content** and **bone mineral density**. Bone mineral content is the total amount of mineral in a bone, whereas bone mineral density is the amount of mineral content in a given area (areal bone mineral density) or volume (volumetric mineral density). Although measures of bone density certainly account for some of the mechanical properties of bone, it does not necessarily tell the whole story.[7] That is because the material properties of the bone can change relative to the bone mass or density. Remember that it is both

> **Bone mineral content** The total amount of mineral in bone
>
> **Bone mineral density** The mineral content in an area or volume of bone

the material properties and the quantity of the material that determine the mechanical properties of a body. And adding a little bit of bone in just the right spot can go a long way in increasing the strength of the bone.[8]

> **Important Point!**   It is not just increasing the bone mineral density, but where that bone is placed, that determines the overall strength of the bone.

Bones seem very well designed for providing structure to the human frame. They appear to have the best combination of stiffness and brittleness to maximize the amount of energy that they can absorb.[6] Bone is **anisotropic**, meaning that its behavior is determined by the direction of forces. For example, bone will not respond to compressive forces the same way in each direction. Similarly, bone is asymmetric. It is stronger in compression than it is in tension, and stronger in tension than it is in shear.

> **Anisotropic** Exhibiting different properties when measured in different directions

> **Important Point!**   Bone is stronger in compression than it is in tension.

Mechanopathology

When a bone breaks, it is called a **fracture**. Acute, or traumatic, fractures occur instantaneously. Stress fractures occur over time. Do not confuse the use of the word *stress* in stress fractures with mechanical stress. Both types of injuries involve stress, it is just the amount that is different.

> **Fracture** The breaking of a bone

Because of the asymmetries associated with bone, you might have guessed that bone is more susceptible to tensile stresses than it is to compressive stresses. This is probably due to the trabeculae. If you think more about it, you would realize how hard it is for a situation to occur when a bone would be pulled apart. How, then, does bone fracture?

If you were to examine a skeleton, you would notice that bones are generally not straight and are not shaped like cylinders. The ends (epiphyses) are generally wider than the middle (diaphysis). The shapes of the bones generally make them susceptible to bending moments. Remember that bending moments create compressive forces on one side of a neutral axis and tensile forces on the other side. Being weaker in tension, this is the side that would develop a fracture first.

Stress fractures occur over time. Bone is dynamic tissue, constantly undergoing modeling (putting down new bone) and remodeling (reabsorbing old bone and laying down new bone). The signal to start these physiological processes is mechanical loading. If you were going to remodel your kitchen, you would have to take out the old cabinets before you put in the new ones. It is the same idea with remodeling bone: The old bone tissue is reabsorbed before the new bone can be laid down. If the frequency of loading is too high, then the signals for bone resorption outpace those for reformation. If you are constantly taking up more tissue than you are putting down, the bone is going to get progressively weaker. This decreases the margin for safety. Eventually, a stress that could be easily tolerated can no longer be tolerated, and the bone breaks.

The Effects of Disuse, Aging, and Exercise

As mentioned earlier, bone is constantly being taken up and laid down. Bones can be heavy, and the body is smart in that it is not going to keep something heavy around that is has to move if it does not have to. So with disuse, bones will get lighter. In general, bones are not going to get thinner, so to get lighter there has to be a decrease in density. General bone loss is called osteopenia. Osteoporosis is a severe loss of bone.

> **Osteopenia**   Lower-than-normal bone mineral density
>
> **Osteoporosis**   Severe decrease in bone mineral density

Can you guess what the effects of a decrease in bone mineral density are? In terms of the mechanical properties, there is a decrease in strength, stiffness, and energy to failure. It is unclear what effects aging and osteoporosis have on the material properties of bone.[7]

Conversely, exercise can have beneficial effects on the mechanical properties of bone, increasing its strength, stiffness, and energy to failure. But you have to pick the right type of exercise. First, the loading has to be dynamic, and it should have a high loading rate.[8] In other words, exercises should be high impact with large rates of deformation. Second, bone formation is limited to the specific sites that are being stressed.[8] This can be a good thing because it means that just a small amount of new bone formation can greatly increase the bone strength in that particular area. But it can also be a bad thing in that it will not protect the bone in other areas or against other types of stresses. This finding would seem to indicate that you would want to be constantly performing novel exercises if you want to increase overall bone health. Finally, bone density seems to be more likely to increase before full skeletal maturity.[8] This means that you would want to increase the density of your bones through exercise while a child, teen, and young adult. It is much harder to put down new bone in your 30s and beyond. Similar to aging and osteoporosis, we know that exercise has a beneficial effect on the quantity of bone and where it is placed, but we do not know how exercise affects the material properties of bone.[7]

> **Important Point!**   To prevent osteoporosis, exercises for bone should be novel, have a high loading rate, and be started while the skeleton is still maturing.

10.4.2   Cartilage

There are three types of cartilage in your body:

- Articular (or hyaline) cartilage, which covers the articular surfaces of bones
- Fibrocartilage, which has some specialized roles
- Elastic cartilage, which is found in the external ear, parts of the nose, and some other places

Here, you are concerned only with articular cartilage.

Function and Structure

Articular cartilage is found where two bones are connected. It serves two important functions. First, it distributes the load transmitted across the joint over a wider area. This decreases the amount of stress on the joint surfaces. Second, it allows for the relative movement of the two opposing joint surfaces with minimal wear and tear. In other words, it decreases the friction, and thus wear, between the two bones. The coefficient of friction between two articular surfaces is extremely low. To give you an idea of how low they are, some coefficients of friction between some common materials is presented in Table 10.3.[9] As you can see, two articular surfaces moving on each other have a lot less friction than ice sliding on ice.

> **Important Point!**   Cartilage increases the area over which a load is distributed and decreases the friction between two articulating bones.

Table 10.3	The Coefficients of Friction for Several Common Objects	
Surfaces	**Coefficient of static friction**	**Coefficient of kinetic friction**
Rubber on dry concrete	1.0	0.8
Wood on wood	0.4	0.2
Steel on steel (lubricated)	0.7	0.6
Ice on ice	0.1	0.03
Synovial joints	0.01	0.01

Cartilage seems to be well designed to accomplish these two tasks, but the anatomy is pretty complex. If you are interested in it, I recommend that you check out one of the more advanced treatments on the subject. For the purposes here, you will lump cartilage into two components: one that is solid and one that is liquid.[10] The solid part will contain mostly collagen and proteoglycans, whereas the fluid part will contain interstitial water with inorganic salts dissolved in it. Think of it as a water-saturated sponge, with the water being (ironically enough) water, and the sponge being the organic solid matrix. That should suffice to understand the mechanics of cartilage.

Mechanics

Articular cartilage is primarily loaded under compression, and the two components (solid and liquid) are what provide its mechanical properties under this type of loading. Fluid flows through the matrix, which stiffens it and helps it support the load. So during compression, it is both the pressure of the fluid and the stiffness of the matrix that transmit the load to the bone[11] and give cartilage its viscoelastic properties (stress-relaxation, creep, hysteresis, and strain rate dependency).

The cartilage not only transmits forces from bone to bone, but it also lubricates the joints. It appears as though all types of lubricating mechanisms (boundary and all three fluid film types) are used by cartilage. The one that is used under a particular loading condition is possibly the one that is most effective in providing lubrication under that condition.[10]

However, cartilage is a little different in that it is permeable: Fluid is flowing into and out of the cartilage as it is compressed (remember the sponge). This has a "self-lubricating" effect.[10] As one surface moves over the other, it pushes fluid out and in front of it. As the load passes over the articular surface, the cartilage becomes less permeable—keeping a layer of fluid between the surfaces.

Mechanopathology

Cartilage seems to be exceptional at performing its tasks, usually producing very little wear and tear over the course of a lifetime. Although other types of injuries can occur to articular cartilage, the most common is osteoarthritis. Osteoarthritis is

> **Osteoarthritis** The progressive degeneration of the articular cartilage and the bone deep to it

a progressive joint disease in which the articular cartilage and subchondral bone (the bone deep to the cartilage) degenerate. Not only does the cartilage decrease in thickness, but it also becomes rougher.

As you can imagine, it can be quite painful and debilitating because the coefficient of friction must increase dramatically.

The exact cause of the osteoarthritis is unknown. Although certainly all the factors that influence how a tissue responds to loading (magnitude, location, direction, duration, frequency, variability, and rate) are involved, it seems likely that the leading cause is a shift in the stress from an area frequently loaded to an area that is loaded less often.[12] Aging, weakness, obesity, malalignment, and injury to ligaments and fibrocartilage of a joint are all risk factors. These can all affect the normal mechanics of the joint, and this alteration of mechanics will stress the articular cartilage in unique ways. Once osteoarthritis is initiated, the degeneration of cartilage can lead to further alterations in mechanics, which then creates a vicious cycle leading to further degeneration.

The Effects of Disuse, Aging, and Exercise

Like most other tissues of the body, cartilage has the ability to adapt, or change, over time. And like most other tissues in the body, the loading that it receives seems to signal these changes to occur. Moderate, repetitive loading appears to best bring about the desired changes of increased proteoglycan content, decreased proteoglycan extractability, and increased cartilage thickness. High-intensity loading, particularly at older ages, brings out the negative effects of decreased collagen network structure, site-specific proteoglycan loss, and reduced cartilage stiffness. Immobilization also leads to a loss in proteoglycan and a reduction in stiffness, but these changes appear to be reversible if the immobilization period is not too long because immobilization does not appear to affect the collagen network structure.[12] The findings show that once again there is an optimal level of loading, below or above which creates negative consequences.

10.4.3 Ligaments

Function and Structure

You have no doubt heard that ligaments connect bone to bone. But that hardly does these important structures justice. A joint is a place where two things are joined together. In this case, it is where two bones are joined together. Unlike the frame of your house, your human frame is designed to move. Different joints allow for varying magnitudes and directions of movements. Ligaments not only provide the support necessary to keep the joints intact, but they also guide certain motions while restricting others.

> **Important Point!** Ligaments hold bones together, restrict certain movements, and guide others.

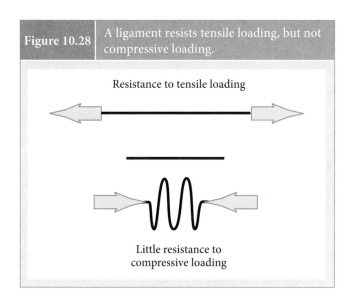

Figure 10.28 A ligament resists tensile loading, but not compressive loading.

Resistance to tensile loading

Little resistance to compressive loading

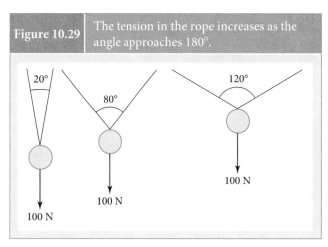

Figure 10.29 The tension in the rope increases as the angle approaches 180°.

20°

80°

120°

100 N

100 N

100 N

To accomplish this task, ligaments are composed primarily of Type I collagen arranged in bundles according to their function. This means that the collagen fibers are not exactly parallel, but they are closely interlaced.[13] Ligaments are very similar in structure to tendons, which connect muscles to bone, but tendons have a higher collagen content, and the bundles are arranged more orderly.[13] Because they are so similar, many texts present ligaments and tendons together. Things are a little bit different here. Because of their importance as part of the muscle–tendon complex, tendons are covered separately.

Mechanics

Ligaments generally sustain tensile loads, usually in one direction. In other words, they have high tensile stiffness and strength, but provide little resistance to compression (Figure 10.28). They may also bear small tensile loads in other directions, depending on their fiber orientation.

Mechanopathology

Injuries to ligaments are called sprains. If ligaments sustain tensile loads and provide little resistance to compression, then it stands to reason that the major mechanism of injury for ligaments is tensile loading. Earlier, you learned that tensile loads occur when an object is being pulled apart in a uniaxial direction. And although that can, and probably does, happen from time to time, a more likely scenario is when the ligament is being forcibly wrapped around part of bone.

> **Sprain** An injury to a ligament that occurs when it is stretched beyond its capacity

To help you understand this, consider a 100 N ball suspended from a rope (Figure 10.29 and Table 10.4).[14] Notice how the tension in the rope increases dramatically as the angle increases from 0° to 160°. At 160°, tension in the rope is almost three times the weight of the ball. From 160° to 180° there is further exponential increase in tension. In fact, at 180° the tension is infinitely high. Now imagine a ligament that is almost straight and already in tension (Figure 10.30). It should not be hard to see how a slight, lateral force (from bending the ligament around a bone) could cause stresses that would easily exceed the ultimate strength of the tendon.

The Effects of Disuse, Aging, and Exercise

Less information appears to be available on the biomechanics of ligaments than on other tissues (like bone and cartilage). Part of the reason for this may be the difficulty in measuring mechanical and material properties of ligaments. Ligaments

Table 10.4	Tension in the Rope as a Function of Joint Angle	
Angle (°)	**Tension (N)**
0	50
20	50.77
40	53.21
60	57.74
80	65.27
100	77.79
120	100.00
140	146.19
160	287.94
179.99	2.86×10^5

Figure 10.30	Tensile loading increases if a ligament is wrapped around a bone.

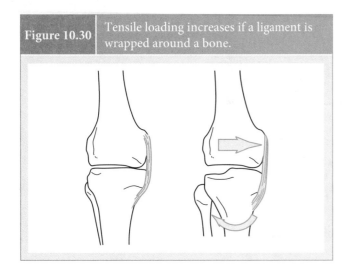

Table 10.5	Key Concepts

- Load
- Deformation
- Stiffness
- Strain energy
- Strength
- Stress
- Strain
- Elastic modulus
- Strain energy density
- Stress relaxation
- Creep
- Hysteresis
- Strain rate dependency
- Lubrication
- Wear
- Failure
- Failure tolerance
- Margin of safety
- Acute injury
- Chronic injury

may not have uniform cross-sections across their lengths, and each ligament may be unique in how it responds to growth, development, and aging. It is difficult to extrapolate findings to different ligaments or the same ligament in different species.[15]

These difficulties notwithstanding, some general trends are associated with ligaments in their biomechanical responses to disuse, aging, and exercise.[15] Long periods of immobilization lead to rather profound decreases in tissue mass, strength, and stiffness. In contrast, exercise can lead to increases in mass, strength, and stiffness—although these increases are pretty modest in comparison to the decreases found with immobilization. In fact, the remobilization period may have to be up to six times longer than the immobilization period to return the ligament to its preimmobilized state. Aging seems to have similar effects of disuse (decreased mass, strength, and stiffness). Not much is known about these effects on the material properties of ligaments.

Section Question Answer

Janice sprained her ankle due to an excessive tensile loading on her ligaments, most probably as they were bent around a bone. Tony broke his leg, probably as a result of his leg being bent and an excessive amount of energy entering his bone. Uncle Joe's osteoarthritis is a degenerative condition that resulted from abnormal loading on his articular cartilage. Each of these injuries was a result of a combination of loading magnitude, location, direction, duration, frequency, variability, and rate that the structure could not handle—albeit in different combinations for each type of injury.

SUMMARY

In this lesson, you learned about the mechanics of materials (Table 10.5). After identifying the different types of load, you saw how a body responds to that load. This response was dictated by the quantity of material, the placement of that material, and its material properties. The loads and their responses were used to develop models of injury, which were applied to various biological tissues. Lubrication and wear were also examined, as well as the critical role that articular cartilage plays in the lubrication of joints.

REVIEW QUESTIONS

1. Define the following terms: load, compression, tension, shear, bending, neutral axis, area moment of inertia, torsion, polar moment of inertia, strength, deformation, ductile, brittle, elastic deformation, plastic deformation, yield point, stiffness, compliance, strain potential energy, toughness, stress, strain, elastic modulus, strain energy density, stress-relaxation, creep, hysteresis, strain–rate dependency, boundary lubrication, fluid film lubrication, squeeze film lubrication, hydrodynamic lubrication, elastohydrodynamic lubrication, wear, interfacial wear, fatigue wear, adhesion wear, abrasion wear, material failure, failure tolerance, margin of safety, acute injury, chronic injury, bone mineral content, bone mineral density, anisotropic, fracture, osteopenia, osteoporosis, osteoarthritis, and sprain.

2. List the factors that determine how an object responds to a load.

3. List the common injuries that occur to bone, cartilage, and ligaments.

4. Describe the differences between the mechanical and material properties of a body.

5. Describe a model for injury.

6. Use the types of loading and models of injury to describe common injuries that occur to bone, cartilage, and ligaments.

7. Describe the effects of disuse, aging, and exercise on the mechanical properties of bone, tendon, and ligaments.

8. Show how a model of injury can be used to reduce injury risk.

9. Describe exercise interventions to prevent maladaptations that occur as a result of aging, disuse, or immobilization.

REFERENCES

1. Whiting WC, Zernicke RF. *Biomechanics of Musculoskeletal Injury.* 2nd ed. Champaign, IL: Human Kinetics; 2008.

2. Allen JH. *Mechanics of Materials for Dummies.* Hoboken, NJ: Wiley Publishing, Inc.; 2011.

3. McGill SM. The biomechanics of low back injury: implications on current practice in industry and the clinic. *Journal of Biomechanics.* May 1997;30(5):465–475.

4. Nordin M, Frankel VH. Biomechanics of bone. In: Nordin M, Frankel VH, eds. *Basic Biomechanics of the Musculoskeletal System.* 2nd ed. Philadelphia: Lippincott, Williams, and Wilkins; 1989:3–30.

5. Whiting WC, Rugg S. *Dynatomy: Dynamic Human Anatomy.* Champaign, IL: Human Kinetics; 2006.

6. Turner CH. Biomechanics of bone: Determinants of skeletal fragility and bone quality. *Osteoporosis International.* 2002;13(2):97–104.

7. Hernandez CJ, Keaveny TM. A biomechanical perspective on bone quality. *Bone.* Dec 2006;39(6):1173–1181.

8. Turner CH, Robling AG. Designing exercise regimens to increase bone strength. *Exercise and Sport Sciences Reviews.* Jan 2003;31(1):45–50.

9. Giancoli DC. *Physics.* 4th ed. Englewood Cliffs, NJ: Prentice Hall; 1995.

10. Mow VC, Proctor CS, Kelly MA. Biomechanics of articular cartilage. In: Nordin M, Frankel VH, eds. *Basic Biomechanics of the Musculoskeletal System.* 2nd ed. Philadelphia: Lippincott, Williams, and Wilkins; 1989:31–58.

11. Cohen NP, Foster RJ, Mow VC. Composition and dynamics of articular cartilage: structure, function, and maintaining healthy state. *Journal of Orthopaedic & Sports Physical Therapy.* Oct 1998;28(4):203–215.

12. Griffin TM, Guilak F. The role of mechanical loading in the onset and progression of osteoarthritis. *Exercise and Sport Sciences Reviews.* Oct 2005;33(4):195–200.

13. Carlstedt CA, Nordin M. Biomechanics of tendons and ligaments. In: Nordin M, Frankel VH, eds. *Basic Biomechanics of the Musculoskeletal System.* Philadelphia: Lippincott, Williams, and Wilkins; 1989:59–74.

14. Hewitt PG. *Touch This! Conceptual Physics for Everyone.* San Francisco: Addison Wesley; 2002.

15. Woo SLY, Abramowitch SD, Kilger R, Liang R. Biomechanics of knee ligaments: injury, healing, and repair. *Journal of Biomechanics.* 2006 2006;39(1):1–20.

Lesson 11

Muscle–Tendon Mechanics

© Maria Teijeiro/OJO Images/Getty Images

LEARNING OBJECTIVES

After finishing this lesson, you should be able to:

- Define the following terms: concentric action, eccentric action, isometric action, stretch-shortening cycle, stiffness, and compliance.
- Describe how the muscle–tendon complex can act like a motor, brake, spring, or strut.
- Trace the flow of energy during concentric, eccentric, and isometric actions.
- Describe the force generated by both the muscle and tendon.
- Describe the muscle–tendon interactions during movement.
- List the factors that affect how much force the muscle–tendon complex can produce.
- Describe and explain how each factor affects the force produced by the muscle–tendon complex.
- Describe the mechanisms of injury for muscle and tendon.
- Describe how injury affects the mechanics of muscle and tendon.

Figure 11.1 Why does an elite long-jumper not take off at a 45° angle as classical mechanics would suggest?

© Maxisport/ShutterStock, Inc.

You know that the takeoff angle that gives a projectile its furthest range is 45°. Yet elite long jumpers (**Figure 11.1**) take off at an angle much less than this (usually between 15° and 27°).[1] Why? Are these elite long jumpers doing something wrong? Should they be jumping with a higher angle of takeoff?

When first learning biomechanics, many of the lessons can be considered applied physics. There is a larger emphasis on the mechanics than there is on the "bio." Specifically, when considering the causes of motion, you learned about the major mechanical principles of impulse–momentum and work–energy. The source of the impulse and work during human (and animal) movement is the muscle–tendon complex, which requires you to interpret these principles in a new light. It is time to put the "bio" in biomechanics and examine some biological principles that govern the movement of living systems.

What you need to realize is that although you are subject to the laws of physics, you are not a machine. Shooting a

| Figure 11.2 | A two-legged hopping maneuver that includes takeoff, airborne, and landing phases. |

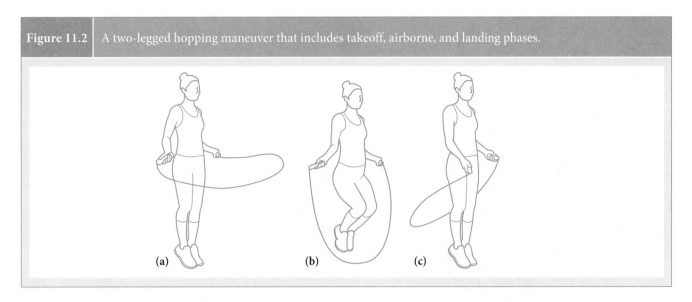

(a) (b) (c)

cannonball at a 45° angle would give you the furthest range, but that does not work when jumping for distance because of your anatomy and physiology. And that brings up two very important points. The first is that form dictates function. How you are built has a tremendous impact on how you move. The second point is that your body is a complex system of interacting parts. Contrary to reductionist thinking, you cannot determine the behavior of the whole system by only looking at the parts. You have to see how the parts are interacting.

Normally, I like to take a "top-down" approach to biomechanical problems. This lesson will be organized a little bit differently. Here, you will see a "bottom-up" approach, where you will look at muscle and tendon individually, and then you will see how they behave together as part of the same system (the muscle–tendon complex). In the next lesson, you will put the muscle–tendon complex into a joint system and see how that alters the behavior yet again. But before examining how the parts interact, you should first appreciate why you have muscle–tendon complexes in the first place and what you need them to do.

11.1 THE FUNCTION OF THE MUSCLE–TENDON COMPLEX (MTC)

Section Question

Although hopping, such as jumping rope (Figure 11.2), is not a normal locomotive behavior for humans, it is a rather "simple" movement that can provide you with a lot of insights into more complex behavior. How many functions of the muscle–tendon complex can you identify during hopping?

In the next section, you will review the individual components that make up the muscle–tendon complex. For now, the important thing to realize is that although the muscle is the active component producing force, it can only control movement by transmitting that force through passive components to the skeletal segments. So you should not think of movement as being regulated by just muscle, but by a muscle–tendon complex. Because that is a mouthful to say, you can just simply refer to it as the MTC.

The fundamental principle of the MTC is that it can only pull, it cannot push. In other words, the force developed in the MTC attempts to bring the two insertion points closer together (Figure 11.3). Which end ultimately moves will depend on the amount of resistance at each segment to which the muscle attaches. Even though the MTC can only develop force in this one direction, it is capable of acting as a motor, brake, spring, or strut (Figure 11.4).[2]

| Figure 11.3 | A muscle–tendon complex (MTC) can only develop force in one direction: toward the center of the MTC. |

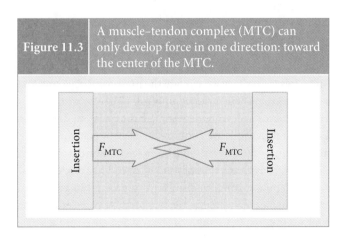

Figure 11.4	The MTC can act like a motor, brake, spring, or strut.

(top left) © sspopov/ShutterStock, Inc., (top right) © Jim Hughes/ShutterStock, Inc., (bottom left) © Vladyslav Danlin/ShutterStock, Inc., and (bottom right) © Guy Shapira, ShutterStock, Inc.

Important Point!   A muscle can only pull, it cannot push.

Important Point!   The MTC can act as a motor, brake, spring, or strut.

Consider the case where hopping is performed only in the vertical direction, and the only movement occurs at the ankle joint. To really exaggerate the movement, imagine that you are on an incline facing upward. (At this point, it may be helpful for you to review some basic gross anatomy and kinesiology of this joint if you are not already familiar with it.) Your primary attention should be directed toward the muscles of the calf, which functionally are referred to as ankle plantar flexors. You are going to examine two muscles in particular, the gastrocnemius and the soleus, which are connected to the foot via the Achilles tendon. They will be considered a single MTC, the triceps surae. For the purposes of this discussion, it is

helpful to think of hopping as occurring in two distinct phases: an upward (propulsive) phase from initiation of upward movement to the instant of takeoff and a downward (braking) phase from the instant of landing to maximum dorsiflexion.

During the propulsive phase, the ankle is plantar flexing and the MTC is shortening. Looking at **Figure 11.5** (from A to B), notice how the two insertion points are further away in part A and are closer together in part B. During the braking phase, the opposite occurs. The ankle dorsiflexes, and the two insertion points move further away from each other (from B to A in Figure 11.5). Because the two insertion points are further apart, the MTC must be lengthening during the braking phase.

Now ask yourself: what is the function of the MTC? First think about what happens during the propulsive phase. To move in the vertical direction and achieve liftoff, the potential and kinetic energies of the body's COM must be increasing. Where does that energy come from? It comes from the MTC. How does it do it? The force generated by the MTC

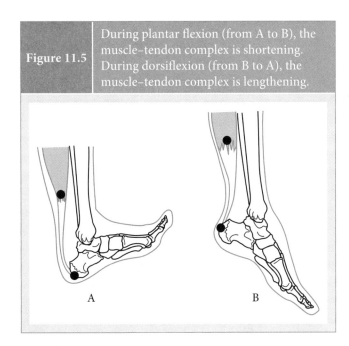

| Figure 11.5 | During plantar flexion (from A to B), the muscle–tendon complex is shortening. During dorsiflexion (from B to A), the muscle–tendon complex is lengthening. |

A   B

Concentric action   An action in which the muscle–tendon complex develops greater force than the external force acting on it and shortens; because the force and the displacement are in the same direction, the MTC is doing positive work—increasing the energy of the skeletal system

and the displacement are in the same direction, so the MTC is shortening and doing positive work. For the MTC to shorten, the force produced by the MTC must be larger than the external force acting on it. Such actions are called concentric actions, and the energy that is generated by the MTC is delivered to the segments. During concentric actions, the MTC is acting as an energy source, or motor.

During the braking phase, the body is losing both potential and kinetic energy. Recall that the energy has to go somewhere, and in this case the energy is going to the MTC. The MTC is still generating force, but it is also lengthening. The force and the displacement are in opposite directions. For this

Eccentric action   An action in which the muscle–tendon complex develops less force than the external force acting on it and lengthens; because the force and the displacement are in the opposite direction, the MTC is doing negative work—decreasing the energy of the skeletal system

to occur, the external forces acting on the MTC must be greater than the force produced by it. This type of action is called an eccentric action. When the force and displacement are in opposite directions, negative work is being performed by the MTC. Energy is moving from the segments to the

MTC. During eccentric actions, the MTC is acting as an energy sink, or brake, by absorbing energy.

Notice how the same MTC (the triceps surae) is controlling both the propulsive and the braking phases, albeit with different actions. During the propulsive phase, the ankle is plantar flexing, and the MTC is acting concentrically. During the braking phase, the ankle is dorsiflexing, and the MTC is acting eccentrically. If both the joint motion and MTC actions are opposite each other (e.g., concentric plantar flexion, eccentric dorsiflexion), then the same MTC is controlling both phases of the movement.

> **Important Point!**   If both the joint motions and the muscle actions are opposite each other, then the same muscle group is controlling the movement.

During eccentric actions, the energy absorbed by the MTC can be briefly stored. If a concentric action immediately follows an eccentric action (as is the case during hopping), the energy stored can be added to the energy produced, enhancing the energy of the concentric muscle action. This involves a process known as the stretch-shortening cycle (SSC), which will be explored in greater detail later in the lesson. During the SSC, the MTC acts like a spring.

Muscle could also generate force, but while neither insertion point is moving. This type of action is called an isometric action. If there is no displacement, the force generated by the MTC must be equal to the external forces acting on it. Because there is no displacement, there is no

Stretch-shortening cycle   A concentric action immediately after an eccentric action; energy stored during the eccentric action contributes to the movement during the concentric action

Isometric action   An action in which the muscle–tendon complex develops a force that is equal to the external force acting on it and does not change its length; because there is no displacement, the MTC is doing no work—the energy of the skeletal system remains unchanged during isometric actions

work being performed by the MTC. However, these isometric actions are no less important as concentric or eccentric actions. Isometric actions are necessary to stabilize segments and transfer energy between different segments.

Think about hopping again. The potential and kinetic energies of all segments (and the COM) are cyclically increasing and decreasing, but the only source of energy is the triceps surae. But what would happen if the triceps surae was the only MTC active during hopping?

It would be very difficult to change the energies of all the segments, and the COM, if the other joints (such as the knee, hips, and spine) were not stabilized but were free to move around. In this case, even though there is not movement at these other joints, there is still a level of muscle activation at each of them. So while the action of the plantar flexors is changing the energy of all the segments in the body, isometric actions of MTCs at other joints are necessary to transfer the energy from the triceps surae to the other segments. The MTCs at the other joints are acting like struts.

There are three actions of the MTC (reviewed in Figure 11.6). There are four functions of the MTC. Three of the functions correspond to three of the actions: During concentric muscle actions the MTC acts like a motor, during eccentric actions the MTC acts like a brake, and during isometric actions the MTC acts like a strut. The fourth function is a combination of two actions: the MTC acts like a spring when an eccentric action is immediately followed by a concentric action. By essentially developing force the same way, the MTC can perform these four functions to produce a wide array of movements.

Section Question Answer

The MTC acts as a motor, brake, strut, and spring via concentric actions, eccentric actions, isometric actions, and the stretch-shortening cycle, respectively.

COMPETENCY CHECK

Remember:

1. List the four functions of the muscle–tendon complex.
2. Define: concentric action, eccentric action, isometric action, and stretch-shortening cycle.

Understand:

1. Match the MTC action with the function of that action.
2. Outline the flow of energy during concentric, eccentric, and isometric actions.

| Figure 11.6 | The three types of muscle actions. (a) when the force produced by the muscle-tendon complex (MTC) is greater than the external force acting on it, the MTC will shorten. (b) When the force produced by the MTC is less than the external force acting on it, the MTC will lengthen. (c) When the force produced by the MTC is equal to the external force acting on it, there is not displacement and no change in length of the MTC. |

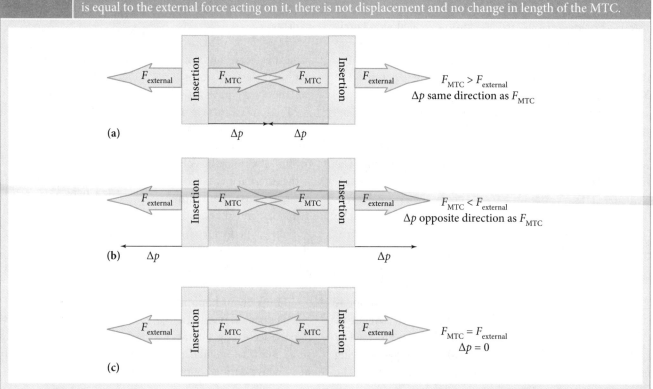

Apply:

1. Give examples of activities that involve concentric actions, eccentric actions, and isometric actions.
2. Give examples of activities that use the stretch-shortening cycle.

11.2 THE INDIVIDUAL COMPONENTS

> **Section Question**
>
> Can a tendon be lengthening while a muscle is shortening?

Now that you understand what the muscle–tendon complex has to do (act like a motor, brake, strut, and spring), it is time to figure out how it does it. You will review some of the basics involved with muscle and tendon, individually. Then you will turn your attention to how they work together to perform the functions outlined earlier, where you will find a few surprises.

Figure 11.7 shows the major structures of the muscle–tendon complex. The major contractile unit of a muscle is the sarcomere. Within the sarcomere, actin (thin) and myosin (thick) filaments are arranged longitudinally. The interaction of the actin and myosin heads is responsible for the muscle action itself. A string of sarcomeres in series is a myofibril, and myofibrils arranged in parallel make up a single muscle fiber. Surrounding the muscle fiber is connective tissue called endomysium. Muscle fibers are grouped together in bundles (also known as fasciculi) and are surrounded by connective tissue called perimysium. As you will learn, the arrangement of the muscle fibers is extremely important in determining the muscle–tendon complex's performance. Collectively, the muscle bundles make up the muscle belly, which is covered by yet another layer of connective tissue called the epimysium. The three layers of connective tissue (perimysium, endomysium, and epimysium), called fascia, are continuous with the tendons on both ends of the muscle belly. Tendons have an intramuscular part (aponeurosis) and an external part that connects the muscle to the bone.

The three layers of connective tissue, aponeurosis, and tendon all have viscoelastic properties that can influence the behavior of the MTC, although the aponeurosis and tendon are believed to be the major contributors.[3] Unless otherwise stated, throughout this lesson, the muscle fibers themselves will be referred to as muscle, all the viscoelastic structures will

| Figure 11.7 | The major structure of the muscle–tendon complex. |

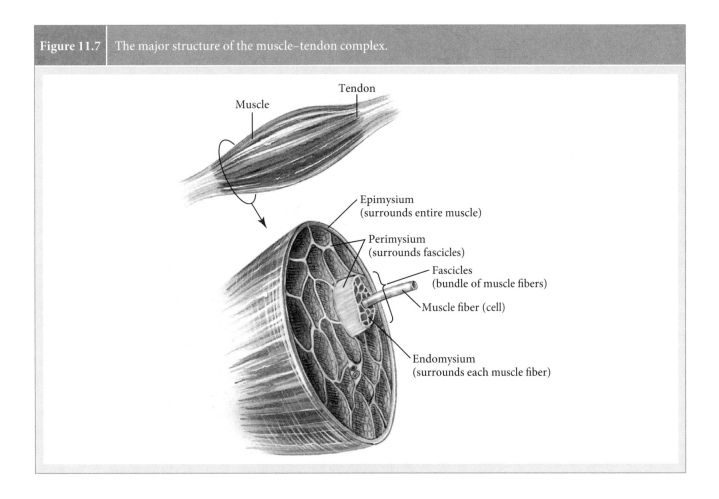

be referred to as tendon, and combined they will be referred to as the muscle–tendon complex.

11.2.1   Muscle

The functional unit of a muscle is the sarcomere (**Figure 11.8**). It is composed of two contractile filaments, actin and myosin, arranged longitudinally. First you should understand how an

individual sarcomere creates force, and then you can appreciate how the sarcomeres work in concert to create muscle force.

We are only interested in the mechanics associated with muscle; the chemical processes will be largely ignored. To describe these actions, let us begin with something you know. Imagine that you are pulling in a rope, hand over hand. First, you would reach out with your right hand, grab onto the rope,

| **Figure 11.8** | The major structure of the sarcomere. |

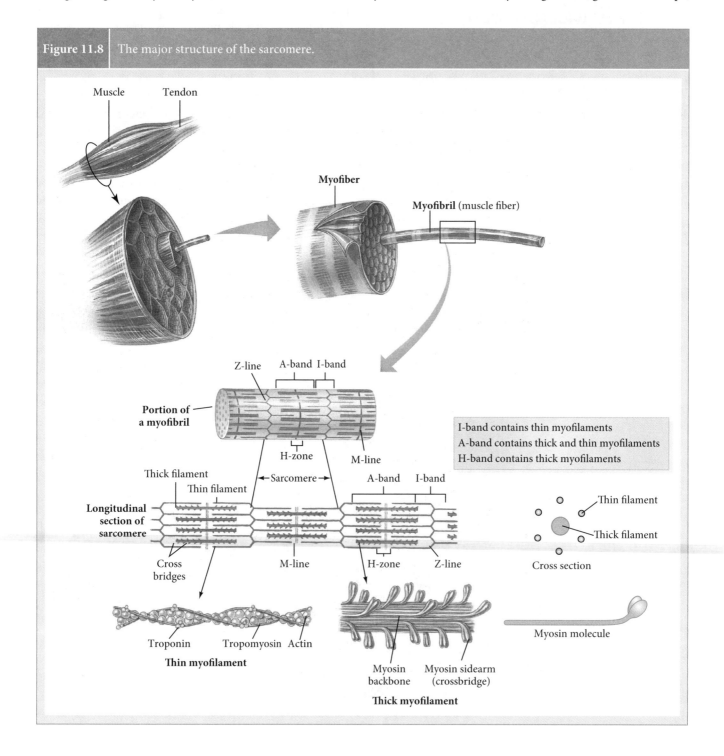

I-band contains thin myofilaments
A-band contains thick and thin myofilaments
H-band contains thick myofilaments

and pull it toward you. That section of the rope slides past you. Then you would reach out with your left hand, grab another section of the rope, and pull it toward you. Next, you would remove your right hand from the rope, reach out again, grab, and pull. The whole process would be repeated in rapid succession to pull in the rope: release, reach out, grab, pull. Technically, you could reach out with both hands at the same time and give the rope a greater tug, but in doing so you would have no tension on the rope while you were reaching out. To maintain some tension on the rope, you would have to alternate hands.

This process is very similar to what happens between the actin and myosin filaments during a muscle action. Think of the rope as the actin filament and you as the myosin filament. Your arms are the myosin cross-bridges. In terms of muscle dynamics, myosin cross-bridges must first attach to the binding site on the actin filament (they are grabbing on the "rope"). Then, the myosin cross-bridge will pivot, pulling on the actin filaments and causing them to slide over the myosin filaments. This is called the power stroke, and its effect is to cause the sarcomere to shorten by a very small amount. The myosin cross-bridge must then detach from the actin site and recock. The whole process—reach out, grab, pull, release—has to be repeated very rapidly for the entire muscle to shorten by a noticeable amount.

To understand how the sarcomeres work in concert to produce force, let us begin with something you know. Think of a 10 kilogram (98.1 N) box suspended by a chain (**Figure 11.9**). If you assume that the chain is massless, then the tension in the chain is equal to the weight of the box (98.1 N). The tension in each individual link would be the same as the tension in the entire chain (98.1N). It does not matter how long the chain is. Adding more links in series (end-to-end) does not change the amount of tension in the chain or in the individual links. But if you put two chains side-by-side (in parallel), then the tension in each chain (and each individual link) is halved (49.05 N).

Now, instead of metal links, think of a chain made up of rubber bands. The same idea applies. Assuming that all the rubber bands in the chain are identical, then the amount of force required to stretch the chain does not change when the rubber bands are arranged in series. However, the amount of force required to stretch the rubber bands increases when they are arranged in a parallel. In addition, think about how far you can stretch the chain of rubber bands. Say you could stretch one rubber band 10 centimeters before it breaks. If you put two rubber bands in parallel, they would still break if the chain is stretched beyond 10 centimeters (although it would take twice as much force to stretch them that far). If you put the rubber bands in series, the chain can now stretch 20 centimeters—10 centimeters for each rubber band (**Figure 11.10**).

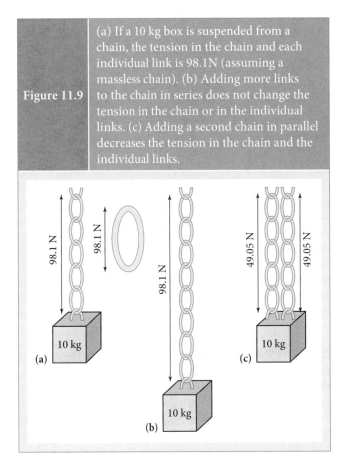

Figure 11.9   (a) If a 10 kg box is suspended from a chain, the tension in the chain and each individual link is 98.1N (assuming a massless chain). (b) Adding more links to the chain in series does not change the tension in the chain or in the individual links. (c) Adding a second chain in parallel decreases the tension in the chain and the individual links.

Now, instead of metal links or rubber bands, think of sarcomeres, which are little force actuators. The sarcomeres can be arranged side-by-side, next to each other. In this case, the sarcomeres are said to be arranged in parallel. When the sarcomeres are arranged end-to-end, they are arranged in

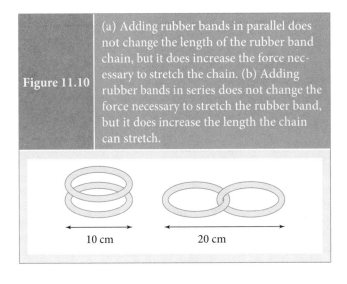

Figure 11.10   (a) Adding rubber bands in parallel does not change the length of the rubber band chain, but it does increase the force necessary to stretch the chain. (b) Adding rubber bands in series does not change the force necessary to stretch the rubber band, but it does increase the length the chain can stretch.

series. It is important to understand that when sarcomeres are arranged in parallel, their force characteristics are additive. That is, the force produced is the sum of the two sarcomeres. However, the length characteristics cannot be added: The length of the two sarcomeres in parallel is not any longer than the individual sarcomeres. The opposite situation occurs when sarcomeres are arranged in series. The force of two sarcomeres in series is the same as the force in an individual sarcomere and cannot be added. The length characteristics of sarcomeres in series can be added.

> **Important Point!**   Muscle architecture determines both the force-producing capabilities and operating range of a muscle.

Muscle architecture refers to how fibers of a muscle are arranged relative to the vector of the force generation (the vector between the two insertion points). The simplest arrangement is called a parallel or fusiform muscle arrangement (**Figure 11.11a**). With this arrangement, all the fibers are aligned parallel to the vector of force generation. And although all the fibers do not necessarily run the entire length of the muscle belly, the fibers are relatively long.

One downside of this arrangement is that only so many fibers can be placed in a certain volume (physical space), which limits the amount of force that a muscle can produce. To increase the amount of force that can be produced by a muscle occupying the same volume, the fibers can be arranged at an angle to the vector of force generation (**Figure 11.11b**). Such an arrangement is referred to as a pennate arrangement, and the angle between the fiber orientation and the vector of force generation is called the pennation angle.

With the parallel arrangement, all the force of the fiber is directed along the vector of the force generation of the muscle. This is not the case with the pennate arrangement. Because of the angle between the fiber orientation and the force vector, a correction factor must be applied, and that correction factor is the cosine of the pennation angle:

$$^{\text{muscle force}}F = {}^{\text{fiber}}F \times \cos\theta_{\text{pennation}} \qquad (11.1)$$

Remember the leading superscript indicates direction. Looking at these arrangements, you might be tempted to say that the parallel arrangement can produce more force because it has a pennation angle of zero, and hence a correction factor of one. Any pennation angle that is greater than zero will have a correction factor of less than one. But you have to consider not only the force of the fibers, but also the number of fibers that are pulling on the tendon: the greater number of fibers with the pennate arrangement more than makes up for the loss of force due to the pennation angle.

To illustrate the differences in muscles, compare the four muscles in **Figure 11.12**. Although the sartorius (SAR) is not a muscle that is part of the calf, it is put in there to show you the dramatic differences between muscle architectures. The sartorius, which crosses both the hip and the knee, has very long muscle fibers and a pennation angle of close to zero. It has one of the smallest pennation angles and longest fiber lengths of any muscle in the lower extremity.[4] Contrast its properties with the soleus (SOL), which has one of the largest pennation angles and shortest muscle fibers of any muscle in lower extremity.[4] The medial (MG) and lateral (LG) gastrocnemius'

Figure 11.11	Two basic muscle architecture types. (a) Parallel muscle type. (b) Pennate muscle type.

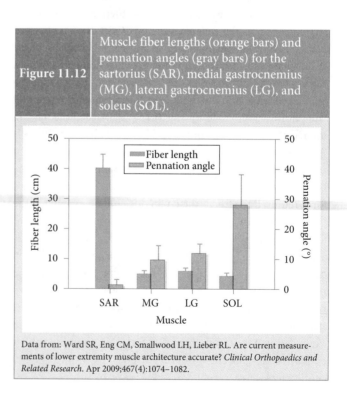

Figure 11.12	Muscle fiber lengths (orange bars) and pennation angles (gray bars) for the sartorius (SAR), medial gastrocnemius (MG), lateral gastrocnemius (LG), and soleus (SOL).

Data from: Ward SR, Eng CM, Smallwood LH, Lieber RL. Are current measurements of lower extremity muscle architecture accurate? *Clinical Orthopaedics and Related Research.* Apr 2009;467(4):1074–1082.

properties fall in between the two extremes for both pennation angle and fiber length.[4]

Although there are other types of fiber arrangements, these two (parallel and pennate) will be the main focus for this lesson. The most important distinction to make is that the parallel muscle will have longer fibers (more sarcomeres in series) and, despite its name, less fibers in parallel. The pennate muscle will have shorter fibers (fewer sarcomeres in series) but more sarcomeres in parallel. The functional implications of these arrangements will be discussed later.

> **Important Point!**   Pennate muscles have more muscle fibers in parallel, but shorter fiber lengths. Parallel muscles have longer fibers, but there are fewer of them in parallel.

11.2.2   Tendon

To be clear, there are many passive elements within the muscle–tendon complex. The biggest include the tendon and aponeuroses, followed by the fascia. There are also elastic elements within the cross-bridge, the actin and myosin filaments, and the structural protein titin. You may learn about these in more advanced textbooks and journal articles. For this lesson, we will lump all these structures together and call them "tendon," even at the risk of the ire of some anatomists and physiologists.

The primary structural component of the passive components is Type I collagen. At their resting lengths, the collagen fibers are not straight, but have a wavy configuration, which has been referred to as a natural crimp.[5] For the fascia, they are arranged more like sheets. The fascia at the various levels converge at the ends of the muscle bellies to form tendons. For tendons, collagen is arranged in parallel bundles similar to the way muscle fibers are in the muscle belly.

The tendons are viscoelastic. The term *viscoelastic* should be reviewed from the lesson on tissue mechanics. What is critical to remember for the discussion here is that tendons have two important (viscous and elastic) properties. Because you are probably more familiar with elastic properties, let us discuss them first.

> **Important Point!**   Tendons have both elastic and viscous properties.

Although the tendons are often thought of (and modeled as) springs, I like to think of them as rubber bands. First, rubber bands provide a force in only one direction—when it is being pulled apart (tension). Unlike a spring, the rubber band offers no resistance when you bring its ends closer

together. Second, the rubber band will only provide resistance after it has been stretched beyond its resting length. If the rubber band is scrunched together, you have to take up the slack before it provides any tension. Third, not all rubber bands offer the same resistance. Some rubber bands are fairly easy to stretch (they are **compliant**), whereas other rubber bands are hard to stretch (they are **stiff**).

> **Compliant**   The opposite of stiffness; a compliant tendon is not very stiff
>
> **Stiffness**   A resistance to change in deformation; an object with greater stiffness requires greater forces for the same amount of deformation

These same qualities are found in the passive elements of the MTC. First, they provide a force only when they are stretched. Particularly in the case of the tendon, the slack must be taken up before the force generated by the muscle can be transmitted to the bone. And the stiffness of the tendon plays a large role in determining the behavior of the MTC.

> **Important Point!**   Tendons only provide force when they are stretched.

Stiffness is the resistance to stretch of the tendon. The stiffer the material, the more force it provides in the direction opposite the stretch. In addition, the more you stretch it, the more force it provides. This is captured by the equation:

$$F_k = k\Delta l \tag{11.2}$$

In this equation, the variable k stands for stiffness and Δl stands for the amount of stretch imposed on the tendon. A stiffer or more elongated tendon will offer more passive resistance. Note that this force can only go in one direction. See **Figure 11.13**. Many factors can decrease tendon stiffness, including aging, pregnancy, immobilization, corticosteroids, and injury.[6]

Second, the viscoelastic elements increase force with the increasing velocity (or rate) of stretch. Think of moving your hand through water. The faster you move your hand, the more force there is exerted by the water. Likewise, the passive elements will also produce force in response to the rate of stretch:[7]

$$F_v = bv \tag{11.3}$$

Where v is the velocity of stretch and b is the proportionality constant. What this means is that force increases with an increased rate of stretch.

The force produced in the passive elements is thus the sum of the force due to both its viscous and elastic properties:[7]

$$F_p = F_k + F_v \tag{11.4}$$

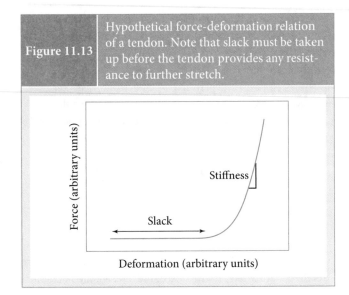

Figure 11.13 | Hypothetical force-deformation relation of a tendon. Note that slack must be taken up before the tendon provides any resistance to further stretch.

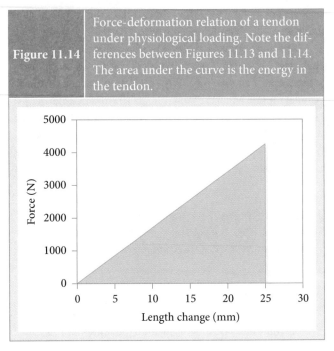

Figure 11.14 | Force-deformation relation of a tendon under physiological loading. Note the differences between Figures 11.13 and 11.14. The area under the curve is the energy in the tendon.

If the actions are pretty slow, the velocity (and F_v) goes to zero. In such cases, the force produced by the passive elements is proportional to their stiffness. As the velocity increases, you would add the F_v component to the total force. Adding the F_v component has the functional consequences of increasing the stiffness, strength, deformation, and energy absorbed by the tendon.[6]

The Energy in a Tendon

Figure 11.13 showed you the force-deformation curve for a hypothetical tendon loaded with an arbitrary amount of force. Theoretically, this is the amount of deformation a tendon could endure before it would rupture. However, tendons rarely deform that much. It is estimated that tendons typically undergo strains of approximately 3% during normal activity,[8] although strains of the Achilles tendon during one-legged hopping have been estimated to be as high as 10%.[9] Even at such high strains, the force-deformation relation during physiological conditions is approximately linear (i.e., a straight line; Figure 11.14).

Because this relationship is approximately linear, you can calculate the area under the curve fairly easily using the formula for a triangle (one-half the base times the height):[10]

$$W_{tendon} = \Delta E_{tendon} = \frac{1}{2}F \times \Delta l \qquad (11.5)$$

The area under the curve represents that energy that is in the tendon. If a tendon is being stretched, that is the energy being stored in the tendon. If the tendon is shortening, it is the amount of energy being released by the tendon. As you will see in the section on the stretch-shortening cycle, the amount of energy stored and released by the tendon is not equal.

Important Point!   The area under the force-deformation curve indicates the amount of energy in the tendon.

11.2.3   Muscle–Tendon Interactions

How the passive components, particularly the tendon, interact with the muscle has an extremely important effect on function. First, a somewhat stiff tendon is necessary to transmit force to the bone. Think of how hard it would be to control a rigid body if there was a very compliant rubber band between it and the motor. However, a compliant tendon has its advantages. It can **decouple** the displacements of the muscle and the MTC.[11] In other words, the displacements of the muscle bundles do not have to be the same magnitude, or even the same direction, as the displacement of the MTC (see Box 11.1). For example, the MTC and tendon can be lengthening while the muscle fibers themselves are shortening. Even if the MTC, muscle, and tendon are all shortening, the velocity of shortening can be different between the muscle and the tendon.[12] With shorter or stiffer tendons, there will be tendency for the MTC and the muscle to have the same displacements (and velocities). With longer or more compliant tendons, there can be a greater degree of uncoupling.[11] Yet if the tendons were too compliant, the muscle would need a greater displacement (and velocity) to transmit force to the bone.

| Decouple   Allow joined subsystems to operate independently |

There seems to be an optimal level of stiffness for tendons to both transmit force to the bone and decouple displacements between the muscle and the MTC.[12] Other implications of the interactions between the active and passive components are discussed later.

Important Point!   Muscle and MTC displacements are not necessarily coupled.

Box 11.1	Applied Research: Using Ultrasound to Determine Decoupled Movements of the Muscle and Tendon

You have probably heard of ultrasound as it applied to prenatal care: It is used to determine the number, gender, and other health aspects of a fetus(es) inside a mother's womb. With ultrasound, high-frequency sound waves are used to produce pictures of inside the body, and it can be used to image muscles, tendons, and other organs. Ultrasound technology was considered a huge breakthrough because it allowed scientists to determine mechanical properties of muscles and tendons in vivo. In this study, investigators used ultrasound to measure both fascicle and tendon length changes during isometric actions of the tibialis anterior during isometric dorsiflexion. They found that during these "isometric" actions, the muscle fascicles shortened and the tendon lengthened by about 14 millimeters. This was one of the first studies to show that the action of the MTC and the muscle fibers are not necessarily the same, and that the tendon can decouple displacements of the muscle and MTC.

Data from: Ito L, Kawakami Y, Ichinose Y, Fukashiro S, Fukunaga T. Nonisometric behavior of fascicles during isometric contractions of a human muscle. *Journal of Applied Physiology*. Oct 1998;85(4):1230–1235.

Section Question Answer

Yes, the muscle and tendon displacements are not necessarily in the same direction all the time. During eccentric actions, the tendon can be lengthening while the muscle length is constant. During isometric actions, the muscle can be shortening while the tendon is elongating. During the stretch-shortening cycle, the muscle may be shortening while the tendon is elongating. And during concentric actions, the tendon can be shortening at a faster rate than the muscle.

COMPETENCY CHECK

Remember:

1. List the components of the muscle–tendon complex.
2. What are the two basic types of muscle architecture?

3. Explain how muscle produces force.
4. Explain how tendon produces force.
5. What are the different properties of tendon that contribute to force?

Understand:

1. What are the differences between a muscle that has sarcomeres arranged in series and a muscle that has sarcomeres arranged in parallel?
2. What are the differences between the forces produced by the elastic and viscous properties of the tendon?
3. How much energy can be absorbed by a tendon?

Apply:

1. When would it be advantageous to have a more compliant tendon?
2. When would it be more advantageous to have a stiffer tendon?

11.3   FACTORS AFFECTING MTC MECHANICS

Section Question

In one experiment,[13] subjects lay prone on a sledge apparatus and "jumped" using only their ankle joints. Three different jump techniques were used. In the squat jump, subjects started from a maximally dorsiflexed position and "jumped" by forceful plantar flexion alone. In the countermovement jump, subjects performed the jumps similar to the squat jump, but started in a maximally plantar flexed position before rapidly moving into the dorsiflexed position and rebounding into the jump. Which technique do you think produced the greatest jump height? Why?

The MTC is biological; it is not a machine. Unlike machines, it will not produce the same amount of force all the time, under every condition. Seven things affect the MTC's force-producing capability (Figure 11.15):

- Physiological cross-sectional area
- Muscle–tendon complex length
- Type of action
- Velocity
- Fiber type
- Recruitment
- Time history of activation

In addition, the MTC demonstrates a great deal of what is called plasticity. That means that it can

Plasticity	The ability to adapt or change

| Figure 11.15 | Factors affecting a muscle's force-producing capabilities. |

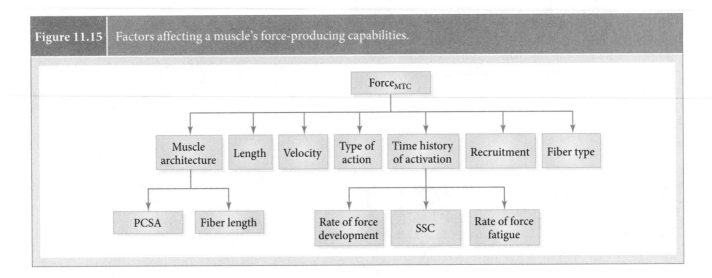

adapt to chronic changes in loading and aging. In this section, you will first explore the factors that influence an MTC's force-producing capability, and then you will examine some of the adaptations to aging, disuse, and exercise.

11.3.1 Physiological Cross-Sectional Area

You have probably heard that someone with big muscles is strong, that a bigger muscle is a stronger muscle, or some other variation of that statement. Statements like these are not 100% accurate. Assume for a second that the conditions are ideal for the production of force. Is this a true statement? Will a bigger muscle produce more force? Not exactly.

It depends on how you define "big." A muscle will occupy a certain volume (space) and have mass, but neither the volume nor the mass are directly related to the force-producing capabilities of the muscle because the sarcomeres in series do not increase the force. It is not the volume, but the cross-sectional area, that determines how much force a muscle can produce. But determining the *anatomical* cross-sectional area only works for a muscle with a parallel arrangement of its fibers. It does not work for a muscle with a pennate arrangement because it does not account for all the fibers producing force (**Figure 11.16**). To account for all the fibers producing force, you must calculate the *physiological* cross-sectional area (PCSA). The techniques and calculations involved in calculating the PCSA are beyond the scope of this lesson, but many good references are available if you are interested in learning more about it.[14]

Important Point! It is the PCSA, and not the anatomical CSA, that determines how much force a muscle can produce.

Two things determine the PCSA: the number of muscle fibers in parallel and the size of those fibers. The number of fibers is related to the muscle architecture: A pennate arrangement will have a larger number of fibers contributing to the force than will a parallel muscle fiber. The number of fibers appears to be genetically determined, and there is no evidence to suggest that you can increase the number of fibers (**hyperplasia**) in a muscle. However, you can increase

Hyperplasia An increase in the number of muscle fibers

| Figure 11.16 | The physiological cross-sectional area only equals the anatomical cross-sectional area for a parallel muscle arrangement. (Top) All of the fibers are accounted for with the parallel muscle arrangement. (Bottom) Only a portion of the fibers are accounted for with a pennate muscle arrangement. |

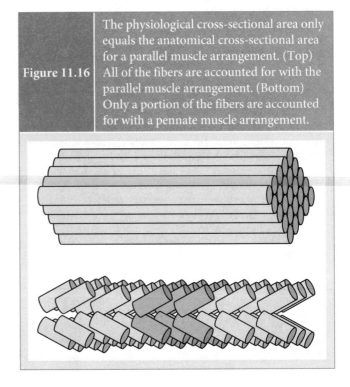

Hypertrophy An increase in the size of the muscle fibers

the size of the muscle fibers (**hypertrophy**) through heavy, progressive resistance exercise.

11.3.2 MTC Length

MTC length is another factor influencing the force it can produce. When considering the MTC length, you must think in terms of sarcomere length, fiber length, and the length of the MTC as a whole.

Sarcomere length is related to the number of actin–myosin cross-bridges that can be formed (**Figure 11.17**). At

approximately the sarcomere's resting length (Figure 11.17b), the maximum number of cross-bridges is formed. When the sarcomere is in its shortened position (the ascending limb, Figure 11.17a), there is a large overlap between the actin and myosin heads, decreasing the number of cross-bridges that can be formed. When the muscle is in a lengthened position (the descending limb, Figure 11.17c), there is also a decrease in the number of cross-bridges that can be formed. A decrease in the number of cross-bridges corresponds to a decrease in the amount of force produced by the muscle fiber.

Remember that the muscle fiber length corresponds to the number of sarcomeres in series, and when sarcomeres are in series, the changes in sarcomere length are additive. For example, say that a muscle fiber length shortens by 1 centimeter. If there are 10,000 sarcomeres in series, then each sarcomere has to shorten by 1 micrometer. If there were only 5,000 sarcomeres in series, each sarcomere would have to shorten by twice (2 μm) as much for the whole fiber to shorten by the same 1 centimeter. Can you guess what would happen if there were 20,000 sarcomeres in series? Every sarcomere would only have to shorten by half as much (0.5 μm) for the same change of 1 centimeter.

Do not assume that every muscle will operate through the entire range in Figure 11.17. Muscles tend to operate in a more narrow range in vivo. Consider the medial gastrocnemius during calf raises and jumping (**Figure 11.18**).[15] Notice how the sarcomere lengths during activities do not encompass the entire area seen with experimental data.

The MTC length is not only determined by the fiber length. The tendons also contribute to the total amount of

Figure 11.17 The force-length curve of an individual sarcomere. (a) the ascending curve, where the actin and myosin overlap too much. (b) The plateau region, where there are an optimal number of actin–myosin interactions. (c) the descending limb, where the actin and myosin are too far apart.

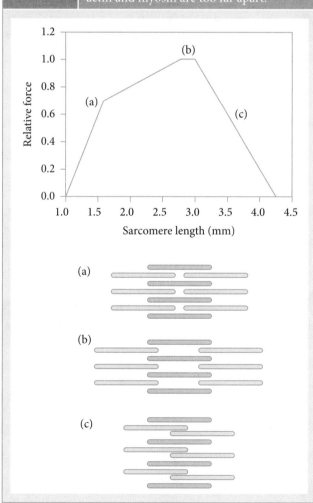

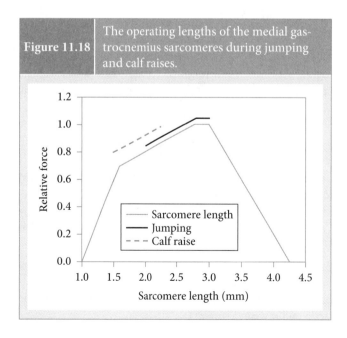

Figure 11.18 The operating lengths of the medial gastrocnemius sarcomeres during jumping and calf raises.

| Figure 11.19 | The force-length curve of a muscle–tendon unit. Muscle (solid black line), tendon (dashed black line), and the muscle–tendon complex (dashed orange line). |

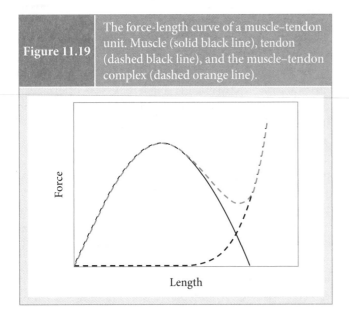

force produced by the MTC, albeit passively. In a shortened position, the tendons produce no additional force. As the MTC lengthens, tension in the tendons increases (think about stretching a rubber band), and the force in the MTC increases (review Figure 11.13).

Two things should be pointed out. First, the total length of the MTC is determined by the length of both the muscle and the tendon. Because both components are involved, a change in length of the MTC does not correspond to a change in length of the muscle. Second, the total force produced by the MTC is determined by both the muscle and the tendon (Figure 11.19).

11.3.3   MTC Action

The force produced by the MTC is also determined by its action. Eccentric actions can produce more force than isometric actions, which can produce more force than concentric muscle actions. Because there is nothing inherent about the descending limb of the force-length curve of a sarcomere (review Figure 11.17), these differences in force production due to MTC action must be viewed with respect to the tendon and contraction dynamics.

First, think about why isometric muscle actions may produce more force than eccentric muscle actions. To assist you, think back to the rope analogy. After every power stroke, one of your hands (the myosin heads) must let go of the rope before it reaches out and grabs another section of the rope. Do you think you can develop more tension on the rope if you just maintained both hands on the rope rather than alternating hand over hand? It is probably the same thing with the actin and myosin filaments.

If that is the case, why can you develop more force eccentrically than isometrically? It is important to remember that just because the MTC is lengthening, it does not mean that the muscle fibers themselves are lengthening. There is growing evidence to suggest that during eccentric actions, the muscle fiber is acting isometrically and the tendons are lengthening. Forces developed passively by the tendons are added to the active force generated by the muscle itself, increasing the total force that can be produced.

Do not confuse the various MTC actions (concentric, isometric, eccentric) with the length-tension relations. An MTC can have an eccentric action in a shortened position, just as it can have a concentric action in a lengthened position.

> **Important Point!**   Do not confuse the type of action with the MTC length.

11.3.4   Velocity

Velocity also affects the force an MTC can produce, but its effect is determined by both the type of action and the fiber length. The relation between force and velocity with the various types of actions is presented in Figure 11.20. You should first verify what was explained earlier: Eccentric actions can produce greater force than isometric actions, which can produce greater force than concentric actions. Second, you should note that during concentric muscle actions, there is a nonlinear, inverse relation between force and velocity. As velocity increases, force decreases and vice versa. You should recognize this from your everyday experience. The speed of movement decreases as you lift heavier objects. Third, notice that eccentric actions do not appear to be as affected by velocity as concentric muscle actions. The most likely reason for this relates to the force produced by the tendon. The tendon can act as a speed buffer. Review Equations 11.2 and 11.3.

| Figure 11.20 | The force-velocity relation of a muscle-tendon complex. |

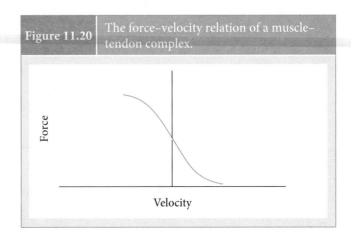

Table 11.1	The Velocities of Shortening for Three Different Fibers With a Different Number of Sarcomeres in Series		
Each fiber is shortening at 2 centimeters per second			
Number of sarcomeres in series	Length change of each sarcomere (μm)		Velocity of shortening (μm/sec)
5,000	2		4
10,000	1		2
20,000	0.5		1

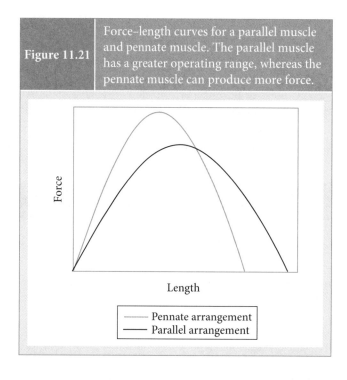

Figure 11.21 Force–length curves for a parallel muscle and pennate muscle. The parallel muscle has a greater operating range, whereas the pennate muscle can produce more force.

Notice that velocity does not have an effect on the elastic properties of the tendon (Equation 11.2), and increasing velocity increases force due to the viscous properties of the tendon (Equation 11.3).

The fiber length also affects the velocity, which also affects the force. In the section on length, you saw muscle fibers that contained differing numbers of sarcomeres (repeated in Table 11.1). Each muscle fiber had to shorten by 1 centimeter. When a fiber had a smaller number of sarcomeres, each sarcomere had to shorten by a greater amount to achieve the goal of 1 centimeter. As demonstrated in Table 11.1, if the shortening had to occur in a specified amount of time (in this case, 0.5 sec), then the fiber with fewer sarcomeres in series had to have a greater velocity of shortening. With a greater velocity of shortening, there is a greater decrement in force production.

Force, Length, Velocity, and Muscle Architecture

In the previous section, it was pointed out that there are two broad categories of muscle architecture: parallel and pennate. Earlier, the generic effects of PCSA, length, and velocity on muscle force were examined. Here, you will see how these various factors interact. The key to understanding this is realizing that pennate muscles usually have relatively short fibers and relatively large PCSAs. Conversely, parallel muscles tend to have relatively long fibers and relatively small PCSAs.

Figure 11.21 shows a general force–length curve for these two types of muscles. With its larger PCSA, the pennate muscle can generate more force. With its longer muscle fibers, the parallel muscle operates over a greater range. So you should not think of one muscle type as being superior to another; they each have their advantages and disadvantages.

Figure 11.22 shows a concentric force–velocity curve for the same two types of muscles. The most striking feature of this graph is that the pennate muscle does not *always* produce more force than the parallel muscle. Certainly the

pennate muscle produces more force at slower contraction velocities. But, because of its shorter fibers, the pennate muscle is more affected by velocity than the parallel muscle. This means that the parallel muscle can operate at higher shortening velocities than a pennate muscle, and after a certain

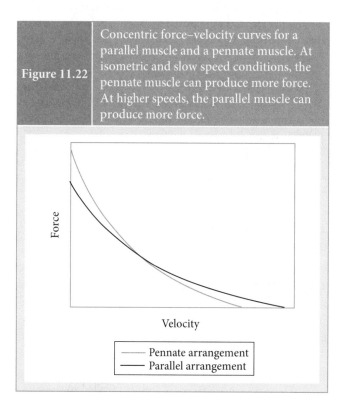

Figure 11.22 Concentric force–velocity curves for a parallel muscle and a pennate muscle. At isometric and slow speed conditions, the pennate muscle can produce more force. At higher speeds, the parallel muscle can produce more force.

velocity, the parallel muscle can actually produce more force than the pennate muscle.

Form dictates function. These muscle arrangements are not arbitrary but very much speak to the functions that these various muscles have to perform. But why have these two types of arrangements? Why not just have one big "super" muscle that has both a large PCSA and long fiber lengths? The answer to that question is in Box 11.2.

| Box 11.2 | Applied Research: Diversity in Muscle Architectural Design |

You have seen how muscles can have two different architectural designs: a parallel muscle with relatively long fibers and small PCSA, and a pennate muscle with relatively short fibers and a large PCSA. What is the purpose of having muscles with these two different designs, particularly when two muscles with different designs but the same function are placed right next to each other? This is the case with the extensor carpi radialis longus (ECRL) and extensor carpi radialis brevis (ECRB) of the wrist. They are located very close to one another, and they both extend the wrist. The ECRB can be described more as a pennate muscle, whereas the ECRL is more of a parallel muscle. If you looked at the force-velocity curves of these two muscles, they would look very similar to the one in Figure 11.19: The ECRB is stronger isometrically and at slower shortening velocities, but the ERCL becomes stronger when velocities exceed 80 millimeters per second. So why have these two distinct muscles rather than one big muscle? Using the architectural design parameters, investigators determined that to have one muscle that could produce the same force and have the same velocity of shortening, this one "supermuscle" would have a mass that was 30% larger than the combined masses of the ERCL and ECRB. They concluded that having two muscles with different architectural properties allowed for the accomplishment of the same tasks but with a lower muscle mass, which is important from an economy point of view.

Data from: Lieber RL, Ljung BO, Friden J. Intraoperative sarcomere length measurements reveal differential design of human wrist extensor muscles. *Journal of Experimental Biology.* Jan 1997;200(1):19–25.

11.3.5   Fiber Type

You may have thought that all skeletal muscle was the same, but this is hardly the case. Muscle fibers can have a variety of different structural and functional characteristics. Originally, fibers were characterized by either their twitch characteristics (fast-twitch, slow-twitch), appearance (red, white), or metabolism (oxidative, glycolytic). Currently, muscle fibers are characterized by the form of their myosin heavy chain protein (called an isoform).

Isoform   A different form of the same protein

Table 11.2	Characteristics of Type I and Type II Muscle Fibers	
Characteristic	**Type I**	**Type II**
Motor neuron size	Small	Large
Nerve conduction velocity	Slow	Fast
Contraction speed	Slow	Fast
Fatigue resistance	High	Low
Fiber diameter	Small	Large

Adapted from: Hunter GR, Harris RT. Structure and function of the muscular, neuromuscular, cardiovascular, and respiratory systems. In: Baechle TR, Earle RW, eds. *Essentials of Strength Training and Conditioning.* 3rd ed. Champaign, IL: Human Kinetics; 2008:3–20.

There are two, rather broad categories of muscle fibers based on their myosin heavy chain: Type I and Type II. Within the Type II classification, there are even more subtypes (IIc, IIa, IIx, IIb) and even intermediary types within the subtypes (IIac, IIax, etc.). All this probably means that muscle fibers fall along a continuum rather than being neatly contained within a particular fiber type. For purposes of this lesson, you will be mainly concerned with the differences between Type I and II fibers. Although some authors downplay the significance that is placed on fiber types,[14] some of the general differences between fiber types are listed in Table 11.2.[16]

Remember that the amount of force a muscle can produce is proportional to the cross-sectional area. The amount of force that can be produced per cross-sectional area is known as the **specific tension**. Although Type I and Type II fibers have been shown to have different specific tensions, they are not as great as you might think. Some investigators have suggested that the specific tension of Type I fiber is ~19 N/cm^2 and the specific tension of Type II fibers is ~24 N/cm^2.[17] Other researchers have suggested that there is no difference in the specific tension of the different fiber types (~22.5 N/cm^2).[14] However, Type II fibers have been shown to produce about 50% more force than Type I fibers.[18]

Specific tension   The amount of force that can be produced per cross-sectional area

Assume for the moment that there is a difference in specific tension between fiber types. Using data on PCSA,[4] fiber type distribution,[19] and specific tension,[17] admittedly (very) rough calculations were made concerning the maximum force-producing capabilities between the gastrocnemius and soleus in Table 11.3. As you can see, even though the gastrocnemius has a much higher percentage of Type II fibers, it still produces about one-third less force than the soleus. The PCSA is the dominant determiner of the force produced.

Table 11.3	Rough Calculations Comparing the Gastrocnemius and Soleus					
	Gastrocnemius			**Soleus**		
	Type I	Type II	Total	Type I	Type II	Total
% of fibers	48	52		88	12	
Total fibers	441216	477984		1549152	211248	
CSA of fiber	2820	3840		2820	3840	
Total CSA	12.4	18.4	30.8	43.7	8.1	51.8
Specific tension	19	24		19	24	
Force produced	236.4	440.5	676.9	830.0	194.7	1024.7

But the effect of fiber type on force is only part of the story, and a very small one at that. Of all the differences between fiber types, the specific tension is probably the least significant.[20] Each fiber type is also affected differently by the shortening velocity. Type II fibers have a maximum shortening velocity that is about 400% of the velocity of Type I fibers. In addition, Type I fibers are more affected (have a steeper decline) by increasing velocity than Type II fibers (Figure 11.23).[18]

Important Point! Specific tension is the least important difference between fiber types.

What is the functional significance of these differences? Recall that power is the product of force and velocity. By plotting the power of the different fiber types (similar to what was done in Figure 11.24), you will see that Type II fibers have a greater (900%) maximum power than Type I fibers, and the load and velocity at which this peak power is achieved is greater in the Type II fibers. These findings suggest that power output is optimized at different speeds by the appropriate recruitment of different fiber types (see later discussion)[18] and may be the most prominent distinction between fiber types.[20]

Type I fibers are more suited for sustained isometric and slow actions because they can develop almost as much force as Type II fibers, but develop their maximum power with their highest efficiency at low velocities, consume less energy, and fatigue very little.[20] Type II fibers are better suited for fast and powerful movements because they can produce

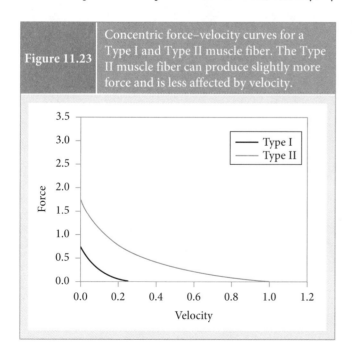

Figure 11.23 Concentric force–velocity curves for a Type I and Type II muscle fiber. The Type II muscle fiber can produce slightly more force and is less affected by velocity.

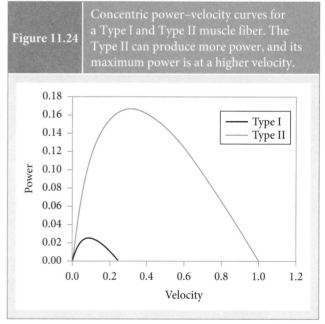

Figure 11.24 Concentric power–velocity curves for a Type I and Type II muscle fiber. The Type II can produce more power, and its maximum power is at a higher velocity.

more power and produce that power with the highest efficiency at high velocities, but they can sustain those outputs for only a short burst.[20] The continuum of muscle fibers allows you to produce a wide array of movements in between these two extremes.

11.3.6   Recruitment

The recruitment of muscle fibers also determines the amount of force the MTC can produce. Although the basic operating unit of a muscle fiber is the sarcomere, the basic operating unit in terms of recruitment is the **motor unit**. A motor unit is an alpha motor neuron and all the muscle fibers that it innervates. Each fiber in a motor unit is of the same fiber type. Indeed, there is good evidence to suggest that a muscle's fiber type is determined by the motor neuron.[21] Recruitment of motor units, and hence the modulation of forces, appears to be governed by two principles: the all-or-none principle and the size principle.

> **Motor unit** A motor neuron and all the muscle fibers it innervates

The all-or-none principle states that an action potential (electrical impulse) does not directly excite muscle fibers, but does so through chemical processes. If the impulse is strong enough to release a sufficient amount of neurotransmitter, then the muscle fibers innervated by that nerve will contract. It is similar to firing a gun:[16] The gun will fire only if the force on the trigger is large enough, and squeezing the trigger harder will not affect the bullet. However, if another action potential reaches the muscle fibers before they relax, then the muscle fibers could contract harder (**Figure 11.25**). This process goes by various names: temporal summation, firing rate, firing frequency, and rate coding.

Force changes via rate coding can only make small changes in the level of force.[14] To accommodate the large range of forces needed in daily activities, you need to recruit various combinations of motor units. This is done through a process that follows the size principle. The size principle states that motor units are recruited in an orderly process from smallest to largest. Type I fibers, which are the smallest, are also the least fatigable and most easily recruited. As more force is needed, increasingly larger Type II motor units are recruited, which leads to a smooth, gradual proportional increase in force. Think about how difficult it would be to control your force output if you recruited motor units in a haphazard fashion, recruiting easily fatigable fibers first, or having wild fluctuations in the force produced by the motor units you recruited, and it makes sense. Of course, this is for a smooth, gradual increase in force. If you need to produce a large amount of force very quickly, you would want to recruit as many motor units as possible nearly simultaneously.

The modulation of force produced by the muscle is governed by these two principles: Force output is changed by

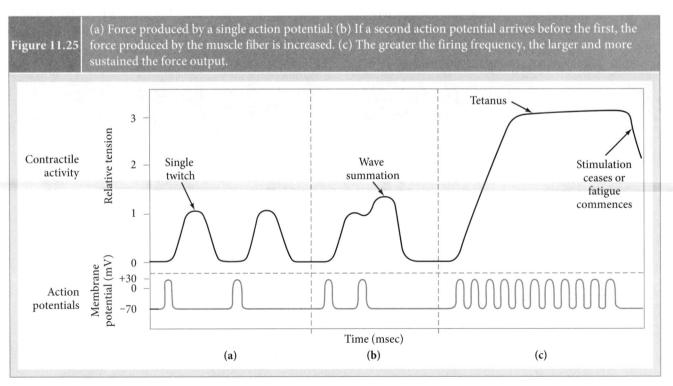

Figure 11.25 (a) Force produced by a single action potential: (b) If a second action potential arrives before the first, the force produced by the muscle fiber is increased. (c) The greater the firing frequency, the larger and more sustained the force output.

either increasing the firing frequency of the activated motor units, or activating more motor units. Both processes lead to increased electrical activity around the muscle. The electrical activity can be measured through a technique called electromyography. See **Box 11.3**.

Important Point!   The magnitude of force is modulated by changing the number of motor units activated and the firing frequency of those motor units.

Box 11.3	Applied Research: Electromyography

A motor unit is activated by an action potential. This action potential involves electrochemical processes. Amplifying and recording the electrical activity of the motor unit action potential is called electromyography (EMG). EMG can be performed non-invasively by using electrodes attached to the surface of the skin (surface EMG) or intramuscularly by inserting an electrode into the muscle belly (fine-wire EMG).

The strength of the EMG signal gives you an indication of the activity level of the muscle by either an increase in firing frequency or the number of motor units activated (it cannot separate the two). When all other factors are equal, an increase in the EMG signal indicates an increase in muscle force. The hard part is getting all other factors to be equal. As you learned, myriad factors affect muscle force. And so it is very hard to determine muscle force from an EMG signal.

That does not mean that EMG is not useful. Quite the contrary. It can tell you if a muscle is active or not during an activity, the timing of that activation, and the relative strength of the activation. It can also provide useful measures of fatigue. You cannot equate the strength of the EMG signal to the force produced by a muscle.

In this investigation, EMG was used to determine the function of the gastrocnemius and soleus during hopping. Interestingly, the gastrocnemius was activated 100 ms before landing and in anticipation of it. The same was not true of the soleus: It seems to be activated in response to landing, as part of the stretch reflex. Both muscles are active as part of the takeoff. Although it was shown that the two muscles are acting differently during the landing of a hop, it could not yet be explained why they would be different. The authors speculated that it may have to do with the fact that the gastrocnemius crosses both the hip and knee, whereas the soleus only crosses the ankle. However, changes in muscle architecture or fiber type composition cannot be ruled out. Clearly, there is a lot more to learn about how muscles function together to produce movement.

Data from: Funase K, Higashi T, Sakakibara A, Imanaka K, Nishihira Y, Miles TS. Patterns of muscle activation in human hopping. *European Journal of Applied Physiology*. Jun 2001;84(6):503–509.

11.3.7   Time History of Activation

The MTC does not develop force instantaneously. Force development takes times. In addition, the force developed by the MTC does not last indefinitely; it will fatigue. Force in a concentric muscle action can also be enhanced by preceding it with an eccentric muscle action.

Rate of Force Development (RFD)

It takes time for a muscle to develop force. It takes approximately 30 to 100 milliseconds from the first signs of electrical activity to the production of measurable tension. This time is often referred to as the **electromechanical delay**.[22] Then it could take up to another 400 milliseconds from the production of measurable tension to the development of maximal tension (**Figure 11.26**).[23]

> **Electromechanical delay**
> The time between the onset of electrical activity at the muscle and production of measurable force

There are several reasons why the production of force takes time. First is the time required for the chemical processes to take place once the action potential reaches the muscle. Then, there is the time it takes for the initial cross-bridge formation. Finally, there is the time needed to take up the slack in the tendon before the force developed in the muscle is transferred to the bone. When you add it all up, you realize that developing force is not an instantaneous process.

RFD becomes very important when you realize that many activities occur in less time than it takes to develop the maximum amount of force. Consider the two muscles in **Figure 11.27**. Muscle A can clearly develop more force than

Figure 11.26	The rate of force development of a muscle. Fmax can take several hundred milliseconds to develop.

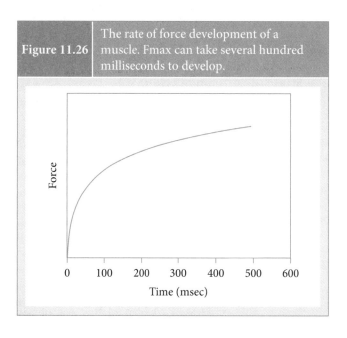

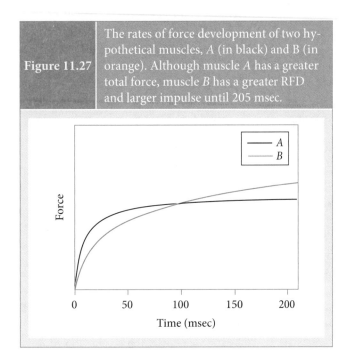

Figure 11.27 The rates of force development of two hypothetical muscles, *A* (in black) and *B* (in orange). Although muscle *A* has a greater total force, muscle *B* has a greater RFD and larger impulse until 205 msec.

Muscle B, in the absolute sense. However, for any activity that is less than 100 milliseconds, Muscle A can develop more force than Muscle B. Even more telling is the mechanical impulse. Look at areas under the curves. For any duration below 205 milliseconds, Muscle B can develop more mechanical impulse than Muscle A. This should give you an indication of just how important RFD is in the performance of human movement.

Important Point!   Rate of force development is important because many activities occur in less time than it takes to develop the maximum amount of force, and it affects the total mechanical impulse.

The Stretch-Shortening Cycle (SSC)

Another important element of the time history of the MTC is the stretch-shortening cycle (SSC). Not all concentric actions preceded by an eccentric action involve an SSC. The SSC is only involved if the following three conditions are met:[24]

1. A well-timed preactivation of the muscle prior to the eccentric action
2. A short, rapid eccentric action
3. An immediate transition from the eccentric action to the concentric action (**amortization phase**)

Amortization phase   The time between the eccentric and concentric actions

However, it is well established that a properly

used SSC can also enhance the force produced by the MTC. There are probably several reasons for this.

First, as you just read, development of force takes time. Say you were going to jump using just your ankles, and you were going to begin from standing. As you are standing, there is minimal activation (and force developed) in your triceps surae. For argument's sake, say you are starting at Point *A* in Figure 11.28, and you have to leave the ground 100 milliseconds after you began plantar flexing. In 100 milliseconds, the force (and impulse) that can be developed by the triceps surae is indicated on the graph as F_A. Now think if you were hopping. Instead of starting from a quiet standing, you were rebounding from the previous hop. During the braking phase, the MTC is developing force during the eccentric action. It is not starting from zero, but say point *B* in Figure 11.25. This has the effect of "preforcing" the MTC.[25] You still had to leave the ground 100 milliseconds after you start plantar flexing, but you begin developing force while you are dorsiflexing. One hundred milliseconds after Point *B*, for the same force–time curve, the force (and impulse) that can be developed is represented as F_B.

Notice that I said, "... for the same force–time curve...." There is good reason to believe that the force–time curve would be enhanced when it is preceded by an eccentric action. Muscle spindles (intrafusal muscle fibers) are sensitive to both the magnitude and rate of stretch. When a muscle is rapidly and forcibly elongated, the muscle spindles are stimulated, evoking increased activation of the musculature and a more forceful contraction. In addition, a portion of the elastic energy stored in the tendon during the eccentric action can be reused during the concentric action. The laws of thermodynamics

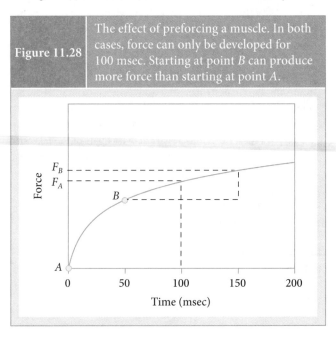

Figure 11.28 The effect of preforcing a muscle. In both cases, force can only be developed for 100 msec. Starting at point *B* can produce more force than starting at point *A*.

tells you that the energy released in the tendon during the concentric action can never be greater (in fact, it cannot even be equal) to the energy that is stored in the tendon during the eccentric action. But a large portion, up to 93%,[26] can be. The percentage not used is lost to heat through the process of hysteresis. The amount that can be reused seems to depend on the length of the amortization phase: The longer the time between the eccentric and concentric actions, the more energy is lost to heat. Finally, the displacements of the muscle and the MTC are uncoupled. During the SSC, the MTC is changing its length rapidly. During the propulsive phase, the muscle can begin shortening before the tendon does. This means that the muscle can be lower than the velocity of the tendon. Because the muscle (Figure 11.17) is greatly affected by, and the tendons' force increases with, shortening velocity, the muscle can operate at a more favorable velocity. The SSC ends up making the tendon act like a catapult or a slingshot:[26] The release of energy during the propulsive phase occurs at a greater rate than it was stored during the braking phase.

> **Hysteresis** The amount of energy lost to heat between loading and unloading

You have just identified four ways the SSC can enhance force output: by preforcing the muscle, by inducing the stretch reflex, by using energy stored in the tendon, and by the catapult effect. The relative contributions of each of these mechanisms have been argued,[23] and there appears to be no clear consensus at the present. However, you should understand that the SSC is important, that there are several mechanisms involved in enhancing force output, and how it is used in the production of human movement.

Important Point! The stretch-shortening cycle enhances force output by preforcing the muscle, by inducing the stretch reflex, by using energy stored in the tendon, and by the catapult effect.

Fatigue

Just as it takes time to develop force in an MTC, the force produced in the MTC cannot be sustained indefinitely. It is subject to fatigue. Fatigue should not be confused with task failure. Fatigue is defined as "any reduction in the force generating capacity of the total neuromuscular system regardless of the force required in any given situation."[27] As you

> **Fatigue** Any reduction in the force-generating capacity of the total neuromuscular system, regardless of the force required in any given situation

can see, this definition implies that fatigue can occur while someone is still able to complete a task. Task failure (not to be confused with material failure), on the other hand, is an inability to continue or complete a desired action.

> **Task failure** The inability to continue or complete a desired action

Important Point! There is a difference between fatigue and task failure.

To see these distinctions, examine **Figure 11.29**. Figure 11.29a is a model for fatigue and task failure during dynamic actions, whereas Figure 11.29b illustrates the same ideas for a sustained isometric action. The black line represents the force the MTC must produce as is required for the task. The orange line represents the force that the MTC is capable of producing, and the slope of the orange line is the rate of force fatigue (RFF).

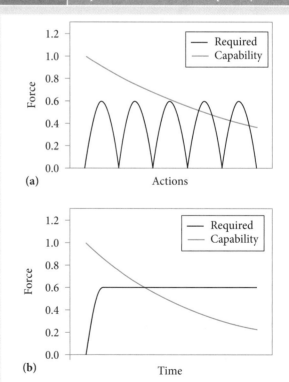

Figure 11.29 A dynamic (a) and static (b) model of task failure. The black lines represent the force requirements of the task. The orange lines represent the force-producing capability of the MTC. Failure occurs when the requirements exceed the capability.

The force capability initially exceeds the requirements, and the task can be completed. Fatigue begins almost immediately;[28] however, the capability still exceeds the requirements, and the task can be completed. With the dynamic actions, the force requirement and capability are approximately equal by the third repetition. By the fourth repetition, the requirements exceed the capability, and task failure occurs. Similarly, for the static action, the RFF will eventually cause the capabilities to fall below the requirements, and the MTC will be incapable of sustaining the action.

There is no single cause of fatigue. Fatigue occurs because one, or several, of the physiological processes involved in the force production of the contractile proteins becomes impaired.[28] There are many potential sites of this impairment within the neuromuscular system, and which site becomes impaired depends on the task being performed. Consult your physiology book or one of the excellent reviews on the topic.[29]

How would you prevent task failure from occurring? There are a couple of possibilities, which are presented in **Figure 11.30**. In Figure 11.30a, you see that the maximum force-producing capability of the MTC was increased, while

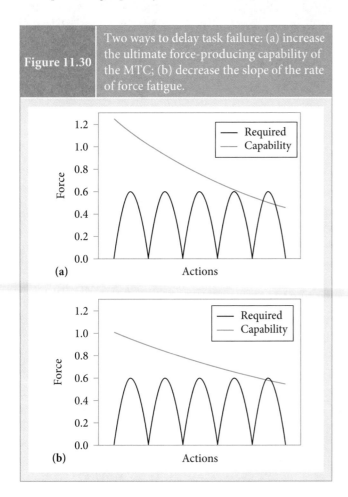

| Figure 11.30 | Two ways to delay task failure: (a) increase the ultimate force-producing capability of the MTC; (b) decrease the slope of the rate of force fatigue. |

the RFF was kept the same as it was in Figure 11.29a. Notice that for the same RFF, four actions could now be performed before task failure occurred. In Figure 11.30b, the maximum force-producing capability was kept the same as it was in Figure 11.29a, but the RFF was decreased. Again, four actions could be performed before task failure occurred. A third option would be to increase both the maximum force-producing capability and decrease the RFF of the MTC (not shown). Finally, the technique could be altered to decrease the force requirement on the MTC. Changing the technique to compensate for the decreased ability of one MTC would ultimately stress another MTC and may not be ideal.

11.3.8   Factors are Interrelated

This section began by listing seven factors that affected the ability of an MTC to produce force. A nice illustration of these factors was presented in Figure 11.15. Although Figure 11.15 looks a lot like the hierarchical models presented elsewhere in this book, I did not describe it as one. Do you know why?

The factors in Figure 11.15 were presented in a mostly vertical arrangement. There were no horizontal connections. This was done to help you learn about each of these factors, but it was a gross oversimplification. Most of the factors also affect one another. For example, you learned how muscle architecture affected both force and velocity. You also saw the interaction between the type of action and velocity. Muscle fiber types interact with velocity, and have an effect on both RFD and RFF. They also affect recruitment. The length of the MTC affects the slack of the tendon; if the tendon slack is removed, then the RFD will be greater. As you can see, very few factors affect the force-producing capability of a muscle independently. In addition, most of these factors are affected by age, disuse, and exercise.

11.3.9   The Effect of Aging, Disuse, and Exercise

The effect of aging, disuse, and exercise on each of the factors described earlier could be a lesson (indeed, an entire textbook) in and of itself. In this section, you are only going to get an overview of some of the mechanical effects. In addition, you need to appreciate a couple of simplifying assumptions. First, because the effects of aging and disuse are fairly similar,[30] unless otherwise noted, they will be considered together. Second, "exercise" is a fairly broad term. For purposes of this discussion, consider exercise to mean heavy resistance exercise.

The effect of aging on the MTC is complicated by the fact that there is generally a decline in physical activity associated with aging. That makes it tough to distinguish inevitable effects of aging from those that are due to a decrease in activity. There are also many reasons for disuse, including spinal cord injury, immobilization due to another type of musculoskeletal injury, or even manned space flight. At the risk of oversimplifying, we will examine some general adaptations.

With aging and disuse, there is an overall decrease in the force-producing capability of the MTC. Just about every factor in Figure 11.15 is responsible for this decline. There is a decline in the PCSA, which is called atrophy. With disuse, the **atrophy** is caused by a decrease in the CSA of the individual fibers. With aging, atrophy is caused by a decrease in both the CSA of the individual fibers and the number of fibers and is known as **sarcopenia**. There is also a decrease in penna-

> **Atrophy**   A decrease in the physiological cross-sectional area
>
> **Sarcopenia**   Age-related decrease in muscle mass

tion angle, but that may be related to the number and size of fibers in parallel. With aging, there is also a decrease in the number of sarcomeres in series. Although this is not as great as the decreased number of sarcomeres in parallel, it will lead to a decrease in shortening velocity Alterations in shortening velocity will also lead to decrease in power. The effect of immobilization is a little more complicated. If an MTC is immobilized in a lengthened position, it appears as though there are an increased number of sarcomeres in series. Similarly, an MTC mobilized in a shortened position results in a decreased number of sarcomeres in series. However, this change in sarcomere number in series with immobilization may be transient: The number of sarcomeres seems to return to the original number after immobilization and a return to normal use. With aging, there does not appear to be a dramatic change in fiber type, but there is a decrease in both recruitment and specific tension. In addition, long-term immobilization appears to result in a selective decrease in the number of Type I fibers.

> **Important Point!**   There are many similarities in the decline of MTC properties between disuse and aging.

The tendon is also affected by aging and disuse.[30] With disuse, there is a rapid decrease in the stiffness and elastic modulus of the tendon. With prolonged disuse, there is also a decrease in the CSA of the tendon. An increase in tendon stiffness would mean that a muscle's sarcomeres would have to shorten more (for a given change in MTC length) or operate in a shortened position (for the same MTC length). Because of the force-length relation (Figure 11.17), this would mean that an MTC would be able to produce less force.

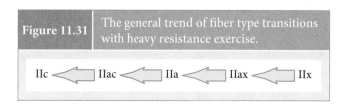

Figure 11.31   The general trend of fiber type transitions with heavy resistance exercise.

But remember that there is also a loss of sarcomeres in series. Theoretically the decrease in tendon stiffness could make up for the loss of sarcomeres in series. Therefore, the overall effects of the changes in the material properties of the tendon are unclear.

As with aging, heavy, progressive resistance exercise can also alter almost every factor in Figure 11.15. There is an increase in the PCSA of the muscle, which is more likely due to an increase in the CSA of the muscle fibers rather than an increase in the number of muscle fibers. Heavy eccentric loading can also increase the number of sarcomeres in series. Force production is also increased through improved recruitment and possibly a change in specific tension. There is also an alteration in muscle fiber types (at least with the Type II isoform). Although you would think that heavy resistance training would make fibers stronger and faster (movement from IIa toward IIx), the opposite actually occurs. In **Figure 11.31**, you see the general direction of the fiber type transformation with heavy resistance exercise. At first, this might not make much sense to you, but what is actually happening is that fibers are moving from being harder to recruit to being easier to recruit. So although each individual fiber may be producing slightly less force, more fibers are being recruited. Various specialized exercises can also make favorable changes in terms of force-velocity, RFD, RFF, and the SSC.

There is less information on the effect of exercise on tendon's material and mechanical properties. Exercise appears to increase the stiffness and Young's modulus. There also appears to be a decrease in hysteresis, but that adaptation appears to be more influenced by the type of exercise. Eccentric exercises and resistance exercises that also incorporate stretching appear to have more of an effect on hysteresis than concentric exercise alone.

> **Section Question Answer**
>
> The section began by asking you which technique would produce a greater jump height: a squat jump or a countermovement jump. In the study, the countermovement enhanced jump height by 14.2 ± 8.9%.[13] The reason why the countermovement enhanced jump height was because it made use of the stretch-shortening cycle.

COMPETENCY CHECK

Remember:

1. Define the following terms: hypertrophy, hyperplasia, isoform, specific tension, motor unit, electromechanical delay, amortization phase, hysteresis, fatigue, task failure, atrophy, and sarcopenia.
2. List the variables that affect the force an MTC can produce.

Understand:

1. Describe how each factor affects the MTC's force-producing capability.
2. Describe the interaction between the type of MTC action and the velocity of movement.
3. Describe the interaction between muscle architecture, force, length, and velocity.
4. Give an example between fatigue and task failure.
5. Describe how variables that affect the force an MTC can produce interact with each other.
6. Which of the variables that affect the force an MTC can produce are modifiable with training?

Apply:

1. How can you improve the ability of the MTC to produce force?
2. Give examples of activities where you use the stretch-shortening cycle.

Figure 11.32 What caused the gasctrocnemius strain? What is the mechanical effect of the injury?

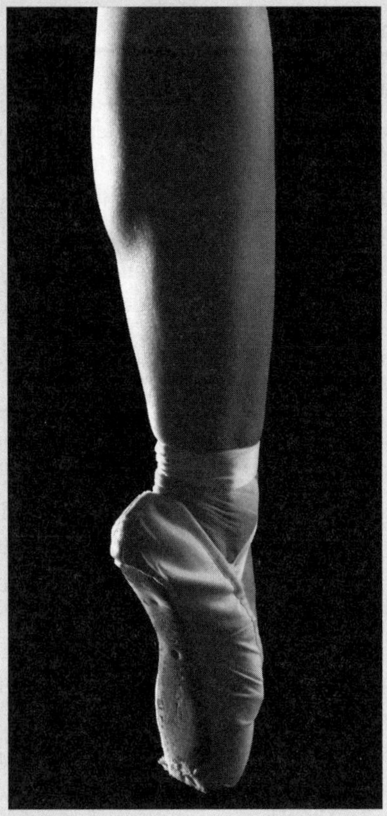

© marby/ShutterStock, Inc.

11.4 INJURY BIOMECHANICS

Section Question

Janice strained her gastrocnemius (**Figure 11.32**), and Justin ruptured his Achilles tendon (**Figure 11.33**). What caused these injuries to happen? What are the mechanical effects of these injuries?

Both the muscle and tendon can be subjected to injury. An injury to the muscle is called a strain (not to be confused with mechanical strain), and the results could be anywhere from a disruption of the myofibrils to a complete tearing of the muscle (**Table 11.4**). An injury of the tendon is referred to as a tendinopathy. Tendons can be injured either acutely or over time, be inflamed or degenerate, and/or be partially or completely torn.

When speaking of injury biomechanics, you probably immediately think about the mechanics that are involved

in producing an injury, or **mechanopathology**. But you might also be interested in how an injury affects the mechanics, called **pathomechanics**. Both topics will be discussed in this section.

> **Mechanopathology** The mechanics that result in an injury
> **Pathomechanics** The mechanics that are a result of an injury

11.4.1 Mechanopathology: The Mechanisms of Injury

Muscle

Although the exact causes of a muscle strain are still unknown, it is commonly accepted that muscle strains are the result of both a stretch and load placed on the muscle.[31] In other words, they occur during eccentric muscle actions. Although there is some debate about whether excessive stress or mechanical strain is the primary culprit, it is probably safe to say that it

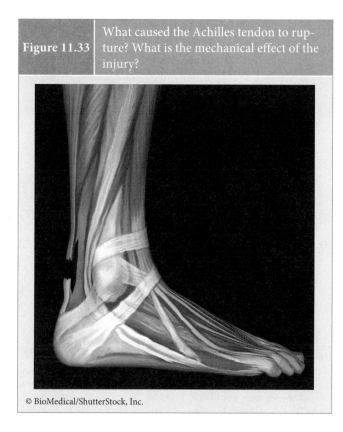

| Figure 11.33 | What caused the Achilles tendon to rupture? What is the mechanical effect of the injury? |

© BioMedical/ShutterStock, Inc.

involves some interaction of the two, which is related to how much energy the MTC has to absorb.

There are many mechanical, biological, and structural theories concerning muscle injury, with no clear consensus on which are correct.[14] Here, you will study a mechanical idea of muscle strain called the "popping sarcomere" hypothesis.[32]

Think about what happens if a myofibril is forced to lengthen, such as landing during a hop. Up to this point, you have probably assumed that each sarcomere increased its length by the same amount. But is that a valid assumption? What if it is not?

Remember that as a sarcomere is stretched beyond its optimal length it gets weaker (the descending limb on Figure 11.17) because there is less of an overlap between the

| Table 11.4 | The Three Classifications of Muscle Strains |

Severity	Description
First degree	Stretching, but minimal tearing (< 5%) of the fibers
Second degree	Partial tearing (5–99%) of the fibers
Third degree	Complete tearing (100%) of the fiber

actin and myosin. Now what if some sarcomeres lengthened more than others? They will be weaker. With continued strain, they will stretch even more than the others because they are weaker, leading to the point where the sarcomere "pops," or is literally torn apart. A torn myofibril is no longer producing any force, so that means that more stress is placed on the remaining myofibrils. If several myofibrils were torn simultaneously, this would put even more stress on the remaining myofibrils. Could this cycle repeat, over and over again, within the space of milliseconds? Could this lead to a torn muscle? Right now, this might be the best mechanical explanation for why a muscle injury occurs.[14]

Tendon

The mechanisms of tendinopathy are also poorly understood.[33] Tendons are strong and are considered to be stronger and able to produce/absorb more energy than muscle.[34] So it seems unlikely that a tendon would rupture before a muscle would. However, tendons also undergo greater strains than the muscle fibers. Repetitive straining of a tendon over time can lead to degenerative changes, which in turn would make the tendon more susceptible to injury (see later discussion). Ruptured tendons show even greater degenerative changes than tendinopathic ones.[35] These findings suggest that even an "acute" rupture of a tendon is not due to greater loads on the tendon, but to a progressive weakening of the tendon over time.

11.4.2 Pathomechanics

Muscle

Although muscle does have the ability to regenerate (produce new muscle tissue), in more severe cases the damage muscle tissue is replaced with a scar.[36] The presence of a scar can have some profound effects on the mechanics of the muscle. It has been suggested that the presence of a scar may not decrease the overall strength of the muscle, but will decrease the optimal length of that muscle.[37] This could lead to a decrease in the operating range of the muscle. In addition, according to the popping sarcomere hypothesis, this muscle (or fibers) will have to operate on the descending limb of the force-length curve, which is what leads to the muscle damage in the first place. This could account for the high incidence of reinjury.

Tendon

Tendinopathy has been shown to alter both the mechanical and material properties of the tendon.[38] Tendinopathy increases the cross-sectional area of the tendon, which you would think would make it stronger. However, both the stiffness and the Young's modulus are lower in the tendinopathic tendon. These findings indicate that the alterations in tissue composition were not favorable. If the tendon is more compliant, then the muscle fibers may have to shorten more for the same change in the length of the MTC. If the muscle fibers are in a more shortened position, they may not be operating at the optimal region of the force-length curve (see Figure 11.15). This could decrease movement economy. In addition, a more compliant tendon would be subjected to higher strains. These higher strains could potentially cause disruptions of the collagen fibers, which would then lead to the degenerative changes in the tendon seen with tendinopathy. This would make the tendon even more susceptible to rupture.

Section Question Answer

The mechanics of injury are still not well understood. Janice likely strained her gastrocnemius during a forceful eccentric action where the weaker sarcomeres "popped," causing her muscle fibers to tear. Justin's repeated use of his Achilles probably led to physiological degradation of the tissue. Eventually, the strength of the Achilles was lower than the forces subjected to it. Both muscle and tendon injuries have a greater potential for reinjury, which highlights the need for proper rehabilitation following injury.

SUMMARY

In this lesson, you learned that it is not just muscle, but the muscle–tendon complex that is responsible for the production and control of movement. The MTC has certain properties when its components interact together that are not present when each component is examined in isolation. Unlike machines, several factors influence how much force an MTC can produce. Many of these properties are adaptable with training, injury, disuse, and aging.

The story is far from over. Just as the muscle behaves differently as part of the MTC, the MTC can behave differently when it is put into an MTC–joint system.

REVIEW QUESTIONS

1. Define the following terms: amortization phase, atrophy, concentric action, compliant, decouple, eccentric action, electromechanical delay, fatigue, hyperplasia, hypertrophy, hysteresis, isoform, isometric action, mechanopathology, motor unit, pathomechanics, plasticity, sarcopenia, specific tension, stiffness, stretch-shortening cycle, and task failure.

2. List the components of the muscle–tendon complex.

3. List the factors that affect the ability of the muscle–tendon complex to produce force.

4. Explain how the muscle acts differently when attached to a tendon compared to when it is in isolation.

5. Outline the differences between a pennate and parallel muscle.

6. Explain the difference between fatigue and task failure.

7. Outline MTC plasticity to exercise, aging, disuse, and injury.

8. Explain how injuries occur and how injuries affect the mechanics of the MTC.

9. Demonstrate how the stretch-shortening cycle is advantageous and when you would use it.

10. Explain how you could increase the force-producing capability of the MTC.

11. Explain how you may decrease injury risk to the MTC.

REFERENCES

1. Linthorne NP, Guzman MS, Bridgett LA. Optimum take-off angle in the long jump. *Journal of Sports Sciences.* Jul 2005;23(7):703–712.

2. Dickinson MH, Farley CT, Full RJ, Koehl MAR, Kram R, Lehman S. How animals move: an integrative view. *Science.* Apr 2000;288(5463):100–106.

3. Muramatsu T, Muraoka T, Takeshita D, Kawakami Y, Hirano Y, Fukunaga T. Mechanical properties of tendon and aponeurosis of human gastrocnemius muscle in vivo. *Journal of Applied Physiology.* May 2001;90(5):1671–1678.

4. Ward SR, Eng CM, Smallwood LH, Lieber RL. Are current measurements of lower extremity muscle architecture accurate? *Clinical Orthopaedics and Related Research.* Apr 2009;467(4):1074–1082.

5. James R, Kesturu G, Balian G, Chhabra AB. Tendon: biology, biomechanics, repair, growth factors, and evolving treatment options. *J. Hand Surg.-Am. Vol.* Jan 2008;33A(1):102–112.

6. Carlstedt CA, Nordin M. Biomechanics of tendons and ligaments. In: Nordin M, Frankel VH, eds. *Basic Biomechanics of the Musculoskeletal System.* Philadelphia: Lippincott, Williams, and Wilkins; 1989:59–74.

7. Yamaguchi GT. *Dynamic Modeling of Musculoskeletal Motion: A Vectorized Approach for Biomechanical Analysis in Three Dimensions.* Boston: Kluwer Academic Publishers; 2001.

8. Zajac FE. Muscle and tendon—properties, models, scaling, and application to biomechanics and motor control. *Critical Reviews in Biomedical Engineering.* 1989 1989;17(4):359–411.

9. Lichtwark GA, Wilson AM. In vivo mechanical properties of the human Achilles tendon during one-legged hopping. *Journal of Experimental Biology.* Dec 2005;208(24):4715–4725.

10. Alexander RM. Storage and release of elastic energy in the locomotor system and the stretch-shortening cycle. In: Nigg BM, MacIntosh BR, Mester J, eds. *Biomechanics and Biology of Movement.* Champaign, IL: Human Kinetics; 2000:19–29.

11. Wakeling JM, Blake OM, Wong I, Rana M, Lee SSM. Movement mechanics as a determinate of muscle structure, recruitment and coordination. *Philos. Trans. R. Soc. B-Biol. Sci.* May 27 2011;366(1570):1554–1564.

12. Wilson A, Lichtwark G. The anatomical arrangement of muscle and tendon enhances limb versatility and locomotor performance. *Philos. Trans. R. Soc. B-Biol. Sci.* May 27 2011;366(1570):1540–1553.

13. Kubo K, Morimoto M, Komuro T, Tsunoda N, Kanehisa H, Fukunaga T. Influences of tendon stiffness, joint stiffness, and electromyographic activity on jump performances using single joint. *European Journal of Applied Physiology.* Feb 2007;99(3):235–243.

14. Lieber RL. *Skeletal Muscle Structure, Function, and Plasticity: The Physiological Basis of Rehabilitation.* 3rd ed. Philadelphia: Wolters Kluwer/Lippincott Williams and Wilkins; 2010.

15. Fukunaga T, Kawakami Y, Kubo K, Kanehisa H. Muscle and tendon interaction during human movements. *Exercise and Sport Sciences Reviews.* Jul 2002;30(3):106–110.

16. Hunter GR, Harris RT. Structure and function of the muscular, neuromuscular, cardiovascular, and respiratory systems. In: Baechle TR, Earle RW, eds. *Essentials of Strength Training and Conditioning.* 3rd ed. Champaign, IL: Human Kinetics; 2008:3–20.

17. Larsson L, Li XP, Frontera WR. Effects of aging on shortening velocity and myosin isoform composition in single human skeletal muscle cells. *American Journal of Physiology-Cell Physiology.* Feb 1997;272(2):C638–C649.

18. Bottinelli R, Pellegrino MA, Canepari M, Rossi R, Reggiani C. Specific contributions of various muscle fibre types to human muscle performance: an in vitro study. *Journal of Electromyography and Kinesiology.* Apr 1999;9(2):87–95.

19. Johnson MA, Polgar J, Weightma. D, Appleton D. Data on distribution of fiber types in 36 human muscles—autopsy study. *Journal of the Neurological Sciences.* 1973 1973;18(1):111–129.

20. Bottinelli R, Reggiani C. Human skeletal muscle fibres: molecular and functional diversity. *Progress in Biophysics & Molecular Biology.* 2000 2000;73(2–4):195–262.

21. Bacou F, Rouanet P, Barjot C, Janmot C, Vigneron P, dAlbis A. Expression of myosin isoforms in denervated, cross-reinnervated, and electrically stimulated rabbit muscles. *Eur. J. Biochem.* Mar 1996;236(2):539–547.

22. Cavanagh PR, Komi PV. Electro-mechanical delay in human skeletal-muscle under concentric and eccentric contractions. *European Journal of Applied Physiology and Occupational Physiology.* 1979 1979;42(3):159–163.

23. Schenau GJV, Bobbert MF, deHaan A. Does elastic energy enhance work and efficiency in the stretch-shortening cycle? *Journal of Applied Biomechanics.* Nov 1997;13(4):389–415.

24. Komi PV, Gollhofer A. Stretch reflexes can have an important role in force enhancement during SSC exercise. *Journal of Applied Biomechanics.* Nov 1997;13(4):451–460.

25. Zajac FE. Muscle coordination of movement—a perspective. *Journal of Biomechanics.* 1993;26:109–124.

26. Alexander RM. Tendon elasticity and muscle function. *Comparative Biochemistry and Physiology a—Molecular and Integrative Physiology.* Dec 2002;133(4):1001–1011.

27. Bigland-Ritchie B, Woods JJ. Changes in muscle contractile properties and neural control during human muscular fatigue. *Muscle Nerve.* 1984 1984;7(9):691–699.

28. Enoka RM, Duchateau J. Muscle fatigue: what, why and how it influences muscle function. *Journal of Physiology-London.* Jan 1 2008;586(1):11–23.

29. Enoka RM, Baudry S, Rudroff T, Farina D, Klass M, Duchateau J. Unraveling the neurophysiology of muscle fatigue. *Journal of Electromyography and Kinesiology.* Apr 2011;21(2):208–219.

30. Narici MV, Maganaris CN. Plasticity of the muscle-tendon complex with disuse and aging. *Exercise and Sport Sciences Reviews.* Jul 2007;35(3):126–134.

31. Lieber RL, Friden J. Mechanisms of muscle injury gleaned from animal models. *Am. J. Phys. Med. Rehabil.* Nov 2002;81(11):S70–S79.

32. Morgan DL. New insights into the behavior of muscle during active lengthening. *Biophysical Journal.* Feb 1990;57(2):209–221.

33. Magnusson SP, Langberg H, Kjaer M. The pathogenesis of tendinopathy: balancing the response to loading. *Nature Reviews Rheumatology.* May 2010;6(5):262–268.

34. Alexander RM. *Principles of Animal Locomotion.* Princeton: Princeton University Press; 2003.

35. Tallon C, Maffulli N, Ewen SWB. Ruptured Achilles tendons are significantly more degenerated than tendinopathic tendons. *Medicine and Science in Sports and Exercise.* Dec 2001;33(12):1983–1990.

36. Jarvinen T, Jarvinen TLN, Kaariainen M, Kalimo A, Jarvinen M. Muscle injuries—biology and treatment. *American Journal of Sports Medicine.* May 2005;33(5):745–764.

37. Brockett CL, Morgan DL, Proske U. Predicting hamstring strain injury in elite athletes. *Medicine and Science in Sports and Exercise.* Mar 2004;36(3):379–387.

38. Arya S, Kulig K. Tendinopathy alters mechanical and material properties of the Achilles tendon. *Journal of Applied Physiology.* Mar 2010;108(3):670–675.

PART III

Joint Level

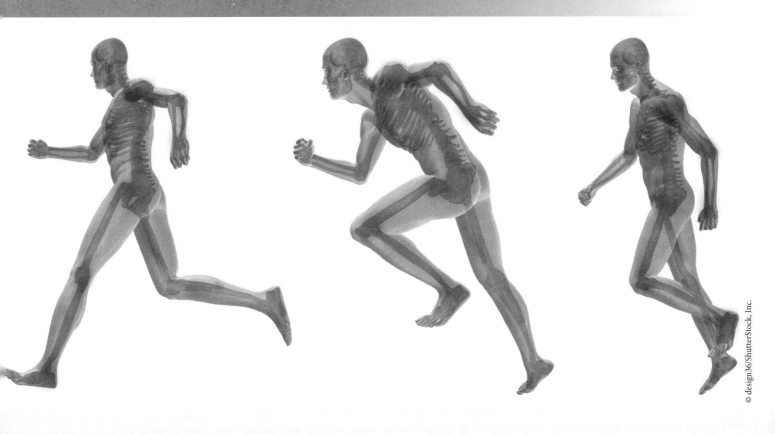

© design36/ShutterStock, Inc.

Lesson 12

Single Joint Concepts

LEARNING OBJECTIVES

After finishing this lesson, you should be able to:

- Define the following terms: degrees of freedom, osteokinematics, arthrokinematics, strength curve, effective mechanical advantage, cocontraction, stability, and quasi stiffness.
- Describe the clinical reference frame.
- Determine on which plane(s) different joint motions typically occur.
- List the six types of diarthroidal joints and the motion(s) associated with each.
- Describe the following types of arthrokinematic motion: rolling, gliding, sliding.
- Describe the effects of a muscle–tendon complex on a joint system.
- Describe the characteristics of a muscle–tendon complex's moment arm, and how changing each will affect its ability to produce torque.
- Describe the effects of a joint system on a muscle–tendon complex.
- Describe how a muscle can act as a motor, brake, spring, or strut.
- Describe how muscles can be cofunctional, antagonistic, or synergistic.
- Describe the difference between static and dynamic stability.

To truly comprehend the principles of biomechanics, you must first master the fundamental concepts of mechanics, which is basically applied physics. While the body is subjected to the laws of physics, the way the various tissues of the body (including the muscle-tendon complex) generates and responds to forces is unique to its biology. So it is important to grasp these biological principles governing movement as well. But the structures of the body do not act in isolation. To truly understand how these parts work, you must see how these parts interact as a system. This lesson is the first step in integration by looking at the interaction of the various parts: how bones interact to allow movement (kinematics), and how muscles interact with joints to control movement (kinetics). It is important to realize that not only do muscles affect joints, but the joint also affects the performance of the muscle. After examining a muscle–joint system with only one muscle, you will then add multiple muscles to the system. The lesson will conclude with a discussion of joint stability.

12.1 CLINICAL REFERENCE FRAMES

> **Section Question**
>
> A woman is walking in a car on a moving train (Figure 12.1). What is the linear velocity of the center of mass of her thigh?

The first thing you must do before you can perform a biomechanical analysis is to establish a frame of reference. In most cases, the frame of reference is fixed and established in relation to Earth. When this is done, it is called a global reference frame. However, in some instances you look at relative motion, that is, one body moving in relation to another one. In such cases, the frame of reference is located in one body or the other. When a frame of reference is located in a body as opposed to Earth, it is called a local reference frame. Both types of reference frames are important. Let us see how they are used when discussing joint motion.

| Figure 12.1 | A woman walking in the car of a moving train. What is the linear velocity of the center of mass of her thigh? |

© BartlomiejMagierowski/ShutterStock, Inc.

You may already be familiar with a clinical reference frame (**Figure 12.2a**), which is a global reference frame. It is very similar to the global reference frame you have been using (**Figure 12.2b**), with some slight modifications. Start with an origin at the COM of the body while standing in the anatomical position and establish axes in the x (anterior/posterior) direction, y (superior/inferior) direction, and z (right/left) direction. The anterior, superior, and right directions will be positive. The clinical reference frame establishes three cardinal planes:

- The sagittal plane—an anterior–posterior plane formed by the x- and y-axes that divides medially/laterally or right/left.
- The transverse plane—a horizontal plane formed by the x- and z-axes that divides the body superiorly/inferiorly.
- The frontal plane—a side-to-side plane formed by the y- and z-axes that divides the body anteriorly/posteriorly.

When movement occurs *in a plane*, the direction of the movement is parallel to the plane. If the movement is a rotation, it occurs about an axis that is perpendicular to that plane:

- Sagittal plane rotation occurs about a medial–lateral axis.
- Frontal plane rotation occurs about an anterior-posterior axis.
- Transverse plane rotation occurs about a superior-inferior axis.

For example, stand with your back to a wall. Movement of your arms up and down along the wall (shoulder abduction and adduction) is occurring about an anterior/posterior axis in the frontal plane. Now turn so that you are perpendicular to the wall, with your right shoulder lightly touching it. Movement of your right arm up and down along the wall (shoulder flexion and extension) is occurring about a medial/lateral axis in the sagittal plane.

Although this type of reference frame is often used clinically, it can also lead to confusion. For example, elbow flexion/extension is said to occur about a medial–lateral axis in the sagittal plane. But if you were to abduct your arm to 90°, that same joint motion would occur about a superior–inferior axis in the transverse plane (**Figure 12.3**). That is because you changed the orientation of the two bones in space (with respect to the global reference frame) without changing the orientation of the two bones in relation to each other (a local reference frame).

In addition, if an axis of rotation is oblique to a cardinal plane, rotation will occur in a plane that is not aligned with a cardinal plane and occurs in two cardinal planes simultaneously. For example, in **Figure 12.4** an oblique axis is offset 45° from the medial/lateral direction (frontal plane). A rotation about this axis will cause simultaneous movement in both the sagittal and transverse planes. In this case, there would be equal amounts of movement in each plane.

> **Important Point!** Rotations occur in a plane about an axis that is perpendicular to that plane. If the axis is oblique, the plane of movement will not be coincident with any of the cardinal planes.

To avoid this confusion, axes and planes should be fixed within segments and use local reference frames (**Figure 12.5**).[1] In this way, if the orientation of the bones changes in space (but no relative motion between the bones occurs), then the reference frames move with the bones. Using the previous example, elbow flexion and extension still occur in the sagittal plane because the orientation of the sagittal plane of the elbow changed with the bones. Likewise, a rotation about an oblique axis is still in a single plane because that plane is rotated.

> **Important Point!** Clinical reference frames should be established for segments and not the body as a whole.

Figure 12.2 | Anatomical planes of movement.

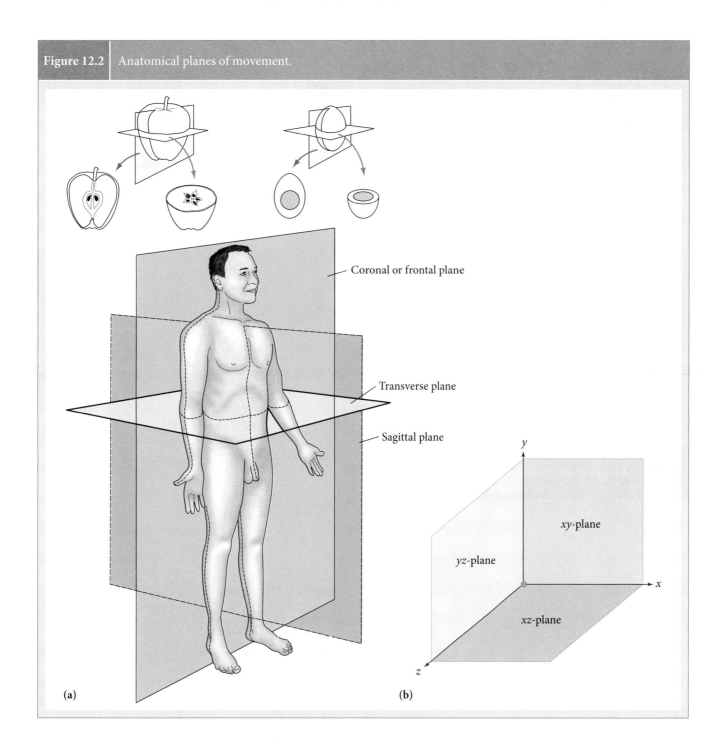

Coronal or frontal plane

Transverse plane

Sagittal plane

xy-plane

yz-plane

xz-plane

(a)

(b)

| **Figure 12.3** | Elbow flexion with the shoulder in neutral (a) and with the shoulder flexed and internally rotated 90° (b). The joint motion is the same, but it occurs in a different plane in the global reference frame. |

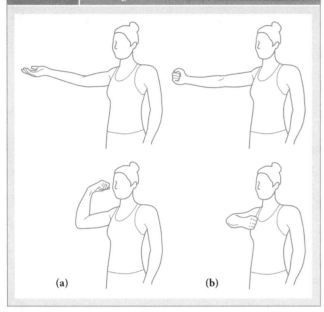

(a)   **(b)**

| **Figure 12.4** | An axis that is offset from a cardinal plane will cause a movement that occurs in more than one cardinal plane. |

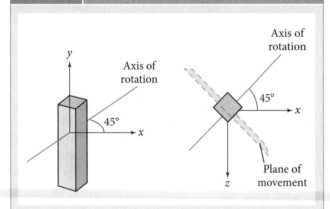

Section Question Answer

In the beginning of this section, you were shown the picture of a woman walking inside a moving train, and you were asked: What is the linear velocity of the center of mass of her thigh? Hopefully, you realize that this is an incomplete question. First, you must know: The velocity of her thigh *in relation to what*? You would get a different "answer" if you talked about the velocity in relation to (fixed) Earth, the moving car, or her (also moving) trunk. Each answer could be technically correct, depending on the question.

| **Figure 12.5** | To avoid problems with a global reference frame, reference frames should be attached locally to bones (segments). |

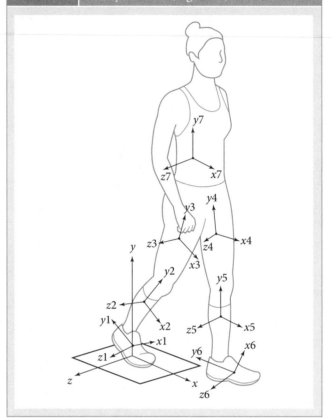

COMPETENCY CHECK

Remember:

1. What planes make up the clinical reference frame?
2. Movement in each plane corresponds to rotations about which axis?

Understand:

1. Which joint motion(s) typically occurs in which planes?
2. Why is it important to establish the frame of reference as a local frame?

Apply:

1. Give an example of a movement in each plane of motion. Reposition your arm or leg and repeat the motion in a different global plane but the same local plane.

12.2 KINEMATICS

Once you have established your frame of reference, you can begin to describe the motion of the body part(s) of interest. But what motion are you interested in? Kinematics can roughly be broken down into two types: **osteokinematics** and **arthrokinematics**. The root words help you identify the motions to which these words are referring. *Osteo* means bone, so *osteokinematics* refers to motions of bones. *Arthro* means joint; *arthrokinematics* refers to motion at the joint surfaces. The type of motion that you examine depends on your level of analysis. If you are interested in gross body movements, you would probably limit your analysis to osteokinematics. However, if you were interested in examining injury mechanisms at joints, you would probably also need to examine the arthrokinematics of that joint. Because you are probably already somewhat familiar with osteokinematic movement, that is a good place to start.

Osteokinematic motion
Rotations of bones

Arthrokinematic motion
Motions at joint surfaces

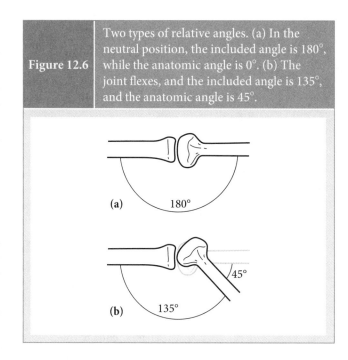

Figure 12.6 Two types of relative angles. (a) In the neutral position, the included angle is 180°, while the anatomic angle is 0°. (b) The joint flexes, and the included angle is 135°, and the anatomic angle is 45°.

(a) 180°

(b) 135°   45°

12.2.1 Osteokinematics

> **Section Question**
>
> Why do you have greater motion at some joints compared to others? Why do certain joints allow movements in one plane but not others? How does this relate to an injury?

It is important to realize that human movement is produced by bones rotating about axes. Even linear motions of the hand (such as pushing or pulling) are accomplished by the coordinated rotations of several joints. When quantifying osteokinematic rotations (called **range of motion**; ROM), two types of measures can be employed. The first is a segment (absolute) angle, which is the angle between the long axis of the bone (segment) and some reference line in the global reference frame (such as the horizontal or vertical). During gait, because it is hard to measure the position of the trunk and the pelvis, the thigh is typically measured with respect to the horizontal.[2] The second type of measurement is a joint (relative) angle, in which the orientation of one bone (segment) is measured relative to the orientation of another bone (segment). There are two subtypes of joint angles: an included angle and an anatomic angle.[3] The included angle is the angle between the long axis of two bones or segments. The anatomic angle is the

Range of motion The amount of rotation, measured in degrees, available at a joint

angle a segment moved from the anatomical position. The difference between these two angles is best illustrated with an example. Consider the two segments attached together in **Figure 12.6a**. The two segments are aligned in this neutral position. So the anatomic angle is 0°, even though they are 180° from each other (the included angle). In **Figure 12.6b**, the joint flexed 45° (anatomic angle), but the segments are 135° apart (included angle).

The decision of which angle is used to represent motion is usually a matter of personal choice and convenience. Anatomic angles are most often used clinically. If you choose to use anatomic angles, it is important that you remember the direction of movement. Using Figure 12.6 as an example, if you measured the segment motion from (a) to (b) as 45° in .01 seconds you could potentially make an error and say the average angular velocity was 4,500° per second when in fact it was −4,500° per second, because the rotation was in the negative direction. This error is often avoided when the included angle is used, but that angle is harder to visualize clinically.

Another point of confusion arises because joint motion is relative motion between two bones. Thus, motion could occur when the distal bone moves with respect to the proximal bone, the proximal bone moves with respect to the distal bone, or both. For example, if the distal segment internally rotates 30° on a fixed proximal segment, or if the proximal segment externally rotates 30° on a fixed distal segment, or if the distal segment rotates internally by 15° while the proximal segment externally rotates 15°, the joint configuration is the same. With a few exceptions that will be noted when appropriate, joint rotation will be synonymous with the rotation of the distal segment—even if that segment is not moving. So if the distal segment internally rotates 30° or the proximal segment externally rotates 30°, the joint will be said to have internally rotated.

> **Important Point!** Joint motion is relative motion between segments and could be caused by movement of either the proximal or distal segment.

Degrees of Freedom

If osteokinematic motion occurs about an axis, it is said to

> **Degrees of freedom**   The number of movements available. Movement must occur in two directions to equal one DOF

have a rotational **degree of freedom** (DOF) in that plane. Think of DOF as having a choice; you do not have a choice if you cannot move, or can only go forward. So motion in two opposite directions will count as one DOF. For example, flexion and extension in the sagittal plane is a single DOF. Because there are three planes of motion, a joint can have up to three rotational DOFs.

Now think about the motions that are available for most joints. Most of the motion available in most joints occurs in the sagittal plane, followed by the frontal and transverse planes. An injury to a joint can occur when motion in a plane exceeds its normal limits, straining the soft tissue structures of the joint. Because the available sagittal plane motion is usually quite large, the limit in this plane is rarely exceeded during normal movements. Comparatively speaking, available motion in the frontal and transverse planes can be quite small, making it easier to exceed the limits in those planes (**Box 12.1**). That is why it is important to understand the DOF of each joint.

Anatomy Meets Mechanics

The available DOFs, and the amount of rotation available at each DOF, are largely determined by the anatomy of the joints. The joint surface of each bone will have two principle directions (**Figure 12.7**). Typically, one bone will have a convex surface, and one bone will have a concave surface, although the amount of curvature each surface has varies greatly between joints. The configuration of these curvatures has a big effect on the amount of motion a joint has in each direction. In addition, the ligaments that hold the bones together will prevent motion in certain directions while

| **Box 12.1** | Applied Research: Planes of Motion and Knee Injuries |

Noncontact injuries occur when an athlete gets hurt, but the injury mechanism does not involve contact with another player or object. Noncontact injuries to the knee, particularly those involving the anterior cruciate ligament (ACL), are a notable problem in athletics. The purpose of this systematic review was to examine the mechanisms of noncontact ACL injuries as reported in 40 peer-reviewed publications. The authors concluded that the mechanism of injury involved limited motion in the sagittal plane, along with motion in the frontal or transverse planes. The potential for injury was greatest when frontal and transverse plane motions were combined.

Data from: Shimokochi Y, Shultz SJ. Mechanisms of noncontact anterior cruciate ligament injury. *Journal of Athletic Training* 2008;43(4):396–408.

guiding motion in others. Together, the bony configuration and the ligamentous constraints determine the function of a particular joint.

Functionally, joints are classified as one of three types: synarthrosis, amphiarthrosis, or diarthrosis. Joints that allow no movement are called synarthroses, whereas joints that allow limited movement are called amphiarthroses. Both of these types of joints would have zero rotational DOF. A "freely movable" joint allows between zero and three DOF, largely depending on the configuration of the two bones making up the joint. These types of joints are called diarthroses and can

| **Figure 12.7** | A joint surface will typically have two principal axes (x- and y-). The surface along each axis is usually convex on one bone and concave on the other. |

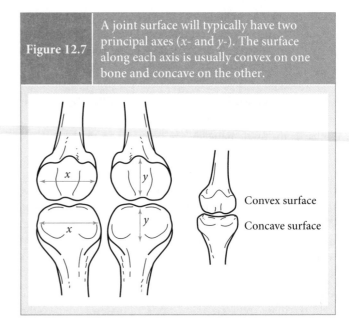

Convex surface

Concave surface

| Figure 12.8 | Different types of diathroidal joints: (a) planar; (b) hinge; (c) pivot; (d) condyloid; (e) saddle; (f) spherical. |

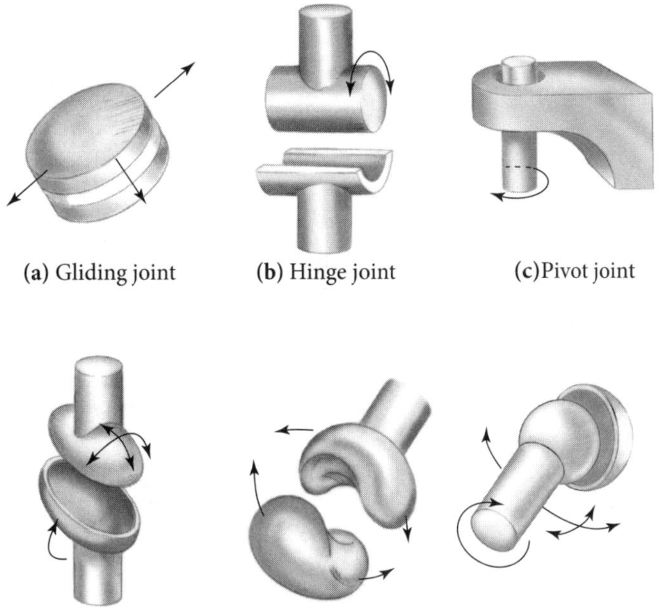

(a) Gliding joint   **(b)** Hinge joint   **(c)** Pivot joint

(d) Condyloid joint   **(e)** Saddle joint   **(f)** Spherical joint

be further classified by the number of degrees of freedom and the type of motion allowed (Figure 12.8):

- **Planar** joints have zero rotational DOF, but do allow for translation between the bones. Because they have no rotational DOF, they have no axis of rotation (AOR).
- **Hinge** joints have one DOF, with an AOR that is closer to the perpendicular of the long axis of the bones. The elbow, ankle, and interphalangeal joints are examples of hinge joints.
- **Pivot** joints also have one DOF, but are different from the hinge joints in that the AOR is closer to the parallel of the long axis of bone. The proximal radioulnar joint is an example of a pivot joint.
- **Condyloid** joints have two DOFs and two AORs. With a condyloid joint, the convex surface in both directions is located in the same bone. The wrist and tibiofemoral joints are examples of condyloid joints.
- **Saddle** joints also have two DOFs (and two AORs). With a saddle joint, one bone has a convex surface in one direction and a concave surface in the other. The first carpometacarpal joint of the thumb is an example of a saddle joint.

- **Spherical** (ball and socket) joints have three DOFs. Technically, there is no AOR, per se, but a center of rotation located in the center of the ball. Movement can occur about any axis that passes through the center, and thus in any plane. Clinically, motion is referenced to the three clinical planes. The hip and glenohumeral joints are examples of spherical joints.

Important Point!   The type of joint determines the number of DOFs and, to a large extent, the amount of motion available in a particular plane.

Section Question Answer

Motion at a joint is determined by its anatomy, particularly its bony configuration and ligamentous constraints. This will limit the planes in which movement occurs and the amount of movement in that plane. If motion exceeds the amount of motion available in a particular plane, injury can result. Typically, the motion available in the frontal and transverse planes are less than the motion in the sagittal plane, so this is where you should look for an injury to occur.

Figure 12.9   Three types of arthrokinematic (joint) motion: (a) gliding, (b) spinning, (c) rolling.

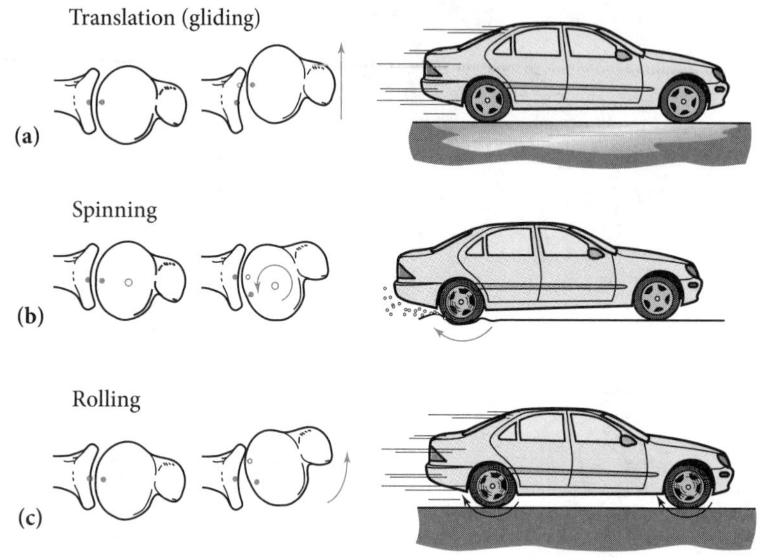

Translation (gliding)

(a)

Spinning

(b)

Rolling

(c)

Adapted from Zuckerman JD and Matsen III FA. Biomechanics of the shoulder, in: *Basic Biomechanics of the Musculoskeletal System*. M Nordin, VH Frankel, eds. Philadelphia: Lippincott, Williams, and Wilkins, 1989, pp 225–248. Figure 12-7, pg. 231.

12.2.2   Arthrokinematics

Section Question

During manual therapy, clinicians are often mobilizing joints to remove restrictions that impede motion. What are they doing?

As the name implies (root word *arthro* = joint), arthrokinematics is the movement of one joint surface relative to another. There are three types of arthrokinematic motion, which are best described with a tire (convex surface) on the ground (for this purpose, think of it as a concave surface):[4]

- **Gliding** (sliding) is pure translation. The same point on the convex surface comes in contact with many points on the concave surface. When you slam on your brakes or skid on ice, the tire is gliding across the road (**Figure 12.9a**).
- **Spinning** is pure rotation. Many points on the convex surface come in contact with the same point on the concave surface. If you ever had your tire stuck in the mud and continued to step on the gas, you have experienced your tire spinning on the ground (**Figure 12.9b**).
- **Rolling** is a combination of gliding and spinning. Many points on the convex surface come in contact with many points on the concave surface. When you

are driving, the tires are normally rolling down the road (**Figure 12.9c**).

With linear translation, every point on a body moves with the same linear velocity. With rotation, the linear velocity is perpendicular to the long axis of the rotating body. The linear velocity of any point during rolling is the vector sum of these two velocities (gliding and spinning; **Figure 12.10**).

As a general rule, when the convex surface is moving on the concave surface, the motion between most joint surfaces should spin. If they did not, then the joint surfaces would not stay congruent for very long. For example, if the convex surface of one bone rolls on the concave surface of another, it will eventually roll off! Looking at Figure 12.10, note that rolling

Figure 12.10   Rolling is a combinaton of spinning and gliding.

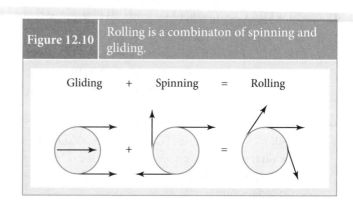

Gliding   +   Spinning   =   Rolling

| Figure 12.11 | If rolling is the sum of spinning and gliding, then the amount of spin is the amount of gliding subtracted from the amount of rolling. In other words, rolling and gliding must be in opposite directions. |

Spinning = Rolling − Gliding

Spinning = Rolling + Gliding

| Figure 12.12 | When the convex surface is moving on the concave surface, rolling and gliding must be in opposite directions. |

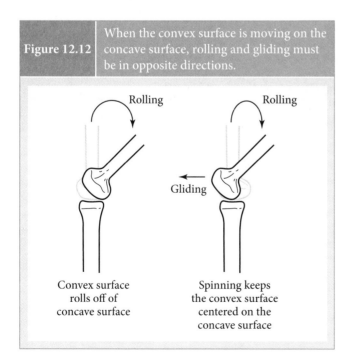

Rolling   Rolling

Gliding

Convex surface rolls off of concave surface

Spinning keeps the convex surface centered on the concave surface

| Figure 12.13 | When the concave surface is moving on the convex surface, rolling and gliding must be in the same direction. |

Pinch here

Gliding

Gap here

Rolling

is a combination of gliding and spinning. So, for the convex surface to spin, rolling and gliding of the convex bone must be in opposite directions (Figures 12.11 and 12.12).

If the concave surface is moving on the convex surface, the concave bone must move in the opposite direction of the convex bone in the previous scenario. This maintains the same *relative* motion between the two bones. If the concave surface was spinning, the joint also would not stay congruent, and there would be a pinching of the joint surfaces in the direction of the spin and a gap in the opposite direction. To stay congruent, the concave surface must translate and rotate on the convex surface. From Figure 12.13, notice that rolling requires the spinning and gliding of the concave bone to be in the same direction.

This is an ideal relation. Oftentimes, a convex bone will not purely spin. The net result is a displacement between the two joint surfaces. Although small displacements may be normal for some joints, excessive amounts may lead to injury. This relation is also the basis for many manual therapies that attempt to mobilize (glide) joints to remove restrictions that prevent normal joint motion: If a bone does not glide properly, the arthrokinematics will not be ideal. Improper arthrokinematics can restrict osteokinematic motion and produce an injury (see Box 12.2). This also explains why stretching may not always increase the ROM of a joint. Stretching will improve the extensibility of the muscle–tendon complex (MTC), but may not target the joint tissues that are restricting arthrokinematic motion. Manual therapy is a specialized skill used by clinicians that is aimed at removing these restrictions.

Section Question Answer

Osteokinematic motion of a joint can be restricted if the arthrokinematic motion of a joint is restricted. Joint mobilizations are often used to improve the arthrokinematic motion of a joint, and thus the osteokinematic motion of that joint.

| Box 12.2 | Applied Research: Joint Mobilizations after Injury to Improve Wrist ROM |

In this case series, six patients who had injury to their wrist underwent treatment that involved both deep heat and joint mobilizations. The author chose joint mobilizations over stretching because he wanted to emphasize the gliding component of wrist motion. The injuries occurred anywhere from 6 weeks to 2½ years prior to treatment, and the deficits in ROM were as large as 45° in flexion and extension. By six treatments, all patients had improved their wrist ROM to within normal limits and maintained 93% of that motion 1 month after the intervention.

Data from: Draper DO. Ultrasound and joint mobilizations for achieving normal wrist range of motion after injury or surgery: a case series. *Journal of Athletic Training* 2010;45(5):486–491.

COMPETENCY CHECK

Remember:

1. Define osteokinematic and arthrokinematic motion, range of motion, and degrees of freedom.
2. List the six types of diarthroidal joints and the motions associated with each.
3. List the three types of arthrokinematic motion.

Understand:

1. What is the difference between an included and an anatomic angle? What are the advantages of using each type?
2. Why is arthrokinematic motion important for proper osteokinematic motion?

Apply:

1. Take a picture of any sporting activity from the web or a magazine. What are the included and anatomic angles of the various joints involved in the movement?

12.3   KINETICS

Section Question

How can Jim lift more weight than George, even though George has more muscle mass?

12.3.1   The Effect of the Muscle–Tendon Complex on a Joint System

A force is required to produce a torque that will rotate a joint about its axis. Forces could be internal (such as muscular forces) or external (such as gravity or contact forces) to the body. Most purposeful movement requires the body's muscular (internal) forces to produce a torque in response to the torque produced by external forces.

Muscle attaches to bone via tendinous insertions. Activation of the muscle produces force, which causes the tendon to pull on the bone. The force produced by the muscle–tendon complex (MTC) will have three effects:

- Compression (or distraction) of the joint.
- Shear across the joint.
- Rotation of the joint.

Each of these can be understood more clearly through the use of an example. **Figure 12.14** is a joint with one DOF. It has a single MTC attached to it. The proximal segment is fixed and does not move. In **Figure 12.15** the proximal bone has been removed, and the MTC force has been broken down into its components that are perpendicular and parallel to the long axis of the distal bone. Depending on the joint configuration, these components can create a shear and/or compression/

| Figure 12.14 | A single joint system with a single muscle attached. |

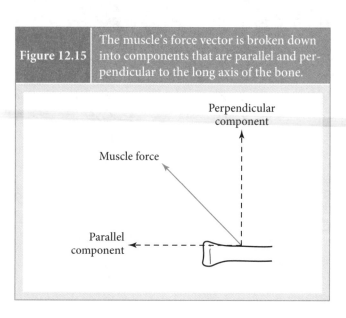

| Figure 12.15 | The muscle's force vector is broken down into components that are parallel and perpendicular to the long axis of the bone. |

Figure 12.16 For this joint configuration, the parallel component creates a compressive force, and the perpendicular component creates a shear force. The effect of the component changes depending on the joint configuration.

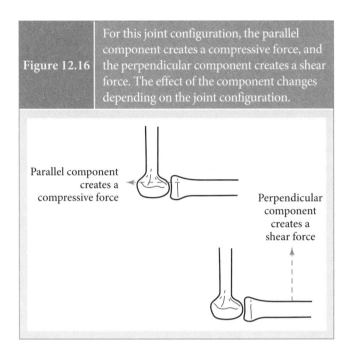

Parallel component creates a compressive force

Perpendicular component creates a shear force

distraction force at the joint. Figure 12.16 shows the effects of these forces for the joint configuration shown. Note that the parallel component creates a compressive force, whereas the perpendicular component creates a shear force. This will change depending on the joint configuration. A force parallel to the joint surface will always create a shear force, and a force perpendicular to the joint surface will always create a compression or distraction force (depending on the direction of the force vector). However, a force parallel to the long axis of a bone is not always perpendicular to the joint surface. In this example, the force perpendicular to the long axis of the bone was parallel to the joint surface.

Important Point! The three effects of a muscle's action on a joint:

1. Compression/distraction.
2. Shear.
3. Rotation.

According to Newton's second law, a net force will induce an acceleration in the direction of the net force. In general, you do not want large translations of the joints; they should stay fairly congruent. Compressive force at a joint is met by an equal and opposite force between bones. Distraction and shear forces are usually negated by forces in the ligaments of the joints.

The third effect of an MTC force is to cause a torque about the joint's AOR. The magnitude and direction of that torque depend on two things: the amount of force produced by the MTC and the MTC's moment arm. Seven factors affect an MTC's ability to produce force, but only two factors determine an MTC's moment arm: the length from the AOR to the tendon's insertion point and the line of pull of the MTC.

Important Point! The amount of torque produced by an MTC will depend on two things:

1. The amount of force produced.
2. Moment arm.

The length from the AOR to the insertion point is fairly straightforward. In Figure 12.17, the AOR is in the center of the convex surface, and the insertion point is where the tendon attaches to the bone. The length, l, is simply measured from the axis of rotation along the length of the bone until insertion point is reached. A longer length usually means that a greater amount of torque can be produced. (Soon you will see why that is not always the case.)

Important Point! Two factors determine the MTC's moment arm:

1. The length from the AOR to the insertion point.
2. The direction of the line of pull.

The MTC's line of pull includes both the sense and the angle it makes from the long axis of the segment. The sense should be obvious: An MTC pulling in a different direction

Figure 12.17 The length, l, is the distance from the joint axis of rotation to the insertion of the MTC on the long axis of the bone.

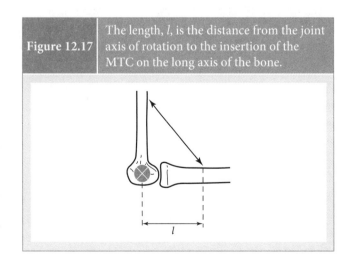

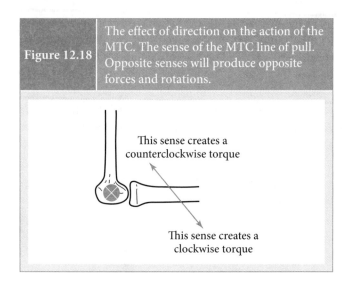

Figure 12.18 The effect of direction on the action of the MTC. The sense of the MTC line of pull. Opposite senses will produce opposite forces and rotations.

This sense creates a counterclockwise torque

This sense creates a clockwise torque

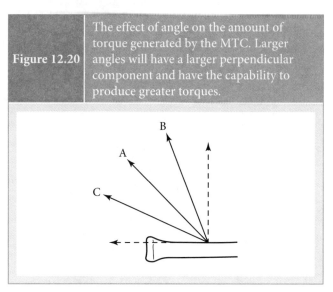

Figure 12.20 The effect of angle on the amount of torque generated by the MTC. Larger angles will have a larger perpendicular component and have the capability to produce greater torques.

will create a torque in a different direction (**Figure 12.18**). The angle, θ_{MTC}, may not be so obvious (**Figure 12.19**). Only the perpendicular component of the force will create a torque (the parallel component will have a moment arm of zero). A larger angle means that the perpendicular component is larger. In **Figure 12.20**, A is the original line of pull of the MTC. B has a greater angle, and thus larger perpendicular component. Conversely, C has a smaller angle, and a smaller perpendicular component.

Important Point! Two factors determine the direction of the line of pull:

1. The sense.
2. The angle of pull from the long axis of the bone.

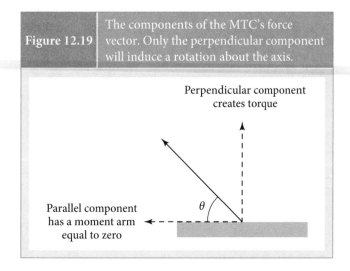

Figure 12.19 The components of the MTC's force vector. Only the perpendicular component will induce a rotation about the axis.

Perpendicular component creates torque

θ

Parallel component has a moment arm equal to zero

Mathematically, the moment arm is going to include both the length, l, and the angle, θ_{MTC}. The appropriate correction factor is sine θ_{MTC} if the angle given is between the long axis of the bone and the MTC's line of pull. If the angle is given perpendicular to the long axis of the bone, the appropriate correction factor is cosine θ_{MTC}. Unless otherwise stated, throughout this text, the angle will always be given from the long axis of the bone. The equation for the length of the moment arm, $^{\perp}d_{MTC}$, is

$$^{\perp}d_{MTC} = l \times \sin\theta_{MTC} \qquad (12.1)$$

It is important to note that the length, l, will have a more profound effect than the angle, θ_{MTC}, because the angle is multiplied by the sine. See **Table 12.1** for an example.

From the preceding discussion, it becomes evident why one man (Jim), with less muscle mass, could lift more weight than another man (George), who has more. **Table 12.2** lists their respective muscle masses and PCSAs for a single muscle joint system. Assuming a specific tension of 22 N/cm², George clearly has the ability to produce more tension within

Table 12.1 The Effect of Doubling the Distance, l, or Doubling the Angle, θ_{MTC}. Doubling the Distance has a More Profound Effect

Condition	l (mm)	θ	$\sin\theta$	$^{\perp}d$ (mm)
1	2	30	.5	1
2	4	30	.5	2
3	2	60	.866	1.73

Table 12.2	Comparing the Maximum Torque-Producing Capability of Two Men, One With a Larger PCSA and the Other With a Longer Moment Arm

The more favorable moment arm leads to a greater torque-producing capability, even though the other man has a greater force-producing capability.

	Jim	George
Muscle mass (g)	31.5	57.6
PCSA (cm^2)	3.8	4.5
Maximum force-producing capability (N)	83.6	99
$^\perp d$ (m)	0.006	0.005
Maximum torque-producing capability (Nm)	0.502	0.495

his muscle. But if Jim has a more favorable moment arm (because his tendon insertion is a bit further away), he could have the advantage in producing more torque, which would be the ultimate factor in how much weight could be lifted.

12.3.2   The Effect of the Joint System on the Muscle Tendon Complex

You should realize that l will never change (unless an orthopedic surgeon is involved). θ_{MTC} will change as a function of the joint angle. Figure 12.21 shows how two muscles, different only in their proximal insertion points, are changing

Figure 12.21	The effect of joint angle on two MTCs with different moment arms. Note that MTC A's moment arm peaks about 90°, while MTC B's moment arm continues to increase throughout the ROM.

as a function of joint angle. How moment arms change as a function of time cannot be predicted and do not follow a general rule; they must be calculated. The important thing to understand is that moment arms do change as the joint angle changes, and this one factor will affect how much torque a muscle can produce.

Recall that torque is the product of force and moment arm. Although it is true that for the same amount of force, a muscle with a larger moment arm will produce greater torque, a muscle may not necessarily be able to produce the same amount of force if the moment arm is increased. Two factors in particular are affected by changing a muscle's moment arm: the length—tension and force—velocity relations.

Consider the length–tension relation of a muscle and then the MTC. It would stand to reason that as a joint is going through its ROM that the length of the muscle will change. When the MTC length and moment arm are combined, the resulting change in the maximum torque as a function of joint angle is known as a **strength curve**.[5] But not all MTCs will go through their entire lengths as part of a joint system, and when combined with the effects of the moment arm, the strength curves will be generally characterized as one of three types (Figure 12.22):[5]

> **Strength curve**   A plot of the maximum torque produced about a joint as a function of joint angle

- Ascending—torque increases as joint angle increases.
- Descending—torque decreases as joint angle increases.
- Ascending–descending—torque first increases and then decreases as joint angle increases.

Because of the myriad factors that determine them, strength curves cannot be calculated or derived but must be determined experimentally. The important point to realize is that the torque about a joint will change as a function of joint angle. The identification of strength curves were the impetus behind the development of various different exercise machines: The hope was that the resistance of the machines would match the human strength and provide an overload of the muscle throughout the entire range of motion (the resistance would be greatest where the greatest torque could be produced and least where the least torque could be produced). However, the design of these machines had limited success in fulfilling their goal (Box 12.3).

> **Important Point!**   The maximum amount of torque that can be produced about a joint will change as a function of joint angle.

Figure 12.22	Types of strength curves. (a) ascending, (b) descending, (c) ascending–descending.

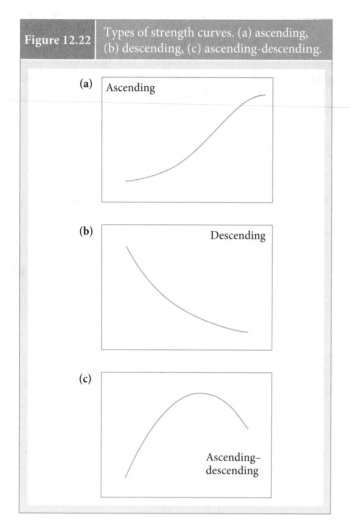

(a) Ascending

(b) Descending

(c) Ascending–descending

Box 12.3	Applied Research: Do the Strength Curves of Machines Match Those of Humans?

Variable-resistance machines are designed so that the amount of resistance varies through the range of motion. The thought is that if the machine's strength curve matches a human, overload will be maximized throughout the ROM—providing a greater overload stimulus. In this study, 10 healthy subjects performed knee extension exercises, and the strength curve was found to be of the ascending–descending type. Torque was lower at increased velocities, but retained the same basic shape. This curve was then compared to the resistance provided by eight different variable-resistance machines. The strength curves of the machines varied greatly, both from one another and from the humans. Different machines displayed ascending curves, descending curves, and ascending–descending curves. Those that did display ascending–descending curves were much flatter than the strength curves of humans, providing greater resistance at the beginning and end ranges of motion. It was concluded that the strength curves of variable-resistance machines do not match those of humans, at least for the knee extensors.

Data from: Folland J, Morris B. Variable-cam resistance training machines: do they match the angle-torque relationship in humans? *Journal of Sports Sciences* 2008;26(2):163–169.

The general type of strength curve (ascending, descending, ascending–descending) appears unaffected by movement speed,[5,6] but the shape of the curve and overall magnitude is greatly influenced by movement speed. At least concentrically, there is an inverse relation between force and velocity. But not every muscle would be affected by velocity in the same way: The location of the muscle's tendinous insertion has a big effect. For a machine or robot, doubling the moment arm should double the amount of torque produced because the force stays constant. But recall that:

$$v = l_r \omega$$

This means that for any given angular velocity, ω, the linear velocity at a point twice as far from the axis of rotation will have twice the linear speed. In other words, a muscle's speed of contraction will be twice as great (it will be shortening twice the distance for the same amount of time). Thus, increasing the moment arm can decrease the ability of a muscle to produce torque as the speed of movement increases.

To see an example of this, return to the example involving Jim and George, but this time switch their moment arms: George's moment arm is .006 meter and Jim's moment arm is .004 meter. You would be tempted to say that George can now create more torque than Jim. To see what happens when you take velocity of movement into account, examine Figure 12.23.

Figure 12.23	Torque-producing capability as a function of angular velocity for two men with different moment arm characteristics.

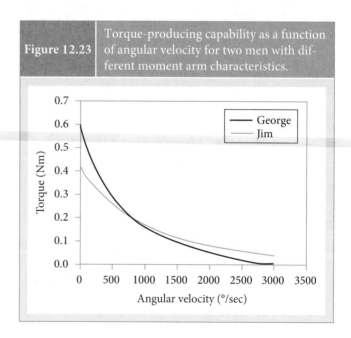

As you probably predicted, George could produce more torque at the lower movement speeds. But as the speed of movement increases, George's advantage diminishes because his muscle is contracting at a higher velocity. At about 900° per second, the two men can produce the same amount of torque. After that, Jim can actually produce more torque than George, even though both his muscle mass and his moment arm are smaller.

This example used a number of simplifying assumptions. It was assumed that the moment arms did not change throughout the movement and that both muscles had the same architectural properties. Both of these could have a pronounced effect on the results,[7] but this example was given just to show you the isolated effect of movement velocity.

COMPETENCY CHECK

Remember:

1. Define: strength curve.
2. What are the three effects of an MTC on a joint?
3. What two things determine an MTC's moment arm?
4. What two things determine an MTC's line of pull?
5. What are the three types of strength curves?

Understand:

1. What happens to the MTC's torque-producing capability if you increase the distance from the AOR to the MTC's insertion point?
2. What happens to the MTC's torque-producing capability if you increase the angle of the line of pull?
3. For each type of strength curve, describe how the torque-producing capability of the muscle changes with a change in joint angle.
4. What happens to the speed of contraction if the MTC's insertion point is further from the axis of rotation? What effect will that have on the ability to produce torque at slow velocities? At fast velocities?

Apply:

1. What is the torque-producing capability of the three muscles in the following table? Assume they all have the same sense, creating a positive torque.

Muscle	Force (N)	Distance (cm)	Angle of Pull (°)	Torque (Nm)
A	220	2	10	
B	154	3	30	
C	396	1.5	20	

12.3.3   How the Muscle–Joint System Interacts with Real-World Loads

Accelerations

Thus far, you have been examining the muscle's ability to produce torque. Now you are in a position to examine the effect of that torque in producing movement. Muscle torque is but one of the torques that are acting on a joint, and you know that:

$$\sum T = I\alpha$$

What other torques are involved? Examine a simple case of performing a dumbbell curl with the upper arm fixed and the forearm moving. To make the illustration simple at the beginning, imagine a situation where the person performing the curl holds the dumbbell statically at several points in the range of motion. Because the movement is being performed in the gravitational field, the force of gravity on the forearm is one of the sources of torque. The contact force of the dumbbell (equal to its weight) is the other source of torque. Examine Figure 12.24, and note that the moment arm of the dumbbell is also changing as a function of joint angle. Specifically, the correction factor is

$$^\perp d_{ext} = l_s \times \sin\theta_s \quad (12.2)$$

where l_s is the length of the segment and θ_s is the segment angle. Similarly, the torque created by the weight of the forearm can be calculated as:

$$^\perp d_g = l_{COM} \times \sin\theta_s \quad (12.3)$$

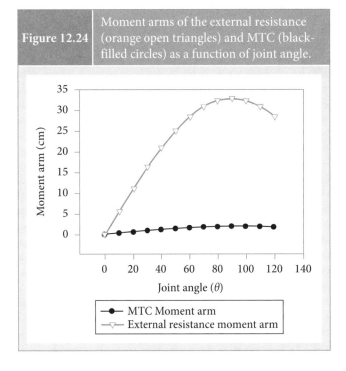

Figure 12.24 Moment arms of the external resistance (orange open triangles) and MTC (black-filled circles) as a function of joint angle.

where l_{COM} is the length from the joint center to the center of mass of the segment and θ_s is the segment angle. Note that the segment angle was used here because the sources of resistance came from a mass being in the gravitational field, making the calculation simpler. Alternatively, you could have calculated the forces perpendicular to the segment. The calculations would have been different, but the answer is the same (see Box 12.4 for details).

It is sometimes useful to compare the moment arm of the muscle to the moment arm of the external resistances. In this example, there are two external resistances: the weight of the segment and the weight of the dumbbell. By dividing the moment arm of the MTC ($^\perp d_{MTC}$) by either the moment

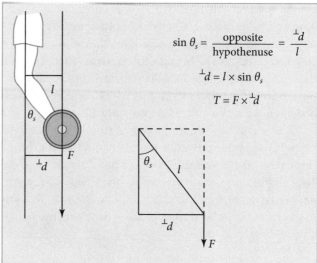

$$\sin \theta_s = \frac{\text{opposite}}{\text{hypothenuse}} = \frac{^\perp d}{l}$$

$$^\perp d = l \times \sin \theta_s$$

$$T = F \times {}^\perp d$$

Note: θ_s is the angle from the vertical

In Figure A, the contact force was broken down into its components, perpendicular and parallel to the long axis of the bone. It is hard to visualize because the angle is obtuse. But the angle, θ_R, is 90° in the negative direction from the segmental angle. So the angle is θ_s minus 90°.

In Figure B, the distance perpendicular from the force vector to the axis of rotation is determined. Conceptually, it is much easier. The advantage of the first method is that it allows you to determine all three effects of the contact force (external resistance): compression, shear, and torque. So sometimes it is worth the extra effort.

Box 12.4	Essential Math—Alternate Methods of Calculating Torque

There are two different ways to calculate the torque about a joint. Although both methods will give you the same result, they are conceptually different.

The first method was shown to you earlier in the lesson. When dealing with MTCs, this method is conceptually easier because the tendon has fixed insertion points, and it is the angle of the line of pull that is changing. With this method, the moment arm remains constant, and the component of the force that is perpendicular to the long axis of the segment is calculated.

In the second method, the force is (hypothetically) fixed, and it is the component of the force vector that is perpendicular distance from the force vector to the axis of rotation that is changing. With this method, the force vector is extended indefinitely, and the perpendicular distance from this vector is calculated.

To see the difference between the two methods, consider the following figure:

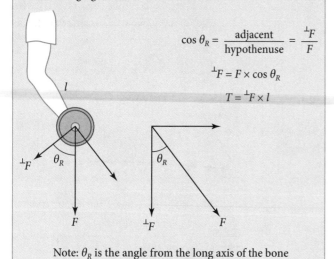

$$\cos \theta_R = \frac{\text{adjacent}}{\text{hypothenuse}} = \frac{^\perp F}{F}$$

$$^\perp F = F \times \cos \theta_R$$

$$T = {}^\perp F \times l$$

Note: θ_R is the angle from the long axis of the bone
$$\theta_R = \theta_s - 90°$$

arm of the segment weight ($^\perp d_g$) or contact force ($^\perp d_{ext}$), you can look at the mechanical advantage (or disadvantage) or the muscle:[8]

$$\text{EMA} = \frac{^\perp d_{MTC}}{^\perp d_{ER}} \qquad (12.4)$$

Where $^\perp d_{ER}$ is the moment arm of the external resistance, either gravity or the external force. If the EMA is greater than 1, the MTC has a mechanical advantage; if it is less than 1 (which it usually is), the MTC is at a disadvantage. For this example, the EMA of the MTC compared to gravity and the dumbbell as a function of joint angle is plotted in Figure 12.25. There are a few important things to note on this graph. First, MTC is at a mechanical disadvantage (EMA < 1) compared to both gravity and the dumbbell. This is not surprising because the moment arm is larger for both the external resistances. This means that to produce the same amount of torque, the MTC must produce a much larger force. Because the MTC's force does not have a single effect, the large muscle forces have the

| Figure 12.25 | Effective mechanical advantage (EMA) of the MTC against the external resistance (top) and gravity (bottom) as a function of joint angle. |

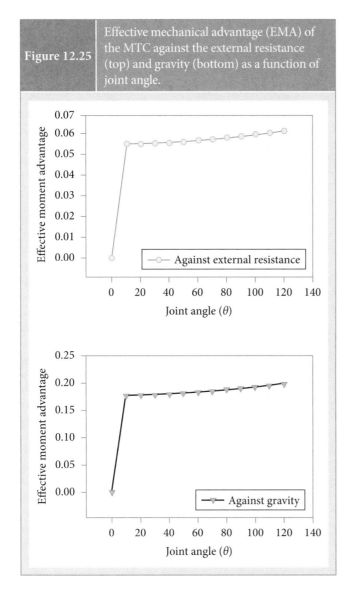

| Figure 12.26 | Arm position during four points of a static dumbbell curl. The moment arm of the external resistance is $l \cdot \sin \theta$. |

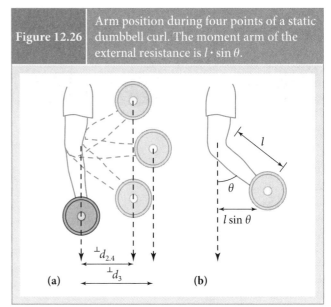

is larger than the moment arm of gravity, and both have the same correction factor ($\sin \theta_s$).

Next, consider the torque required by the muscle to hold the dumbbell in place at the various points in the ROM. If the forearm is held statically, you know that the sum of the torques must be zero. You also know that there are three torques: MTC, gravity (weight of the segment), and the contact force (force provided by the dumbbell). Looking at Figure 12.26 and using the right-hand rule, you should be able to determine that gravity and the contact force produce a clockwise (or negative) torque. To cancel those out, the MTC torque must be positive, which you can also verify with the right-hand rule. Algebraically, we can summarize this as follows:

$$\sum T = I\alpha$$
$$\sum T = 0$$
$$T_{\text{MTC}} - T_{\text{g}} - T_F = 0$$
$$T_{\text{MTC}} = T_{\text{g}} + T_F$$

Graphically, the contributions of each of these torques is presented in Figure 12.27. Note the sum of the torques in each case is zero.

Thus far, the discussion has centered around a very particular case: when the sum of the torques are zero, and no movement is occurring. If the sum of the torques is not zero, then there will be an acceleration—in the direction of the net torque. Note that this is the *net* torque, and not necessarily the

potential to create large compressive and shear forces across the joints. Second, the EMA has much different shape than the moment arm of the MTC graph: note how it is essentially flat throughout the range of motion. Any increase in the moment arm of the MTC is offset by the increase in the moment arm of the external resistances. This helps to explain "sticking points," those points in the range or motion that are hard to overcome when lifting a resistance. They cannot be explained by the moment arm of the MTC alone; oftentimes they can occur where the moment arm of the MTC is larger (but where the EMA is either similar, or even lower) than other points in the ROM. Third, the mechanical disadvantage is greater for the dumbbell than the segment weight, but the shapes of the curves are identical. Again, this should be unsurprising because the moment arm of the external force

Figure 12.27	Net torque (orange closed diamonds) is a function of the torques created by the dumbbell (black open squares), weight of the forearm (gravity; black closed circles), and the MTC (orange open triangles).

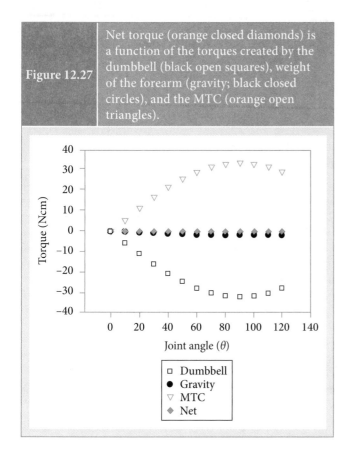

Figure 12.28	Angular position (top), velocity (middle), and acceleration (bottom) during the upward phase of a biceps curl.

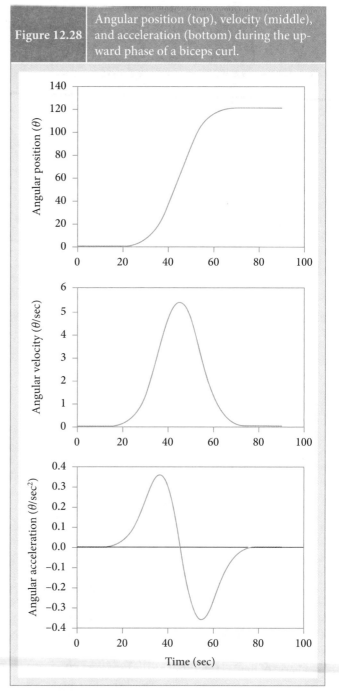

MTC torque, and that the net torque could be in the opposite direction of the net velocity (displacement).

To illustrate these ideas, consider the dumbbell curl again, this time with movement. For ease of discussion, consider separate "up" and "down" phases and neglect the weight of the forearm. During the "up" phase, the angular displacement is approximately 120° (0–120° anatomic angles). The angular velocity and acceleration curves in Figure 12.28 should look very familiar to you by now. Examine the acceleration curve more closely. If you subtract the acceleration caused by the dumbbell from the net acceleration, you can determine the acceleration caused by the MTC. Alternatively, you could have calculated the net torque from the acceleration and then subtracted the external torque to determine the MTC torque, which was done in Figure 12.29. You will note that in this case the net and MTC torques are in the same direction.

Now look at the "down" phase (Figures 12.30 and 12.31). During this part of the movement, the net acceleration is less than that caused by gravity. For this to occur, the MTC torque must be in the opposite direction of the net torque. In this case, the elbow flexors are acting eccentrically.

Energy Exchanges

Section Question

There are three types of muscle actions: concentric, eccentric, and isometric. Why are muscles capable of having three types of actions?

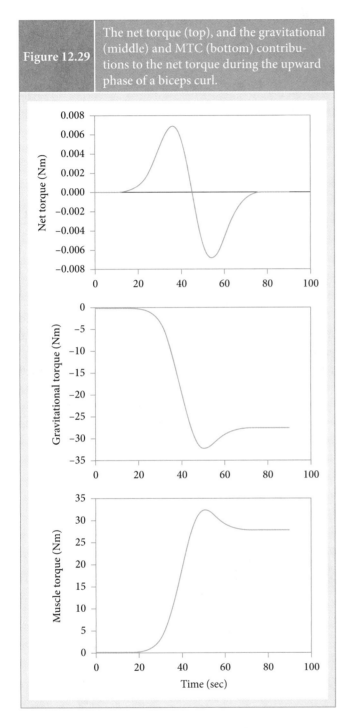

Figure 12.29 The net torque (top), and the gravitational (middle) and MTC (bottom) contributions to the net torque during the upward phase of a biceps curl.

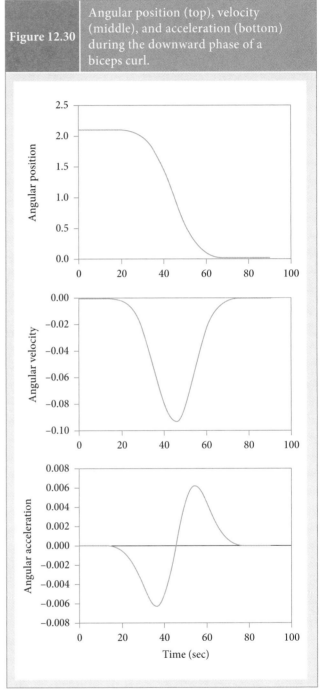

Figure 12.30 Angular position (top), velocity (middle), and acceleration (bottom) during the downward phase of a biceps curl.

Up to this point, you have been examining the MTC by looking at its forces (torques) and resulting accelerations. Additional insights can be gathered using work-energy methods. Using these methods, you will see that the MTC has the potential to function as a motor, brake, spring, or strut,[9] facilitating energy exchanges between the MTC and the segments of the skeletal system.

Important Point! The MTC can act as a motor, brake, spring, or strut.

The most obvious function of the MTC is to act as a motor, or energy source, by generating energy and delivering that energy to the segments. If the muscle torque and the joint

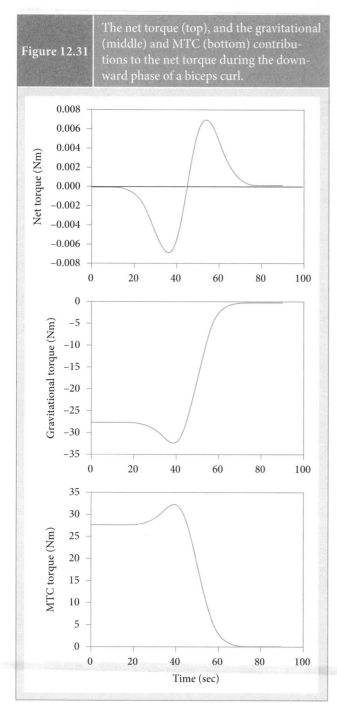

Figure 12.31 The net torque (top), and the gravitational (middle) and MTC (bottom) contributions to the net torque during the downward phase of a biceps curl.

torque and the joint angular displacement are in opposite directions. The sign of the work is negative, and the muscle is lengthening. This means that muscles are decreasing the mechanical energy of the skeletal system (hence the negative sign) by absorbing energy via eccentric muscle actions. An example here would be your abdominal muscles absorbing energy as you lowered yourself down from the sit-up position.

Remember that the first law of thermodynamics tells you that energy cannot be created or destroyed; it is only transformed. The chemical energy that you ingest is stored in the body and can later be converted to mechanical energy via the metabolic pathways that produce ATP. But when the MTC absorbs energy, where does it go? Part of that energy can be stored in the elastic components of the MTC, but only for a brief amount of time, where it can be reused. Any unused energy will ultimately be dissipated as heat.

So how is that energy reused? The energy stored in the elastic components of the MTC is analogous to the energy stored in a rubber band. When the rubber band is let go, it snaps! The energy stored in the rubber band is converted to kinetic energy. Likewise, if an eccentric muscle action is immediately followed by a concentric muscle action (known as the stretch-shortening cycle), the energy in the concentric phase is enhanced as the energy stored in the elastic components is added to the energy generated by the contractile components. In this way, muscles are acting as springs. The use of the elastic properties of the MTC is an important part of human function (see **Box 12.5**).

Finally, muscle can act as a strut, transferring energy from one segment to another via an isometric muscle action. In this

Box 12.5	Applied Research: The Springlike Action of MTCs Enhance Performance

In this study, six subjects were asked to jump while lying on a sliding sledge that was angled 17° from the horizontal. This apparatus allowed the jump to be performed using only the ankle joint. In the first condition, subjects started in a maximally dorsiflexed position (which allowed for no elastic recoil of the tendon) and pushed off into maximal plantar-flexion. In the second condition, subjects started in a maximally plantar-flexed position before dropping into maximal dorsiflexion, and then immediately rebounded into maximal plantar-flexion. This condition allowed for significant elastic recoil of the tendon.

The second condition enhanced jump height by an average of 14%. The investigators concluded that the improved performance was attributable to both an increase in muscle activation (the stretch reflex) and elastic recoil of the tendon.

Data from: Kubo K, Morimoto M, Komuro T, Tsunoda N, Kanehisa H, Fukunaga T. Influences of tendon stiffness, joint stiffness, and electromyographic activity on jump performances using single joint. *European Journal of Applied Physiology* 2007;99(3):235–243.

angular displacement are in the same direction, the work of the MTC is positive. Positive work means that the muscle is generating energy and delivering that energy to the segments via a concentric muscle action. Thus, the energy of the skeletal system is increasing (which is why the work is positive). The simple case of performing a sit-up against gravity is a good example of positive work being done by abdominal muscles.

The MTC can also act as a brake, or energy sink, by absorbing energy from the segments. This happens when the muscle

<table>
<tr><td>Figure 12.32</td><td>Potential energy of the lower body becomes kinetic energy of the lower body. Via isometric muscle action of the abdominals, the kinetic energy of the lower body is transferred to the upward body. Kinetic energy of the upper body becomes potential energy of the upper body as it changes its height above the ground.</td></tr>
</table>

(a)   (b)   (c)   (d)

case, the muscle is not generating or absorbing energy, and the amount of energy decreased in one segment is increased in another. So the total work (change in energy of the system) is zero, but there is still a torque produced by the muscle. The only way to have torque and not have work is if the angular displacement is zero, hence the isometric muscle action. As a simple example (**Figure 12.32**),[10] imagine that you are lying on your back with your hips flexed to 90°. Your legs, being some height above the ground, have a certain amount of potential energy. If you let your legs drop to the ground, but did not allow your hips or spine to extend, your upper body would rotate upward. In this case, by isometric contraction of your hip flexors, you transferred energy from your lower body to your upper body.

Much of the research in biomechanics involves calculating the torque and/or work done by the MTCs of various joints. Because these calculations take up a huge portion of the biomechanics literature, you should understand how they are calculated and what they mean. This is discussed in **Box 12.6**.

<table>
<tr><td>Box 12.6</td><td>Interpreting Research: The Net Joint Torque</td></tr>
</table>

The net joint torque, or net joint moment, (T_{NJ}) is such a ubiquitous construct in biomechanics research that you should become very acquainted with it. One of the essential problems in

biomechanics is determining the force and/or torque requirements during an activity. With advanced technologies (such as kinematic MRI and fluoroscopy) it is possible to determine a muscle's moment arm accurately, but only for tasks constrained by the use of such equipment. But it is nearly impossible to directly measure the amount of force that a given muscle produces, so biomechanicians are required to estimate it.

The basic gist of the method is to accurately measure the positions and orientations of the various segments of the body, and from this determine the linear and angular velocities and accelerations. Segmental masses and moments of inertia data are obtained from published sources. Known forces and moments are then calculated for each segment. More complicated versions of the following equation are then used to determine the net joint torque:

$$\sum T_{\text{known}} + T_{NJ} = I\alpha$$
$$T_{NJ} = I\alpha - \sum T_{\text{known}}$$

The sum of all torques is equal to the product of the moment of inertia and the angular acceleration. All of the torques include the known torques (due to gravity, joint reaction forces, etc.) and the torque created by the muscle (which is unknown). After subtracting the known torques from the product of the moment of inertia and the angular acceleration, the remainder is the net joint torque. The T_{NJ} is the net of all muscles capable of producing a torque around that joint. It is not representative of any one muscle. Oftentimes, it is equated with the torque required of a group of muscles (such as the knee extensors), but this assumes that there is no cocontraction of antagonistic muscles. This is often a faulty assumption, but it is the best estimation of muscular demand that presently exists.

Other variables are then derived from the T_{NJ}:

Net Joint Torque Impulse: This is the integration of the T_{NJ} with respect to time, or the area under the T_{NJ}-time curve. It represents the T_{NJ}'s contribution to the change in angular momentum of the segment.

Net Joint Torque Work: This is the integration of the T_{NJ} with respect to angular displacement, or the area under the T_{NJ}-angular displacement curve. It represents the T_{NJ}'s contribution to the change in energy of the system. A positive sign means that the T_{NJ} is adding energy to the system and is interpreted as a concentric muscle action. A negative sign means the T_{NJ} is absorbing energy from the system and is interpreted as an eccentric muscle action.

Net Joint Torque Power: This is the change in T_{NJ} work with respect to the change in time. It is alternately calculated as the product of the T_{NJ} and the joint angular velocity. It represents the rate at which the T_{NJ} is generating or absorbing energy. The sign $(+/-)$ is interpreted the same as it is for T_{NJ} work.

COMPETENCY CHECK

Remember:

1. What is the effect of a net torque?
2. How is an effective mechanical advantage calculated?
3. What two things determine an MTC's line of pull?
4. What are the three types of strength curves?

Understand:

1. What is the implication of an effective mechanical advantage less than 1?
2. What happens to the MTC's torque-producing capability if you increase the angle of the line of pull?
3. For each type of strength curve, describe how the torque-producing capability of the muscle changes with a change in joint angle.
4. Give an example of an MTC acting like a motor, a brake, a spring, and a strut.

Apply:

1. What is the torque-producing capability of the three muscles in the following table? Assume they all have the same sense, creating a positive torque.

Muscle	Force (N)	Distance (cm)	Angle of Pull (°)	Torque (Nm)
A	220	2	10	
B	154	3	30	
C	396	1.5	20	

2. Assume muscles A and C create a flexion torque, and muscle B creates an extension torque. Describe the energy changes and actions of the muscles if:

A. The joint rotates from 0° to 90° of flexion.
B. The joint rotates from 90° of flexion back to 0°.
C. The joint stays fixed at 45°.

Note: For each, assume that the MTC torque is constant, and take each muscle as a separate case (that is, they are not activated simultaneously on the same muscle).

12.3.4   Multiple Muscles and Single Joint Systems

Section Question

Why do you have more than one muscle that can serve the same role (such as flex the elbow)?

No joint has a single muscle crossing it; several muscles cross any given joint. Sometimes, those muscles will have the same effect on a joint. In those cases, the muscles are said to be cofunctional.[11] Other muscles will have opposite effects on a joint. Those muscles are called antagonists. Muscles working together to produce a movement are synergistic.[11] As you will see, not all synergistic muscles are cofunctional, but all cofunctional muscles are synergistic.

> **Cofunctional** Muscles that have the same effect on a joint or group of joints
>
> **Antagonist** Muscles that have the opposite effect on a joint
>
> **Synergist** Muscles that work together to produce a movement

To begin, think about why you have more than one muscle that is cofunctional. Obviously, two muscles could produce more force than one muscle of the same size, but why not just have one larger muscle? One obvious answer is redundancy. If something were to happen to that one muscle (or its nerve), another muscle could potentially take over and still allow at least some movement of the joint in a particular direction. But two cofunctional muscles rarely have the same muscle architectural or moment arm characteristics, and that has some interesting functional consequences.[12]

Consider two cofunctional muscles, A and B. Muscle A has a larger PCSA and a larger moment arm than muscle B. All else being equal, you would expect muscle A to produce more torque than muscle B. And in a static situation, you would probably be correct. If you are considering higher movement velocities, the same argument may not hold up. First, you know that muscle A, with its shorter muscle fibers, is going to be more affected by velocity than muscle B. Second, you know that because muscle A has a longer moment arm than muscle B, it must have a higher linear (contraction) velocity. Combined together, muscle A is going to be much more affected by movement speed than muscle B, and muscle B may actually be able to produce more torque than muscle A at higher movement speeds. By having two cofunctional muscles with different mechanical properties, the same task can be performed at greater velocity extremes than if there was only one muscle.[12]

You also have muscles crossing both sides of a joint's AOR and thus will produce torque in opposite directions. This is generally a good thing; you want to be able to move a joint in two directions. These muscles are generally referred to as antagonists, and although this name implies an adversarial relationship, it should not.

Cocontraction occurs when both prime movers and antagonists are activated at the same time. From a movement perspective, this is considered a bad thing because one muscle (the antagonist) is, in essence, "fighting" the

> **Cocontraction** Activation of a prime mover and an antagonist simultaneously

other (the prime mover). If a task requires a prime mover to create a torque of 50 Nm, but an antagonist is generating 20 Nm of torque, then the prime mover must create a torque of 70 Nm to net 50 Nm. This makes movement inefficient because it will require more metabolic energy to complete the same task. However, some researchers have argued that a certain amount of cocontraction is a good thing in terms of enhancing joint stability and decreasing injury risk (see the next section).

Antagonistic muscles can also be cofunctional, particularly if they are activated sequentially (rather than simultaneously) and the movement speed is rapid. For example, say you wanted to rapidly extend your elbow from a flexed position. The movement will start with zero velocity and end with zero velocity, so you know the overall change in momentum (and thus impulse) must also be zero. You certainly do not want to rely on the ligaments and bones to stop the movement—you could hyperextend your elbow and hurt yourself. In these situations, you would want the elbow extensors to rapidly extend your elbow during the first part of the movement and your elbow *flexors* to activate during the latter part of the movement to bring the forearm to a stop. In this case, the elbow extensors are acting as a motor, and the elbow flexors are acting as a brake.

In many movements such as this, the acceleration phase is larger in both duration (time) and displacement. This is advantageous for generating impulse and energy, but also puts the antagonists at a disadvantage. Those muscles must generate a larger amount of force to reduce the same amount of impulse or absorb the same amount of energy because the time and displacement are smaller during the deceleration phase.

> **Section Question Answer**
>
> You have more than one muscle that can perform the same role for a number of reasons. First, if something were to happen to one muscle (say nerve damage causing paralysis), you could potentially still perform the movement (as long as the other nerves were intact). Second, due to differences in muscle architecture and moment arm length, an MTC will be able to produce more torque at a certain velocity than others. Having more than one muscle performing the same action allows you to operate over a wide velocity spectrum.

> **COMPETENCY CHECK**
>
> **Remember:**
> 1. Define the different muscle roles: cofunctional, synergist, and antagonist.
> 2. Define: cocontraction.
>
> **Understand:**
> 1. Explain why you have more than one muscle to perform the same role.
> 2. Explain why the term *antagonist* is really a misnomer.
>
> **Apply:**
> 1. During resistance exercise, are you training specific muscles or movements?

12.4 JOINT STABILITY

> **Section Question**
>
> Mary's physical therapist said that her injury was the result of an unstable joint. What did she mean?

Stability and instability are important concepts in joint mechanics. Unfortunately, there is not a universally accepted definition of instability; several have been used in the literature, including[13]

> **Stability** The ability of a system to produce a reference position or trajectory in the presence of a disturbance

- Excessive and occasionally uncontrolled range of motion resulting in joint dislocation.
- Small, abnormal movement in an otherwise normal range of motion, which may result in pain due to "impingement" at the joint.

- A small amount of force necessary to move a joint through its range of motion (or low joint stiffness).

A better definition might be that a stable joint is one that always gives responses that are appropriate to the stimulus. In other words, the joint does what it is supposed to do, without exposing the joint structures to damaging stress. It should be clear that because we often require a joint to do different things in different situations, the definition of joint stability will be context dependent. In most cases, the context is in response to some disturbance while attempting to maintain some reference position (static stability) or trajectory (dynamic stability) goal.

For example, consider an arbitrary joint held together by one ligament (Figure 12.33a). There is a large separation between the two segments so that you can see what is going on with the ligament. Suppose the goal of the task is simply to transmit forces across the segments (in the direction of the arrows in Figure 12.33b), so that you do not have any displacement (movement). The joint will have static stability if it stays aligned, it does not move, and the forces are transmitted from one segment to the next across the joint without any damage.

Now imagine another force is applied to one of the segments, at right angles to the long axis of the segment (a shear force; Figure 12.33c). Such a disturbing force is called a perturbation. If the perturbation is of a small enough

> **Perturbation** A force causing a disturbance or deviation in a system

magnitude, the contact forces between the joint surfaces will be larger than the perturbation—preventing any movement. In addition, the ligament can provide a restraining force that, in addition to the contact forces, will prevent motion from occurring. These two forces (contact forces and ligamentous forces) make up a passive subsystem of joint stability.[14]

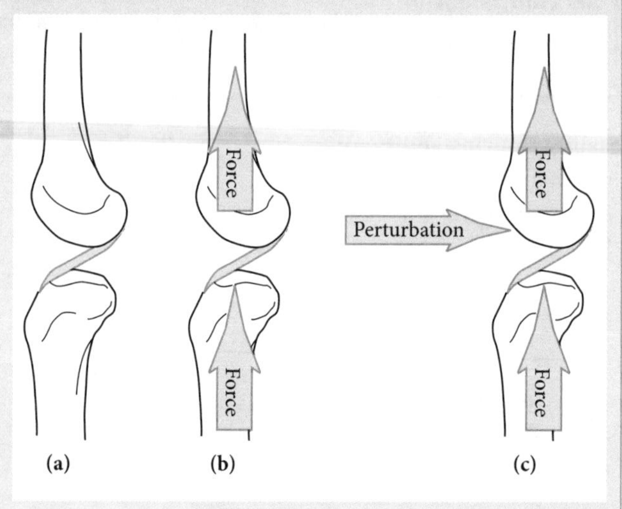

If the perturbation is large enough, it could actually cause a displacement of one of the segments. There are a few consequences when this occurs. First, the congruency between the segments is altered, decreasing the amount of contact area, which in turn increases the stress on the joint surfaces remaining in contact. Second, the direction of the contact forces may be altered. Third, there is now a shear across the joint surfaces, which will increase the strain on the ligament. As the strain increases, so too does the stress on the ligament (Figure 12.34).

From this perspective, static stability can be equated with what is called joint stiffness: The stiffer the joint, the greater the static stability because a larger perturbation force is required to move the segments out of alignment. To distinguish joint stiffness from the property of a material body also called stiffness, a q is used instead of a k:[15]

$$q = \frac{F}{\Delta p} \tag{12.5}$$

That is, a joint with high stiffness would require a large force to displace a segment by a small amount. It would appear that high stiffness is desirable. How, then, can the stiffness of the joint be increased?

You already know that the passive subsystem is responsible for stiffness: the contact forces and tensile forces of the ligaments. Increasing these forces will increase stiffness. The muscle–tendon complex (MTC) can also increase the stiffness of the joint and has been referred to as the active subsystem of joint stability.[14] MTCs can increase stiffness in the following ways. First, activating the muscles that cross a joint will increase the compressive (normal) forces on the joint surfaces. Increasing the normal forces will increase the maximum potential frictional force. Second, MTCs rarely follow a path that is parallel to the long axes of bone and perpendicular to joint surfaces; they run obliquely to both. The component of an MTC's line of pull that is in the opposite direction of the disturbing force can also resist that perturbation. These concepts are illustrated in Figure 12.35. You can now see that a displacement of the joint surfaces will occur if there is an inequity between the disturbing force and resistance force:

$$\Delta p = F_{disturbing} - F_{resisting} \tag{12.6}$$

Figure 12.33 (a) A joint held together by a single ligament (space between bones exaggerated for effect). (b) A force is transmitted from one bone to the other. (c) A perturbation force can alter force transmission between bones.

Figure 12.34 | (a) A perturbation can cause a shear force across a joint. (b) A shear force can increase strain on a ligament.

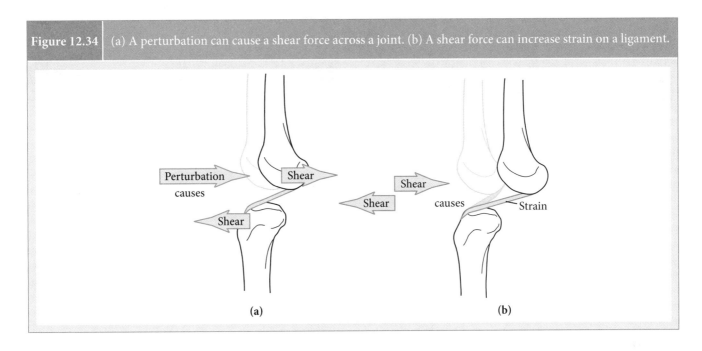

(a)   (b)

Figure 12.35 | The force of a muscle–tendon complex can negate a perturbation by increasing compressive forces across a joint (increasing the frictional force) and creating a shear force in the opposite direction of the perturbation.

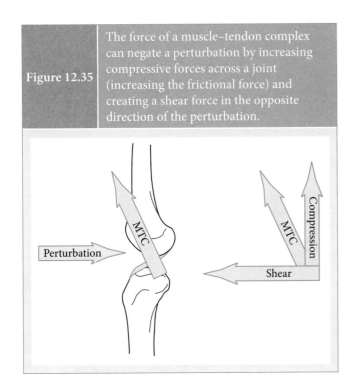

It follows that if the disturbing force is equal to the resistive force, then the displacement will be zero. And the following components make up the resistive force:

$$^x F_{resistive} = {}^N F_{contact} + F_f + {}^x F_{ligaments} + {}^x F_{MTC} \quad (12.7)$$

the normal force, the friction force, the force provided by the ligaments and capsular structures, and the force of the muscle–tendon complex (all in the direction opposite the disturbing force).

In this example, the joint surfaces are relatively flat so as to not have any appreciable normal forces that are in opposition to the perturbation. A good example of this would be the knee. A relatively deep joint, like the hip, will have a normal force in just about any direction a perturbation can occur (minus, of course, a pure distraction).

What happens if the perturbation causes a displacement? A restoring force would then be necessary to return the alignment of the joint. If not, then the joint is said to be unstable. If so, the joint is stable in a mechanical sense.

A perfect response would be the segment immediately returning back to its original position; however, this rarely happens. Usually the segment will oscillate around before returning to its original position. The amount of time it takes to return to its alignment is a measure of the system's performance (Figure 12.36).

Static stability means segments maintain a certain position and orientation, so that a joint maintains a certain alignment in the presence of a perturbation. In such instances, a large amount of stiffness would be advantageous. But joint stiffness does not come without a price. First, stiffness increases the compressive forces across a joint. Second, stiffness costs in terms of the metabolic energy required to maintain muscle activation. Finally, you cannot equate stability with stiffness in every context (Box 12.7).

| Figure 12.36 | In the presence of a perturbation, a system will oscillate around, before returning to, the reference position. If the system returns to the intended position, it is said to have static stability. |

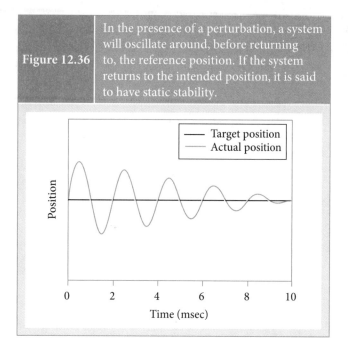

| Figure 12.37 | In the presence of a perturbation, a system will oscillate around, before returning to, the intended trajectory. If the system returns to the intended position, it is said to have dynamic stability. |

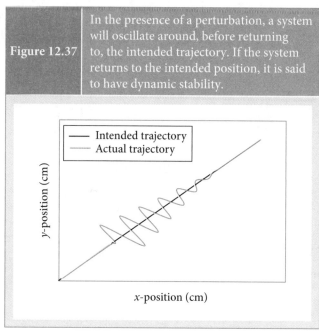

In contrast, dynamic stability is the ability to maintain a given trajectory (as opposed to position) in the presence of a perturbation (**Figure 12.37**).[16] As you could probably have guessed, in such cases measures of stiffness are inadequate

because the joint is moving. Excessive cocontraction will cause unnecessary fatigue and decrease the accuracy of movement.[17] In such cases, the focus shifts to robustness and performance.[16]

Robustness would indicate how large a disturbance the system could tolerate and still return to its intended trajectory. **Performance** measures how far away from its intended trajectory a system will move and how quickly it returns to that trajectory following the disturbance. You want joints that can respond to a large variety of disturbances quickly and accurately. This requires not only joint structures and muscles that have the physical capacity to handle such disturbances, but also a nervous system that can quickly detect and respond to such disturbances.

> **Robustness** The magnitude of the disturbance that a system could tolerate and still return to its intended position or trajectory
>
> **Performance** How large a deviation, and how quickly a system returns to its intended position or trajectory following a disturbance

---

| Box 12.7 | Applied Research: Conflicting Thoughts on Cocontractions and Joint Stability |

In this study, subjects were asked to keep their wrists in a neutral position while a motor produced extensor torques that resembled either a constant, elastic, or unstable load. Compared to the constant and elastic perturbations that required wrist flexor activity, the unstable perturbations required significant cocontraction of all the wrist muscles. It was concluded that "cocontraction was the most effective way to provide the necessary postural stability to counteract an unstable load."

In contrast, another study required subjects to balance on an unstable seat in both a "normal" condition, and with increased trunk muscle coactivation. Increased coactivation actually decreased performance in this case. These authors suggest that there are cases where increased coactivation is advantageous (for example, when receiving a body check in hockey), but there are other times when fine motor control requires a more supple joint.

Study 1 Data from: De Serres SJ, Milner TE. Wrist muscle activation patterns and stiffness associated with stable and unstable mechanical loads. *Experimental Brain Research* 1991;86(2):451–458.

Study 2 Data from: Reeves NP, Everding VQ, Cholewicki J, Morrisette DC. The effects of trunk stiffness on postural control during unstable seated balance. *Experimental Brain Research* 2006;174(4):694–700.

Section Question Answer

Defining a joint as "unstable" is not very clear. What Mary's therapist probably means is that a small perturbing force can cause the two joint surfaces to slide away from one another, increasing the demand on the ligaments, which could result in injury.

COMPETENCY CHECK

Remember:

1. Define the following terms: stability, perturbation, robustness, and performance.

Understand:

1. Where do the forces that stabilize a joint come from?

Apply:

1. What could you do to improve the performance and robustness of joint system in the presence of a perturbing force?

SUMMARY

In this lesson, you learned how joints and muscles interact to create joint systems using the key concepts in Table 12.3. You learned the joint motion is not just bones rotating about axes, but also motion that occurs between two joint surfaces. Muscle–tendon complexes create forces, and those forces can create not only a rotation, but also shear and compression/distraction forces as well. Not only do MTCs affect joints, but the location of the MTC in relation to a joint also affects its performance, most notably due to the length–tension and force–velocity relations. Another way of analyzing MTC function was to examine its ability to act like a motor, brake, spring, or strut. When more than one MTC crosses a joint (and this happens for every joint), those MTCs often take on complementary roles. Finally, you learned that MTCs not only move joints, but also stabilize them. You will later learn to apply these concepts to specific joints in the lower extremity, trunk, and upper extremity.

Table 12.3	Key Concepts

- Clinical reference frame
- Degrees of freedom
- Osteokinematics and types of joints
- Arthrokinematics: gliding, spinning, and rolling
- Strength curves
- Muscle-induced accelerations and energy exchanges
- Functional relations between muscles
- Joint stability

REVIEW QUESTIONS

1. List the three clinical reference frames. In which plane typically does the largest amount of motion occur? What is the implication for the other planes of movement?
2. How many degrees of freedom does a condyloid joint have? Where are the axes of rotation?
3. If a convex surface is rolling anteriorly on a concave surface, then which direction must it be sliding? Why?
4. What are the three effects of a muscle–tendon complex on a joint?
5. What is the effect of doubling the distance between the axis of rotation and the tendon insertion point? Would the effect be the same if you doubled the angle of the line of pull? Why or why not?
6. True or False: An MTC can produce constant torque throughout a joint range of motion. Why or why not?
7. What is the effect of doubling the joint angular velocity on a MTC's contraction velocity?
8. Give an example of an MTC acting like a motor, a brake, a spring, and a strut.
9. Give an example of when two MTCs would be co-functional, synergists, and antagonists.
10. When would cocontraction be beneficial? When would it be detrimental?
11. When can stability be equated with quasi stiffness? When is it inappropriate to do so?
12. Assuming the neutral position (0°) is when the two bones are parallel (180°) from each other. Complete the following table:

Anatomic Angle (°)	Included Angle (°)
0	180
30	
45	
90	
120	

13. The ankle has a neutral position (0°) where two bones are at right angles to each other (90°). Complete the following table:

Anatomic Angle (°)	Included Angle (°)
	90
	5
	10
	95
	120

14. Assuming the neutral position (0°) is when the two bones are parallel (180°) from each other. Complete the following table:

Parameters	Case 1	Case 2	Case 3
Joint angle (°)	30	45	90
d (cm)	2	3	1.5
θ_m (°)	10	20	35
Muscle force (N)	200		
Shear force (N)		469.846	
Compressive force (N)		171.01	
Torque (Nm)			9.5

15. For each of the following, determine: (1) the change in energy; (2) if the energy was absorbed, generated or transferred by the muscle; and (3) the type of muscle action (concentric eccentric, or isometric).
    a. Flexors create a constant torque of .5 Nm while the joint is fixed at 30° of flexion.
    b. Extensors create a constant torque of 6 Nm while the joint moves from 0° to 70° of flexion.
    c. Flexors create a constant torque of 1.3 Nm while the joint moves from 30° of flexion to 70° of flexion.

REFERENCES

1. Cappozzo A, Catani F, Della Croce U, Leardini A. Position and orientation in-space of bones during movement—anatomical frame definition and determination. *Clinical Biomechanics.* Jun 1995;10(4):171–178.

2. Perry J. *Gait Analysis: Normal and Pathological Function.* Thorofare: SLACK Incorporated; 1992.

3. Zatsiorsky VM. *Kinematics of Human Motion.* Champaign, IL: Human Kinetics; 1998.

4. Zuckerman JD, Matsen III FA. Biomechanics of the shoulder. In: Nordin M, Frankel VH, eds. *Basic Biomechanics of the Musculoskeletal System.* Philadelphia: Lippincott, Williams, and Wilkins; 1989:225–248.

5. Kulig K, Andrews JG, Hay JG. Human strength curves. *Exercise and Sport Sciences Reviews.* 1984;12(1):417–466.

6. Knapik JJ, Wright JE, Mawdsley RH, Braun J. Isometric, isotonic, and isokinetic torque variations in 4 muscle groups through a range of joint motion. *Physical Therapy.* 1983;63(6):938–947.

7. Lieber RL, Ljung BO, Friden J. Intraoperative sarcomere length measurements reveal differential design of human wrist extensor muscles. *Journal of Experimental Biology.* Jan 1997;200(1):19–25.

8. Biewener AA. Scaling body support in mammals—limb posture and muscle mechanics. *Science.* Jul 7 1989;245(4913):45–48.

9. Dickinson MH, Farley CT, Full RJ, Koehl MAR, Kram R, Lehman S. How animals move: an integrative view. *Science.* Apr 2000;288(5463):100–106.

10. Zatsiorsky VM. *Kinetics of Human Motion.* Champaign, IL: Human Kinetics; 2002.

11. Zajac FE, Neptune RR, Kautz SA. Biomechanics and muscle coordination of human walking—Part I: Introduction to concepts, power transfer, dynamics and simulations. *Gait & Posture.* Dec 2002;16(3):215–232.

12. Lieber RL, Friden J. Functional and clinical significance of skeletal muscle architecture. *Muscle Nerve.* Nov 2000;23(11):1647–1666.

13. Flanagan SP, Kulig K. Biomechanica. In: Placzek JD, Boyce DA, eds. *Orthopaedic Physical Therapy Secrets.* 2nd ed. St. Louis: Mosby Elsevier; 2006:12–25.

14. Panjabi MM. The stabilizing system of the spine. 1. Function, dysfunction, adaptation, and enhancement. *Journal of Spinal Disorders.* Dec 1992;5(4):383–389.

15. Latash ML, Zatsiorsky VM. Joint stiffness—myth or reality. *Human Movement Science.* Dec 1993;12(6):653–692.

16. Reeves NP, Narendra KS, Cholewicki J. Spine stability: the six blind men and the elephant. *Clinical Biomechanics.* Mar 2007;22(3):266–274.

17. Gribble PL, Mullin LI, Cothros N, Mattar A. Role of cocontraction in arm movement accuracy. *Journal of Neurophysiology.* May 2003;89(5):2396–2405.

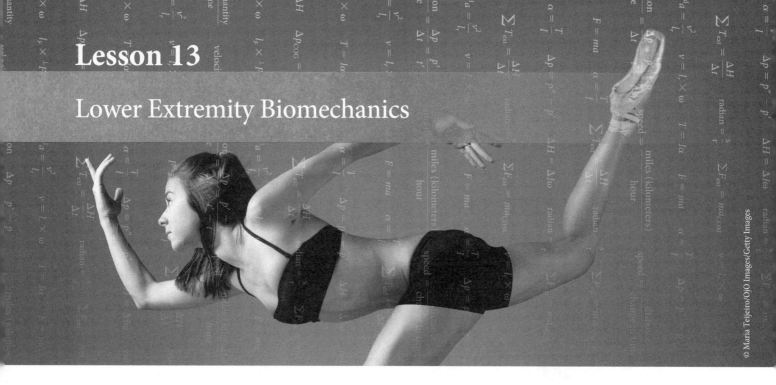

Lesson 13

Lower Extremity Biomechanics

LEARNING OBJECTIVES

After finishing this lesson, you should be able to:

- Identify the major joints of the lower extremity.
- For each joint, determine which type of joint it is, how many rotational degrees of freedom it has, the motions that are available at that joint, and the "normal" ranges of motion.
- For each joint, describe the arthrokinematic motions of the proximal segment moving on the distal segment and the distal segment moving on the proximal segment.
- Describe the torques that can be produced by the muscles crossing each joint.
- Determine the net acceleration of a segment based on the torques produced by external forces and muscle–tendon complexes.
- Explain how the orientation of a segment affects the torque requirements of the muscle–tendon complex.
- Describe how the foot interacts with the ground.
- Give examples of how biomechanical analyses aid in the understanding of lower extremity injuries.

INTRODUCTION

Before embarking on this lesson, you should have a good handle on how a generic joint, controlled by an MTC, works. In this lesson, you will learn about the various specific joint systems in the body. Most of the discussion will center on joints that have rotational degrees of freedom and are involved in the production of movement. Your exploration will begin from the ground up. You will start with the foot and see how it interacts with ground. The interphalangeal joints of the toes will be ignored. The intertarsal joints will, for the most part, be represented as the arch of the foot with two struts and a spring tying them together. The ankle (subtalar and talocrural joints), knee (tibiofemoral and patellofemoral joints), and hip will all be discussed in turn. Here, you will see many mechanics concepts put into "action."

13.1 THE FOOT AND ANKLE COMPLEX

> **Section Question**
>
> Trisha, a high-school cross country runner, has been diagnosed with a stress fracture of the second metatarsal (Figure 13.1). What muscles can she train with exercise to help prevent the reoccurrence of this injury?

13.1.1 Function and Structure

Before studying any system, you first need to determine its function or purpose. What is the purpose of the foot and ankle complex? Why is there a foot at the end of your leg at all? The foot and ankle complex serves several important functions. First, it provides a base of support (BOS) to maintain balance. Remember that the base of support is determined by the contact area(s) with the ground and that in quiet standing the center of gravity (COG) cannot extend beyond the BOS. If there was no foot at all, the BOS would be very small, and balance (at least while standing on the ends of your two legs) would be precarious at best. Second, the foot–ankle complex serves as an energy source, an energy sink, and is involved with energy exchanges. Remember that muscle–tendon complexes generate and absorb energy, and to do so they must generate force while displacing (i.e., perform work). The foot

| **Figure 13.1** | What muscles can she train to prevent the reoccurrence of a metatarsal stress fracture? |

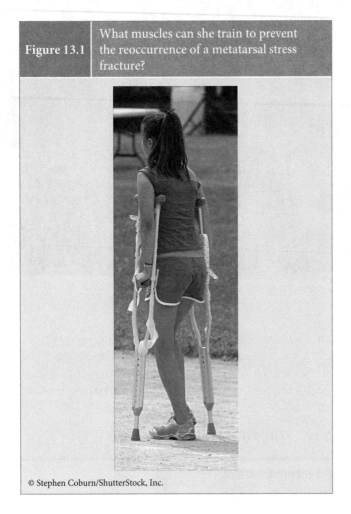

© Stephen Coburn/ShutterStock, Inc.

and ankle complex will transfer force and energy generated from the more proximal segments to the ground or other object the foot is in contact with (bicycle pedal, ball, etc.) and vice versa. Third, the foot and ankle complex extends the time you are in contact with the ground. This is important in a number of different activities.

> **Important Point!** The foot and ankle complex provides a base of support; is important in energy production, generation and transfer; and extends the time you are in contact with the ground.

There are many different ways you could analyze the components of the foot and ankle complex. You could look at the 25 joints and the 28 bones that construct them (**Figure 13.2**). You could look at the three regions common in the clinical literature:[1] the forefoot (metatarsals and phalanges), midfoot (navicular, cuboid, and three cuneiforms), and hindfoot (talus and calcaneus). Each of these ways of looking at it may be appropriate, depending on what it is that you are trying to

analyze. In the interest of brevity and at the risk of simplicity, here you will concentrate on the metatarsal–phalangeal (MTP) joints, the arches of the foot, and the ankle complex.

13.1.2 The Metatarsal–Phalangeal Joints

The MTP joints are the joints that connect your toes to your foot. They are formed between the convex head of each metatarsal and the concave proximal end of each proximal phalange. These joints are important because they support the body, provide traction, and control the forward motion of the center of mass (COM) during propulsion.[2] They are condyloid joints, meaning they have two rotational degrees of freedom (DOF): flexion/extension and abduction/adduction. Although many people have done extraordinary things with their feet (check out the book or movie *My Left Foot*), you will be predominately interested in the flexion/extension DOF for many activities you will be studying later.

The axis of rotation for this movement runs through the metatarsal heads. Typical ROM value for flexion is 20° and for extension is 80°.[3] Muscle–tendon complexes (MTCs) on the plantar side of the axis will produce a flexor torque, whereas MTCs on the dorsal side will produce an extensor torque (**Figure 13.3**). The muscles are presented in **Table 13.1**. In general, the MTPs can generate more flexor torque than extensor torque. This should clue you to the fact that they have much greater demands during normal activity.

> **Important Point!** MTCs passing on the plantar side of the MTP axis produce a flexor torque. MTCs passing on the dorsal side produce an extensor torque.

Most of the torque requirements for the MTP joint come from its interaction with the ground. Because the inertial effect of the phalanges is so small, torque at the MTP joints is considered negligible until the center of pressure (COP) of the ground reaction force (F_{GRF}) moves anterior (distal) to the joint.[4] After that, the MTP joints produce a flexor torque. Because this flexor torque occurs while the toes are extending, the MTP MTCs absorb energy. The MTP joints do not appear to actively flex during push-off, so they do not appear to generate energy during walking, running, or jumping.[4,5] However, they may play an important role with the arches of the foot, which is explained next.

13.1.3 The Arches of the Foot

The three arches in each foot are critical in allowing the foot and ankle complex to perform its functions. That is because the arches allow the foot to be supple at some times and rigid at others. Think about what would happen if your foot was

| Figure 13.2 | Major bones and joints of the foot: (a) lateral view; (b) medial view. |

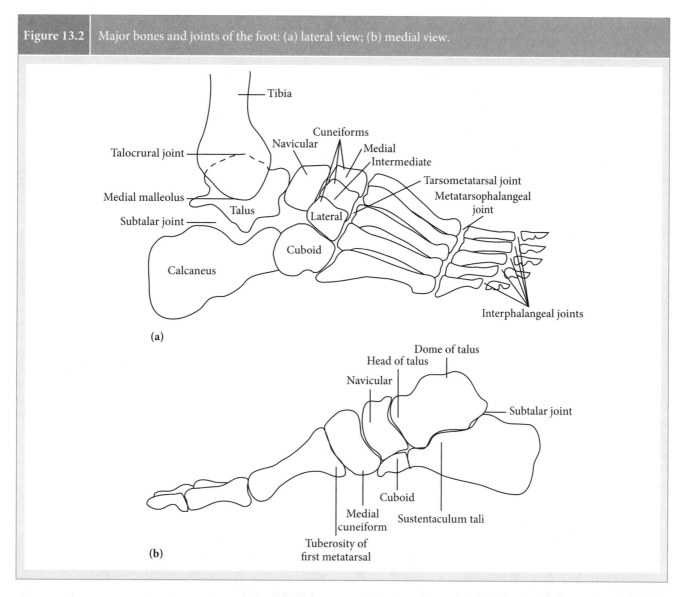

(a)

(b)

| Figure 13.3 | The MTP joints. Muscles dorsal to the axis of rotation will produce an extensor torque. Muscles plantar to the axis of rotation will produce a flexor torque. |

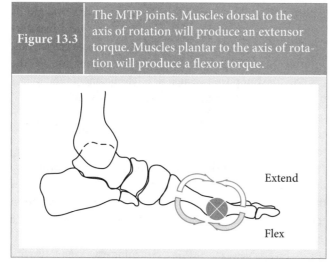

Table 13.1	Muscles Acting on the MTP Joints
Dorsal (extensor torque)	Extensor digitorum longus Extensor hallucis longus Extensor digitorum brevis Extensor hallucis brevis
AOR: Through the metatarsal heads	
Plantar (flexor torque)	Flexor digitorum longus Flexor hallucis longus Flexor digitorum brevis Flexor hallucis brevis Flexor digiti minimi Lumbricals Interossei

always rigid. If you were standing on uneven terrain, your BOS would be very small because it would only involve a small portion of the foot that was actually in contact with the ground. So you want to have a supple foot. This is also important in absorbing energy when the foot makes contact with the ground. Remember that energy exchanges involve work, and work requires both a force and a displacement. A supple foot can absorb energy by allowing movement between the bones. So having a supple foot is a good thing, but is it always a good thing?

Now think about what happens when you want to push off the ground. Remember Newton's third law. For you to move up and forward, there needs to be a net force moving you up and forward. That force is the reaction force, which comes from you pushing downward and backward against the ground. If the foot was very supple, it would not be very effective at allowing this to occur. To give you an opposite example, think about what happens when the terrain is very supple—like running on the beach. It is a lot harder to run on the beach than on a firm surface because some of the force is actually moving the sand away from you. The same thing happens if your foot is supple. It would not be ideal for an effective push-off. During push-off, it would be helpful if the foot could lock into place. This is why there are arches of the foot: they allow the foot to alternate between being supple and being rigid.

> **Important Point!**   The arches allow the foot to be either supple or rigid, depending on the task requirements.

The joints created by the bones of the midfoot (most notably the talonaviuclar, calcaneocuboid, and tarsometatarsal joints) form three arches within the foot: a transverse arch across the width of the foot and medial and longitudinal arches running along the length of the foot. Rather than thinking of them as three separate arches, they can be described as a single twisted osteoligamentous plate (**Figure 13.4**).[1] The anterior portion of the plate is at the metatarsal heads, is oriented horizontally, and is in full contact with the ground. The posterior portion of the plate is the calcaneous and is oriented vertically. The twist of the plate gives the arches their shape. Weight bearing tends to untwist the plate, causing the arch to flatten. During unloading, the weight rebounds to its original shape.

Several ligaments are responsible for giving the arch its shape. These include the plantar calcaneonavicular ligament, the interosseous talocalcaneal ligament, and the long and short plantar ligaments. Muscles also contribute to arch of the foot, although their function occurs more during movement than just quiet standing.[1] The tibialis posterior, flexor

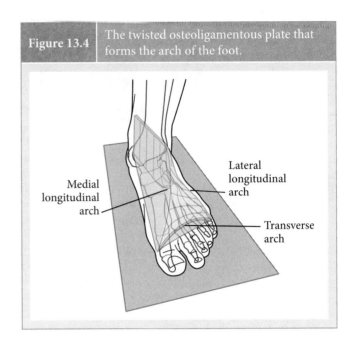

Figure 13.4   The twisted osteoligamentous plate that forms the arch of the foot.

Medial longitudinal arch

Lateral longitudinal arch

Transverse arch

digitorum longus and brevis, flexor hallucis longus and brevis, and peroneus longus may all play a role. However, the most important structure as far as the arch is concerned may be the plantar fascia (aponeurosis).

From a mechanical perspective, it may be helpful to think of the arches of the foot as a **truss** with two **struts** and a spring acting as a **tie rod** (**Figure 13.5**).[6] The triangular-shaped truss is one of the strongest geometric shapes often used in bridging and building construction. The anterior strut (or leg of the triangle) is formed by the navicular, cuboid, cuneiforms, and metatarsals. The posterior strut is formed by the talus and calcaneous. Although some authors consider just the plantar fascia as the tie rod,[1] for your purposes include all the ligaments and MTCs that hold the arch in place. Because that is a bit of a mouthful to continually say, the ligaments, MTCs, and plantar fascia in the dorsum of the foot will be referred to as the "spring" of the foot in this text.

Body weight pushes down on the apex of the strut, flattening the triangle and stretching the spring. During this time, the arch flattens somewhat. As body weight is removed from the struts when the COP passes anterior to the midfoot, the spring recoils. The arch returns to its normal height. This spring is capable of storing and returning elastic energy, with

> **Truss**   A structure composed of one or more triangular units
>
> **Strut**   A structural unit designed to resist compressive forces
>
> **Tie rod**   A slender structural unit that is designed to resist tensile forces

Figure 13.5	The arches of the foot as a truss with two struts and a spring as a tie-rod.

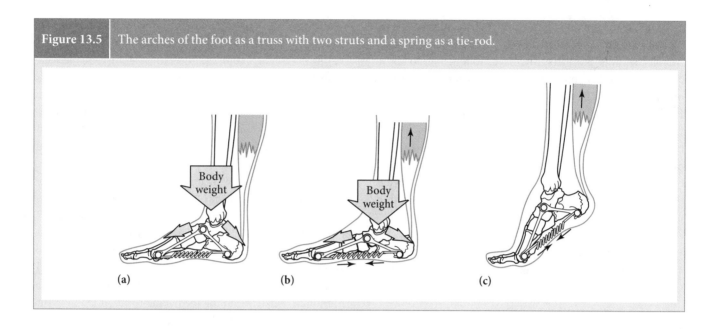

(a)   (b)   (c)

a fairly small hysteresis loop.[7] It is estimated that the spring can return 78% percent of the energy absorbed by it, which is equivalent to about half the energy returned via the Achilles tendon.[7] This means the spring of the foot does have an impact on movement economy.

The spring of the foot has one other important function: It ties the rearfoot and forefoot together.[6] It acts to buttress the arch[6] and transmit forces from the Achilles tendon out to the metatarsal heads (**Figure 13.6**).[8,9] This is easy to visualize if you stand up and perform a heel raise: as soon as your gastrocnemius/soleus shortens, you go up on your toes.

The importance of the spring to proper foot functioning should not be underestimated. Stretching of the spring during the braking phase allows the arch to drop and energy to be dissipated. Tensioning of the spring during the propulsive phase stiffens the foot and allows forces to be transmitted from the Achilles to the forefoot. In addition, energy absorbed by the spring during the braking phase may be reused during the propulsive phase.

Important Point!   Proper foot functioning is accomplished by modulating the stiffness of the spring of the foot.

You may now begin to realize the importance of the MTP joints during a push-off. It may seem inefficient to have energy absorbed by the MTP MTCs that is not reused because the MTPs do not flex during push-off.[4,5] However, before making that conclusion you should keep in mind a basic philosophical postulate articulated by Latash:[10] "If something in the design of the human body looks suboptimal, it is likely that we have missed something important." And the important thing that you may have missed is that extension of the MTP joints increases tension on the spring of the foot.[11]

Figure 13.6	The force of the Achilles tendon is transferred to the forefoot via the spring of the foot.

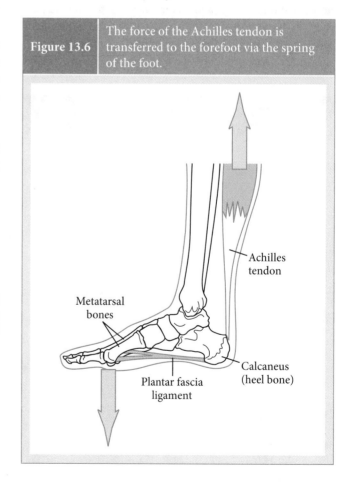

Achilles tendon

Metatarsal bones

Calcaneus (heel bone)

Plantar fascia ligament

During the braking phase, the gastrocnemius/soleus will typically be acting eccentrically—so there is tension in the Achilles tendon. But the spring needs to lengthen to absorb energy during this phase. In addition, during the propulsive phase the spring needs to shorten and the foot needs to stiffen. How is that accomplished? It is accomplished through MTP extension! Extending the MTP joints increases the stiffness of the spring of the foot.[11] Increased stiffness of the spring pulls on the calcaneous and locks the foot in place so that this can occur. So rather than being "wasted," the energy absorbed by the MTP MTCs during extension is directed back toward the plantar fascia.

> **Important Point!**   Stiffness of the spring of the foot is modulated by the coordination of the MTP joints and ankle complex.

13.1.4   The Ankle Complex

The ankle complex consists of two joints: the subtalar joint and the talocrural joint (Figure 13.7). The subtalar (below the talus) joint is formed by the talus and calcaneous. It should be noted that the anterior and medial facets of the subtalar joint share a capsule with the talonavicular joint, and if the talus is moving on a fixed calcaneous, then movement is also occurring at the talonavicular joint. The subtalar joint also has one

Figure 13.7	The ankle complex consists of the talocrural joint and the subtalar joint.

rotational DOF: pronation and supination. The axis of rotation for the subatalar joint is also an oblique axis, meaning that pronation and supination occur in all three planes. The majority of movement is inversion/eversion in the frontal plane. Because inversion/eversion is the only motion that can be measured clinically at the subtalar joint, sometimes these are the only components that are discussed. Typical ranges of motion for the subtalar joint are 35° of inversion and 20° of eversion.[3]

The talocrural joint is formed from the talus, tibia, and fibula. Mechanically, it has been likened to a mortise, but I like to think of the tibia and fibula that act like an adjustable (concave) wrench that clamps on to the (convex) talus (Figure 13.8). It is a hinge joint with one rotational degree of freedom. The axis runs through the distal poles of the malleoli. Because the pole of the lateral malleolus is more distal and slightly posterior to the medial malleolus, it is an oblique axis that is not quite in the sagittal plane. Typical ranges of motion for the talocrural joint are 50° of plantar flexion and 20° of dorsiflexion.[3]

The subtalar joint is an important link between the leg and foot. In some respects, the talus has been described as a **ball bearing**.[1] The purpose of a ball bearing is to support both axial and radial loads. The axial load would be the weight of the body pressing

> **Ball bearing**   A component that separates moving parts and takes a load

down on the lower leg. This load has to be redirected out in a radial direction toward the foot. The talus is designed to do just that, transmitting the load from the lower leg anteriorly toward the navicular and posteriorly toward the calcaneus.

Figure 13.8	The talocrural joint acts like an adjustable wrench with the tibia and fibula clamping onto the talus.

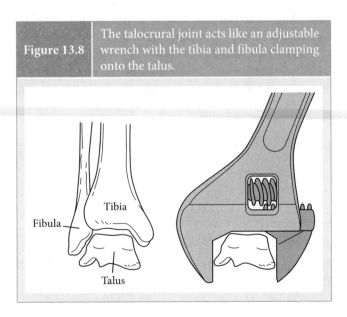

| Figure 13.9 | The ligaments of the ankle complex: (a) the deltoid medial, (b) the lateral ligaments. |

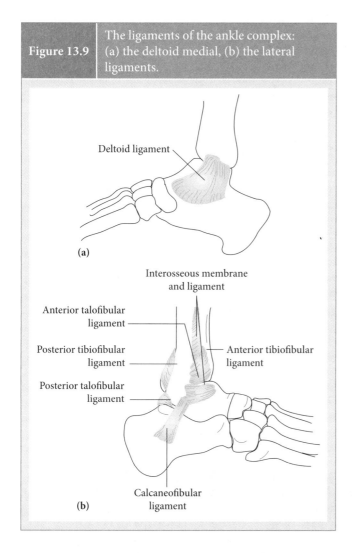

(a)

(b)

Deltoid ligament

Interosseous membrane and ligament

Anterior talofibular ligament

Posterior tibiofibular ligament

Anterior tibiofibular ligament

Posterior talofibular ligament

Calcaneofibular ligament

| Figure 13.10 | The subtalar joint. Muscles that pass medial to the axis of rotation will produce a supinator torque, and those that pass lateral to it will produce a pronator torque. |

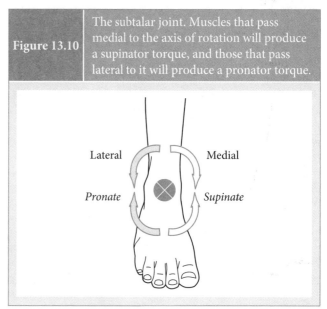

Lateral   Medial

Pronate   Supinate

| Table 13.2 | The Muscles Crossing the Subtalar Joint |

Medial (supinator torque)	AOR: Through the talar head	Lateral (pronator torque)
Gastrocnemius Soleus Tibialis posterior Tibialis anterior Flexor hallucis longus Flexor digitorum longus Plantaris		Peroneus longus Peroneus brevis Peroneus tertius

| Figure 13.11 | The talocrural joint. Muscles that pass anterior to the axis of rotation will produce a dorsiflexor torque, and those that pass posterior to it will produce a plantar flexor torque. |

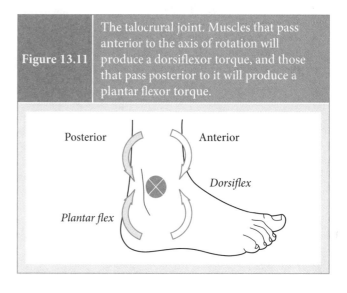

Posterior   Anterior

Dorsiflex

Plantar flex

Ligaments that hold the ankle complex together are located on the medial and lateral sides (Figure 13.9). There are four ligaments on the medial side (anterior tibiotalar, tibionavicular, tibiocalcaneal, and posterior tibiotalar) that are collectively known as the deltoid ligament. Its major function is to check or restrain eversion. The three ligaments on the lateral side (anterior talofibular, calcaneofibular, and posterior talofibular) check or restrain inversion.

MTCs that cross medial to the subtalar axis will produce a supination torque, and those that pass lateral to it will produce a pronation torque (Figure 13.10). These muscles are listed in Table 13.2. The maximum supinator torque is greater than the pronator torque. MTCs that cross anterior to the talocrural joint axis will produce a dorsiflexion torque, whereas MTCs that are posterior to the axis will produce a plantar flexion torque (Figure 13.11). These muscles are listed in Table 13.3. The plantar flexor torque tends to be much stronger than the dorsiflexor torque.

Table 13.3	The Muscles Crossing the Talocrural Joint
Anterior (dorsiflexor torque)	Tibialis anterior Extensor hallucis longus Extensor digitorum longus Peroneus tertius
AOR: Through the malleoli	
Posterior (plantar flexor torque)	Gastrocnemius Soleus Peroneus longus Peroneus brevis Flexor hallucis longus Tibialis posterior Flexor digitorum longus Plantaris

Every MTC that crosses the talocrural joint also crosses the subtalar joint. So these MTCs have actions at both joints. By bisecting the two axes, you can create four functional quadrants (**Figure 13.12**). MTCs in quadrant 1 produce a dorsiflexion torque about the talocrural joint and a supination torque about the subtalar joint. MTCs in quadrant 2 produce a dorsiflexion torque about the talocrural joint and a pronation torque about the subtalar joint. MTCs in quadrant 3 produce a plantar flexion and supination torque, whereas MTCs in quadrant 4 produce a plantar flexion and pronation torque. As you may have guessed, the MTCs in quadrant 3 can produce the most torque, and the muscles in quadrant 2 can produce the least amount of torque.

13.1.5 Toes, Foot, and Ankle Interactions with the Ground

Although the foot can be involved in handling instruments or striking objects (e.g., soccer balls), its arguably most important function is in its interaction with the ground. Although other external forces are involved (the force due to gravity, inertial forces), for the purposes of this discussion they will be ignored, and the torque produced by the ground reaction forces (F_{GRF}) and muscles will be highlighted. A major consideration is where the center of pressure (COP) of the F_{GRF} is located in relation to the ankle joint center (JC_{ankle}).

During the braking phase, the initial contact can either be in the midfoot, forefoot, or rearfoot. If, during initial contact with the ground, the COP is posterior to the JC_{ankle}, then the $^yF_{GRF}$ will produce a plantar flexion torque (**Figure 13.13**). If this external torque ($T_{external}$) is not counteracted by an MTC torque (T_{MTC}) the foot will have a large clockwise ($-$) angular

Figure 13.12	Functional quadrants of the ankle complex. Muscles in quadrant 1 produce a dorsiflexion torque about the talocrural joint and a supination torque about the subtalar joint. Muscles in quadrant 2 produce a dorsiflexion torque about the talocrural joint and a pronation torque about the subtalar joint. Muscles in quadrant 3 produce a plantar flexion and supination torque, whereas muscles in quadrant 4 produce a plantar flexion and pronation torque.

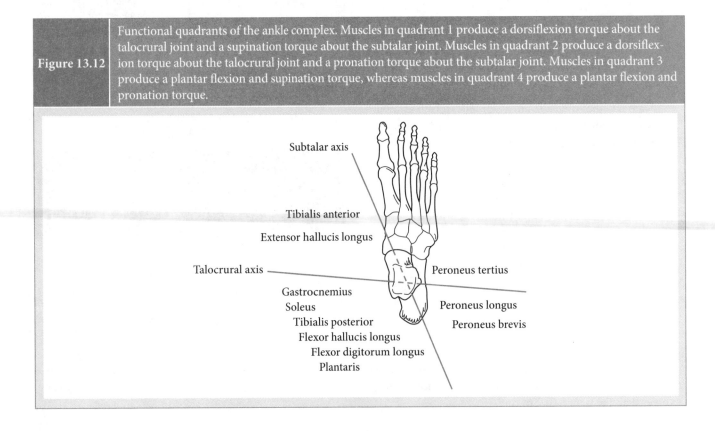

Figure 13.13	When initial contact is made posterior to the JC_{ankle}, the F_{GRF} will produce a plantar flexion torque that will accelerate the foot unless the MTCs produce a countertorque.

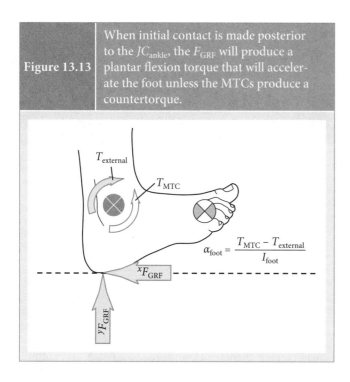

acceleration (α_{foot}) and will probably forcibly "slap" the ground. The opposite situation occurs when the initial contact with the ground is anterior to the JC_{ankle} (**Figure 13.14**). When this happens, $^{y}F_{GRF}$ produces a dorsiflexion torque, and the foot will rotate in the counterclockwise (+) direction.

Figure 13.14	When initial contact is made anterior to the JC_{ankle}, the F_{GRF} will produce a dorsiflexion torque that will accelerate the foot unless the MTCs produce a countertorque.

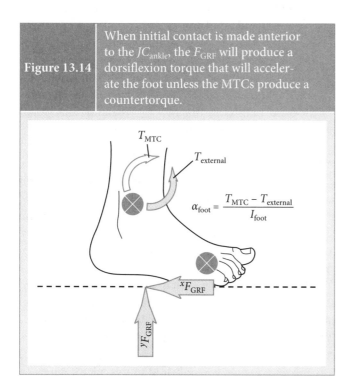

MTC torques can either act in the same or opposite direction of the torque produced by external forces. The actual α_{foot} is determined by the sum of these torques. During the braking phase, the muscles are usually decreasing α_{foot}. Can you imagine how damaging it could be to the foot if they did not? To decrease α_{foot}, T_{MTC} must be in the opposite direction of $T_{external}$. This means that the MTCs are acting eccentrically, absorbing energy that has been transferred to them from the segments.

You should also note the effect of $^{x}F_{GRF}$ in both cases: it produces a negative (plantar flexion) torque. With a rearfoot contact, this torque adds to the torque created by $^{y}F_{GRF}$. With a forefoot contact, it is in the opposite direction. Because the $^{y}F_{GRF}$ is usually so much greater than the $^{x}F_{GRF}$, the $^{y}F_{GRF}$'s effect tends to dominate. However, in all cases, α_{foot} is determined by the sum of the external and MTC torques.

> **Important Point!** The acceleration of the segment is determined by the sum of the MTC and external torques, as well as the moment of inertia of the segment.

A similar situation exists during the propulsive phase (**Figure 13.15**). During push-off, the COP is almost always anterior to the JC_{ankle}. And you should remember that the $^{x}F_{GRF}$ is directed anteriorly during the propulsive phase. The F_{GRF} in both directions will be producing a dorsiflexion torque. During this time, the talocrural joint will be plantar flexing, and the foot will be rotating in the clockwise direction. T_{MTC} will be a plantar flexor torque, generating energy

Figure 13.15	During push-off, the MTCs of the ankle must produce a net plantar flexor torque.

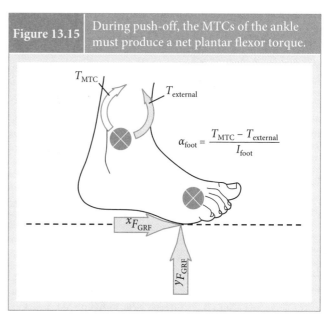

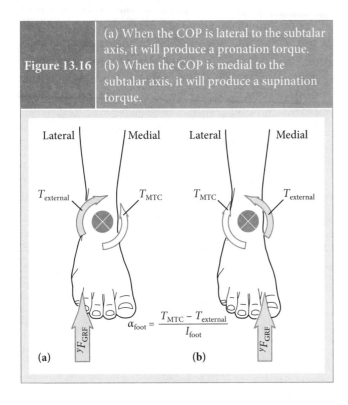

Figure 13.16 (a) When the COP is lateral to the subtalar axis, it will produce a pronation torque. (b) When the COP is medial to the subtalar axis, it will produce a supination torque.

$$\alpha_{\text{foot}} = \frac{T_{\text{MTC}} - T_{\text{external}}}{I_{\text{foot}}}$$

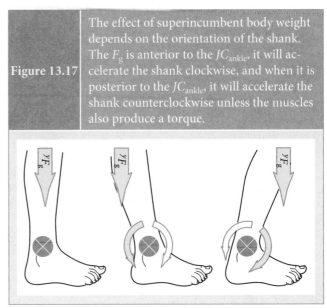

Figure 13.17 The effect of superincumbent body weight depends on the orientation of the shank. The F_g is anterior to the JC_{ankle}, it will accelerate the shank clockwise, and when it is posterior to the JC_{ankle}, it will accelerate the shank counterclockwise unless the muscles also produce a torque.

and delivering it to the segments to increase their kinetic and potential energies.

These types of analysis are not limited to the sagittal plane. Again, ignore all the external forces except the F_{GRF}, and specifically the $^yF_{\text{GRF}}$. If the COP is lateral to the subtalar axis (**Figure 13.16a**), the foot will rotate in the clockwise direction—causing the foot to pronate. If the COP is medial to the subtalar axis (**Figure 13.16b**), the foot will rotate in the counterclockwise direction, causing the foot to supinate. Again, the actual α_{foot} will depend on the net torque determined by the sum of T_{external} and T_{MTC}.

Remember that a joint is made up of two segments and that a joint torque has an effect on each segment. In the preceding paragraphs, you saw how the ankle muscular torques affected the rotations of the foot. Now see how they affect movement of the tibia and fibula, which are collectively referred to as the **shank**.

To see these effects examine **Figure 13.17**. Use a very simple model where the foot is flat on the floor and the only external force is the **superincumbent body weight**. If the shank is oriented vertically, the external force (F_g) will not produce any torque about the ankle joint. If the shank is rotated counterclockwise from this position,

> **Shank**   The part of the leg from the knee to the ankle
>
> **Superincumbent body weight**   The body weight that is lying on top of the segment(s) of interest

F_g will create a plantar flexion torque and want to continue to rotate the shank in a counterclockwise direction. If the shank is rotated clockwise from this position, F_g will create a dorsiflexion torque and want to continue to rotate the shank in a clockwise direction. In addition, the moment arm of F_g increases with an increase in rotation in either direction.

If the goal was to stand upright and not move or fall over, T_{MTC} is necessary to counteract the torque produced by F_g. For example, if the shank were rotating in the clockwise direction you would fall flat on your face unless T_{muscle} was a plantar flexor torque to arrest the movement. This plantar flexor torque would transfer energy from the segments to the MTCs, where it is absorbed. If the plantar flexor torque was large enough, then it could start to rotate the shank counterclockwise. In this case, the MTC is generating energy and delivering to the segments via concentric actions.

13.1.6   Injuries to the Foot and Ankle

Numerous injuries can occur to the foot and ankle. A detailed accounting of all of them is impossible within the space of this lesson. Here you will consider two of the most common: stress fractures of the metatarsals and inversion ankle sprains. They provide good illustrations of how improper mechanics can lead to injury.

Stress fractures are a result of microtrauma to bone, and stress fractures of the metatarsals are fairly common. As explained earlier in the lesson, the ankle MTCs control the acceleration of the foot after initial contact via eccentric actions. An uncontrolled foot landing can lead to a "hard slap" of the ground, which could certainly lead to this microtrauma. In addition, a "hard" landing of the ground due to a large effective mass of the foot can lead to the same result. These are rather straightforward explanations. But there is another interesting cause of stress fractures.

The spring of the foot is responsible for transferring forces from the Achilles tendon to the forefoot. Decreasing the tension in the spring, either by fatiguing the muscles or cutting the plantar fascia, significantly increases the strains on the metatarsals.[12] In addition, during push-off, decreasing the tension in the spring by the same mechanisms shifts the ground reaction force and contact area from the toes to the metatarsals and from the medial metatarsals to the more lateral ones.[13] The plantar fascia is a passive tissue, and if it is stretched to its plastic region, there is not much you can do to restore it to its original length. However, the muscles underneath the plantar fascia can be changed. These results suggest the strength and endurance of these muscles is important in the prevention of stress fractures of the metatarsals.

Inversion ankle sprains, as the name suggests, occur when there is too much inversion motion at the ankle. This can happen when your foot lands in a hole or on another player's foot, or when you step off of the curb incorrectly. With inversion, the lateral ankle ligaments are placed on tension. These tensile forces are exacerbated by the ligaments bending about the lateral malleolus (Figure 13.18). If the strains on the ligaments exceed their yield point, microtears occur. Greater stresses mean greater strains, which means more tearing occurs and results in a more severe sprain. Remember that when

a tissue exceeds its yield point, a plastic deformation occurs. This means the ligament will not return to its original length, and the joint will be lax. Greater laxity means there is a greater chance for more reinjury, and a person who has had a lateral ankle sprain can enter a viscous cycle where they always have greater ankle instability.

Inversion ankle sprains are much more common than eversion ankle sprains. There are a number of reasons for this. First, the lateral malleolus extends further distally than the medial malleolus. This provides a bony block against excessive eversion. Second, the deltoid ligaments appear stronger than the lateral ankle ligaments. This is no doubt due to the fact that they are stressed more often during daily activity. Third, when the foot makes contact with the ground, the body is already anticipating it to pronate. To control pronation, the muscles that control pronation eccentrically (the supinators) activate prior to contact. If the body were to wait to activate these muscles until after the foot made contact with the ground, it would be too late because the response would be too slow. The body is not anticipating the subtalar joint to invert with foot contact and therefore the pronators are not primed to respond to this motion.

Section Question Answer

Trisha's stress fracture is a result of too much stress on the second metatarsal. This could be because her foot is hitting the ground too hard. Assuming that she is a rearfoot striker, the contact between the foot and the ground is controlled by the torque produced by the dorsiflexor MTCs acting eccentrically. These muscles have been shown to be particularly susceptible to fatigue during running.[14] In addition, the spring of the foot transmits forces from the rearfoot to the forefoot. As mentioned previously,[12] fatigue of the MTP flexors also increases stress on the metatarsal heads during push-off. Therefore, a comprehensive exercise program would include improving the strength and endurance of both the talocrural dorsiflexors and MTP flexors.

| Figure 13.18 | With a lateral ankle sprain, excessive inversion bends the ligament, increasing the tension on it. |

Ankle sprain

Sprained lateral ligament

Normal   Inversion

COMPETENCY CHECK

Remember:

1. What bones are involved in the metatarsal–phalangeal joints, subtalar joint, and talocrural joint?
2. What structures are involved in the "spring" of the foot?
3. What type of joint is the metatarsal phalangeal joint? How many rotational degrees of freedom does it have? What motions are available? What is considered to be the "normal" ranges of motion?

4. What type of joint is the talocrural joint? How many rotational degrees of freedom does it have? What motions are available? What is considered to be the "normal" ranges of motion?

Understand:

1. Describe how the spring of the foot operates during braking and propulsive phases.
2. During initial contact, the MTCs of the ankle complex are acting eccentrically. Which muscle groups are acting eccentrically during a rearfoot contact? Which muscle groups are acting eccentrically during a forefoot contact?
3. Describe the arthrokinematic motions of the talocrural joint for plantar flexion and dorsiflexion when: (a) the talus is moving on the tibia/fibula, and (b) the tibia/fibula are moving on the talus.
4. Determine the resultant acceleration of the foot segment. The moment of inertia is 0.04 kg·m². Ignore any inertia effects.

$T_{external}$	T_{MTC}	α_{foot}
65	−80	
30	−20	
−20	10	

Apply:

1. How does improving the strength and endurance of the dorsiflexors and MT extensors decrease injury risk?
2. How can improving the strength of the dorsiflexors and plantar flexors improve balance?

13.2 KNEE COMPLEX

Section Question

When choosing an exercise to increase strength of the quadriceps, should you choose a knee extension exercise (Figure 13.19) or a squat (Figure 13.20)?

You just learned how the foot and ankle complex interact with the ground. As you will learn, the hip, with its large muscle mass, is a major energy producer of the lower extremity. The knee is the mechanical link between the two.

Important Point! The knee is the link between the foot that interacts with the ground and the hip that is the major torque producer of the lower extremity.

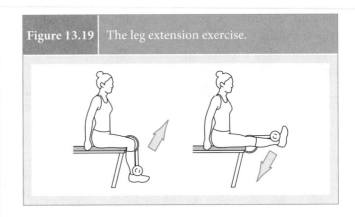

Figure 13.19 The leg extension exercise.

Figure 13.20 The squat exercise.

The knee complex is actually comprised of two joints: the tibiofemoral and the patellofemoral joint (Figure 13.21). The tibiofemoral joint is formed between the convex femoral condyles and the concave tibial condyles. The fibula does not articulate with the femur. The tibiofemoral joint is a bicondyloid joint that has two rotational DOFs: flexion/extension and internal/external rotation. The axis of rotation for

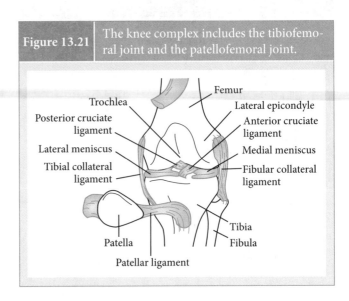

Figure 13.21 The knee complex includes the tibiofemoral joint and the patellofemoral joint.

Table 13.4	The Major Ligaments of the Knee and the Motions They Restrain

Ligament	Movement restrained
Anterior cruciate	Anterior translation
	Valgus
	Varus
	Internal rotation
Posterior cruciate	Posterior translation
Lateral collateral	Varus
	Lateral rotation
Medial collateral	Valgus
	Lateral rotation
	Anterior translation

flexion/extension is through the femoral epicondyles. Just like the talocrural joint, it is an oblique axis, meaning that motion technically occurs in all three cardinal planes. The axis of rotation for internal/external rotation is just medial to the centroid of the tibial shaft. Typical ranges of motion for the knee are 0° of extension, 145° of flexion, 20° of internal rotation, and 30° of external rotation.[3] The patellofemoral joint is a plane synovial joint with no rotational DOF. It acts as both a spacer and lever for the extensor mechanism of the tibiofemoral joint. Its role will be explained after the discussion of the tibiofemoral joint.

13.2.1   The Tibiofemoral Joint

Four major ligaments hold the tibia and femur together and help guide their motion. These ligaments are summarized in Table 13.4. In addition, two fibrocartilaginous rings, called menisci, also serve important functions summarized in Table 13.5. It is important to understand how the menisci and cruciate ligaments work together to produce the arthrokinematic motion at the knee. This will be illustrated by examining a knee flexing by the femur moving on a fixed tibia. The other motions (knee extension by a femur moving on a fixed tibia, knee flexion and extension by a tibia

Table 13.5	The Functions of the Menisci

Reduction of compressive stress between the tibia and femur
Lubrication of the joint surfaces
Prevention of synovial capsule impingement
Contribution to joint stability
Assistance with gliding motion of tibia and femur

moving on a fixed femur) can be reasoned by following a similar thought process.

Following the concave–convex pattern, flexion of the knee is caused by a posterior rolling and anterior gliding of the convex femur on the concave tibia. As the femur rolls posteriorly, tension is increased in the anterior cruciate ligament (ACL). This tension will pull the femur anteriorly on the tibia in an attempt to bring the two ends of the ACL to their resting length. The ACL is assisted in doing this by the menisci.

The menisci are attached to the tibia, but they move with the femur. As the femur rolls posteriorly, a shear force is created on the posterior horns of the menisci. According to Newton's third law, a posterior force on the menisci by the femur is met with an equal, anterior force on the femur by the menisci. Similar to an anteriorly directed ground reaction force propelling you forward, an anteriorly directed meniscal reaction force will move the femur forward. Muscle forces can also have an effect on the arthrokinematics of the knee joint (discussed later), but the passive forces of the ACL and posterior horns of the menisci guide the femur as it rolls during knee flexion. Can you outline how the other actions are guided?

Important Point!   The cruciate ligaments and the menisci have important roles in the arthrokinematic movements of the knee joint.

Before going over the translational effects of the MTCs, let us review the rotary effects of the MTCs at the knee. MTCs crossing anterior to the tibiofemoral axis will produce an extensor torque, and those that pass posterior to it will produce a flexor torque (Figure 13.22). These muscles are listed in Table 13.6. The extensors are stronger than the flexors. Theoretically, there are four sets of MTCs that can produce an external rotation torque and four sets of MTCs that could produce an internal rotation torque (Figure 13.23). However, the anterior MTCs are not oriented to produce any kind of rotation torque, leaving only the posterior MTCs to do so. Of those, only the two in blue are anatomically possible. These muscles are listed in Table 13.7.

Because the anterior MTCs do not produce a rotational torque, and because some posterior MTCs produce an internal rotation torque while others produce an external rotation torque, the knee MTCs can be organized into three functional areas (Figure 13.24). MTCs in the first area only produce an extension torque. MTCs in the second area produce a flexion and internal rotation torque, whereas MTCs in the third area produce a flexion and external rotation torque.

Figure 13.22	The tibiofemoral joint. MTCs anterior to the joint axis of rotation will produce an extensor torque, and MTCs posterior to it will produce a flexor torque.

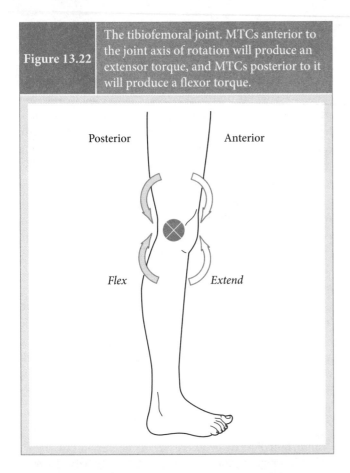

Figure 13.23	MTCs that can theoretically rotate the knee. Of all the possibilities, only the two in blue are anatomical possibilities.

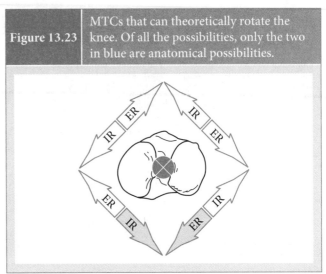

Table 13.7	Muscles Producing Internal and External Rotator Torques at the Knee

Medial (internal rotator torque)	AOR: Through the shaft of the tibia	Lateral (external rotator torque)
Semimembranosus		Biceps femoris
Semitendinosus		
Gracilis		
Sartorius		
Popliteus		

then the external force (F_g) is a flexion torque that needs to either be overcome or controlled by the MTC torque. If the foot were moving through the air during a leg curl exercise (**Figure 13.25b**), then F_g is producing an extension torque that needs to either be overcome or controlled by the MTC torque.

Table 13.6	Muscles Producing Flexor and Extensor Torques at the Knee

Anterior (extensor torque)	AOR: Through the femoral epicondyles	Posterior (flexor torque)
Rectus femoris		Biceps femoris
Vastus medialis		Semimembranosus
Vastus lateralis		Semitendinosus
Vastus intermedius		Gracilis
		Sartorius
		Popliteus
		Gastrocnemius

Figure 13.24	MTCs of the tibiofemoral joint grouped in functional areas.

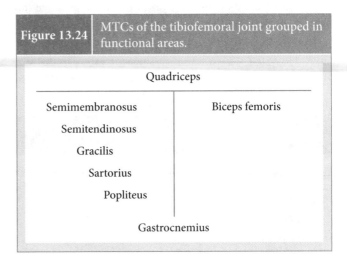

Important Point!   MTCs anterior to the tibiofemoral axis will produce an extensor torque. MTCs posterior and lateral to the axis will produce a flexor and external rotator torque. MTCs posterior and medial to the axis will produce a flexor and internal rotator torque.

MTC torques affect both the shank and thigh. If the foot is moving through the air (such as when kicking a ball or performing a knee extension exercise, **Figure 13.25a**),

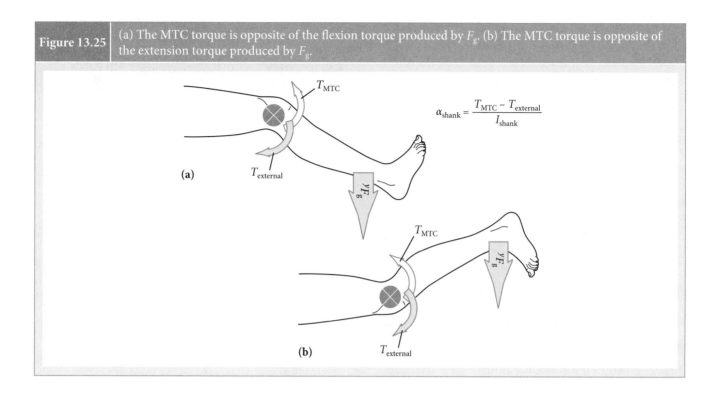

Figure 13.25 (a) The MTC torque is opposite of the flexion torque produced by F_g. (b) The MTC torque is opposite of the extension torque produced by F_g.

$$\alpha_{shank} = \frac{T_{MTC} - T_{external}}{I_{shank}}$$

Forces superior to the knee will also create torque demands about the knee joint. If you ignore the inertial forces, and only consider superincumbent body weight, you will see that the external torque produced by F_g is dependent on its location relative to the knee joint center (JC_{knee}). If F_g is posterior to JC_{knee}, then it will produce a flexion torque. Conversely, if F_g is anterior to JC_{knee}, it will produce an extension torque. These ideas are displayed in Figure 13.26.

All the MTCs crossing the tibiofemoral joint will produce a compressive force across the joint. Traditionally, it is believed that the hamstrings would produce a posterior force on the tibia, and the quadriceps would produce an anterior force on the tibia. The gastrocnemius, attached to the posterior femur, would produce a posterior force on the femur. This would make the quadriceps and gastrocnemius antagonistic to the ACL and agonistic with the PCL. Similarly, the hamstrings would be antagonistic to the PCL and agonistic to the ACL. This has led to the popular, clinical belief that cocontraction of the hamstrings is a protective mechanism for the ACL.

Although generally true, interpret these types of blanket statements with care. First, the effect of any muscle force on joint shear has to be determined with the segment orientations in mind. Although a muscle may create shear in the same direction throughout the ROM, other times it may not.

Figure 13.26 (a) If F_g is posterior to the knee joint center, it will create a flexion torque. (b) If F_g is anterior to the knee joint center, it will create an extension torque.

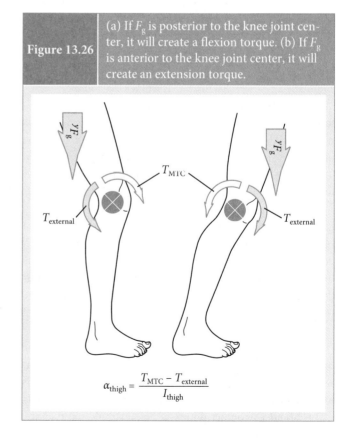

$$\alpha_{thigh} = \frac{T_{MTC} - T_{external}}{I_{thigh}}$$

And even if the shear force maintains the same direction, the magnitude of it may change. The shear forces provided by the three muscles mentioned earlier (quadriceps, hamstrings, and gastrocnemius) all seem to decrease with increased knee flexion angles.[15] Second, although increasing the hamstrings may provide a protective, posterior shear force to assist the ACL, it would also increase the demands on the quadriceps for the same amount of torque. Say the net torque had to be an extensor torque of 10 Nm. This is produced by the quadriceps. If cocontraction of the quadriceps produced a 2 Nm flexor torque, then the quadriceps demand would increase to 12 Nm to produce the same net effect of 10 Nm. This strategy would only work if the hamstrings were producing more shear than the quadriceps for the same amount of torque. Finally, the net joint shear force (or shear force on the ACL) is determined by *all* forces, moments, and the segment orientations. In situations such as a squat, lunge, or step-up, the tibial shear force is actually in the posterior direction at greater knee flexion angles.[16]

13.2.2 The Patellofemoral Joint

The patella is a sesamoid bone that lies anterior to the tibiofemoral joint. The bone itself articulates with the femur, but not the tibia. Superiorly, it attaches to all four quadriceps muscles via the quadriceps tendon. Inferiorly, it attaches to the tibia via the patellar tendon (sometimes referred to as the patellar ligament). It is a plane synovial joint that has no rotational DOFs. Although the patella may offer protection to the anterior aspect of the knee, it has two main mechanical functions. One of them should be fairly obvious to you. The other may not be.

To understand both functions, it is necessary to follow the tension developed in the quadriceps. This tension is transmitted to the patella via the quadriceps tendon. From the quadriceps tendon it is transmitted to the patellar ligament. The torque created by the patellar ligament is altered by the width and orientation of the patella.

First, the patella acts like a spacer. It does this by changing the patellar ligament's line of pull, increasing the moment arm of the patellar ligament. Think about what would happen if you did not have a patella. The line of pull would be from its distal insertion on the tibial tuberosity to the femoral epicondyles. The thickness of the patella changes this, enhancing the torque created by the force in the patella. Hopefully, it is not too difficult for you to see this function.

The second function is a bit trickier. If the force in the quadriceps tendon and the force in the patellar ligament were identical, the patella would just act as a spacer. This happens with very small knee flexion angles. But the patella pivots about its contact points. Using a simple static model

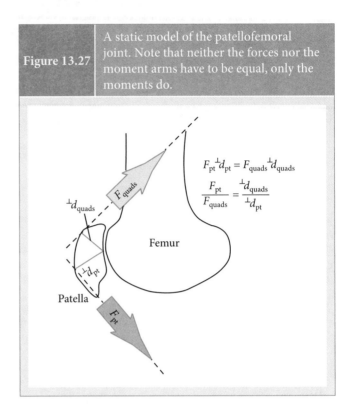

| Figure 13.27 | A static model of the patellofemoral joint. Note that neither the forces nor the moment arms have to be equal, only the moments do. |

$$F_{pt} {}^\perp d_{pt} = F_{quads} {}^\perp d_{quads}$$

$$\frac{F_{pt}}{F_{quads}} = \frac{{}^\perp d_{quads}}{{}^\perp d_{pt}}$$

(**Figure 13.27**),[17] you can see that the moment arms of the quadriceps and patellar ligament are different as the patella pivots about the contact point. Thus, the orientation of the patella changes the magnitude of the force in the patellar tendon as a function of the quadriceps force. As the knee flexes, the patella reduces the amount of force transmitted from the quadriceps to the patellar tendon.[17]

13.2.3 Injuries to the Knee

Injuries to the knee are common in physical activities. These include ligamentous sprains, meniscal tears, osteoarthritis, and many more. Here you will learn about the mechanical causes of some representative injuries. This discussion should be thought of as an overview rather than an exhaustive treatment of these topics.

First you will examine ligamentous injuries, and the ACL will be used as an example. Because the ACL attaches to the posterior part of the distal femur and anterior part of the proximal tibia, its primary function is to prevent or restrain the tibia from moving anterior on the femur. Another, equally valid, way of looking at its function is to think of it as checking posterior movement of the femur on the tibia. But for now, just consider the femur fixed and the tibia moving.

If the tibia moves anterior, tension will develop in the ACL. The two ends of the ACL are furthest apart when the knee is fully extended, and they move closer together as the knee flexes. This means that the ACL has greater strains placed on it when the knee is fully extended, and these strains decrease as the knee flexes. Enough anterior displacement of the tibia could cause the ACL to rupture if those strains exceed the yield point. But this rarely happens.

Ligaments receive tensile stress when they are bent. This should give you an indication that the ACL is somehow bent when it gets injured. And that is what happens when the tibia is not only displaced anteriorly but is also internally rotated. The ACL is lateral to the PCL. When the knee internally rotates, the ACL wraps around the PCL. This can have a dramatic increase on the tensile stress in the ACL and is considered the key mechanism for ACL sprains.[18] If a valgus or varus torque is added to the anterior displacement force and internal rotation torque, even greater stresses can develop. So ACL sprains occur due to a combination of loading of an anteriorly directed force, an internal rotation torque, and a valgus or varus torque.[18]

The meniscus plays an important part in the normal arthrokinematic movements of the tibiofemoral joint. Recall that the menisci attach to the tibia, but move with the femur. During arthrokinematic motion, the menisci are subjected to shear strains. These strains are exacerbated with internal or external rotation. Similar to the mechanism of ACL injuries, meniscal tears occur when there is a rotation of the femur on the tibia.

Patellofemoral pain syndrome (PFPS) is an injury to the patellofemoral joint (hence the name), not the tibiofemoral joint. Stress is the key to understanding PFPS. Remember that stress is force per unit area. Here the force is the patellofemoral joint reaction force, which has contributions from both the quadriceps force and the patellar ligament force. The area is the contact area between the patella and the femur. As the knee flexes from 0° to 60°, the patella moves inferior in the trochlear groove of the femur. The trochlear groove is deeper inferiorly than it is superiorly, and the contact area increases from 0° to 60°, after which it stays about the same with deeper knee flexion angles.

PFPS is caused by excessive stress, which may be a result of a force that is too high or a contact area that is too small. The general trend seems to be that the problem is too little contact area. In general, PFPS seems to be exacerbated at low flexion angles. This makes sense because the contact area increases with higher knee flexion angles. So something is occurring at these low flexion angles to decrease the contact area. One thought was that there was an imbalance in muscle forces between the vastus medialis and vastus lateralis. The stronger vastus lateralis would move the patellar laterally and decrease its contact area with the trochlear groove. However, this hypothesis has been refuted.[19] Current thinking is that

the problem is not with a patella moving on the femur, but the femur moving underneath the patella.[20] Like the ACL, this finding suggests that the problem may not be the knee itself.

Section Question Answer

To answer any question about exercise prescription, you must apply your knowledge of biomechanical principles. There could be lots of different aspects from which to make comparisons between leg extension and squat exercises, but for now focus on just the biomechanics of the knee joint. Three areas should be of particular interest: the torque demands at the knee joint, shear forces at the knee, and patellofemoral stress.

In terms of torque demands, each exercise has unique torque-angle relations. With the knee extension exercise, external torque demands are greatest at zero degrees of knee flexion (full extension) and decrease with knee flexion angles. The opposite situation occurs with the squat exercise: The external torque is least at full extension and increases as the knee flexion angles increase (up to a point). These relations are displayed in Figure 13.28. They suggest that both leg extensions and squats should be prescribed to strengthen the knee extensors through the full range of motion.

Shear forces may be a concern when performing lower extremity exercises. During the leg extension exercise, shear forces are in the anterior direction during the end range of extension (< 25°) and in the posterior direction at knee flexion angles greater than 25°.[21] This means that there is stress on the ACL near full extension and stress on the PCL at greater flexion angles. The amount of shear force depends on the distance

Figure 13.28 The torque-angle relations of the leg extension and squat exercises.

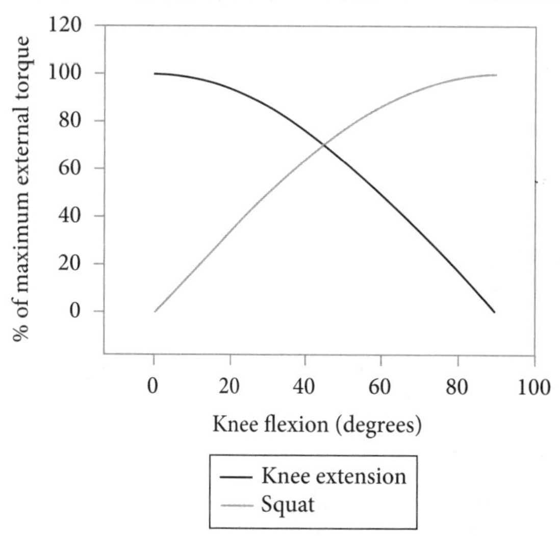

the external resistance is placed from the JC_{knee}: The further away from the joint the resistance is placed, the greater the load on the ACL.[22] During the squat exercise, shear forces are directed in the posterior direction—putting tension on the PCL.[21]

Finally, patellofemoral stress is also different between the exercises. During the leg extension exercise, forces in the quadriceps are greatest at full extension. At full extension, the contact area between the patella and femur are also the least. This would suggest that patellofemoral stress is greatest at full extension during the leg extension and progressively decreases as the flexion angle increases. The relation between patellofemoral stress and knee joint angle during the squat is not as clear-cut. As mentioned previously, knee extensor force increases with increased flexion angle, but the contact area also increases from 0° to 60°, after which it stays fairly constant. Research has shown that patellofemoral stress increases linearly with increasing flexion angle.[23] This finding is not surprising from 60° to 90° because the contact area is constant, whereas the force increases with the external demand. It is somewhat surprising from 0° to 60° because contact area is increasing. However, the increase in force is larger than the increase in contact area.[23]

Taken together, what do these results mean? Both the squat and the leg extension exercise have advantages and drawbacks. If the goal is to train the quadriceps throughout the full range of motion, then both exercises should be employed. Having strains on the ACL and PCL is not necessarily a bad thing. It is needed to improve their strength. But if strain on the cruciates is a concern, such as during the early parts of rehabilitation, then squats should be performed for ACL patients and leg extensions for PCL patients. Finally, for patients with PFPS, a combination of the two exercises would decrease stress on the patellofemoral joint. Employing leg extensions from 45° to 90° of knee flexion, and squats from 0° to 45° of flexion may be ideal. The key is to match the biomechanics of the exercise with the person for whom you are prescribing the exercise.

COMPETENCY CHECK

Remember:

1. What type of joint is the tibiofemoral joint? How many rotational degrees of freedom does it have? What motions are available? What are considered to be the "normal" ranges of motion?

Understand:

1. Describe the arthrokinematic motions at the tibiofemoral joint for both flexion and extension with the tibia

moving on the femur and the femur moving on the tibia. Discuss how each of these motions is assisted by the cruciate ligaments and menisci.
2. Draw a free body diagram of the knee. Explain how the quadriceps, hamstrings, and gastrocnemius create shear forces.
3. Explain why superincumbent body weight creates a large external torque with greater knee flexion angles.

Apply:

1. Use the concepts of knee biomechanics to show how you can decrease injury risk or prescribe exercise.

13.3 HIP

Section Question

You wish to reduce the bone-on-bone forces of a patient following hip-replacement surgery. What advice would you give him concerning walking with a cane (Figure 13.29) or carrying a suitcase (Figure 13.30)?

Figure 13.29	What advice do you give a patient walking with a cane?

© Ljupco Smokovski/ShutterStock, Inc.

Figure 13.30	What advice do you give a patient walking with a suitcase?

© Gert Johannes Jacobus Vrey/ShutterStock, Inc.

Figure 13.31	Ligaments of the hip.

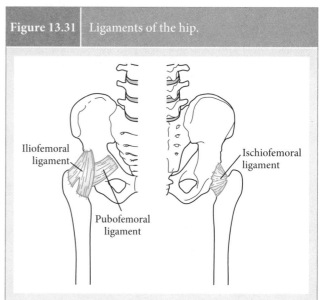

Table 13.8	Functions of the Ligaments of the Hip

Ligament	Movement restrained
Iliofemoral	Extension
	External rotation
Pubofemoral	Abduction
	Extension
	External rotation
Ischiofemoral	Internal rotation
	Extension
	Adduction

The hip, or coxofemoral, joint is a ball-and-socket (spherical) joint, with the femoral head as the ball and the acetabulum of the pelvic bone as the socket. Remember that spherical joints permit rotation about any line passing through the center of the sphere, so the hip can rotate about any line passing through the center of the femoral head. This is probably a good thing because the talocrural and tibiofemoral axes are oblique and not aligned with each other. Motion at the hip is usually expressed as three rotational DOFs: flexion/extension, ab/adduction, and internal/external rotation. It is important to remember that the clinical names of the rotations at the hip in the sagittal plane are the reverse of those at the knee: A counterclockwise rotation of the shank is extension at the knee, whereas a counterclockwise rotation of the thigh is flexion at the hip. Typical ranges of motion for the hip are 120° of flexion, 20° of extension, 45° of abduction, 30° of adduction, 40° of internal rotation, and 40° of external rotation.[3]

Three relatively strong ligaments keep the head of the femur inside the acetabulum (Figure 13.31). They are the iliofemoral, pubofemoral, and ischiofemoral ligaments. The movements they restrain are listed in Table 13.8. In addition, an acetabular labrum deepens the socket and keeps the femoral head firmly inside the acetabulum.

MTCs that cross anterior to the femoral head will produce a flexion torque, and those that pass posterior to it will produce an extension torque (Figure 13.32). These muscles are listed in Table 13.9. Like the knee, the extensors are stronger than the flexors. MTCs that are medial to the axis will produce an adductor torque, whereas muscles lateral to it will produce an abductor torque (Figure 13.33). These muscles are listed in Table 13.10. Unlike the knee, there are MTCs in multiple directions that can produce an external rotation torque and MTCs in multiple directions that could produce an internal rotation torque (Figure 13.34). These muscles are listed in Table 13.11.

You should realize that this is a gross simplification of the muscle function. At the hip more so than any other joint that you have studied thus far, MTCs can change their line of pull depending on the joint position. For example, the hip adductors

| **Figure 13.32** | The hip joint. MTCs anterior to the joint axis of rotation will produce an extensor torque, and MTCs posterior to it will produce a flexor torque. |

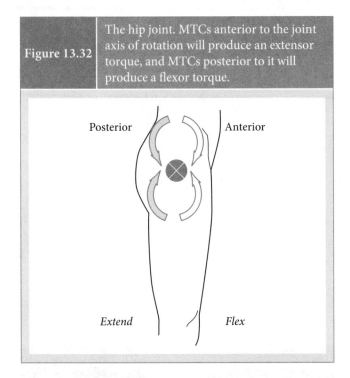

| **Figure 13.33** | The hip joint. MTCs medial to the joint axis of rotation will produce an adductor torque, and MTCs lateral to it will produce a abductor torque. |

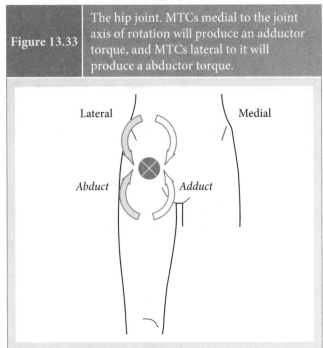

Table 13.9	Muscles Producing Flexor and Extensor Torques at the Hip	
Anterior (flexor torque)	AOR: Through the femoral head	**Posterior (extensor torque)**
Iliacus		Gluteus maximus
Psoas major		Biceps femoris (long head)
Tensor fascia latae		Semimembranosus
Sartorius		Semitendinosus
Rectus femoris		Adductor magnus

Table 13.10	Muscles Producing Adductor and Abductor Torques at the Hip	
Medial (adductor torque)	AOR: Through the femoral head	**Lateral (abductor torque)**
Pectineus		Gluteus medius
Adductor brevis		Tensor fascia latae
Adductor longus		Gluteus minimus
Adductor magnus		
Gracilis		

(such as the pectineus, adductor longus, brevis, and magnus) can all produce a flexor torque at the hip from an extended position to about 40° to 50° of hip flexion.[1] After the distal insertions are superior to the proximal insertions, the muscles produce extensor torques. As another example, the piriformis produces an external rotator torque until the hip is flexed to about 90°. After that, it switches and produces an internal rotator torque. Similar examples exist with other muscles.[1,24]

Important Point! MTCs can change the direction of the torque they produce depending on the orientation of the hip.

As with the other joints, you can think about the effect of a joint torque on either the proximal or distal segment. Actually, as you will see in the lesson on multijoint concepts, muscles can actually cause rotations of segments they do not

| **Figure 13.34** | MTCs that can theoretically rotate the hip. |

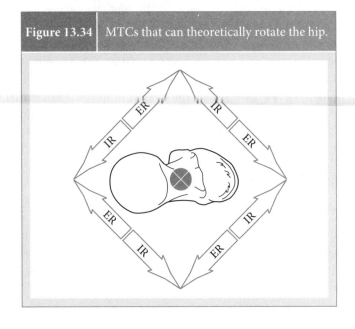

Table 13.11	Muscles Producing Internal and External Rotator Torques at the Hip	
Internal rotator torque	**External rotator torque**	
Pectineus	Gluteus maximus	
Adductor brevis	Piriformis	
Adductor longus	Gluteus medius	
Tensor fascia latae	Gluteus minimus	
	Obturator internus	
	Gemellus superior	
	Obturator externus	
	Gemellus inferior	
	Sartorius	

even cross! But for now, we are going to keep it simple. You should be able to articulate the effect of a hip muscle torque on producing a rotation of the thigh. You can see if you understand it as part of the review at the end of the lesson. Here you will see the effect on rotations of the pelvis.

As with the other joints, ignore the inertial forces and just focus on the effect of the superincumbent body weight. We will examine this effect in both the sagittal and frontal plane while standing on one leg. If F_g is anterior to the hip joint center (JC_{hip}), it will produce a flexion torque; if it is posterior, it will produce an extension torque (**Figure 13.35**). Similarly, if F_g is medial to the JC_{hip}, it will produce an adduction

Figure 13.35	(a) If F_g is anterior to the hip joint center, it will create a flexion torque. (b) If F_g is posterior to the hip joint center, it will create an extension torque.

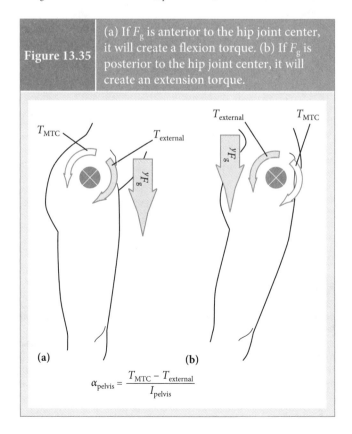

(a)   **(b)**

$$\alpha_{\text{pelvis}} = \frac{T_{\text{MTC}} - T_{\text{external}}}{I_{\text{pelvis}}}$$

Figure 13.36	(a) If F_g is medial to the hip joint center, it will create an adduction torque. (b) If F_g is lateral to the hip joint center, it will create an abduction torque.

torque, and if it is lateral, it will produce an abduction torque (**Figure 13.36**).

The actual acceleration of the pelvis (α_{pelvis}), and corresponding energy changes, depends on the magnitude and direction of the muscle torques. If the muscle torque is equal and opposite of the external torque, then there will be no α_{pelvis} and no changes in energy of the system. If T_{MTC} is opposite in direction and greater in magnitude than T_{external}, then α_{pelvis} will be in the same direction as T_{MTC}. The energy of the system will be increased as energy generated by the muscles is transferred to the segments. If T_{MTC} is opposite in direction and smaller in magnitude than T_{external}, then α_{pelvis} will be in the same direction as T_{external}. However, the energy of the system will be decreased as it is transferred from the segments and absorbed by the muscles.

Section Question Answer

Due to the relative strength of its structures, injuries to the hip are fairly rare. Usually, faulty mechanics at the hip manifest themselves as injuries at other joints. However, total hip replacement surgeries are sometimes required. In such cases, it is prudent to reduce the bone-on-bone forces at the hip. Biomechanical analyses can aid with quest.

Consider the use of a cane. Should the cane be held on the same side of the hip surgery, or on the opposite side? Let us say a person weighs 222 pounds, which means that the head, arms, and trunk weigh about 150 pounds. The person can push down on the cane with a force of 22 pounds. Thus, superincumbent body weight (F_g) is equal to 128 pounds.

For the same numbers, assume that the moment arm from the JC_{hip} to the COM of the head, arms, and trunk (HAT) is 4 inches, then the torque created by F_g is (128 × 4) 512 inch-pounds. That means the hip abductors have to produce the same 512 inch-pounds torque in the opposite direction. If the hip abductors have a moment arm of 2 inches, then the force of the hip abductors would be (512/2) 256 pounds. Adding the force of the abductors to the superincumbent body weight, you can estimate that the bone on bone force would be (128 + 256) 384 pounds.

If the cane is moved to the opposite side, its effect on superincumbent body weight is the same: 128 pounds of force and a torque of 512 inch-pounds. However, on the opposite side, the cane would create an abduction torque. If its moment arm is 8 inches, it would create an abduction torque of (22 × 8) 176 inch-pounds. This torque is subtracted from the adduction torque of F_g, so the torque produced by F_g is (512 − 176) 336 inch-pounds. That means the hip abductors have to produce the same 336 inch-pounds torque in the opposite direction. If the hip abductors have a moment arm of 2 inches, then the force of the hip abductors would be (336/2) 168 pounds. Adding the force of the abductors to the superincumbent body weight, you can estimate that the bone-on-bone force would be (128 + 168) 296 pounds. This represents a 22% reduction in the bone-on-bone forces from using the cane in the opposite hand and a 34% reduction from using no cane at all.

The same reasoning can be applied to carrying a suitcase. If the patient were carrying a suitcase that weighed 22 pounds, the 176 inch-pounds of torque would now be an adduction torque, not an abduction torque. This is added to the torque produced by the F_g (150 × 4 = 600) for a total of 776 inch-pounds. The abductor force would be (776/2) 388 pounds and the bone-on-bone force would be (150 + 22 + 388) 560 pounds. This shows you how much of a compressive load muscular forces put on a joint. It is not a good idea to carry a suitcase if you are trying to decrease the compressive forces across a joint!

COMPETENCY CHECK

Remember:

1. What type of joint is the hip joint? How many rotational degrees of freedom does it have? What motions are available? What are considered to be the "normal" ranges of motion?

Understand:

1. Give examples of the hip motion with the pelvis moving on the hip and the hip moving on the pelvis.
2. Explain the effect of $T_{external}$ and T_{MTC} at the hip on α_{thigh}.

Apply:

1. Use the concepts of hip biomechanics to show how you can decrease injury risk or prescribe exercise.
2. Use the concepts of hip biomechanics to show how you can decrease forces on the hip while carrying a suitcase or using a cane.

SUMMARY

In this lesson, you began to apply mechanical and biological principles to specific joints in the lower extremity. The majority of this lesson related to the torque production of the MTCs about the joint axes of the lower extremity. The shear forces and compressive forces of the MTCs were also illustrated at the knee and hip joints, respectively. You also saw how biomechanists view injuries. After you are familiar with each individual joint, you can then turn your attention to how they work together in the production of movement patterns.

REVIEW QUESTIONS

1. Identify the major joint of the lower extremity.
2. For each joint, determine which type of joint it is, how many rotational degrees of freedom it has, the motions that are available at that joint, and the "normal" ranges of motion.
3. For each joint, describe the arthrokinematic motions of the proximal segment moving on the distal segment and the distal segment moving on the proximal segment.
4. Describe the torques that can be produced by the muscles crossing each joint.
5. Explain how the orientation of a segment affects the torque requirements of the muscle–tendon complex.
6. Describe how the foot interacts with the ground.
7. Explain the effect of $T_{external}$ and T_{MTC} at the hip on α_{thigh}.
8. Give examples of how biomechanical analyses aid in the understanding of lower extremity injuries.

REFERENCES

1. Levangie PK, Norkin CC. *Joint Structure and Function: A Comprehensive Analysis.* 3rd ed. Philadelphia: F. A. Davis; 2001.

2. Rolian C, Lieberman DE, Hamill J, Scott JW, Werbel W. Walking, running and the evolution of short toes in humans. *Journal of Experimental Biology.* Mar 1 2009;212(5):713–721.

3. Flanagan SP. Mobility. In: Miller T, ed. *NSCA's Guide to Tests and Assessments.* Champaign, IL: Human Kinetics; 2012:275–294.

4. Stefanyshyn DJ, Nigg BM. Mechanical energy contribution of the metatarsophalangeal joint to running and sprinting. *Journal of Biomechanics.* Nov-Dec 1997;30(11–12):1081–1085.

5. Stefanyshyn DJ, Nigg BM. Contribution of the lower extremity joints to mechanical energy in running vertical jumps and running long jumps. *Journal of Sports Sciences*. Feb 1998;16(2):177–186.

6. Kim W, Voloshin AS. Role of plantar fascia in the load-bearing capacity of the human foot. *Journal of Biomechanics*. Sep 1995;28(9):1025–1033.

7. Ker RF, Bennett MB, Bibby SR, Kester RC, Alexander RM. The spring in the arch of the human foot. *Nature*. Jan 8 1987;325(6100):147–149.

8. Erdemir A, Hamel AJ, Fauth AR, Piazza SJ, Sharkey NA. Dynamic loading of the plantar aponeurosis in walking. *Journal of Bone and Joint Surgery—American Volume*. Mar 2004;86A(3):546–552.

9. Cheung JTM, Zhang M, An KN. Effect of Achilles tendon loading on plantar fascia tension in the standing foot. *Clinical Biomechanics*. Feb 2006;21(2):194–203.

10. Latash ML. *Synergy*. Oxford: Oxford University Press; 2008.

11. Carlson RE, Fleming LL, Hutton WC. The biomechanical relationship between the tendoachilles, plantar fascia and metatarsophalangeal joint dorsiflexion angle. *Foot & Ankle International*. Jan 2000;21(1):18–25.

12. Donahue SW, Sharkey NA. Strains in the metatarsals during the stance phase of gait: implications for stress fractures. *Journal of Bone and Joint Surgery—American Volume*. Sep 1999;81A(9):1236–1244.

13. Hamel AJ, Donahue SW, Sharkey NA. Contributions of active and passive toe flexion to forefoot loading. *Clinical Orthopaedics and Related Research*. Dec 2001(393):326–334.

14. Reber L, Perry J, Pink M. Muscular control of the ankle in running. *American Journal of Sports Medicine*. Nov-Dec 1993;21(6):805–810.

15. Fleming BC, Renstrom PA, Ohlen G, et al. The gastrocnemius muscle is an antagonist of the anterior cruciate ligament. *Journal of Orthopaedic Research*. Nov 2001;19(6):1178–1184.

16. Heijne A, Fleming BC, Renstrom PA, Peura GD, Beynnon BD, Werner S. Strain on the anterior cruciate ligament during closed kinetic chain exercises. *Medicine and Science in Sports and Exercise*. Jun 2004;36(6):935–941.

17. Yamaguchi GT, Zajac FE. A planar model of the knee joint to characterize the knee extensor mechanism. *Journal of Biomechanics*. 1989;22(1):1–10.

18. Markolf KL, Burchfield DI, Shapiro MM, Shepard ME, Finerman GAM, Slauterbeck JL. Combined knee loading states that generate high anterior cruciate ligament forces. *Journal of Orthopaedic Research*. Nov 1995;13(6):930–935.

19. Powers CM. Rehabilitation of patellofemoral joint disorders: a critical review. *Journal of Orthopaedic & Sports Physical Therapy*. Nov 1998;28(5):345–354.

20. Powers CM, Ward SR, Fredericson M, Guillet M, Shellock FG. Patellofemoral kinematics during weight-bearing and non-weight-bearing knee extension in persons with lateral subluxation of the patella: A preliminary study. *Journal of Orthopaedic & Sports Physical Therapy*. Nov 2003;33(11):677–685.

21. Escamilla RF, Fleisig GS, Zheng NG, Barrentine SW, Wilk KE, Andrews JR. Biomechanics of the knee during closed kinetic chain and open kinetic chain exercises. *Medicine and Science in Sports and Exercise*. Apr 1998;30(4):556–569.

22. Zavatsky AB, Beard DJ, O'Connor JJ. Cruciate ligament loading during isometric muscle contractions—a theoretical basis for rehabilitation. *American Journal of Sports Medicine*. May-Jun 1994;22(3):418–423.

23. Wallace DA, Salem GJ, Salinas R, Powers CM. Patellofemoral joint kinetics while squatting with and without an external load. *Journal of Orthopaedic & Sports Physical Therapy*. Apr 2002;32(4):141–148.

24. Neumann DA. *Kinesiology of the Musculoskeletal System*. 2nd ed. St. Louis: Mosby Elsevier; 2010.

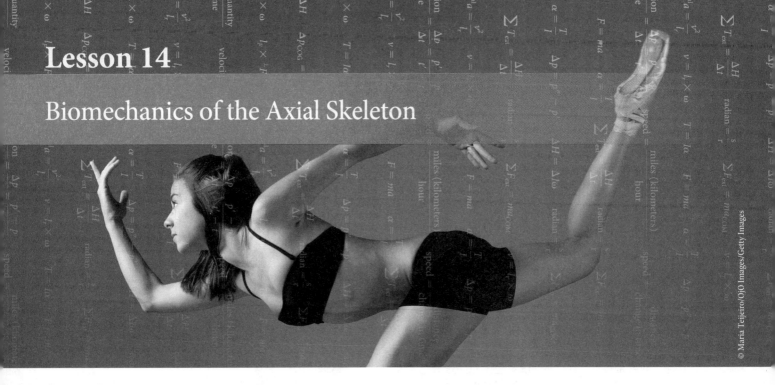

Lesson 14

Biomechanics of the Axial Skeleton

LEARNING OBJECTIVES

After finishing this lesson, you should be able to:

- Define the following terms: pelvic rotation, pelvic tilt, functional spinal unit, joint stiffness, intra-abdominal pressure, and buckling.
- List the components of the functional spinal unit.
- List the components of the intervertebral disk.
- List the functions of the sacroiliac joint.
- List the functions of the intervertebral disk.
- Explain how the zygapophysial joints guide the motion of the functional spinal unit.
- Explain where an MTC has to be located to produce torque on the spine in various directions.
- Describe the origin of shear and compressive forces acting on the spine.
- List available ranges of motion in each region of the spine.
- List the muscles acting on the spine and the torque(s) they produce.
- Demonstrate ways to decrease loading on the spine.
- Explain why you can produce more torque in the extension than in flexion in the craniocervical region.
- Explain why the MTCs in the cervical region have to produce a constant extensor torque in an upright position.
- Discuss the controversy surrounding intra-abdominal pressure and its role in spinal stiffness and compression loading of the lumbar spine.
- List the factors that determine if a column buckles.
- Explain the role of the MTCs in preventing buckling.
- Describe a buckling injury that occurs at the thoracolumbar spine and at the craniocervical spine.
- Explain the mechanisms for a traumatic brain injury.

INTRODUCTION

The spine, ribs, and skull collectively make up the axial skeleton. In this lesson, you will learn predominately about motions at the spine. There will be some discussion of the sacroiliac (SI) joint, as well as biomechanics of traumatic brain injury to the skull. Mechanics involving the respiration (involving the costovertebral joints) and mastication (involving muscles of the temporomandibular joint), although important, will not be discussed. Detailed treatments of those topics are available elsewhere[1,2] if you are interested.

14.1 BASIC FUNCTION AND STRUCTURE

14.1.1 The Sacroiliac (SI) Joint

The joint above the hip joint is the sacroiliac (SI) joint (**Figure 14.1**). The sacrum is part of the axial skeleton, whereas the pelvic bones are part of the appendicular skeleton. The SI joint, then, connects the axial skeleton to the lower extremity (appendicular skeleton).

As with any joint or movement, a discussion of biomechanics should begin with a discussion of the purpose or function. Three functions have been described for the SI joint.[3] First, it must support the loads from the spine above. Second, it must transfer those forces laterally to the pelvis. In this respect, it is similar to the ball-bearing talus that transmits the longitudinal loads of the shank radially to the foot. Newton's third law states that for every force there is an equal and opposite reaction force. So the SI joints also transmit loads from the pelvis to the spine (**Figure 14.2**). It could be argued that a solid bone (i.e., a fused sacrum

| Figure 14.1 | The sacroiliac joint. |

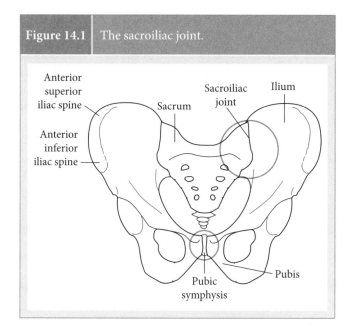

and pelvic girdle) would better serve both these functions. But is should be appreciated that there is movement of the pelvis, however slight. And this fact highlights the third function of the SI joint: It acts to relieve stress caused by a twisting pelvis.[3]

Important Point!   The purpose of the SI joint is to support the load of superincumbent body weight from above, transmit forces between the spine and pelvis, and relieve stress caused by a twisting pelvis.

During hip extension, tension in the strong ligaments of the hip rotates the pelvis anteriorly. Conversely, during hip flexion, tension in the hamstrings rotates the pelvis posteriorly. During activities such as gait, hip flexion and extension are antisymmetric: When one hip is flexing, the other hip is extending. This means that the two halves of the pelvis are rotating in opposite directions (Figure 14.3). The force can be absorbed by the extremely strong ligaments of the SI joint and thus relieve the sacrum of the torsional stress created by these movements.

To avoid confusion, movement of the pelvis with respect to the sacrum will be called a pelvic rotation. Movement of the entire pelvic girdle (pelvis and sacrum) with respect to the hip and spine will be called a pelvic tilt.

Pelvic rotations with respect to the sacrum are very small (2° or less).

> **Pelvic rotation**   Movement of a pelvic bone in relation to the sacrum
>
> **Pelvic tilt**   Movement of the entire pelvic girdle with respect to the hips and spine

Oblique axes for the movements are different and go through the pubic symphysis. There are no muscles that control the SI joint. Movements that do occur are a result of either the forces of the trunk acting on the pelvis, or muscles pulling on the pelvic bone. We will summarize this very brief discussion of the SI joint by saying that the SI joint does not have a rotational degree of freedom that is under voluntary control. Its purpose is to transmit forces from the spine to the pelvis, and it is designed to relieve torsional stresses associated with small rotations of the pelvis that are a part of normal movement.

| Figure 14.2 | Forces are transmitted between the sacrum and the pelvis through the SI joint. |

| Figure 14.3 | The two halves of the pelvis rotate in opposite directions during gait. |

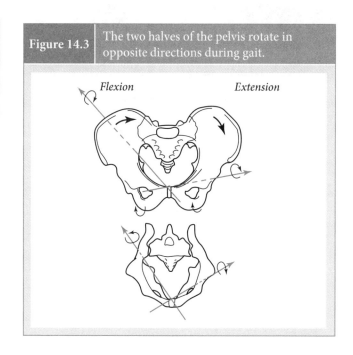

Important Point!   Rotations of the SI joint are very small, and there are no muscles that control them.

14.1.2   Functional Spinal Unit

The spinal column is generally regarded as having five sections: craniocervical, thoracic, lumbar, sacrum, and coccyx. The sacrum was discussed previously, and the coccyx will not be covered in this lesson. The remaining sections of this lesson will focus on the lumbar, thoracic, and cervical regions.

The basic motion element of the spinal column is the functional spinal unit. In this section, you will learn about the common features and terminology of the functional spinal unit. In the subsequent sections, you will learn about the individual differences in the functional spinal units that give each region its unique characteristics.

A **functional spinal unit** consists of two vertebrae and the intervertebral disk between them (**Figure 14.4**). As such, it really consists of three joints: the interbody joint and the two (right and left) zygapophysial joints. The functional spinal unit is named according to the two vertebrae that are involved, with the superior vertebra mentioned first. So the functional spinal unit of L4 and L5 would be referred to simply as L4–L5. If you see two vertebrae mentioned by themselves (such as L4–L5), assume that it means the L4–L5 functional spinal unit.

> **Functional spinal unit**
> Two adjacent vertebrae and the intervertebral disk

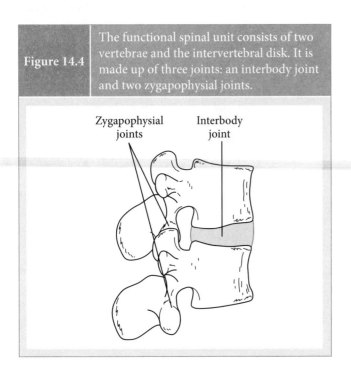

| **Figure 14.4** | The functional spinal unit consists of two vertebrae and the intervertebral disk. It is made up of three joints: an interbody joint and two zygapophysial joints. |

Zygapophysial joints   Interbody joint

The interbody joint consists of the vertebral bodies and the intervertebral disk. In between the disk and vertebral body is a cartilaginous endplate. It is classified as a symphysis joint. The disk serves the two important functions. First, it allows for an even distribution of load between the two vertebral bodies. Second, it provides space between the two vertebral bodies so that the upper vertebra can tilt without coming into contact with the lower vertebra. To accomplish the first task while performing the second, the disk must be deformable as well as strong.

Important Point!   The intervertebral disk evenly distributes the load between two vertebral bodies while allowing them to glide around one another without touching.

To accomplish these tasks, the disk acts like a tire.[4] The tire consists of two parts: the rubber tube and the air inside it. Air pressure inside the tube separates the rim from the road, but also allows the tire to deform if it is going over something like a small pebble. Also note that when you apply a compressive load to a tire, it bulges at its sides. But the more air you have inside the tire (the greater the pressure), the less the tire bulges. The disk has some important similarities, as well as a major difference.

The intervertebral disk also has two parts (**Figure 14.5**). The annulus fibrosus is analogous the rubber of the tire. It is made of concentric rings of fibrocartilage. However, the fibers of the ring are not parallel to the disk, but are at an angle. In addition, the orientation of every other ring is in the opposite direction. This means that the annulus fibrosus was designed to resist tensile forces from all directions. (Remember, compressive forces on the disk cause it to bulge at the side, putting tensile stress on the annulus fibers, which helps it maintain its height under load.)

Where your tire has air inside it, the disk has a substance called the nucleus pulposus. Some people have used the analogy of nucleus pulposus being like the jelly inside a jelly doughnut. The nucleus pulposus is what pressurizes the disk,

| **Figure 14.5** | The intervertebral disk consists of an annulus fibrosus and a nucleus pulposus. |

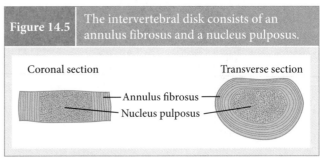

Coronal section   Transverse section

Annulus fibrosus
Nucleus pulposus

giving it its height. Also remember that the greater pressure inside the disk means less bulging (and less stress) on the annulus fibrosus.

Motions that can occur at the functional spinal unit are based on the motion of the anterior body of the superior vertebra. Flexion and extension occur when the spine bends anteriorly and posteriorly, respectively, about a medial–lateral axis. Lateral flexion (also called lateral bending) is a side-bending motion where rotation occurs about an anterior–posterior axis. Right lateral flexion occurs when the right shoulder moves toward the right pelvis or the right pelvis moves toward the right shoulder. Rotation is a twisting about the longitudinal axis. Turning toward your right side would be rotation right. If you were hanging from a bar and twisted so that your anterior pelvis was facing to the left, it would still be rotation right because the relative position of the segments would still be the same. Although somewhat of an oversimplification, consider the axis of rotation for all motions to pass through the superior edge of the inferior vertebra, about half the distance between its anterior and posterior borders (Figure 14.6). The interbody joints theoretically allow rotational motion in all three planes, but the two zygapophysial joints strongly influence the allowable motion of the functional spinal unit.

The zygapophysial joints are formed by the articulation of the superior articular processes of the inferior vertebra with the inferior articular processes of the superior vertebra. You will also sometimes see them called "facet joints" or "apophysial joints" and the name spelled "zygapophyseal" or "apophyseal." In each case, they are talking about the same thing. They are plane synovial joints. They act like railroad tracks,[2] guiding the motion of the functional spinal unit. This is easy to see if you visualize the articular processes being oriented completely in the sagittal or transverse planes. If they are oriented in the sagittal plane, they will guide flexion and extension while blocking rotation. Likewise, if they are oriented in the transverse plane, they will allow guide rotation and block extension. In reality, the articular processes' orientation is not coincident with a single plane, but the zygapophysial joints determine the motions available at the functional spinal unit.

Important Point!   The zygapophysial joints strongly influence the motion of the functional spinal unit.

Motion in a region (thoracolumbar spine, craniocervical spine) is dependent on the motion of every functional spinal unit in the region. Each functional spinal unit does not have the same amount of motion, and motion in a region is not simply the sum of the motion in each unit. For example, several of the cervical functional spinal units are actually slightly extended during full flexion of the neck.[5] Depending on your level of analysis, you may be interested in this level of detail or simply the gross movement of an entire region. In this lesson, with a few exceptions, you will be examining gross motion at a region as opposed to detailed motion at each functional spinal unit.

Important Point!   Movement in a spinal region is a result of complex interactions between the motions of each functional spinal unit in the region.

Like any other joint, the torque produced by a muscle–tendon complex (MTC) acting on the spine is dependent on its line of pull relative to the axis of rotation (Figures 14.7 to 14.9). MTCs that cross anterior the flexion–extension axis will create a flexor torque, whereas MTCs that pass posterior to it will produce an extensor torque. MTCs lateral to the anterior-posterior axis will produce a lateral flexor torque. Because MTCs can only pull and cannot push, they always produce a lateral flexor torque on the same side (ipsilateral) where they are located. In other words, MTCs to the right of the axis will always produce a right lateral flexor torque; they cannot produce a left lateral flexor torque. As with external and internal rotation of the hip, there is at least a theoretical possibility that

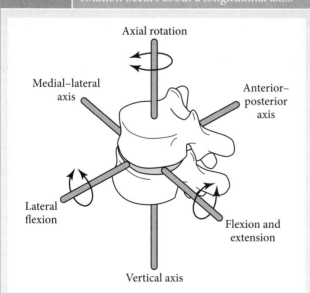

| Figure 14.6 | The axes of rotation for the functional spinal unit. Flexion–extension occurs about a medial–lateral axis, lateral flexion occurs about an anterior–posterior axis, and rotation occurs about a longitudinal axis. |

Axial rotation

Medial–lateral axis

Anterior-posterior axis

Lateral flexion

Flexion and extension

Vertical axis

Figure 14.7	MTCs that cross anterior the flexion–extension axis will create a flexor torque, whereas MTCs that pass posterior to it will produce an extensor torque.

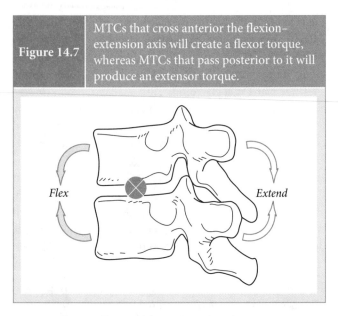

Figure 14.8	MTCs lateral to the anterior–posterior axis will produce a lateral flexor torque.

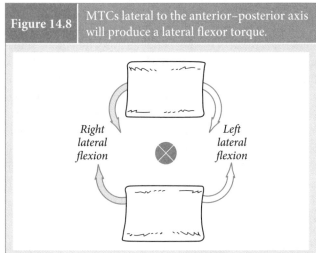

Figure 14.9	Several MTCs can produce a rotator torque to a given side of the body. Rotation right (RR) is shown here.

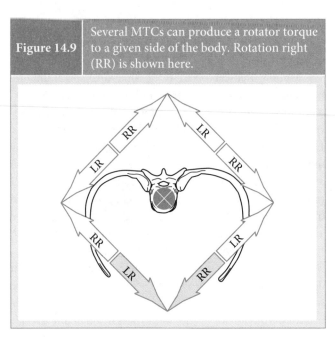

As an example, consider a case where a functional spinal unit is oriented at 45° from the vertical in the global reference frame (Figure 14.10a). Assume that it is L4–L5, and there is a superincumbent force of F_g acting on it. Neglect the mass of the functional spinal unit. In the local frame of the functional spinal unit (Figure 14.10b), this force would be resolved into its x- and y-components. Because the angle was 45°, both components would be equal to $0.707F_g$ in the positive x- and negative y-directions. This is equivalent to an anterior shear

several MTCs can produce a rotator torque to a given side of the body. Specific MTCs acting on a region of the spine will be discussed later in their respective sections.

Loading on the spine is a result of the superincumbent bodyweight, the weight of any carried objects, inertia forces/torques of these masses, and muscular forces. The spine almost always has a compressive load applied to it. Depending on the orientation of the spine, it may also have shear loads as well. The amount of shear load borne by the passive structures of the functional spinal unit is highly dependent on the muscles forces acting on the spine.

> **Important Point!** Loading on the spine is a result of: the superincumbent bodyweight, the weight of any carried objects, inertial forces of these masses, and muscular forces.

Figure 14.10	Forces acting on a functional spinal unit (a) in the global frame and (b) in the local frame. The forces produce compression and shear stresses on the functional spinal unit.

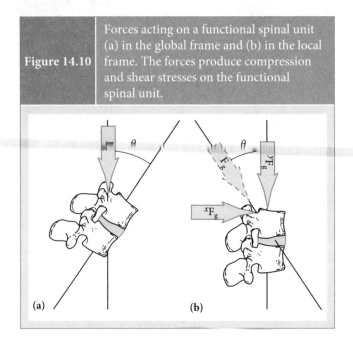

and a compressive load. To maintain static equilibrium, there would need to be equal and opposite forces in the negative x- and positive y-directions. The force in the y-direction would come from the reaction force of the inferior vertebra. Force in the x-direction could come from both muscle and passive structures of the spine.

COMPETENCY CHECK

Remember:

1. Define the following: pelvic rotation, pelvic tilt, and functional spinal unit.
2. List the three joints that make up the functional spinal unit.
3. List the two parts that make up the intervertebral disk.

Understand:

1. List the functions of the sacroiliac joint.
2. List the functions of the intervertebral disk.
3. Explain how the zygapophysial joints guide motion of the functional spinal unit.
4. Explain where an MTC has to be located to produce torque in various directions on the spine.

Apply:

1. Explain how you could decrease compressive and shear loading on a functional spinal unit.

14.2   REGION-SPECIFIC MECHANICS

14.2.1   Thoracolumbar Spine

> **Section Question**
>
> Several workers at a shipping company have developed low back pain. You have been called in for a consultation (Figure 14.11). What advice can you give these workers?

The lumbar and thoracic regions are distinct. However, in practice, it is hard to separate the motions of the two. That is because gross spinal motion is usually determined by the movement of the shoulders relative to the movement of the pelvis. This involves both the lumbar and thoracic regions. For that reason, both regions will be discussed in this section. And although there are differences within each functional spine unit of each region, things will be simplified here so you will only be learning about the aggregate characteristics of each region.

The lumbar spine consists of five vertebrae. Because T12–L1 and L5–S1 are similar to the other functional spinal

Figure 14.11 What advice can you give workers at a shipping company to help prevent injuries to the low back?

© Yuri Arcurs/ShutterStock, Inc.

units in the lumbar region, you can say there are six functional spinal units and 18 joints in the lumbar spine. There are 12 thoracic vertebrae forming 12 functional spinal units with 36 joints. The thoracic vertebrae articulate with the ribs, which also have an impact on their motion. In addition, MTCs attaching to the ribs and pelvis will create torque about the spine because the ribs attach to the spine at the costovertebral and costotransverse joints.

> **Important Point!**   MTCs attaching to the ribs and pelvis will create torque about the spine because the ribs attach to the spine at the costovertebral and costotransverse joints.

Collectively, the thoracolumbar spine is capable of movement in all three planes, although there are varying contributions from each of the regions. These ranges of motion are summarized in Table 14.1.[2] The varying degrees of

	Region		
Motion	**Lumbar**	**Thoracic**	**Thoracolumbar**
Flexion	45	40	85
Extension	15	25	40
Lateral flexion	20	25	45
Rotation	5	35	40

Table 14.1 Motions Available at the Lumbar Spine, Thoracic Spine, and Thoracolumbar Region

Table 14.2	MTCs Producing a Flexor and Extensor Torque in the Thoracolumbar Region		
Anterior (flexor torque)	**AOR: Through the vertebral bodies**	**Posterior (extensor torque)**	
Rectus abdominis		Spinalis	
External oblique		Longissimus	
Internal oblique		Iliocostalis	
Psoas major		Multifidus	
		Quadratus lumborum	

Table 14.3	MTCs Producing a Lateral Flexor Torque in the Thoracolumbar Region	
Lateral (flexor torque)	**AOR: Through the vertebral bodies**	
External oblique		
Internal oblique		
Iliocostalis		
Longissimus		
Quadratus lumborum		

motion are largely due to the orientation of the zygapophysial joints in both regions and the ribs in the thoracic region. The muscles that create torque in these directions follow the general pattern for the functional spinal unit (see Figures 14.7 to 14.9). These muscles are listed in Tables 14.2 to 14.4. Similar to the lower extremity, muscles that create in more than one plane can be grouped into functional quadrants (Figure 14.12).

The amount of torque that is required of the MTCs depends on both the orientation of the spine in the gravitational field and any other external forces/torques (including inertial ones). For a very simple example, take a look at what happens when body position changes (Figure 14.13). The lever arm of the COM_{HAT} is some distance, l_{COM}, from your

Table 14.4	MTCs Producing a Contralateral and Ipsilateral Torque in the Thoracolumbar Region
Ipsilateral rotator torque	**Contralateral rotator torque**
Internal oblique	External oblique
Iliocostalis	Multifidus

point of interest (say L5–S1). The external torque about L5–S1 would be zero if the COM_{HAT} was located directly above the axis of rotation. Because the actual COM_{HAT} lies anterior to the vertebral column, this is a bit of a fantasy—but stick with it. Assuming the spine stays rigid and flexion occurs at the

Figure 14.12	The functional quadrants for MTCs acting on the thoracolumbar region.

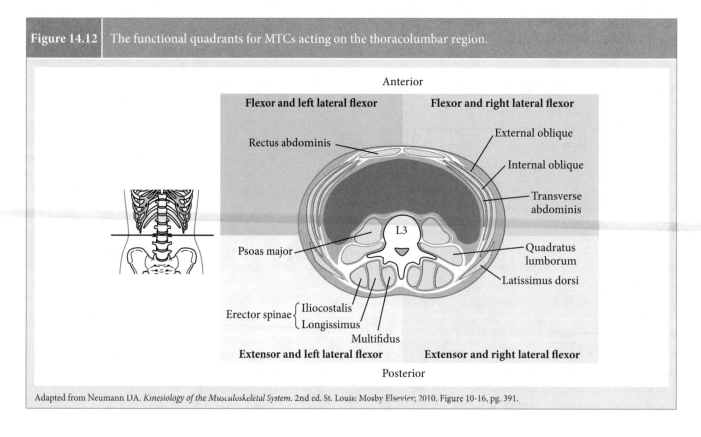

Anterior

Flexor and left lateral flexor   **Flexor and right lateral flexor**

Rectus abdominis
External oblique
Internal oblique
Transverse abdominis
L3
Psoas major
Quadratus lumborum
Latissimus dorsi
Erector spinae { Iliocostalis / Longissimus
Multifidus

Extensor and left lateral flexor   **Extensor and right lateral flexor**

Posterior

Adapted from Neumann DA. *Kinesiology of the Musculoskeletal System*. 2nd ed. St. Louis: Mosby Elsevier; 2010. Figure 10-16, pg. 391.

| Figure 14.13 | The external torque is determined by F_{gHAT}, the hip flexion angle, and l. |

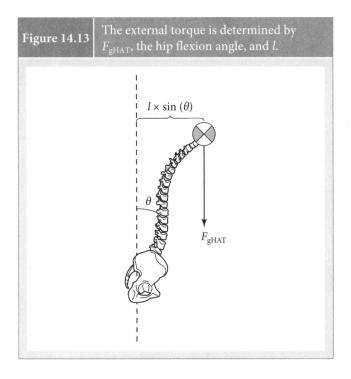

$l \times \sin(\theta)$

θ

F_{gHAT}

hips, the hip flexion angle will change the orientation of the trunk with respect to the vertical axis. To determine the magnitude of the change, you would use the formula:

$$T_{external} = F_{gHAT} \times l_{COM} \times \sin\theta_{hip} \quad (14.1)$$

Assuming F_{gHAT} and l_{COM} are constants, $T_{external}$ is a function of the hip flexion angle. The sine of any angle increases from 0° to 90°, so you should be able to see how the external demands at L5–S1 change as a function of hip flexion angle. If this was a static situation, then a constraint is that the angular acceleration is zero:

$$\sum T = I\alpha = I \times 0 = 0 \quad (14.2)$$

And because only the MTC and external forces are producing torques:

$$T_{MTC} + T_{external} = 0 \quad (14.3)$$

So it follows that:

$$T_{MTC} = -T_{external} \quad (14.4)$$

Because θ_{hip} is the only variable that can change $T_{external}$, it is also the only variable that determines the torque demands on the lumbar extensors.

If the man was flexing or extending his hips in this position, then you would also have to include the inertia torque when determining the external demand at L4–L5:

$$T_{MTC} = T_{inertial} - T_{external} \quad (14.5)$$

Remember that

$$T_{inertial} = I\alpha_{HAT} \quad (14.6)$$

Where I is the moment of inertia taken about L5–S1 and α_{HAT} is the angular acceleration of the HAT. Care must be taken to ensure that this is the acceleration of the trunk flexing or extending and not the trunk moving in space due to hip motion. This could also create a torque (called an interactive torque), but that is something to be considered with multi-joint systems.

Another frequent situation where loads are placed on the lumbar spine is when you are carrying an object. Say the same man is carrying a 10 kilogram package. Even if he is standing upright, the package will create at torque about the lumbar spine (in this case, use L4–L5). If he is holding the package close to his chest (Figure 14.14a), then the external torque would be

$$T_{external} = F_{g_{package}} \times d_{trunk} \quad (14.7)$$

Where d_{trunk} is the distance from the spine to the front of his chest. If the man holds the package some distance away from trunk, d_{arm}, as in Figure 14.14b, this distance must be added to the d_{trunk}, and the equation 14.7 becomes

$$T_{external} = F_{g_{package}} \times (d_{trunk} + d_{arm}) \quad (14.8)$$

You can begin to appreciate how holding the package away from your body increases the torque demands on the spine because it increases the external torque caused by the package.

Not only are there torque demands on the spine, but there are also forces creating loads on the spine as well. The head, arms, and trunk (HAT) are approximately 60% of total body weight.[6] This means that an 80 kilogram man will have ~471 Newtons of compressive force on the lower lumbar spine during quiet standing, assuming no muscle activation at all. In many (if not most) cases, there are muscle activations, particularly when carrying a load.

> **Important Point!**   You should always look for torques, axial forces, and shear forces acting on the spine.

To see this effect, return to the example of the man holding the package. These numbers are very rough; you should not take them as being exact, but only to give you a flavor for what is going on. The effective mechanical advantage (EMA) of the MTC compared to the package is roughly 0.3[7] when the package is held close to the chest. This means

Figure 14.14 | The external torque (a) when the package is held close to the body, and (b) when it is held at arm's length.

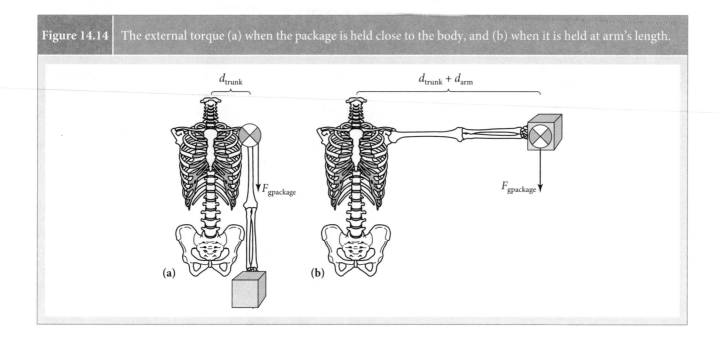

(a)   (b)

that the spinal erector force would have to be three and one third times greater than the weight of the package. For a 10 kilogram package, that is 327 Newtons of force! If the package is held at arm's length, the EMA drops to ~0.08. That means that the extensor MTCs have to generate almost 13 times as much force (1275 Newtons for a 10 kilogram package). Remember, the force of the MTC creates a compressive penalty on the spine. Using these very rough numbers, the compressive forces on the spine are four times greater when you hold the same package at arm's length compared to close to the chest.

Compressive forces are not the only loads on the spine. Consider again the man standing, just with his body weight, at an angle from the vertical. Draw a free-body diagram (Figure 14.15a) and then rotate it so that the reference frame is a local one in the spine (Figure 14.15b).

Assume that the system is in static equilibrium, so the following constraints are met:

$$^{x}a = 0$$
$$^{y}a = 0$$
$$\alpha = 0$$

Let us focus on the x-direction. If the force due to gravity is

$$^{x}F_{g} = F_{g} \times \sin\theta \qquad (14.9)$$

then this is an anterior shear force on the spine. To maintain static equilibrium, the equation of motion becomes

$$^{x}F_{g} - {}^{x}F_{internal} = 0 \qquad (14.10)$$

And:

$$^{x}F_{internal} = {}^{x}F_{g} \qquad (14.11)$$

where $^{x}F_{internal}$ represents a posterior shear force necessary to counteract the anterior shear force and maintain equilibrium. It is equal to:

$$^{x}F_{internal} = {}^{x}F_{g} = F_{g} \times \sin\theta_{hip} \qquad (14.12)$$

Figure 14.15 | The F_{gHAT} in the global frame and in the local frame. In the local frame, F_{gHAT} produces a compressive and shear force.

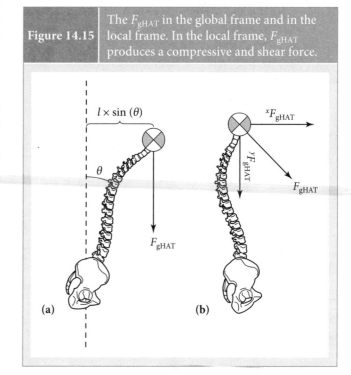

(a)   (b)

showing you that the amount of shear is dependent on the superincumbent bodyweight and the amount of hip flexion. Once again, the sine of an angle increases from 0° to 90°. Assuming that superincumbent body weight does not change, the amount of anterior shear that the spine system is subjected to increases with hip flexion angles.

Section Question Answer

Low back disorders are complex and do not have many easy answers. However, one simple principle is to decrease the load on the lumbar spine. As you just learned, both external and internal forces on the spine increase when the spine is oriented at an angle to the gravitational field and when loads are held further away from the spine. Maintaining an upright posture and keeping external loads close to the trunk are two ways to decrease the load on the spine.

COMPETENCY CHECK

Remember:

1. List the available ranges of the motion of the thoracolumbar spine.
2. List the muscles that act on the thoracolumbar spine and the torque(s) that they produce.

Understand:

1. Explain why loads increase on the thoracolumbar spine with an increased angle of orientation in the gravitational field.
2. Explain why loads increase on the thoracolumbar spine when an external load is carried further away from the spine.

Apply:

1. Demonstrate techniques for lifting and carrying that will decrease the load on the thoracolumbar spine.

14.2.2   The Craniocervical Region

Section Question

When designing a helmet (Figure 14.16), why do you want to place the center of mass of the helmet posterior to the center of mass of the head?

| Figure 14.16 | When designing a helmet, why place the center of mass of the helmet behind the center of mass of the head? |

© gosphotodesign/ShutterStock, Inc.

Functionally, the head can be considered a platform that houses the sensory apparatus for "hearing, vision, smell, taste, and related lingual and labial sensations."[5] It is also where the nose and mouth are, which provide oxygen and nourishment to the body. The head must be able to scan the environment and be directed toward objects of interest. The function of the cervical spine, then, is to support the head as well as move and orient it in three-dimensional space. Four parts of the craniocervical region are necessary for it to perform these functions: the cradle (the atlas; C0–C1), the axis (C1–C2), the root (C2–C3) and the column (remaining cervical functional spinal units).[5] These parts are presented in Figure 14.17.

Important Point!   The function of the cervical spine is to support the head and orient it in three-dimensional space.

The cradle consists of the rather deep concave facets of C1 articulating with the convex occipital (sometimes referred to as C0) condyles. The major motions at C0–C1 are flexion

Figure 14.17	The four parts of the cervical spine: cradle, axis, root, and column.

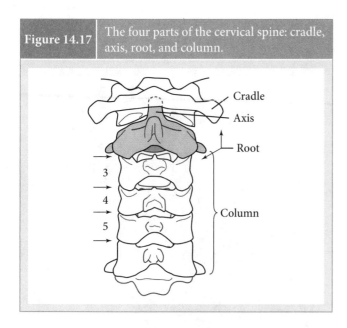

Table 14.5	MTCs Producing a Flexor and Extensor Torque on the Cervical Region		
Anterior (flexor torque)		AOR: Through the vertebral bodies	**Posterior** (extensor torque)
Scalenes Sternocleidomastoid Longus colli			Trapezius Longissimus Semispinalis Splenius cervicis Levator scapulae

Table 14.6	MTCs Producing a Lateral Flexor Torque in the Cervical Region		
Lateral (lateral flexor torque)		AOR: Through the vertebral bodies	**Lateral**
Semispinalis Splenius Sternocleidomastoid Trapezius Longus colli Scalenes			

Table 14.7	MTCs Producing a Contralateral and Ipsilateral Rotation Torque in the Cervical Region	
Ipsilateral rotator torque		**Contralateral rotator torque**
Longissimus		Multifidus Rotators Semispinalis Sternocleidomastoid Splenius

Table 14.8	MTCs Producing a Flexor and Extensor Torque on the Cranial Region		
Anterior (flexor torque)		AOR: Through the vertebral bodies	**Posterior** (extensor torque)
Longus capitus Rectus capitus anterior Stylohyoid			Trapezius Longissimus capitis Semispinalis capitis Splenius capitus Rectus capitis posterior

and extension. Flexion occurs according to the convex–concave pattern: anterior rolling and posterior gliding of the convex C0 on the concave C1. Lateral bending and rotation are severely limited. C1 acts as a washer,[5] distributing the load from the skull to the vertebral column below. The atlas of C1 articulates with the dens of C2, with the alar ligament holding the two in position. Few MTCs act directly on these functional spinal units. MTCs that attach to the skull transfer torques inferiorly to C0–C1 and C1–C2.[5] C2–C3 is different from the rest of the cervical functional spinal units. It appears as though the orientation and location of the facets act to anchor C2 into C3.[5] The remaining functional spinal units act as column on which the head and rest of the vertebrae sit.

Important Point! MTCs attaching to the skull transfer torques inferiorly to the upper cervical vertebrae.

The cervical spine is capable of movement in all three planes. Typically, there are approximately 50° of flexion, 60° of extension, 45° of lateral bending in each direction, and 80° of rotation in each direction.[4] MTCs of the cervical region are usually referred to as either capital MTCs (MTCs that have their superior attachment on the skull) or cervical MTCs (that have their superior attachment on the cervical vertebrae). The cervical MTCs that produce torque in these directions follow the general pattern for the functional spinal unit (see Figure 14.7). These MTCs are listed in Tables 14.5 to 14.7. The capital MTCs can be thought of as acting on an axis of rotation that goes through the center of the foramen magnum (Tables 14.8 and 14.9). MTCs that are anterior to the medial–

Table 14.9	MTCs Producing a Lateral Flexor Torque in the Cranial Region	
Lateral (lateral flexor torque)	**AOR: Through the vertebral bodies**	**Lateral**
Rectus capitus Semispinalis capitis Splenius capitis Sternocleidomastoid Trapezius Longus capitis Stylohyoid		

Table 14.10	MTCs Producing a Contralateral and Ipsilateral Rotation Torque in the Cranial Region	
Ipsilateral rotator torque		**Contralateral rotator torque**
Rectus capitis posterior major Oblique capitis inferior Splenius capitis		Sternocleidomastoid

lateral axis will produce a flexor torque, whereas MTCs that are posterior to it will produce an extensor torque. MTCs that are lateral to the anterior–posterior axis produce a lateral flexor torque. This can create four functional quadrants:

- Quadrant I—extensor and left lateral flexor torque
- Quadrant II—extensor and right lateral flexor torque
- Quadrant III—flexor and right lateral flexor torque
- Quadrant IV—flexor and left lateral flexor torque

These quadrants are represented in Figure 14.18. Capital MTCs producing a rotation torque do not follow as clear-cut a pattern and are presented in Table 14.10.

When comparing the flexors to the extensors, it should be noted that there are more extensors than there are flexors, the extensors tend to have a greater physiological cross-sectional area than the flexors, and the extensors have a longer moment arm than the extensors. In addition, the moment arm of the flexors decrease as the neck extends. This has implications for whiplash injuries, which will be discussed later in this lesson.

Figure 14.18	The functional quadrants for MTCs acting on the cervical region.

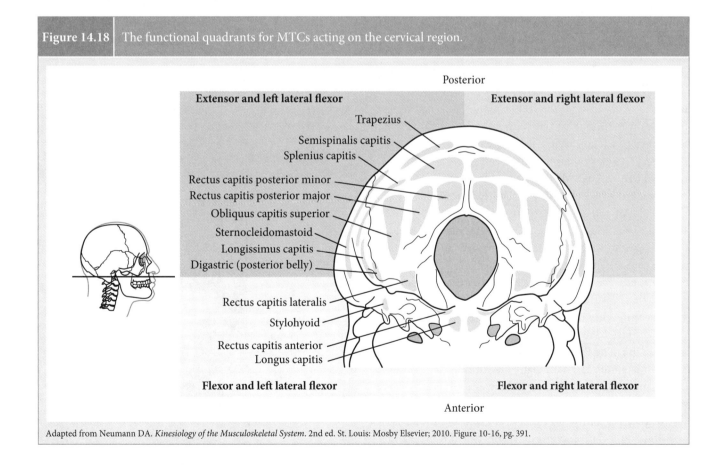

Adapted from Neumann DA. *Kinesiology of the Musculoskeletal System.* 2nd ed. St. Louis: Mosby Elsevier; 2010. Figure 10-16, pg. 391.

Important Point!   You can produce greater extensor torque than flexor torque in the craniocervical region.

The head weighs approximately 7% of total body weight.[6] During standing, the COM_{head} is anterior to the cervical spine.[8] This means that there is always a flexion moment on the cervical spine that must be counteracted by the cervical extensors to keep the head from flexing. Like any other body region, the actual acceleration of the head would depend on the external torque, the torque of the MTCs, and the moment of inertia of the head:

$$\alpha_{head} = \frac{T_{MTC} + T_{external}}{I_{head}} \qquad (14.13)$$

As you will see, the inertia of the head has some big implications for injuries. If the external torque and the MTC torque are in the same direction, then the head will obviously accelerate in that direction. If the MTC torque and external torque are in opposite directions, three outcomes are possible. If the MTC torque is greater than the external torque, then the head will accelerate in the direction of the MTC torque. Likewise, if the external torque is greater than the MTC torque, the head will accelerate in the direction of the external torque. If the MTC torque and the external torques are equal, the head will not accelerate and will be in equilibrium.

It is interesting to see the kinds of loads that the cervical spine is subjected to as a result of balancing the weight of the head. This is another simplified example, and the numbers should be interpreted with caution. Let us examine two situations: one involving the capital extensors (Figure 14.19) and the other involving the cervical extensors (Figure 14.20).[9]

As mentioned previously, the head weighs approximately 7% of total body weight. However, the weight of the head also does not fluctuate tremendously, either. So, for people who weigh less, this percentage is higher and for people who weigh more, this percentage is less.[8] But for this example, use 7%. For the same 80 kilogram man discussed earlier, that is ~550 Newtons of compressive force on the cervical spine due to the head alone. And remember that you would also have to add the compressive forces due to the musculature. The EMA of the capital extensors in a neutral position is about 2. That means that the capital extensors have a mechanical advantage and have to generate a force of about half the weight of the head to counteract the extensor torque. The EMA of the cervical extensors in the same position is 0.5. The cervical extensors are at a mechanical

Figure 14.19 External torque about the cranial axis of rotation produced by the weight of the head. The cranial extensors are at a mechanical advantage.

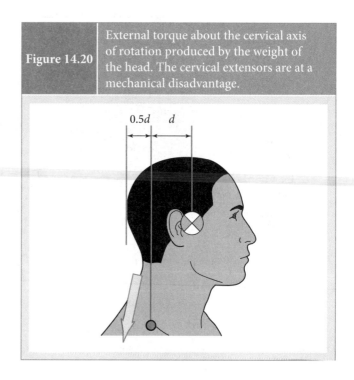

disadvantage and thus have to supply a force that is approximately twice the weight of the head to maintain a neutral position. Thus, the compressive loads on the lower cervical spine are about three and one half times the weight of the head. Certainly these loads are very tolerable. But the cervical region was not designed to tolerate high loads, something you will soon see.

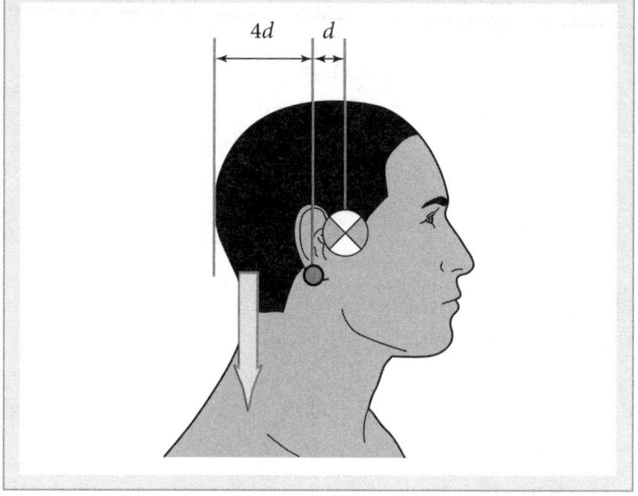

Figure 14.20 External torque about the cervical axis of rotation produced by the weight of the head. The cervical extensors are at a mechanical disadvantage.

The center of mass of the head is anterior to the axes of rotation—both at the cervical region and at the cranial region. This produces a flexion torque while standing upright that must be countered by the craniocervical extensors. If the COM$_{helmet}$ is anterior to the COM$_{head}$, then both would create flexor torques about the axes of rotation.[8] This could lead to undue fatigue of craniocervical extensors. It is more beneficial to place the COM$_{helmet}$ as posterior as possible to ease the demand on the craniocervical extensors.

COMPETENCY CHECK

Remember:

1. List the ranges of motion available at the craniocervical region.
2. List the MTCs that cross the craniocervical region and the torques they produce.

Understand:

1. Why can you produce more torque in extension than in flexion in the craniocervical region?
2. Why do the MTCs in the craniocervical region have to produce a constant extensor torque in an upright position?

14.3   SPINAL INJURIES

Section Question

What can you tell football players to help them reduce the potential for injuries to the neck (**Figure 14.21**)?

Injuries can occur to any structure in the body, and the spinal column is no exception. Different types of stresses can be applied to various structures such as bone, ligament, cartilage, and muscle. The individual structural components of the spinal column are no different. However, when looking at the spinal column as a whole, there is a new mode of failure you should be familiar with: buckling. To understand buckling, you need to appreciate the concepts of stiffness and stability. A brief review of these important ideas, and their application to the spine, is in order.

Figure 14.21   What can you tell football players to help protect their necks from injury?

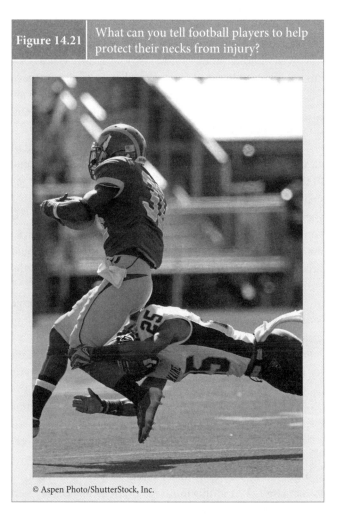

© Aspen Photo/ShutterStock, Inc.

14.3.1   Basic Concepts

Spinal Stiffness, Intra-abdominal Pressure, and Stability

The concepts about joint stiffness and stability both relate to how a joint system responds to a disturbing force or torque. Throughout this entire discussion, the disturbance can be a force or a torque (or both). If the disturbing force or torque is greater than the resistive force or torque provided by the system, then the joint will displace. Joint stiffness relates the disturbing force or torque to the displacement. Remember that it is not exactly the same idea as the material property of stiffness, so you should distinguish it as joint (or quasi) stiffness to be precise with your language. Greater joint stiffness means that the same disturbing force or torque produces less displacement of the system or the elements within it.

All the joint structures contribute to joint stiffness. In the case of the spine, this includes the bony articulations, disks, cartilage, ligaments, joint capsules, and MTCs. The MTCs contribute to stiffness via both reflex and voluntary mechanisms. Intrinsic joint and muscle properties respond very rapidly. And although they exhibit viscoelastic properties, they are not involved in the voluntary regulation of joint stiffness. That role is reserved for the MTCs. The MTCs can absorb the greatest amount of energy, and it is preferential to have them do so over other structures such as ligaments.

Stiffness of the spine may be among the most difficult of all the joints to both comprehend and practice. This is due to the length of the spine and the sheer number of joints involved. In the lumbar spine alone, it has been estimated that varying degrees of stiffness are required in 36 dimensions.[10] If you add the thoracic spine, the number climbs to 108! With so many dimensions that must be precisely controlled, it should become apparent that no one, single muscle can stiffen the entire spine.

The intervertebral disk can, and does, bear a fairly large load. The load borne by the disk is a function of the pressure inside it (much like the pressure inside a tire). This has led many to speculate that pressure inside the abdominal cavity (called intra-abdominal pressure; IAP) can bear some of the load placed on the spinal column in the lumbar region (there are no load-bearing anatomical structures anterior to the

> **Intra-abdominal pressure** Pressure inside the abdominal cavity

spine).[11] Others contend that muscle activation is required to increase IAP, and this muscle activation can actually lead to an increase in compressive force across the spine.[12] Yet another argument is that IAP is merely a by-product of the muscle activity necessary to stiffen the spine.[13] It seems that the role of IAP in spinal unloading is far from resolved, as different models produce different results.[14] Whatever the ultimate determination of the effect of IAP is on the compressive forces in the spine, it is safe to say that IAP increases with muscle activation, the IAP/muscle activation increases the stiffness of the spine, and only low levels of muscle force are necessary to sufficiently stiffen the spine in most situations.[10]

However, spinal stiffness should not be equated with stability.[15] Rather than simply resisting a perturbation, it may sometimes be advantageous to yield to it and then "bounce back." For example, imagine any activity where you have to maintain balance on an unstable surface. In one particular study, subjects did this by sitting on an unstable surface.[16] When subjects were asked to increase stiffness of their spine, they were more likely to lose their balance. This finding highlights the fact that stability is a very intricate concept, and biomechanists have a lot more work to do before fully understanding it.

Buckling

There are many different types of injuries that can affect the spinal column. Injuries to the lumbar spine appear to be the most common, whereas injuries to the cervical region are potentially the most serious. Although injuries to the thoracic region can occur, they are somewhat rare. In this section you will learn about a common mechanism of injury (buckling) and then see how it is applied in a few different scenarios.

The spine can be thought of as a column. All columns are under compressive forces. If the column is short, then material failure occurs when the stresses exceed the strength of the material. If the column is many times longer than it is wide, then buckling can occur well before the stresses equal the strength of the mate-

> **Buckling**   Deformation of a column

rial. The middle of the column moves (anteriorly, laterally, posteriorly) and shortens under a load that it can no longer support (Figure 14.22). This is easy to demonstrate: Simply take an index card and stand it upright between your finger and a firm surface. Now press down, and watch the index card buckle under the force of your finger.

| Figure 14.22 | Buckling of column. |

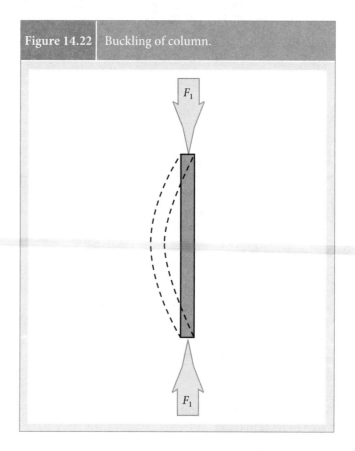

A simplified formula to show you the maximum force a general column can withstand before buckling is

$$F_{\text{buckling}} = \frac{YI_{\text{area}}}{l^2} \qquad (14.14)$$

Where Y is the elastic modulus, I_{area} is the area moment of inertia, and l is the length of the column. Although this equation may not be directly applicable to the human spine, it illustrates a couple of important points. The first thing you should notice is that the length term is squared, so increasing the length of the column decreases the force before buckling by a factor of four. You really cannot do anything about the length of the spinal column anyway, nor can you change the area moment of inertia. But you can increase the force before buckling by increasing the stiffness of the spine.

Another important point to realize is that a force that is off-center (an eccentric force, not to be confused with an eccentric MTC action) will decrease the strength of the column because it will start to bend the column immediately. Where this becomes really important is with the relative position of the vertebrae within the spinal column.

Recall that your spine is not a straight column, but has a gentle curvature to it. Load is transmitted through the spinal column with this natural curvature. Buckling can occur when this natural alignment is not kept. In this way, a centered force at the ends of the column can become an eccentric force at one of the functional spinal units—inducing the column to deform at that particular functional spinal unit (Figure 14.23). When this occurs, depending on the direction of the buckling, certain parts of the functional spinal unit will be subjected to compressive stresses while the opposite side will be subjected to tensile stresses.

14.3.2 Thoracolumbar Injuries

With the concepts of buckling in mind, you can begin to appreciate how injuries to the spinal column occur. First, begin with an injury to the thoracolumbar region and the familiar example of bending down to pick up an external load. When performed by keeping the lumbar spine in a relatively neutral position (or within a few degrees from full flexion at each functional spinal unit), very little stress is placed on the ligaments of the lumbar spine.[17] Almost all the demands are taken up by the muscles. You will also remember that this position allows for the compressive forces to be centrally transmitted across the vertebral bodies. However, if just one functional spinal unit goes into full flexion, the spinal column can buckle at the position. Injury can then result to soft tissue structures around that functional spinal unit.

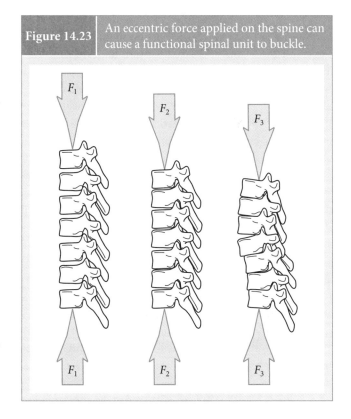

Figure 14.23 An eccentric force applied on the spine can cause a functional spinal unit to buckle.

How can this happen? MTCs like the iliocostalis and longissimus traverse the length of the lumbar spine. The multifidus, on the other hand, only traverses a couple of functional spinal units. It has been hypothesized that the erector spinae produce a gross torque over the entire lumbar spine, while the multifidus produces torque at specific segments, fine-tuning the position of each functional spinal unit.[10] This type of spinal buckling can thus occur if, by fatigue or errors in motor control, the larger erector spinae are highly activated while the multifidus has a low activation, or when muscle activations of all muscles are low.[18]

The importance of muscle activation in protecting the spine cannot be overemphasized. But it has to be the right amount at the right time by the right muscles. Too much activation can be extremely fatiguing and may make the spine unresponsive to perturbations. However, too little activation can also be problematic because the time required to activate and develop force within an MTC is relatively slow. Nowhere is this more evident than with whiplash injuries.

14.3.3 Craniocervical Injuries

Whiplash was originally thought to be an overextension injury, but here you will see how it is more appropriate to think of it as a buckling injury. Both the head and trunk have inertia, and the neck is what links the two. If the neck is stiff, the head and trunk will move as one. If not, they are moving as two separate masses that are linked together. During impact, the trunk is not only projected forward but also upward.[19] This means that there is an eccentric force acting on the neck from below. The craniocervical region as a whole does not exceed its physiological limit rotating into extension as the trunk moves upward and forward relative to the head, but the lower cervical functional spinal units are extending around an abnormal axis while the upper ones are flexing.[19] The anterior portions of the functional spinal units are abnormally separated, while the posterior elements are jammed together. All this takes place within 150 milliseconds after impact.[19] Even if the muscles try to activate to stiffen the neck at impact, they are just too slow to do much about it.

Whiplash shows you what happens when the trunk causes a buckling from below, but while the head is free to move. Much more disastrous consequences can occur if the head is struck axially from above and does not move. Catastrophic injuries can occur when the head gets "stuck," and inertia of the trunk collapses on top of the head, buckling the neck. This can happen when a player "spears" another player or object with his helmet, or when the head strikes the ground as when diving into shallow water. The injuries sustained to the craniocervical spine are not the result of the initial impact of the head or of the neck going into hyperflexion or hyperextension subsequent to the impact. Rather, they occur when the head strikes an object and momentarily stays stationary while the trunk continues to move with a certain amount of inertia. In such cases, the inertia of the trunk buckles the spine—causing some functional spinal units to flex while others extend.[20] This is very similar to what you saw in whiplash injuries. The major differences are that the inertial force of the trunk is directed more axially, and the head is not free to move. These differences cause greater compressive forces and more complex buckling patterns, leading to potentially more harmful injuries with this mechanism.

14.3.4   Traumatic Brain Injuries

Of course, a blow to the head does not always result in a neck injury. And an impact to the head can lead to a head (brain) injury either in the presence or absence of a neck injury. It is not easy to determine which structures are injured due to a particular impact; it involves a complex interaction of the two objects (head and whatever it is that it is impacting) as well as the direction and magnitude of the resulting force. You will not be finding the answer to this question (if such an answer exists) in this section. Rather, you will just be learning about the basic mechanics of injuries involving the brain that do not produce any injuries to the skull. The mildest form of this injury is a concussion, whereas more serious ones include

epidural hematomas, subdural hematomas, and intracerebral hemorrhaging.

How can an impact injure the brain without injuring the skull that protects it? The key is to understand that the brain: (1) is not a solid mass, but a large group of axonal connections interspersed with vascular structures, and (2) sits inside a skull filled with cerebrospinal fluid. These facts mean that one part of the brain can be moving relative to another, and the brain can be moving relative to the skull. Traumatic brain injuries occur by one of these two mechanisms, and the head does not even have to be hit for these to happen.

In one case, the brain makes an impact with the inside of the skull. When this happens, a compressive injury to the point of contact can occur. Think about what happens if you have a ball in a wagon and then move the wagon really quickly. The ball strikes the back of the wagon. If a stationary head is struck by a moving object, the same thing can happen. The skull can move and strike the brain inside it. If a moving head suddenly comes in contact with a massive stationary object (like the ground or a goalpost), then the brain can get bounced around inside the skull like a pinball. In this case, the moving brain can hit the skull, and this can occur on the opposite side of the initial impact.

In the other case, one part of the brain moves relative to another part of the brain. With acceleration/deceleration of the head, particularly with rotational accelerations in any plane of movement, shear stresses occur. This can cause a disruption or rupture of axons or vessels. A good example of this would be when a boxer gets caught with a hook punch to the jaw. Note that the punch impacts with the mandible and not skull itself. This induces a large acceleration about the longitudinal (vertical) axis. Because the brain is not a solid mass, portions further from the vertical axis can be moving at a different rate than portions closer to the vertical axis. This will put a tensile stress on the axons and vessels and can literally pull them apart.

Section Question Answer

The craniocervical region is injured due to buckling of the column. There are several things that a football player can do to help protect himself. First, he can ensure that he tackles using the correct technique where he makes contact with his shoulder pads instead of "spearing" with his helmet. Second, he can make sure that he is wearing the correct helmet/neck roll combination that will help limit movement of the neck. Third, prior to impact he needs to "bull" the neck by cocontracting the neck and trapezius MTCs. Finally, he needs to ensure that the MTCs of the neck have adequate strength and endurance to do this repeatedly.

COMPETENCY CHECK

Remember:

1. Define the following terms: joint stiffness, stability, intra-abdominal pressure, and buckling.
2. List the factors that determine if a column buckles.

Understand:

1. Explain the role of the MTCs in preventing buckling.
2. Describe a buckling injury that occurs at the thoracolumbar region and at the craniocervical region.
3. Explain the mechanisms for a traumatic brain injury.

Apply:

1. Discuss ways to prevent injuries to the spine.

SUMMARY

In this lesson, you learned about the biomechanics of the axial skeleton, particularly the spine. In terms of rotations and torques, the spine is similar to any other joint: MTCs act some distance away from the axis of rotation. The net effective torque, determined by summing the external and MTC torques, determines the acceleration of the spine. When examining injury mechanisms, the individual structures of the spine respond similarly to like tissues elsewhere in the body. However, the spine as a whole should be thought of as a column, with buckling being the major mode of injury. Any mechanism that causes an injury to the craniocervical region can also cause a traumatic brain injury. Brain injuries occur because the brain is not a solid mass, and it floats inside the skull. Movement of one portion of the brain relative to another causes shear stresses, whereas the skull impacting the brain causes compressive stresses.

REVIEW QUESTIONS

1. Define the following terms: pelvic rotation, pelvic tilt, functional spinal unit, joint stiffness, intra-abdominal pressure, and buckling.
2. What are the components of the functional spinal unit?
3. What are the components of the intervertebral disk?
4. What are the functions of the sacroiliac joint?
5. What are the functions of the intervertebral disk?
6. How do the zygapophysial joints guide the motion of the functional spinal unit?

7. Explain where an MTC has to be located to produce torque on the spine in various directions.
8. Where do shear and compressive forces acting on the spine originate?
9. What are the available ranges of motion in each region of the spine?
10. List the muscles acting on the spine and the torque(s) they produce.
11. How can you decrease loading on the spine?
12. Why can you produce more torque in the extension than in flexion in the craniocervical region?
13. Why do the MTCs in the cervical region have to produce a constant extensor torque in an upright position?
14. Discuss the controversy surrounding intra-abdominal pressure and its role in spinal stiffness and compression loading of the lumbar spine.
15. What are the factors that determine if a column buckles?
16. How do the MTCs prevent buckling?
17. List a buckling injury that occurs at the thoracolumbar spine and at the craniocervical spine.
18. What are the mechanisms for a traumatic brain injury?

REFERENCES

1. Levangie PK, Norkin CC. *Joint Structure and Function: A Comprehensive Analysis*. 3rd ed. Philadelphia: F. A. Davis; 2001.
2. Neumann DA. *Kinesiology of the Musculoskeletal System*. 2nd ed. St. Louis: Mosby Elsevier; 2010.
3. Bogduk N. *Clinical Anatomy of the Lumbar Spine and Sacrum*. 4th ed. Edinburgh: Elsevier; 2005.
4. Adams MA, Dolan P. Spine biomechanics. *Journal of Biomechanics*. Oct 2005;38(10):1972–1983.
5. Bogduk N, Mercer S. Biomechanics of the cervical spine. I: Normal kinematics. *Clinical Biomechanics*. Nov 2000;15(9):633–648.
6. de Leva P. Adjustments to Zatsiorsky-Seluyanov's segment inertia parameters. *Journal of Biomechanics*. Sep 1996;29(9):1223–1230.
7. Jorgensen MJ, Marras WS, Granata KP, Wiand JW. MRI-derived moment-arms of the female and male spine loading muscles. *Clinical Biomechanics*. Mar 2001;16(3):182-193.
8. Zatsiorsky VM. *Kinetics of Human Motion*. Champaign, IL: Human Kinetics; 2002.
9. Oatis CA. *Kinesiology: The Mechanics and Pathomechanics of Human Movement*. Philadelphia: Lippincott, Williams, and Wilkins; 2004.
10. McGill SM. *Low Back Disorders: Evidence-Based Prevention and Rehabilitation*. 2nd ed. Champaign, IL: Human Kinetics; 2007.
11. Daggfeldt K, Thorstensson A. The role of intra-abdominal pressure in spinal unloading. *Journal of Biomechanics*. Nov-Dec 1997; 30(11–12):1149–1155.
12. Nachemson AL, Andersson GBJ, Schultz AB. Valsalva maneuver biomechanics—effects on lumbar trunk loads of elevated intra abdominal pressures. *Spine*. 1986;11(5):476-479.
13. Marras WS, Mirka GA. Intra-abdominal pressure during trunk extension motions. *Clinical Biomechanics*. Jul 1996;11(5):267–274.

14. Stokes IAF, Gardner-Morse MG, Henry SM. Intra-abdominal pressure and abdominal wall muscular function: Spinal unloading mechanism. *Clinical Biomechanics*. Nov 2010;25(9):859–866.

15. Reeves NP, Narendra KS, Cholewicki J. Spine stability: the six blind men and the elephant. *Clinical Biomechanics*. Mar 2007;22(3):266–274.

16. Reeves NP, Everding VQ, Cholewicki J, Morrisette DC. The effects of trunk stiffness on postural control during unstable seated balance. *Exp. Brain Res*. Oct 2006;174(4):694–700.

17. Cholewicki J, McGill SM. Lumbar posterior ligament involvement during extremely heavy lifts estimated from fluoroscopic measurements. *Journal of Biomechanics*. Jan 1992;25(1):17–28.

18. Cholewicki J, McGill SM. Mechanical stability of the in vivo lumbar spine: implications for injury and chronic low back pain. *Clinical Biomechanics*. Jan 1996;11(1):1–15.

19. Bogduk N, Yoganandan N. Biomechanics of the cervical spine Part 3. minor injuries. *Clinical Biomechanics*. May 2001;16(4):267–275.

20. Nightingale RW, McElhaney JH, Richardson WJ, Best TM, Myers BS. Experimental impact injury to the cervical spine: relating motion of the head and the mechanism of injury. *Journal of Bone and Joint Surgery—American Volume*. Mar 1996;78A(3):412–421.

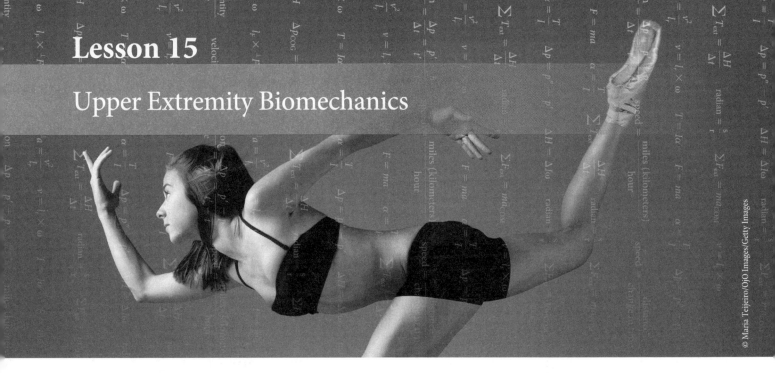

Lesson 15

Upper Extremity Biomechanics

LEARNING OBJECTIVES

After finishing this lesson, you should be able to:

- Identify the major joints of the upper extremity.
- For each joint, determine which type of joint it is, how many rotational degrees of freedom it has, the motions that are available at that joint, and the "normal" ranges of motion.
- For each joint, describe the arthrokinematic motions of the proximal segment moving on the distal segment and the distal segment moving on the proximal segment.
- Describe the torques that can be produced by the muscles crossing each joint.
- Determine the net acceleration of a segment based on the torques produced by external forces and muscle–tendon complexes.
- Explain how the orientation of a segment affects the torque requirements of the muscle–tendon complex.
- Compare and contrast the precision and power grips.
- Give examples of how biomechanical analyses aid in the understanding of upper extremity injuries.

INTRODUCTION

In this lesson, you will examine the mechanics of the upper extremities. You do a lot more with your hands than your feet, and this means that the upper extremities are more complex than the lower extremities. Most people are not subjecting their upper bodies to large contact forces, but the velocities with which the arms move can still subject the structures to high loads. In this lesson, you will examine them, starting with the shoulder and working distally to the fingers.

15.1 THE SHOULDER COMPLEX

Section Question

Mark is a water polo player. In activities such as throwing (Figure 15.1), he places great demands on his shoulder in terms of both mobility and stability. How does the shoulder accomplish these seemingly contradictory tasks? How can Mark improve his performance and reduce his risk of injury?

The shoulder complex is presented in Figure 15.2. The shoulder is not a single joint but a complex of three joints and one "articulation." The shoulder is capable of very large ranges of

Figure 15.1	How does the water player's shoulder satisfy the competing demands for mobility and stability?

© muzsy/ShutterStock, Inc.

Figure 15.2	The shoulder complex consists of the sternoclavicular joint, acromioclavicular joint, scapulothoracic articulation, and glenohumeral joint.

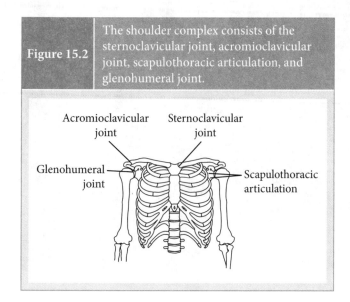

Figure 15.3	The sternoclavicular joint is a saddle joint with two degrees of freedom. Elevation and depression occurs about an anterior–posterior axis that goes through the sternal end of the clavicle; protraction and retraction occurs about a longitudinal axis that goes through the sternum.

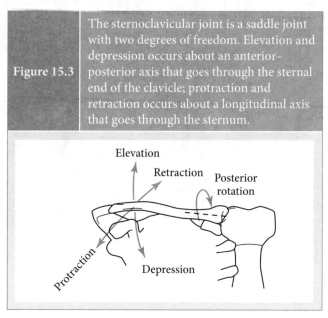

motion, which are necessary to be able to place the hand in a large variety of places in space. To achieve this mobility, the joint cannot have very strong ligamentous or bony constraints. In activities such as throwing, the shoulder is also subjected to large velocities, accelerations, and forces. These activities require the shoulder to be stable and transfer large forces from the trunk to the arm. Mobility and stability are generally on opposite ends of a continuum: A more stable joint is less mobile, and a more mobile joint is less stable. To achieve these competing demands, the three joints and the articulation along with the muscles play a key role in modulating the stability and mobility of the shoulder complex. Before examining how they work together, it is important to have an understanding of each of the components in isolation.

15.1.1   Function and Structure

As you will soon see, what you normally think of as "shoulder motion" is really a composite motion of three joints: the sternoclavicular, acromioclavicular, and glenohumeral joints. We can simplify this a bit by discussing motion of the scapulothoracic motion and glenohumeral motion because scapulothoracic motion is the combination of sternoclavicular and acromioclavicular motion.

15.1.2   The Sternoclavicular Joint, Acromioclavicular Joint, and Scapulothoracic Articulation

Proper placement of the glenoid fossa of the scapula is imperative for proper functioning of the shoulder. To correctly place the scapula on the thorax, motion must occur

at sternoclavicular and acromioclavicular joints. The sternoclavicular joint (Figure 15.3) is a saddle joint with two degrees of freedom. Elevation and depression occur about an anterior–posterior axis that goes through the sternal end of the clavicle. For elevation and depression, the clavicle is convex and the sternum is concave. Typical ranges of motion are 45° of elevation and 10° of depression. Protraction and retraction occur about a longitudinal axis that also goes through the sternum. With saddle joints, the members are convex in one direction and concave in the other, so the clavicle is concave, and the sternum is convex. Typical ranges of motion are about 30° in each direction.

The acromioclavicular joint is considered a plane, gliding joint, but it does have a single degree of freedom: upward/downward rotation. This rotation occurs about an anterior–posterior axis that goes through the distal end of the clavicle (Figure 15.4). The acromioclavicular joint has up to 30° of upward rotation. In addition, small rotations occur about a vertical axis (winging) and medial–lateral axis (tipping). Winging and tipping are thought to make microadjustments of the position of the glenoid fossa. These small motions are beyond the scope of this lesson. If you are interested, you should consult references on functional anatomy.[1,2]

Motion of the scapula in relation to the thorax is due to a combination of motions occurring at the sternoclavicular and acromioclavicular joints (Figure 15.5). I will introduce the fundamental concept of adding joint rotations together here. It is really just an extension of the relative motion that is part of linear kinematics. For example, to determine the relative

Figure 15.4 The axis of rotation for the acromioclavicular joint is an anterior–posterior axis that goes through the distal end of the clavicle.

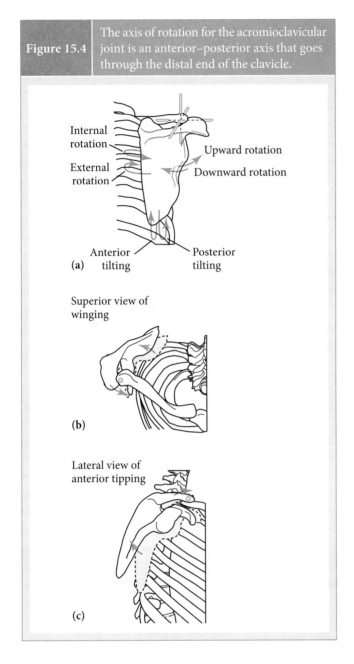

(a)

Internal rotation
External rotation
Upward rotation
Downward rotation
Anterior tilting
Posterior tilting

Superior view of winging

(b)

Lateral view of anterior tipping

(c)

velocity between two bodies, *a* and *b*, in the global reference frame, you use the equation:

$$^b v_a = {}^{global} v_a - {}^{global} v_b$$

Rearranging the equation gives you:

$$^{global} v_a = {}^{global} v_b + {}^b v_a$$

This equation is interpreted as the velocity of body *a* in the global reference frame is equal to the velocity of *a* relative to *b* plus the velocity of *b* in the global reference frame. If instead of velocity you use angle and substitute trunk, clavicle, and

scapula for global, *a*, and *b*, respectively, you get

$$^{trunk}\theta_{scapula} = {}^{trunk}\theta_{clavicle} + {}^{clavicle}\theta_{scapula} \quad (15.1)$$

Rotation of the clavicle relative to the trunk is the sternoclavicular motion, and rotation of the scapula in relation to the clavicle is the acromioclavicular motion. As you will see, raising the arm overhead (**shoulder elevation**) requires proper scapulothoracic motion.

> **Shoulder elevation**
> Raising the arm overhead in any plane

Important Point! Proper shoulder function requires scapulothoracic motion.

As with any other joint, the action of an MTC on the scapulothoracic articulation is dependent on its line of pull in relation to the axis of rotation. The scapulothoracic articulation is a little bit different than other joints because you have to consider muscles that attach to both the clavicle and the scapula. For example, the serratus anterior and rhomboids will create a protraction and retraction torque, respectively, about the sternoclavicular joint, but their attachments are on the scapula and not the clavicle. Forces are transmitted from the scapula to the clavicle to produce torque. In addition, the force vectors for upward/downward rotation can be a little tricky. In **Figure 15.6a**, you see MTCs that produce an upward rotator torque positioned on all sides of the scapula. Muscles that produce a downward rotator torque (**Figure 15.6b**) can be located right next to them, with a different orientation. MTCs acting on the scapulothoracic articulation are listed in **Tables 15.1** to **15.3**.

Table 15.1	MTCs Producing an Elevator and Depressor Torque on the Scapulothoracic Articulation
Elevator torque	**Depressor torque**
Upper trapezius	Lower trapezius
Levator scapulae	Pectoralis minor
Rhomboids	Subclavius

Table 15.2	MTCs Producing a Protractor and Retractor Torque on the Scapulothoracic Articulation
Protractor torque	**Retractor torque**
	Middle trapezius
Serratus anterior	Rhomboids
	Lower trapezius

Table 15.3	MTCs Producing an Upward and Downward Rotator Torque on the Scapulothoracic Articulation

Upward rotator torque	Downward rotator torque
Serratus anterior	Rhomboids
Upper trapezius	Pectoralis minor
Lower trapezius	

Important Point! To determine the torque-producing capability of an MTC on the scapulothoracic articulation, you must examine MTCs attaching to both the clavicle and scapula and pay attention to their line of pull.

15.1.3 The Glenohumeral Joint

Formed by the articulation of the convex humeral head and the concave glenoid fossa, the glenohumeral joint is a spherical and can rotate about any axis passing through the center of the humeral head. It is typically expressed as having three degrees of freedom: flexion/extension, ab/adduction, and internal/external rotation (Figure 15.7). Typical ranges of motion of the glenohumeral joint are 120° flexion, 60° extension, 120° abduction, 0° adduction, 70° internal rotation, and 90° external rotation.[3]

Remember that these are the ranges of motion for the humerus rotating with respect to the scapula. If you wanted to know the motion of the humerus with respect to the trunk,

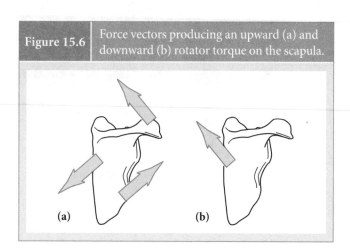

Figure 15.6	Force vectors producing an upward (a) and downward (b) rotator torque on the scapula.

(a)   (b)

you would have to add the rotation of the scapula with respect to the trunk to the rotation of the humerus with respect to the scapula:

$$^{trunk}\theta_{humerus} = {}^{trunk}\theta_{scapula} + {}^{scapula}\theta_{humerus} \qquad (15.2)$$

To show you an example of how this works, consider what is called the "typical" shoulder range of motion of 180° of abduction. Typical motion of the humerus with respect to the scapula is 120°. Typical motion of the scapula is 60° (which, incidentally, is made up of 30° from the sternoclavicular joint and 30° from the acromioclavicular joint). Therefore the total amount of abduction is

$$^{trunk}\theta_{humerus} = {}^{trunk}\theta_{scapula} + {}^{scapula}\theta_{humerus}$$
$$= 60° + 120° = 180° \qquad (15.3)$$

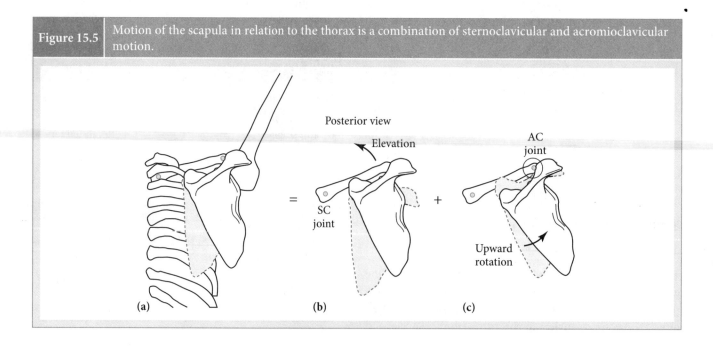

Figure 15.5	Motion of the scapula in relation to the thorax is a combination of sternoclavicular and acromioclavicular motion.

Posterior view

Elevation

AC joint

=   SC joint   +   Upward rotation

(a)   (b)   (c)

Figure 15.7	The axes of rotation for the glenohumeral joint. Flexion/extension occurs about a medial–lateral axis, ab/adduction about an anterior–posterior axis, and internal/external rotation about a longitudinal axis.

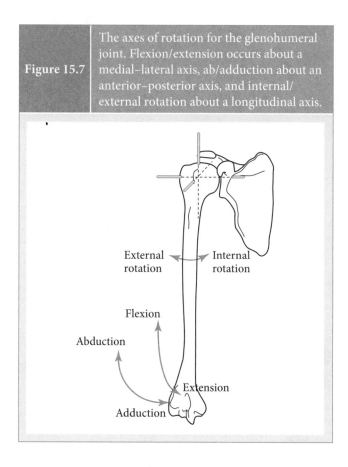

Figure 15.8	MTCs anterior to the axis of rotation will produce a flexor torque; MTCs posterior to the axis of rotation will produce an extensor torque.

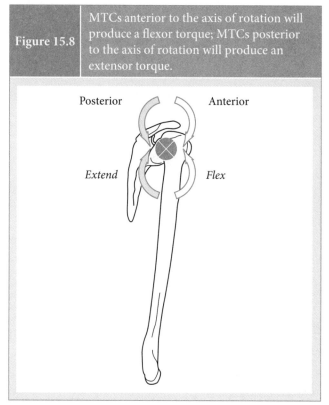

This should demonstrate to you that, although often overlooked, movement of the scapula is of vital importance for proper shoulder functioning. During shoulder flexion, the scapulothoracic motion must be protraction, elevation, and upward rotation. Similarly, during abduction, the scapulothoracic motion must be retraction, elevation, and upward rotation.

Important Point!   Proper movement of the shoulder requires motion of all three joints: glenohumeral, acromioclavicular, and sternoclavicular.

Returning specifically to the glenohumeral joint, MTCs anterior to the medial–lateral axis will produce a flexor torque, whereas MTCs posterior to it will produce an extensor torque (Figure 15.8). These muscles are listed in Table 15.4. MTCs that are lateral to an anterior–posterior axis will produce an abductor torque, and MTCs medial to this axis will produce an adductor torque (Figure 15.9). These muscles are listed in Table 15.5. As with the hip, there are several theoretical locations for MTCs producing an internal or external rotator torque (Figure 15.10). These muscles are listed in Table 15.6.

When analyzing the effects of the torque produced by a given MTC group, you first have to establish if the arm or the

Table 15.4	Muscles Producing Flexor and Extensor Torques About the Glenohumeral Joint

Anterior (flexor torque)	AOR: Through the humeral head	Posterior (extensor torque)
Anterior deltoid Pectoralis major (Clavicular portion) Coracobrachialis Biceps brachii		Latissimus dorsi Teres major Posterior deltoid

Table 15.5	Muscles Producing Abductor and Adductor Torques About the Glenohumeral Joint

Medial (adductor torque)	AOR: Through the humeral head	Lateral (abductor torque)
Latissimus dorsi Teres major Pectoralis major (Sternal portion)		Middle deltoid Anterior deltoid Supraspinatus

trunk is accelerating. For example, compare a push-up exercise versus a bench press (Figure 15.11). With the bench press, the hand is holding a load that will ultimately be accelerating.

Figure 15.9	MTCs that are lateral to an anterior–posterior axis will produce an abductor torque, and muscles medial to this axis will produce an adductor torque.

Lateral   Medial

Abduct   *Adduct*

Figure 15.10	MTCs that can theoretically produce a rotation torque about the glenohumeral joint.

ER   IR   IR   ER   ER   IR   ER   IR

Table 15.6	MTCs Producing Internal and External Rotator Torques About the Glenohumeral Joint

Internal rotator torque	**External rotator torque**
Pectoralis major	Infraspinatus
Latissimus dorsi	Teres minor
Teres major	

Figure 15.11	(a) Bench press; (b) push-up.

(a)

(b)

With the push-up, the hand is fixed on the ground, and it is the trunk that is accelerating. In the first case, the MTCs of the glenohumeral joint would affect the acceleration of the arm, whereas in the second case the MTCs of the glenohumeral joint would affect the acceleration of the trunk.

Let us examine this concept further by looking at two movements: a dumbbell lateral raise and an Iron Cross (Figure 15.12). First, analyze the dumbbell lateral raise in a static condition. Once again, you start with the equation of motion:

$$\sum T = I\alpha$$

And list the constraints:

$$\alpha = 0$$

Neglecting the mass of the arm, you will have an external torque due to the weight of the dumbbell. You should recognize by now that this torque will be determined by the weight of the dumbbell, the length of the arm, and the shoulder

| Figure 15.12 | (a) Dumbbell lateral raise; (b) the Iron Cross. |

(a) © boyan1971/ShutterStock, Inc. (b) © tankist276/ShutterStock, Inc.

abduction angle:

$$T_{\text{dumbbell}} = F_{\text{gdumbbell}} \times l_{\text{arm}} \times \sin\theta \qquad (15.4)$$

You will also have a torque produced by the MTCs. With these two torques and the identified constraint, the equation becomes

$$\sum T = T_{\text{MTC}} - T_{\text{dumbbell}} = 0 \qquad (15.5)$$

And rearranging the equation and making the appropriate substitution:

$$T_{\text{MTC}} = T_{\text{dumbbell}} = F_{\text{gdumbbell}} \times l_{\text{arm}} \times \sin\theta \qquad (15.6)$$

If dumbbell was accelerating and you wished to determine the magnitude and direction of that acceleration, after the appropriate substitutions you would have

$$\alpha_{\text{arm}} = \frac{T_{\text{MTC}} - T_{\text{dumbbell}}}{I_{\text{arm}}} \qquad (15.7)$$

where the moment of inertia is the inertia of the arm-dumbbell system taken about the shoulder joint center.

The Iron Cross is a bit more difficult to analyze because two arms are involved in the movement. To simplify things and illustrate the point here, let us perform a qualitative analysis. There are three forces acting on the body: a reaction force at each ring, plus the weight of the COM_{body}. Hopefully, you can see that if the body was in static equilibrium, then both glenohumeral joints have to produce an adductor torque because the weight of the body is creating an abduction torque. If the COM_{body} is accelerating, then you must account for both the sum of the torques and the moment of inertia of the body taken about the appropriate point.

15.1.4   Functional Classification of the Shoulder Musculature

The preceding discussion examined the MTCs of the shoulder complex in relation to their anatomical classification. It is important that you know them as such. But I find that it is helpful to understand the MTCs according to a functional classification as well. Functionally, the muscles of the shoulder complex are classified as: protectors, pivoters, positioners, and propellers.[4]

Important Point! Functionally, the muscles of the shoulder complex are classified as: protectors, pivoters, positioners, and propellers.

The glenohumeral joint is the most mobile joint in the body. This means that its motions cannot be constrained by bony architecture or strong ligamentous constraints. Indeed, the glenoid fossa is very shallow, and the glenohumeral ligament complex is fairly weak. This means that MTCs must prevent the glenohumeral joint from dislocating or subluxing. Any MTC crossing the glenohumeral joint has the potential to perform this function. However, these MTCs can also produce a torque, which may be counterproductive to the movement being performed.[5]

The glenohumeral protectors function to keep the humeral head centered on the glenoid fossa. The rotator cuff MTCs (supraspinatus, infraspinatus, teres minor, subscapularis) and the long head of the biceps brachii are ideally suited to perform this task. First, they have fairly small moment arms and will not create large torques that have to be counteracted. Second, notice that these MTCs are arranged so that they are pulling the humeral head inside the glenoid fossa from various directions (**Figure 15.13**). The rotator cuff MTCs have fairly large PCSAs and short fiber lengths.[6] This means that they can produce a relatively large amount of force, but operate over a fairly short range. In addition, large activations are not required to keep the humeral head inside the glenoid fossa,[7] but they must constantly do it. This would suggest that endurance of these MTCs is more important than their absolute strength.

> **Important Point!**   The rotator cuff and long head of the biceps brachii act as glenohumeral protectors, keeping the humeral head centered in the glenoid fossa.

Proper shoulder function requires that the glenoid fossa is positioned in the direction of the humeral movement

Figure 15.13 The MTCs of the glenohumeral protectors are arranged in such a way as to provide force vectors centering the humeral head inside the glenoid fossa.

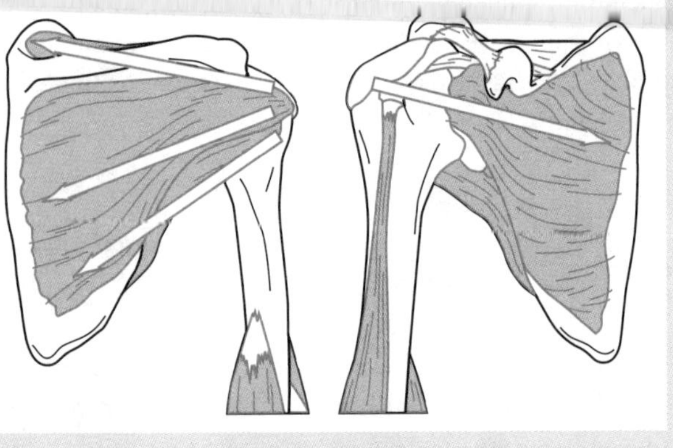

(protracted when the humerus is anterior to the body, retracted when the humerus is lateral to the body, and elevated and upwardly rotated when the humerus is elevated). This is the role of the scapular pivoters. Muscles that move the scapula (Tables 15.1 to 15.3) are in this category. If the scapula is not positioned correctly, then abnormal motion between the scapula and humerus may occur, or the contact forces between them may be off-center. Both of these things could lead to injury.

> **Important Point!**   Muscles attaching to the scapula are scapular pivoters, and their role is to position the glenoid fossa.

The deltoids are the humeral positioners. Their function is to position the humerus in relation to the thorax. It is interesting to note the position of the humerus relative to the thorax, for many common throwing activities in sport are similar (**Figure 15.14**).[8] It appears as though positioning the arm in space during throws is achieved by abducting the arm to 90° and then laterally flexing the thoracolumbar spine accordingly. This is not to say that it will happen for all activities, and there are clearly times when the arm will be elevated beyond 90°. The deltoids have a relatively large fiber length, meaning that they can operate over the large length ranges required by the mobility of the shoulder.[5]

Finally, the propellers transfer torque from the trunk to the arm. These muscles include the pectoralis major, latissimus dorsi, and teres major. These are the relatively large muscles that attach from the trunk to the humerus. Although some authors contend that they function more to transfer energy than produce it,[4] their energy generating capabilities should not be overlooked.

15.1.5   Injuries to the Shoulder Complex

Numerous injuries can occur to the shoulder complex. This is due to the competing demands placed on it during the myriad activities in sport and life. Numerous injuries can occur to the shoulder, but many of them have a common mechanism.

Ideally, the humeral head should stay roughly centered on the glenoid fossa during all activities. This becomes more important as the amplitude of movement and the forces across the joint become larger. Movement of the humeral head out of the center of the glenoid fossa, particularly in the superior or anterior directions, can lead to a number of injuries.

Figure 15.14   The angle between the humerus and thorax is similar across many different throwing activities. The position of the humerus in space is achieved by flexing the trunk.

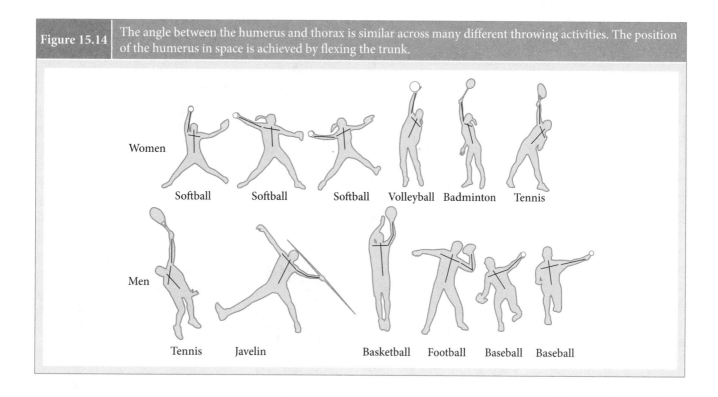

Important Point!   Many injuries occur because the humeral head moves out of the center of the glenoid fossa. This happens because of impairments of either the glenohumeral protectors and/or scapular pivoters.

If the humeral head migrates superiorly, it can hit against the coracoacromial arch that forms a roof above it. When the humeral head and coracoacromial arch make contact, they will pinch (or impinge) on the structures in between them: the supraspinatus tendon, the tendon of the biceps brachii long head, and subacromial bursa. This leads to a condition known as shoulder impingement syndrome.

Though the humeral head could also potentially move posteriorly or inferiorly, the most common direction would be anteriorly. Due to the relative weakness of the glenohumeral ligament complex, the humerus could dislocate, sublux, and potentially tear the labrum surrounding the glenoid fossa.

Although these injuries are different and distinct, the mechanism is the same: a displacement of the humeral head. Because it is the role of the protectors (rotator cuff and long head of the biceps brachii) to prevent this from happening, it is easy to think that these injuries are a result of a weak rotator cuff. As previously mentioned, due to the relatively large PCSAs, the rotator cuff muscles do not need a large amount of activation to stabilize the humeral head. Endurance of the

rotator cuff may certainly be a factor, but another intriguing factor can contribute to these injuries: inadequate movement of the scapula.

Remember that the glenoid fossa should be directed in the movement of the humerus, and that movement of the humerus in relation to the thorax is determined by movement of the scapula in relation to the thorax and movement of the humerus in relation to the scapula. If there is inadequate movement of the scapula, there may be excessive movement of the humerus to make up for the deficiency. This excessive movement can pull the humerus away from its central location within the glenoid. Because the rotator cuff has a relative short fiber length, this movement away from the central location will bring to the descending limb of the force-length curve—making them less effective in performing their role. In addition, by being lengthened outside their normal operating range, the muscles are more likely to be strained and torn.

Injuries to the shoulder provide a very good example of the interplay between two joints (in this case, the scapulothoracic articulation and the glenohumeral joint) and how a deficiency at one joint (the scapulothoracic) can manifest itself as an injury at another (the glenohumeral). These mechanopathologies can lead to such diverse injuries as impingement syndrome, joint instabilities, and muscle strains, and it points to the need to constantly guard against limiting yourself to a single joint during analysis of movement.

Section Question Answer

Mobility and stability do have competing demands; the more mobile a joint is, the less stable it is (and vice versa). The shoulder complex balances these demands by having several joints involved in its movement. The key is the scapula: Movement of the scapula accounts for one third of the total overhead arm movement whereas rotation of the scapula makes it easier for the glenohumeral protectors (rotator cuff and long head of the biceps brachii) to keep the humeral head centered on the glenoid fossa. Mark should focus on improving the strength and endurance of the glenohumeral protectors and scapular pivoters to improve his performance and reduce his risk of injury.

COMPETENCY CHECK

Remember:

1. Define shoulder elevation.
2. What joints and articulations make up the shoulder complex?
3. What bones are involved with each joint and articulation?
4. What type of joint is the glenohumeral joint? How many rotational degrees of freedom does it have? What motions are available? What are considered to be the "normal" ranges of motion?
5. Which muscle–tendon complexes can produce torque at the glenohumeral joint or scapulothoracic articulation?

Understand:

1. Describe the functional classification of the MTCs of the shoulder complex.
2. Would there be a greater torque requirement of the shoulder MTCs if they were accelerating the humerus or the trunk? Why?
3. Compare and contrast the shoulder complex to the hip joint.

Apply:

1. When strengthening the MTCs to prevent a shoulder injury, what should be the order or priority for the different functional groups?

15.2   THE ELBOW AND FOREARM

Section Question

Would it be easier to tighten or loosen a screw (Figure 15.15)? Why?

| Figure 15.15 | Why is it easier to tighten a screw than it is to loosen it? |

© djem/ShutterStock, Inc.

15.2.1   Structure and Mechanics of the Elbow and Forearm

The elbow functions as another link in the upper extremity chain. Kinematically, the addition of this link allows for the hand to reach a greater variety of positions in space. Kinetically, the elbow acts as another energy source and sink. The forearm allows you to orient your hand from a pronated (palm down) to a supinated (palm up) position and everywhere in between.

The elbow joint consists of the articulations between the humerus and radius (humeroradial joint) and the humerus and the ulna (humeroulnar joint). The proximal radioulnar joint is also within the joint capsule, which is part of the forearm. The other part of the forearm is the distal radioulnar joint. The elbow is a hinge joint with a single degree of freedom: flexion and extension. The axis of rotation goes through the humeral epicondyles (Figure 15.16). Typical ranges of motion are for 140° flexion and for 0° extension. The forearm also has a single degree of freedom: supination and pronation.

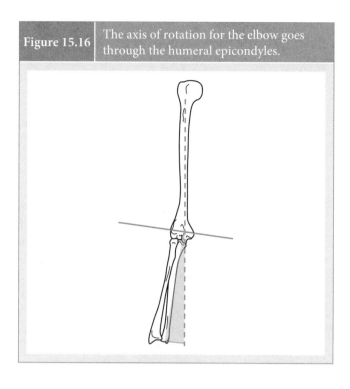

Figure 15.16 The axis of rotation for the elbow goes through the humeral epicondyles.

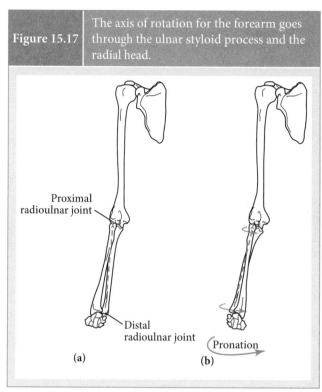

Figure 15.17 The axis of rotation for the forearm goes through the ulnar styloid process and the radial head.

Proximal radioulnar joint

Distal radioulnar joint

Pronation

(a)

(b)

During these motions, the radius rotates about ulna, which does not move during these motions. The axis of rotation of the forearm is oblique, running from the ulnar styloid process through the radial head (**Figure 15.17**). Typical ranges of motion are 80° for pronation and 80° for supination.

Important Point! The radius rotates around the ulna during pronation and supination. The ulna does not move during these motions.

Like the knee, two collateral ligaments limit valgus and varus motion while guiding flexion and extension of the elbow (**Figure 15.18**). Unlike the knee there are no cruciate ligaments inside the capsule, but the configuration of the bones (particularly the trochlear notch of the ulna) makes the elbow a fairly stable joint. The ulnar collateral ligament, also called the medial collateral ligament, limits valgus motion, whereas the radial (lateral) collateral ligament limits varus motion. As you will see, the elbow is more commonly subjected to valgus forces, which leads to higher injury rates of the ulnar collateral ligament.

For the elbow, MTCs passing anterior to the axis of rotation will produce a flexor torque, whereas MTCs with a line of pull posterior to it will produce an extensor torque (**Figure 15.19**). These muscles are listed in **Table 15.7**. Because the radius rotates during pronation and supina-

Figure 15.18 The ligaments of the elbow joint.

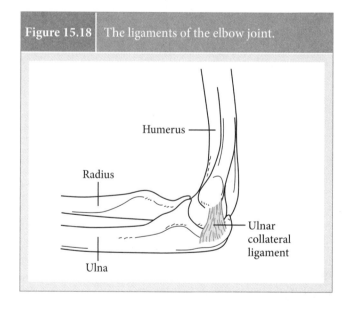

Humerus

Radius

Ulnar collateral ligament

Ulna

tion, MTCs that produce a torque about the elbow and are attached to the radius are affected by elbow position. Those MTCs that produce a torque about the elbow and are attached to the ulna are not affected by forearm position. This idea will be explored further in the next paragraph. MTCs

Table 15.7	MTCs Producing Flexor and Extensor Torques About the Elbow		
Anterior (flexor torque)		**AOR:** Through the humeral epicondyles	**Posterior (extensor torque)**
Biceps brachii Brachialis Brachioradialis			Triceps brachii Anconeus

Table 15.8	MTCs Producing Supinator and Pronator Torques About the Forearm

Supinator torque	Pronator torque
Biceps brachii Supinator Brachioradialis	Pronator teres Pronator quadratus Brachioradialis

Figure 15.19	MTCs anterior to the elbow axis of rotation produce a flexor torque; MTCs posterior to it produce an extensor torque.

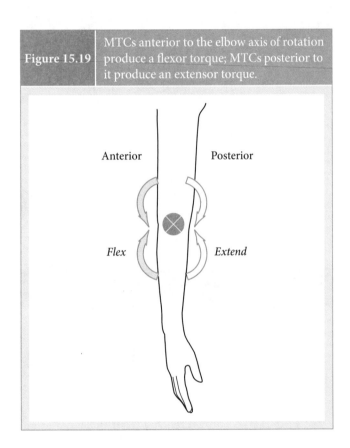

torque is unaffected by forearm position. Of the MTCs that produce a flexor torque about the elbow, the biceps brachii and brachioradialis are attached to the radius. The brachialis is not and therefore is unaffected by forearm position. The biceps brachii produces both a flexor and a supinator torque. If the forearm is not fully supinated, it cannot produce its maximum torque about the elbow. The brachioradialis produces a flexor torque and can produce a pronator torque or supinator torque, depending on forearm position. Due to its line of pull, the brachioradialis always wants to return the forearm to the neutral (thumbs-up) position. Therefore, it cannot produce its maximum torque about the elbow unless the forearm is in the neutral position.

The role of the elbow flexors has been acknowledged by bodybuilders for a long time. If you want to train the biceps as an elbow flexor, you should perform curls with a supinated grip. To emphasize the brachioradialis, you would do hammer curls with a neutral grip. And while the brachialis is unaffected by forearm position, the amount of torque generated by the biceps and brachioradialis is decreased with a pronated grip. Thus, more emphasis is placed on the brachialis with reverse curls.

15.2.2   Injuries to the Elbow Joint

Due to its congruency, the elbow is a fairly stable joint. And certain "elbow" injuries, such as tennis elbow, are really due to forces at the wrist (which will be covered in the next section). Here, you will look at a mechanism that tends to get the elbow in trouble: valgus loading.

Recall that the elbow joint has a single rotational degree of freedom: flexion/extension. It does not have an abduction/adduction degree of freedom. That does not mean that there is not a valgus/varus load on the elbow. In fact, it is quite the contrary. This is best understood if you abduct your shoulder to 90° and then externally rotate it to 90° (Figure 15.20). This is sometimes referred to as the 90/90 position, for obvious reasons. Now imagine that you are holding a mass, such as a ball, in your hand.

that produce a supinator and pronator torque about the elbow are listed in Table 15.8. More torque can be produced in supination compared to pronation. Screws and bolts are threaded to take advantage of this fact (unless you happen to be left-handed, and then you are at a disadvantage).

Important Point!   MTCs that produce a torque about the elbow and are attached to the radius are affected by forearm position; those attached to the ulna are not.

All the MTCs producing an extensor torque about the elbow are attached to the ulna. This means that elbow extensor

Figure 15.20 | The 90/90 position of the shoulder.

© Ahturner/ShutterStock, Inc.

Figure 15.22 | Changing the elbow angle changes the moment arm of the ball. If the elbow angle is zero, then there is no valgus torque on the elbow.

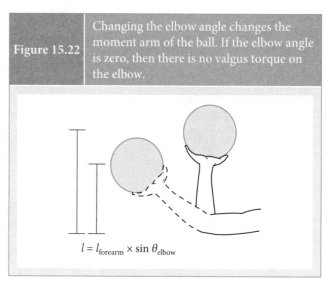

$$l = l_{forearm} \times \sin \theta_{elbow}$$

Consider what would happen if you were to externally rotate your shoulder so that the forearm was slightly passed the vertical. Drawing a free-body diagram of the forearm in this position, as was done in **Figure 15.21**, is helpful in understanding the loads placed on it. For this analysis, ignore the mass of the forearm and just consider the mass of the ball, F_{gball}. By now, you should be able to calculate that the torque about the elbow:

$$T_{ball} = F_{gball} \times {}^{y}l_{forearm} \times \sin \theta \qquad (15.8)$$

where θ is the angle between the forearm and the vertical. The torque created by the ball is a valgus torque. It should be apparent that a ball with a larger mass, a longer forearm, or a larger angle from the vertical will all contribute to a larger valgus torque at the elbow.

If the elbow is flexed to 90°, then the entire length of forearm should be used in the equation. If the elbow if flexed more or less than 90°, then the "length" of the forearm must be corrected for this angle. Can you explain why?

In this situation, you have to think in three dimensions. You wish to know the distance from the elbow joint center to the center of mass of the ball in the vertical (y) direction (**Figure 15.22**). To determine this, you should apply the formula:

$$ {}^{y}l_{forearm} = l_{forearm} \times \sin \theta_{elbow} \qquad (15.9)$$

Knowing that the sine of 90° is 1, you should be able to realize that if the elbow is flexed to 90°, then the entire length of the forearm contributes to the moment arm of the ball. At all other angles the length will be less. At 0° of elbow flexion, the length of the forearm in the y direction is zero, and thus the valgus torque on the elbow created by the ball is also zero.

Figure 15.21 | A free-body diagram of the forearm. Valgus torque is determined by the angle from the vertical, the length of the forearm, and the weight of the ball.

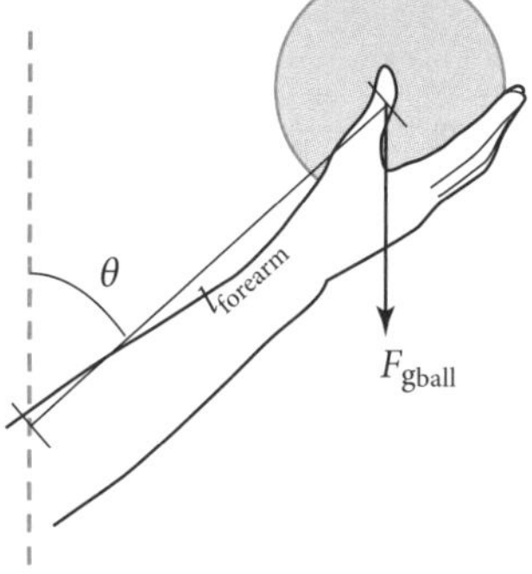

θ

$l_{forearm}$

F_{gball}

Section Question Answer

Most screws are threaded such that, when you are facing the screw, it moves toward you (loosens) when it is turned counterclockwise and moves away from you (tightens) when it moves clockwise. This is known as a right-handed thread because it follows the right-hand rule. But the design also gives a person who is naturally right-handed an advantage. Because you can produce more torque in supination than you can in pronation, it will be easier to tighten the screw (if you are right-handed) than it will be to loosen it. Unless, of course, you are left-handed: then it would be easier to loosen the screw than it would be to tighten it.

COMPETENCY CHECK

Remember:

1. What type of joints are the elbow and forearm? How many rotational degrees of freedom do they have? What motions are available? What are considered to be the "normal" ranges of motion?

Understand:

1. Describe how forearm position affects the ability to produce torque at the elbow.
2. Compare and contrast the knee and elbow joints.

Apply:

1. Use the concepts of elbow biomechanics to show how you can decrease injury risk or prescribe exercise.

15.3   THE WRIST AND HAND

> **Section Question**
>
> Mary is a collegiate athlete who has been recently diagnosed with tennis elbow (**Figure 15.23**). Why is this really an injury of the wrist? What recommendations can you make to Mary to help her?

With the wrist and hand, the full purpose of the upper extremity can be realized. Although many sporting activities involve throwing, not everyone participates in those sports.

| **Figure 15.23** | Why is tennis elbow an injury of the wrist? |

© Mai Techaphan/ShutterStock, Inc.

On the other hand (pun intended!), everyone uses their hands in almost every activity of daily living. Seen in this way, the shoulder and elbow joints can be thought of as getting the hand "pretty close" to where it needs to be to grab or manipulate objects. The forearm and wrist joints orient the position of the palm. The wrist also makes fine-tuned adjustments for the position of the hand. Finally, the wrist and hand manipulate objects—either by taking hold of them (prehensile movements) or pushing/pulling on them without holding onto them (nonprehensile movements).[9]

15.3.1   The Wrist

The wrist joint is comprised of the articulation of the concave radius and ulna with the convex proximal carpal row (scaphoid, lunate, triquetrum). It is a condyloid joint with two degrees of freedom: flexion/extension and radial/ulnar deviation. For both sets of movements, the axis of rotation passes through the capitate (**Figure 15.24**). For flexion/extension this is a radial–ulnar axis, and for radial/ulnar deviation it is a palmar–dorsal axis. Typical ranges of motion are 80° for flexion, 70° for extension, 20° for radial deviation, and 30° for ulnar deviation.[3]

MTCs that are palmar to the flexion/extension axis will produce a flexor torque, whereas MTCs dorsal to this axis will produce an extensor torque (**Figure 15.25**). These muscles are listed in **Table 15.9**. MTCs that are radial to the radial/ulnar deviation axis will produce a radial deviation torque, whereas

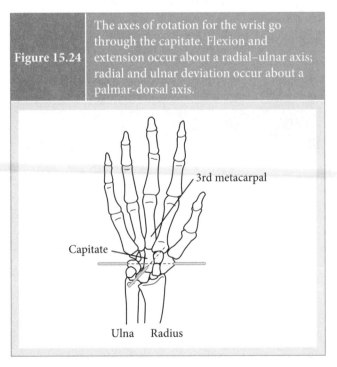

| **Figure 15.24** | The axes of rotation for the wrist go through the capitate. Flexion and extension occur about a radial–ulnar axis; radial and ulnar deviation occur about a palmar-dorsal axis. |

3rd metacarpal

Capitate

Ulna   Radius

Figure 15.25	MTCs that are palmar to the flexion/extension axis will produce a flexor torque, whereas MTCs dorsal to this axis will produce an extensor torque.

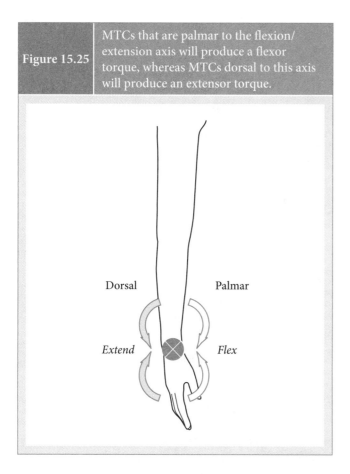

Figure 15.26	MTCs that are radial to the radial/ulnar deviation axis will produce a radial deviation torque, whereas MTCs with a line of pull that is ulnar to it will produce an ulnar deviation torque.

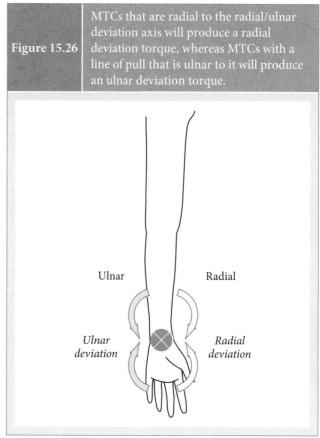

MTCs with a line of pull that is ulnar to it will produce an ulnar deviation torque (Figure 15.26). These muscles are listed in Table 15.10. As you have seen several times already with MTCs that have actions in two planes, functional quadrants can be created for the wrist (Figure 15.27).

To see the intricate relationship between internal and external moments at the wrist, examine the relatively simple act of tapping a nail into a wall with a hammer. For the

moment, assume that only the wrist is involved in the action. Also assume that the hand is massless and that the moment of inertia for the hammer is taken about the wrist joint. The torque of the wrist MTCs must be equal to the inertial torque minus the torque due to gravity:

$$\sum T = I_{hammer}\alpha_{hammer} \qquad (15.10)$$

$$T_{MTC} + T_{external} = I_{hammer}\alpha_{hammer} \qquad (15.11)$$

$$T_{MTC} = T_{inertial} - T_{external} \qquad (15.12)$$

Table 15.9	Muscles Producing Flexor and Extensor Torques About the Wrist	
Anterior (flexor torque)	**AOR: Through the capitate**	**Posterior (extensor torque)**
Flexor carpi radialis		Extensor carpi radialis longus
Flexor carpi ulnaris		Extensor carpi radialis brevis
Flexor digitorum superificialis		Extensor carpi ulnaris
Flexor digitorum profundus		Extensor digitorum
Palmaris longus		

Table 15.10	Muscles Producing Radial and Ulnar Deviation Torques About the Wrist	
Radial (radial deviation torque)	**AOR: Through the capitate**	**Ulnar (ulnar deviation torque)**
Flexor carpi radialis		Flexor carpi ulnaris
Extensor carpi radialis longus		Extensor carpi ulnaris
Extensor carpi radialis brevis		

Figure 15.27 | Functional quadrants for the wrist MTCs.

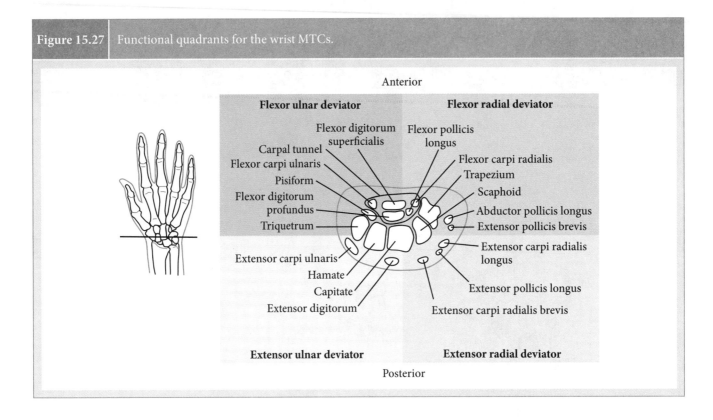

Anterior

Flexor ulnar deviator   **Flexor radial deviator**

Flexor digitorum   Flexor pollicis
superficialis   longus
Carpal tunnel
Flexor carpi ulnaris   Flexor carpi radialis
Pisiform   Trapezium
Flexor digitorum   Scaphoid
profundus   Abductor pollicis longus
Triquetrum   Extensor pollicis brevis
                                              Extensor carpi radialis
Extensor carpi ulnaris   longus
Hamate
Capitate   Extensor pollicis longus
Extensor digitorum   Extensor carpi radialis brevis

Extensor ulnar deviator   **Extensor radial deviator**

Posterior

In **Figure 15.28**, the rotation of the hammer toward the wall is in the positive direction, and rotation away from the wall is in the negative direction. From position *A* to position *B*, the only external torque is the weight of the hammer, and it is in the negative direction. This means that the MTCs must produce enough positive (ulnar deviation) torque to overcome the external force and produce the desired acceleration. At the instant the hammer is at position *B*, the torque produced by the weight of the hammer is zero, and the MTC torque is equal to the inertial torque. From position *B* to the instant before impact at point *C*, the weight of the hammer is producing an external torque, but this time in the positive direction. It is assisting with the acceleration of the hammer.

At impact, position *C*, there is a contact force between the hammer and the nail that produces a torque in the negative direction. This requires a large positive torque, or the hammer that would radially deviate the wrist. To prevent this from happening, the MTCs must produce an equal and opposite positive force (ulnar deviation torque).

From position *A* to position *C*, the hammer had a positive angular velocity from start to impact, after which it immediately went to zero. On the return, from position *C* back to position *A*, the hammer will initially speed up in the negative direction (a negative acceleration) produced by a radial torque

Figure 15.28 | The action of a hammer striking a nail on a wall. (*A*) The start position; (*B*) when the hammer is completely vertical; and (*C*) when the hammer strikes the nail.

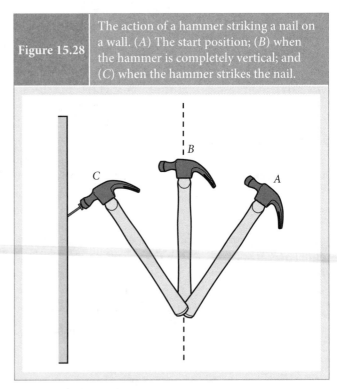

of the MTCs. Incidentally, from position *C* to position *B*, the weight of the hammer is producing a positive torque, which must be overcome as well. From position *B* to position *A*, the weight of the hammer is once again producing a negative torque. However, at some point, the hammer's angular velocity must be decreasing so that the hammer comes to a zero angular velocity at point *A*. This requires a positive torque from the MTC, which is produced by the MTCs producing an ulnar deviation torque acting eccentrically.

Throughout this example, the MTCs were referred to as a single muscle group, either those producing an ulnar deviation torque or those producing a radial deviation torque. Examining Figure 15.27, you will note that there is no such muscle group: MTCs that produce either a radial or ulnar deviation torque also produce a flexor or extensor torque. To keep the hammer on course in a pure plane, the internal flexor and extensor torques must cancel to produce an effective torque of zero in those directions.

Although this is a relatively simple example, and you can probably do it with ease after some practice, think about all the torques that must be accounted for. First, there is the torque produced by gravity, which is nonlinear, decreasing in the negative direction from point *A* to point *B*, and increasing in the positive direction from point *B* to point *C*. Second, there is the torque produced by impact with the nail. Third, there is the inertial torque that must increase and then decrease, particularly from point *C* to point *A*. Finally, there are the internal torques in directions other than the direction of movement that must be counterbalanced. An impressive amount of torques for such a simple task!

15.3.2 The Thumb and Fingers

The four fingers have a condyloid joint where the proximal phalanges articulate with the metacarpals (the metacarpal–phalangeal joints) and hinge joints where the middle phalanges articulate with the proximal (proximal interphalangeal joints) and distal (distal interphalangeal joints) phalanges. The metacarpal phalangeal joints have two degrees of freedom, flexion/extension and ab/adduction, with the axis of rotation for each degree of freedom passing through the metacarpal head. Typical ranges of motion are 90° for flexion, 20° for extension, 25° for abduction, and 0° for adduction.[3] The interphalangeal joints are hinge joints with a single degree of freedom (flexion–extension) passing through the head of the more proximal phalange. Typical ranges of motion 100° for flexion and 10° for extension.[3] MTCs that have a line of pull that are palmar to the axes of rotation will flex the metacarpal–phalangeal joints and interphalangeal joints, whereas MTCs that have a line of pull that are dorsal to the axes of rotation will extend the metacarpal–phalangeal joints and interphalangeal joints.

The MTCs with a line of pull lateral to the carpometacarpal joints will ab/adduct them.

The thumb is different from the other fingers in a couple of respects. First, there are only two phalanges, and so only one interphalangeal joint. The interphalangeal joint is similar to the other interphalangeal joints. The metacarpal phalangeal joint of the thumb is a hinge joint with a single degree of freedom: flexion and extension. Whereas the other carpo-metarcarpal joints are plane joints with no rotational degrees of freedom, the first carpo-metacarpal joint is a saddle joint with two degrees of freedom: flexion/extension and ab/adduction. The axis of rotation for flexion/extension goes through the trapezium, and the axis of rotation for ab/adduction goes through the base of the first metacarpal. Typical ranges of motion for the first carpo-metacarpal joint are 15° for flexion, 75° for extension, and 70° for abduction.

Because the thumb is rotated approximately 90° from the rest of the fingers, thumb motion is also described differently than the rest of the fingers. Flexion and extension occur in the plane of the palm, whereas ab/adduction occurs in a plane perpendicular to it. The "palmar" side of the thumb is where the pad is, and the "dorsal" side is where the nail is. Any MTC with a line of pull palmar to the axis of rotation going through trapezium will flex the first carpo-metacarpal joint, and any MTC with a line of pull dorsal to it will extend. The "ulnar" side of the thumb is closest to the index finger. Any MTC with a line of pull on the ulnar side of the first metacarpal will abduct the first carpo-metacarpal joint, whereas any MTC with a line of pull on the radial side will adduct it.

Prehension: Gripping, Pinching, and Grasping

Prehension is the act of grasping or seizing something. In many movements, you can say that this is the

> **Prehension** The act of grasping or seizing something

ultimate purpose of the entire extremity: to be able to reach out and grab something and then do something with it once you have it, whether it is feeding yourself, combing your hair, or driving a car. It does not much matter if the hand can reach the object if you cannot manipulate that object once the hand reaches it.

The basic mechanics of holding an object include the need to balance the forces and torques. In this section, you will examine how the fingers and thumb generate those forces. There are many ways to hold onto an object, based on both the amount of finger flexion and the area in contact with the palm. This discussion will be limited to the precision grip, or pinch

(Figure 15.29a), and power grip, or grasp (Figure 15.29b).[9] The differences between the two depend on:[10]

- Area of contact within the hand.
- Number of fingers involved in the activity.
- Amount of finger flexion.
- Position of the thumb.
- Position of the wrist.

The precision grip, or pinch, is used for precise manipulation of small, relatively light objects. It involves contact between the object and the distal phalanges of the thumb and one or more fingers (although it typically involves just the index finger or index middle fingers) on opposite sides of the object. The fingers remain relatively fixed, and the thumb pushes against them. Precision of movement occurs because the thumb can apply a force/moment on the object being held.

The power grip, or grasp, is used for manipulation of heavier objects, and its purpose is to resist any system of forces acting on the object being held.[9] It involves contact between the object, phalanges, and palm. The fingers are more flexed, and the thumb reinforces the fingers, acting as a buttress in the direction opposite the palm.

Part of the differences in the force produced between the two grips lies with the role of the thumb. With the precision grip, the thumb is producing the force that grips the object. But another part of the differences in force produced lies with the moment arms of the reaction forces produced by the object being gripped.[10] Figure 15.30 compares the moment arms of the reaction force about the metacarpal-phalangeal joint with the power grip (Figure 15.30a) and the precision grip (Figure 15.30b). You will notice that the moment arm for object with the power grip ($^{\perp}d_1$) is fairly small, meaning that it will not produce a large torque about the metacarpal-phalangeal joint. In contrast, the moment arm for the object with the precision grip ($^{\perp}d_2$) is comparatively quite large.

15.3.3　Injuries to the Wrist

When discussing injuries to the elbow, I mentioned that tennis elbow was really an injury of the wrist. Now it is time to explain why that is. Tennis elbow (the medical term is lateral epicondlyasia) is a painful condition that involves the lateral

| Figure 15.29 | (a) The precision grip; (b) the power grip. |

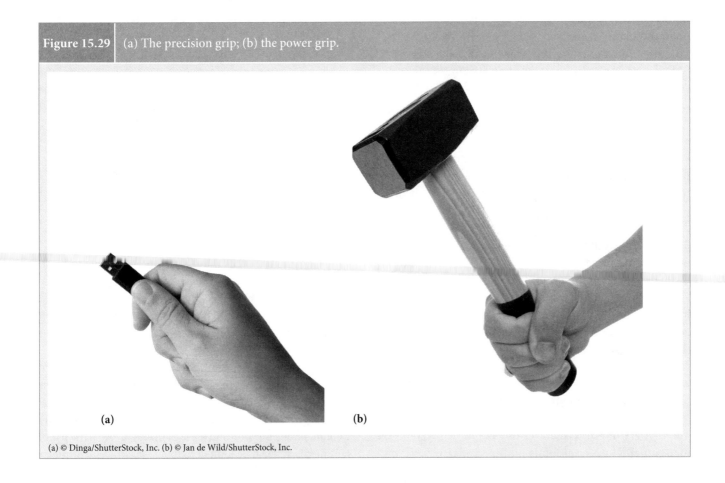

(a)　　　　　　　　　　　　　　　　　　(b)

(a) © Dinga/ShutterStock, Inc. (b) © Jan de Wild/ShutterStock, Inc.

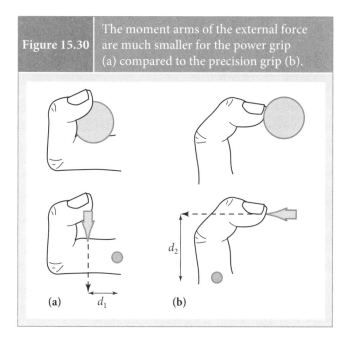

Figure 15.30 The moment arms of the external force are much smaller for the power grip (a) compared to the precision grip (b).

(a) d_1

(b) d_2

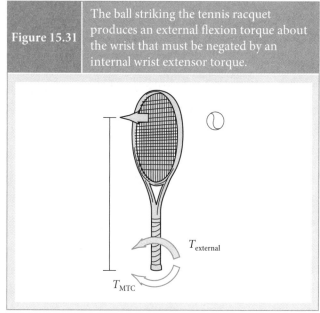

Figure 15.31 The ball striking the tennis racquet produces an external flexion torque about the wrist that must be negated by an internal wrist extensor torque.

$T_{external}$

T_{MTC}

epicondyle of the humerus. The muscles that attach to the lateral epicondyle include the extensor carpi radialis longus and brevis, and extensor carpi ulnaris. These MTCs produce negligible torque at the elbow because of extremely small moment arms (the axis of rotation for flexion/extension of the elbow is through the humeral epicondyles). Although they can play an important role in stabilizing the elbow, it is their function at the wrist that causes the problem. It is called "tennis" elbow for a reason, and it happens because of the backhand stroke.

Let us take a top-down approach to looking at a tennis backhand. A ball with a certain mass (m_{ball}) has a fairly large velocity (v_{ball}, estimated to be about 60–70 meters per second). The ball impacts with the racquet. This can be modeled as a collision, with the racquet having a certain velocity ($v_{racquet}$) and an effective mass of the racquet-arm system. Neither the ball nor the strings of the racquet are perfectly rigid, and both will deform on impact. The goal of the backhand stroke is to impart a large momentum to the ball, changing its course and returning the ball to the opponent with a high velocity. This requires that the racquet have a high velocity and a large effective mass.

There will be equal and opposite forces acting on the ball and the racquet. During the backhand, the force imparted on the racquet by the ball will create a flexion torque at the wrist. To effectively transfer momentum to the ball, you obviously do not want the wrist to flex on impact. Therefore, you need the MTCs to create an equal and opposite wrist extensor torque (**Figure 15.31**).

The relatively high velocity of the ball means that it has a relatively high momentum, which in turn means a relatively high impact force on the racquet. Coupled with the relatively high moment arm produced by the racquet (Figure 15.28), the ball produces a fairly high torque about the wrist. The moment arm of the wrist extensors is much smaller. Being at a mechanical disadvantage, the forces are that much higher. The repeated exposure to high forces with high loading rates is what leads to tennis elbow.

Section Question Answer

There are several ways to decrease the risk of tennis elbow. First, improving the strength and endurance of the wrist extensors would mean that they are working at a smaller percentage of their capacity, and there would be a larger margin of safety. Second, using a shorter racquet would decrease the moment arm of the ball and decrease the demand on the wrist extensors. Finally, decreasing the tension on the strings would decrease the force transmitted to the racquet. Of course, this also means that performance would decline, as the velocity of the ball coming off the racquet would be less.

COMPETENCY CHECK

Remember:

1. Define prehension.
2. What type of joint is the wrist? How many rotational degrees of freedom does it have? What motions are available? What are considered to be the "normal" ranges of motion?

3. What type of joints are the metacarpalphlangeal and interphalangeal joints? How many rotational degrees of freedom do they have? What motions are available? What are considered to be the "normal" ranges of motion?

Understand:

1. Compare and contrast the ankle to the wrist.
2. Compare and contrast a typical finger to the thumb.
3. Compare and contrast the precision and power grips.

Apply:

1. When would you use the precision grip over the power grip and vice versa?
2. Why is tennis elbow an injury of the wrist instead of the elbow? What can you do to prevent it?

SUMMARY

With this lesson, you learned about the mechanics of most of the individual joints of the upper extremities. With the exception of some single joint exercises, rarely does one joint act in isolation. With certain joint complexes, you began to appreciate how important it is for joints to work together. Joints work together to produce many of the movements associated with activities of daily living and sport.

REVIEW QUESTIONS

1. What are the major joints of the upper extremities?
2. For each joint: what type of joint is it? How many rotational degrees of freedom does it have? What motions are available at that joint, and what are the "normal" ranges of motion?
3. Describe the torques that can be produced by the muscles crossing each joint.
4. How does the orientation of a segment affect the torque requirements of the muscle–tendon complex?
5. Compare and contrast the precision and power grips. Describe situations where you would use each.
6. Give examples of how biomechanical analyses aid in the understanding of upper extremity injuries.

REFERENCES

1. Levangie PK, Norkin CC. *Joint Structure and Function: A Comprehensive Analysis.* 3rd ed. Philadelphia: F. A. Davis; 2001.
2. Neumann DA. *Kinesiology of the Musculoskeletal System.* 2nd ed. St. Louis: Mosby Elsevier; 2010.
3. Flanagan SP. Mobility. In: Miller T, ed. *NSCA's Guide to Tests and Assessments.* Champaign, IL: Human Kinetics; 2012:275–294.
4. Jobe FW, Pink M. Classification and treatment of shoulder dysfunction in the overhead athlete. *Journal of Orthopaedic & Sports Physical Therapy.* Aug 1993;18(2):427–432.
5. Veeger HEJ, van der Helm FCT. Shoulder function: The perfect compromise between mobility and stability. *Journal of Biomechanics.* 2007;40(10):2119–2129.
6. Ward SR, Hentzen ER, Smallwood LH, et al. Rotator cuff muscle architecture—Implications for glenohumeral stability. *Clinical Orthopaedics and Related Research.* Jul 2006(448):157–163.
7. McQuade KJ, Murthi AM. Anterior glenohumeral force/translation behavior with and without rotator cuff contraction during clinical stability testing. *Clinical Biomechanics.* Jan 2004;19(1):10–15.
8. Fleisig GS, Barrentine SW, Escamilla RF, Andrews JR. Biomechanics of overhand throwing with implications for injuries. *Sports Medicine.* Jun 1996;21(6):421–437.
9. Napier JR. The prehensile movements of the human hand. *Journal of Bone and Joint Surgery-British Volume.* 1956;38(4):902–913.
10. Oatis CA. *Kinesiology: The Mechanics and Pathomechanics of Human Movement.* Philadelphia: Lippincott, Williams, and Wilkins; 2004.

PART IV

Limb Level

© Stefan Scharr/Shutterstock, Inc.

Lesson 16 Multijoint Concepts

Lesson 16

Multijoint Concepts

LEARNING OBJECTIVES

After finishing this lesson, you should be able to:

- Define the following terms: angle-angle diagram, interaction torque, chain configuration, compensatory motion, coordination, coupled motion, motion path, motor redundancy, movement pattern, and reach area.
- Determine the position of the end effector of a multijoint chain.
- Identify the reach area when there are restrictions to the motions of the joints in the chain.
- Describe how motor redundancy and compensatory motion can be both helpful and harmful in completing tasks.
- Describe the basic movement patterns and the benefits and limitations of each.
- Interpret angle-angle diagrams.
- Calculate and interpret coupling angles.
- Explain how interaction torques occur.
- List the different ways energy can be transferred in a multijoint system.
- Explain the roles of biarticular muscles.
- Relate joint torques to endpoint forces.
- Explain why fast movements use proximal-to-distal sequencing.

INTRODUCTION

Joints do not work in isolation; they are usually part of a larger system. In this lesson, you will learn the basic concepts of how the various segments and joints interact. You will then put it all together. You may have heard multijoint movement referred to as a kinematic or kinetic chain. Unfortunately, this terminology is a bit ambiguous and will not be used here. But because the terms are so pervasive in kinesiology, you should be familiar with them. See **Box 16.1**.

To elucidate the concepts of a multijoint system, you will be looking at the same chain throughout this lesson. To simplify the math, the chain will consist of two links, each with a length of one. Also, the analyses will be limited to two-dimensional, planar movement. Many of the ideas presented in this lesson are based on more advanced treatments of the subjects,[1-3] but are simplified here so that you get the basic ideas.

Many of the examples in this lesson will use a standard, two-link chain (**Figure 16.1**). Two links, or segments, form a joint. For this lesson, each joint will be a hinge joint with one rotational degree of freedom. The end of one segment, the beginning of the next segment, and the joint between them are all located at the same position in space. Segmental angles are absolute angles relative to the positive x-axis in the global reference frame. Joint angles are relative angles between two segments. Two or more joints create a chain. Joints and segments will be numbered from the most proximal to the most distal, starting with 1. To conduct an analysis, you must start with a segment angle. This segment angle will be referred to as joint "zero," with segment "zero" being the negative y-axis in the global frame. At the end of the chain is the end effector. Many times, the end effector is the hand or foot that interacts with the environment. However, the end effector can also be the body's center of mass. This lesson will start with a discussion of the kinematics before moving on to the kinetics of multijoint systems.

Box 16.1	Multijoint Terminology

Discussing multijoint concepts can be confusing because there is no universal, consistent use of some terms. A link, or segment, is a single element of a chain. These two terms are often used interchangeably (as is done in this lesson), and there appears to be no harm in doing so. Two links connected by a joint is called a kinematic pair, and two or more kinematic pairs create a kinematic chain. Notice that the descriptor of the chain is kinematic because you are describing the motion between links in a chain.

In engineering and robotics, a distinction is made between a chain that is open and one that is closed. With an open kinematic chain, one end of the chain is free to move (i.e., it is not attached to anything). Both ends are fixed with a closed kinematic chain (see the following figure). It is important to realize that with an open kinematic chain, each joint can move independent of the other joints. This is not the case with a closed kinematic chain: Movement of one joint will cause the other joints to move in a predictable manner. This should be evident for the closed kinematic chain in the figure.

Applying these terms to human movement is precarious because both ends of the chain are never fixed. However, some authors will state the chain is open when the distal end (hand or foot) is moving and is closed when it is fixed. For example, a squat would be a closed chain activity, whereas a knee extension exercise would be an open chain. Problems arise when you compare a push-up to a bench press, or a pull-up to a pull-down, or even a squat to a leg press. For each of these pair of activities, the distal end is fixed for one, but it is moving for the other. Yet the mechanics of each exercise within the pair are very similar.

A better way might be to characterize exercises by the consequences of the definitions rather than the actual descriptors. For example, an open kinematic chain movement occurs when each joint is free to move independent of the others, whereas a closed kinematic chain occurs when movement at one joint creates movement at another (movements are coupled). Under these revised definitions, both activities in each pair would be considered closed kinematic chains. Although this may appear to resolve the conflict, in some movements joint motion may be coupled even if there are not any physical constraints causing them to be so. It may be the most effective way to perform the activity (examples will be provided in this lesson), so constraints are caused by the task.

Things can be even more confusing when you see the terms *open* and *closed kinetic chains*. As you know, *kinematics* refers to the description of motion without concern for the forces that are causing the motion, whereas *kinetics* refers to the causes of motion (forces and torques). In engineering and robotics literature, the term *kinetic chain* is not used. In human movement literature, many authors will use the term *kinetic chain*, often interchangeably with *kinematic chain*. *Kinematic chain* is a more apt descriptor because the distinction between open and closed is being made on the motion, not the forces causing the motion. That is not to say that forces are not transmitted through the chain (they are). And although many roboticists try to minimize these forces, humans capitalize on them. But even if you accept the revised definitions proposed, these forces exist in both open and closed kinematic chains.

Because of the ambiguity that exists with the terms *open* and *closed kinematic* and *kinetic chains*, these terms will not be used in this text. I feel that they do not add much to our understanding of human movement. In contrast, the dynamic patterns of flexion/extension and swing/whip described later will provide a more useful classification of human movement.

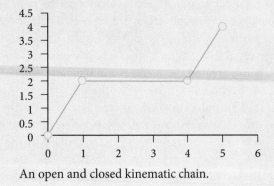
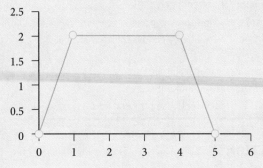

An open and closed kinematic chain.

Figure 16.1	A serial, two-link chain.

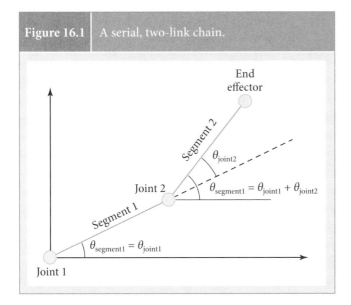

16.1 KINEMATICS

Section Question

You are teaching an elementary-grade-school class some fundamental sports skills. In one lesson, you are going to teach them to throw a ball overhead. In the next lesson, you will have them pitch a softball underarm. In the final unit, you will have them throw darts. How are these skills similar, and how are they different (Figure 16.2)?

16.1.1 Chain Configuration and Reach Area

The **chain configuration** refers to the orientation of each segment relative to each other and the global reference frame. Usually, the chain is configured in a certain way so that the end effector is

> **Chain configuration** The orientation of each link relative to each other and the global reference frame

Figure 16.2	What do these three throwing motions have in common?

(left) © JJ pixs/ShutterStock, Inc., (middle) © Aspen photo/ShutterStock, Inc., (right) © Alexander Chelmodeev/ShutterStock, Inc.

located at a particular point in global space. Say you wanted to reach out and grab a cup or kick a ball. The location of the cup or ball is where your hand or foot needs to be. The chain (in these cases, the arm or leg) is configured in such a way as to get the hand or foot there.

Before beginning your analysis, you must pick a frame of reference. Logical places for the origin could be either at the beginning of the chain or at the location of an object where you wish the end effector to be. For the analyses that follow, unless otherwise stated, the origin will be at the beginning of the chain, which will be at the center of Joint 1. Also, for these analyses, Segment 1 will start with an orientation of $-90°$ in global space. This may seem like it is making the examples more complicated (and it is), but it is also making them more realistic because the femur or humerus is normally oriented $-90°$ in global space while standing.

To determine the position of the end of Segment 1 (which is also the location of Joint 2 and the beginning of Segment 2) in global space, you need to know both the length of Segment 1 and the segment angle. (You would also need to know the location of Joint 1 if it was not at the origin of the global reference frame.) The length of Segment 1 is fixed, so the only variable is the angle of Segment 1. Recall that the segment angle is found by knowing the relation between the joint and global axes and the joint angle:

$$^{global}\theta_{segment1} = {}^{global}\theta_{reference} + {}^{reference}\theta_{segment1} \quad (16.1)$$

If the orientation of Segment 1 was coincident with the positive x-axis, then the segment and joint angles would be the same. Because they are not, you must adjust for it. For example, while standing with the humerus in a resting position (Figure 16.3a), the humerus reference system is rotated -90 from the global reference system. Thus:

$$^{global}\theta_{humerus} = {}^{global}\theta_{thorax} + {}^{thorax}\theta_{humerus}$$
$$= -90° + 0° = -90° \quad (16.2)$$

If the shoulder was flexed to $90°$ (Figure 16.3b), then the segment angle would become

$$^{global}\theta_{humerus} = {}^{global}\theta_{joint} + {}^{joint}\theta_{segment}$$
$$= -90° + 90° = 0° \quad (16.3)$$

To determine the position of Joint 2 in global space, you would then use the following formulae:

$$^{x}p_{J2} = l_{segment1} \times \cos(\theta_{segment1}) \quad (16.4)$$
$$^{y}p_{J2} = l_{segment1} \times \sin(\theta_{segment1}) \quad (16.5)$$

To determine the end of Segment 2 (which is also the location of either Joint 3 and the beginning of Segment 3, or the end effector), you need to know the location of Joint 2, the length of Segment 2, and the orientation angle of Segment 2. You previously calculated the location of Joint 2, and the length of Segment 2 is fixed. So you need to determine the angle of Segment 2, which is

$$^{global}\theta_{segment2} = {}^{global}\theta_{segment1} + {}^{segment1}\theta_{segment2} \quad (16.6)$$

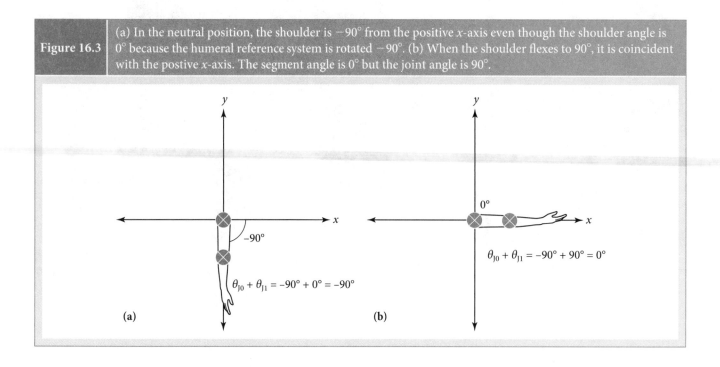

Figure 16.3 (a) In the neutral position, the shoulder is $-90°$ from the positive x-axis even though the shoulder angle is $0°$ because the humeral reference system is rotated $-90°$. (b) When the shoulder flexes to $90°$, it is coincident with the positive x-axis. The segment angle is $0°$ but the joint angle is $90°$.

$$\theta_{J0} + \theta_{J1} = -90° + 0° = -90°$$

(a)

$$\theta_{J0} + \theta_{J1} = -90° + 90° = 0°$$

(b)

And substituting Equation 16.1 for $^{\text{global}}\theta_{\text{segment1}}$ gives you:

$$^{\text{global}}\theta_{\text{segment2}} = {}^{\text{global}}\theta_{\text{reference}} + {}^{\text{reference}}\theta_{\text{segment1}} + {}^{\text{segment1}}\theta_{\text{segment2}}$$
$$(16.7)$$

In **Figure 16.4a**, both the elbow and shoulder are in a neutral position. The orientation of the forearm in the global reference frame is

$$^{\text{global}}\theta_{\text{forearm}} = {}^{\text{global}}\theta_{\text{reference}} + {}^{\text{reference}}\theta_{\text{humerus}} + {}^{\text{humerus}}\theta_{\text{forearm}}$$
$$= -90° + 0° + 0° = -90°$$
$$(16.8)$$

When the shoulder is in neutral and the elbow is flexed to 90° (**Figure 16.4b**), then the orientation of the forearm is

$$^{\text{global}}\theta_{\text{forearm}} = {}^{\text{global}}\theta_{\text{reference}} + {}^{\text{reference}}\theta_{\text{humerus}} + {}^{\text{humerus}}\theta_{\text{forearm}}$$
$$= -90° + 0° + 90° = 0°$$
$$(16.9)$$

You do not normally think in terms of segment angles, but joint angles. So it is more convenient to write it that way. The following notation will be used for joint angles:

$$\theta_{J0} = {}^{\text{global}}\theta_{\text{reference}} \qquad (16.10)$$
$$\theta_{J1} = {}^{\text{reference}}\theta_{\text{segment1}} \qquad (16.11)$$
$$\theta_{J2} = {}^{\text{segment1}}\theta_{\text{segment2}} \qquad (16.12)$$

Hopefully, you can start to recognize the pattern that is emerging. Every segment angle is just the sum of all the joint angles (remember that in the absence of a leading superscript, you assume it is in the global space):

$$\theta_{\text{segment1}} = \theta_{J0} + \theta_{J1} \qquad (16.13)$$
$$\theta_{\text{segment2}} = \theta_{J0} + \theta_{J1} + \theta_{J2} \qquad (16.14)$$

With this new notation and assuming J1 is at the origin, you can rewrite the location of Joint 2 using joint coordinates:

$$^{x}P_{J2} = l_{\text{segment1}} \times \cos(\theta_{J0} + \theta_{J1}) \qquad (16.15)$$
$$^{y}P_{J2} = l_{\text{segment1}} \times \sin(\theta_{J0} + \theta_{J1}) \qquad (16.16)$$

The location of the end effector is found in a similar fashion, but you must start at the location of J2:

$$^{x}P_{\text{end effector}} = {}^{x}P_{J2} + l_{\text{segment2}} \times \cos(\theta_{J0} + \theta_{J1} + \theta_{J2}) \qquad (16.17)$$
$$^{y}P_{\text{end effector}} = {}^{y}P_{J2} + l_{\text{segment2}} \times \sin(\theta_{J0} + \theta_{J1} + \theta_{J2}) \qquad (16.18)$$

The equations may look a little scary, but they are not that bad. Just remember that the process just reiterates for every link. You just add the position of the joint and the additional joint angle at every step.

Now that you know how to find the endpoint of a chain, you can begin to ask questions involving the use of the end effector. Again, if you are interested in grabbing a cup or kicking a ball, you would want to know things like: Can I reach the cup or ball? How can the chain be configured to reach the cup or ball?

The first task is to determine if the hand or foot can even reach that position in global space. This is normally called the **reach area**, or workspace. Do not get

> **Reach area**   The space in which the end effector can be positioned

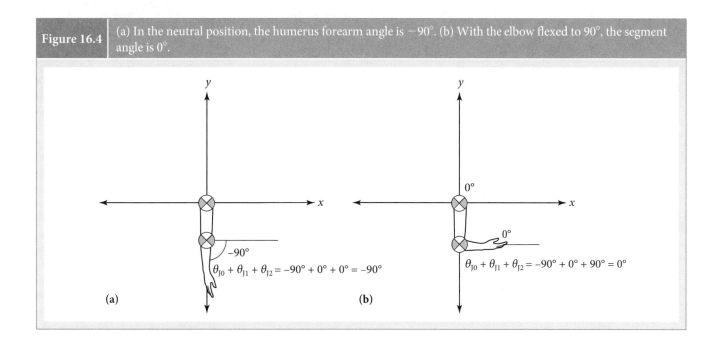

Figure 16.4   (a) In the neutral position, the humerus forearm angle is −90°. (b) With the elbow flexed to 90°, the segment angle is 0°.

thrown off by the name, the reach area is just as applicable to kicking (or any other multijoint activity) as it is to reaching. It just defines the area that the end effector can reach.

You could go through the iterative process outlined earlier, repeat it for every imaginable chain configuration, and create a cloud of points. But that would take a long time and get boring very quickly. For an easier way to determine the reach area, you must know the location of the origin of the chain, the lengths of the links, and the ranges of motion (ROM) of each joint angle. If a chain has two links of equal length and each joint can rotate 360° (Figure 16.5a), then the reach area of this chain is a circle centered at Joint 1 with a radius that is equal to the sum of the two lengths. If Segment 2 is shorter than Segment 1, then the reach area is a donut with the radius of the hole the difference between their two lengths (Figure 16.5b).

Most joints do not allow for 360° of rotation. What happens when the ROM is restricted? Instead of a circle, the reach area is defined by a set of circular sectors (see Box 16.2) that is defined by the beginning and end ranges of motion of each joint. This is shown when Joint 1 has an ROM from 0° to 180°, with 0° being coincident with the negative y-axis. Because the maximum length of the chain occurs when the angle of Joint 2 is zero, first construct a circular sector with a radius of the sum

of the lengths from −90° to 90° (Figure 16.6a). If Joint 2 is free to rotate 360°, then a circle with a radius of Link 2 should be added to each end of the circular sector (Figure 16.6b). Half the circle is located inside the reach area already established and can, for the time being, be ignored. Other restrictions at Joint 1 will just change the location and length of the circular sector. In Figure 16.7, the ROM for Joint 1 is 0° to 135°.

Restrictions at Joint 2 do not have the same effect as Joint 1. If Joint 2 can achieve full extension, then the reach area will still be the circle with the radius the sum of the two lengths. If Joint 2 cannot achieve full extension, then the circle will have a smaller radius. Examples are provided in Figure 16.8. The outer ring shows the reach area when Joint 2 is restricted to 45° to 180°. The inner ring shows the reach area when Joint 2 is restricted to 90° to 180°. Notice how much smaller the radius of the circle is when there is a larger joint restriction (compare the radii of both with the reach area in Figure 16.4).

The last two examples illustrated what happens if either Joint 1 or Joint 2 had a restricted ROM, but the other joint did not. Figure 16.9 shows you the combined effect of having both joints restricted. Joint 1 is restricted to 0° to 135°, while Joint 2 is restricted to 45° to 120°. Again, a joint angle of 0° at Joint 1 corresponds to a segment orientation of −90° in global

Figure 16.5	(a) If the two segments are the same length and each joint can rotate 360°, then the reach area is a circle defined by the sum of the two lengths. (b) If Segment 2 is shorter than Segment 1, then the reach area is shaped like a donut. The area inside the smaller circle cannot be reached.

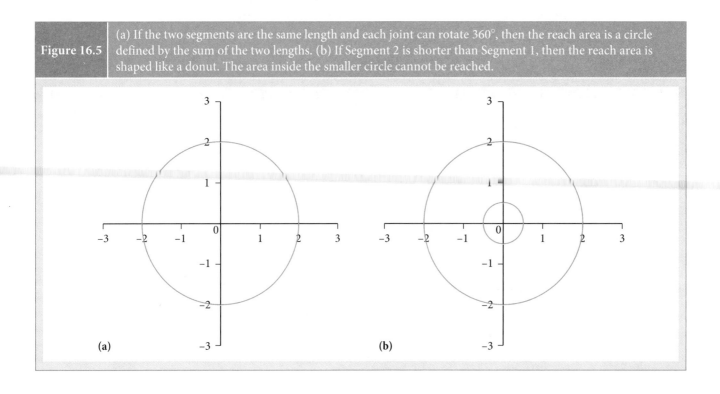

Box 16.2 | Essential Math: Circular Sector

A circular sector is the area enclosed by two radii and an arc. If the two radii are in the same location, the area is a circle. If the two radii are 180° apart, the area is a semicircle. If the radii are less than 180°, the area is a wedge. Examples are shown in the figure.

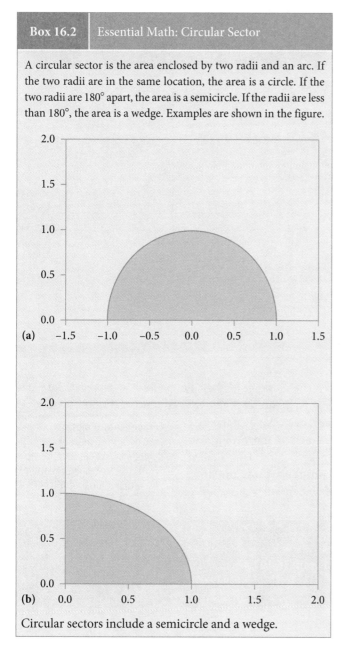

(a)

(b)

Circular sectors include a semicircle and a wedge.

Figure 16.6 | The reach area for a two-segment chain when Joint 1 is restricted to 0° to 180°.

(a)

(b)

space. Notice the additional area that cannot be reached by the end of this chain.

If the two joints have a large amount of mobility, then within the reach area the end effector can arrive at a point with two possible configurations. This is illustrated in Figure 16.10. Notice that, although both configurations are theoretically possible, you would be hard pressed to find a joint in the human body that had the range of motion of Joint 2 in the figure. If a typical range of motion is between 0° and 180°, then a certain point can be reached by only one configuration.

Points can be reached outside the reach area, or by a different configuration inside the reach area, if one or more additional joints and links are added to the chain. With three links, for example, the reach area is increased, and there are numerous chain configurations that reach the same point (Figure 16.11). It should be no surprise, then, that both the upper and lower extremity chains each have three joints and three links.

There are several other possibilities if, for whatever reason, a particular configuration could not be used to reach a point. This attribute has gone by many names, but probably the most

Figure 16.7	The reach area for a two-segment chain when Joint 1 is restricted to 0° to 135°.

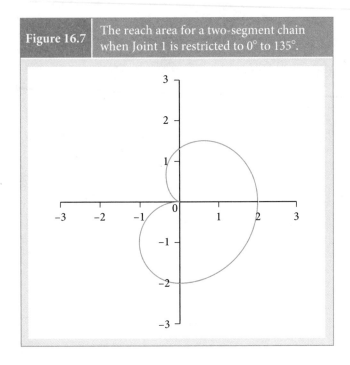

Figure 16.9	The reach area when Joint 1 is restricted to 0° to 135°, while Joint 2 is restricted to 45° to 120°.

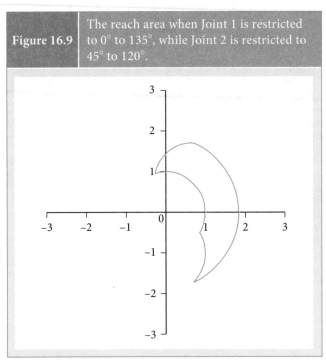

> **Motor redundancy**   The ability to perform the same task in multiple ways

common is **motor redundancy** (other terms include the more positive-sounding motor abundance and the more negative sounding motor degeneracy). In general, this is a good thing. If there was not a certain amount of redundancy in the system, then your end effector would fail to reach

a point if just one joint did not perform as expected. But if one joint had a decrease in ROM or torque production, there could be a corresponding increase at one or more joints to make up for it. This is called **compensatory motion**. Compensatory motion is a

> **Compensatory motion**   Increased motion at one degree of freedom to make up for decreased range of motion at another degree of freedom

Figure 16.8	The reach area for a two-segment chain when Joint 2 is restricted (A) to 45–180 and (B) from 90–180.

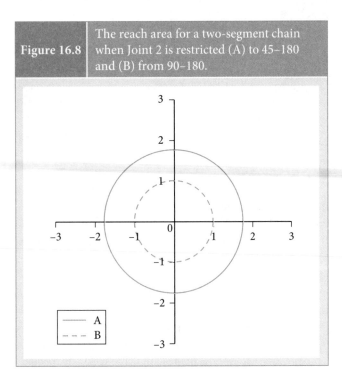

Figure 16.10	If both joints are unrestricted, there are two chain configurations that can reach the same point in space. One of the configurations is not usually physiologically possible.

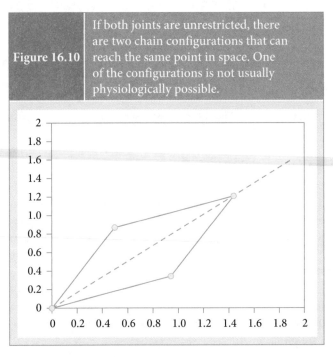

Figure 16.11 If there are three links in the chain, there are a large number of configurations that can reach the same point.

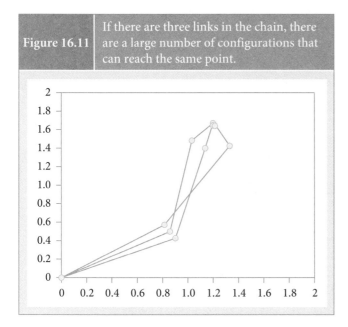

Figure 16.12 To push something with the upper extremity requires the upper arm and forearm to rotate in opposite directions.

Photo: © Andresr/ShutterStock, Inc.

great thing as long as the demands are well within the capacity of the compensatory joint(s). However, it can become injurious if the demands are at the limits or exceed the capacities of the compensating joints. In this way, a deficiency at one joint can manifest itself as an injury at another joint.

16.1.2   Patterns of Movement

Although understanding the chain configuration is important, in biomechanics you are usually more interested in how things change. In other words, how does the multilink chain move? If you think about it, this can be a very daunting task; there are literally thousands of ways the body moves. However, things can be simplified if you realize that there are typical movement patterns, and most movements are variations of one of these patterns.

A **motion path** is the trajectory of movement by any point of interest. Motion paths can refer to the linear motion of a point or the angular motion of a rigid body. A **movement pattern** is any recognizable spatial and temporal regularity, or any interesting relation, between moving bodies.[4] In other words, movement patterns look at two or more motion paths. For example, if you notice that one segment rotates clockwise while another rotates counterclockwise or if one segment begins to rotate after another segment has completed its rotation, you have uncovered a pattern. Human beings are

Motion path   The trajectory of movement by any point of interest

Movement pattern   Any recognizable spatial and temporal regularity, or any interesting relation, between moving bodies

not machines, and so we cannot expect these movement patterns to be exact every single time. There will always be variations in the way a movement is being performed. However, basic patterns are helpful in understanding how you move.

Many descriptors have been used to describe patterns of movement in a multilink chain. Each of them are good, but can also be somewhat confusing if they are taken too literally. When you push something with your hand, your arm and forearm are rotating in opposite directions (**Figure 16.12**). The rotation of the two segments in opposite directions causes your hand to follow a nearly linear motion path. This is a common movement pattern that you see not only in the upper extremity. When you "push" off the ground, your thigh and shank are rotating in opposite directions. Because of these common features, some authors will refer to this as a pushlike movement pattern.[5] The basic idea is that the two segments are rotating in opposite directions, and the end effector is following a nearly linear motion path.

When you pull something in toward you, these same two criteria are met. The motions are just reversed. Similarly, the eccentric (or down) phase of a bench press or squat would also fall into the category of a pushlike pattern. It is somewhat grammatically unappealing to call a movement when you are pulling something into you a pushlike pattern, but it is also nice to have a short, descriptive term to relate an entire class of movements that share similar features. Some authors have separated these into push patterns and pull patterns,[6] and that seems a bit more appealing. But during the eccentric phase of a bench press or squat, you are pushing, albeit with a force that is lower than that of gravity pulling the barbell to the center of the earth.

Others have referred to this as a countercurrent movement pattern because the two segments are rotating in opposite directions, whereas others will say that the two segments are simply antiphase. These terms are probably more accurate, but less descriptive. At the risk of confusing these patterns of motion with joint motion, they will be referred to as extension or flexion patterns.[1]

> **Extension pattern**   A movement pattern with the segments rotating in opposite directions. The end effector follows a linear motion path as it moves away from the origin of the chain.
>
> **Flexion pattern**   A movement pattern with the segments rotating in opposite directions. The end effector follows a linear motion path as it moves toward the origin of the chain.

An **extension or flexion pattern** occurs when the segments in the chain are simultaneously rotating in opposite directions (**Figure 16.13**). When this occurs, the end effector has a more or less linear motion path. With an extension pattern, the distance between the origin of the chain and the end effector is increasing. With a flexion pattern, the distance between the origin of the chain and the end effector is decreasing. Collectively, they will be referred to as flexion/extension patterns.

Because we usually think in terms of joint rather than segment motion, it is helpful to think about the joint motions

Table 16.1	Anatomical Motions Associated With Flexion and Extension Patterns of the Upper and Lower Extremities		
		Pattern	
		Extension	**Flexion**
Upper extremity	Shoulder	Flexing	Extending
	Elbow	Extending	Flexing
Lower extremity	Hip	Extending	Flexing
	Knee	Extending	Flexing

associated with the flexion and extension patterns. These are listed in Table 16.1. You have to be careful because nomenclature for the knee is different than the elbow: an anterior rotation of the elbow is called flexion, whereas that same anterior rotation of the knee is called extension.

The opposite of a flexion/extension pattern occurs when the segments are rotating in the same direction (**Figure 16.14**). These patterns have been called throwlike patterns,[5,6] concurrent patterns, in-phase, and rotations. Although throwlike seems like a good descriptor, this pattern is also seen with kicking. You may also throw something with an extension pattern. Here, when segments are rotating in the same direction, they will be called **swing patterns** (if the movement speed is submaximal) or **whip patterns** (if the movement is maximal). Collectively, they will be referred to as swing/whip patterns.

> **Swing pattern**   A movement pattern with the segments rotating in the same direction with a submaximal velocity. The end effector follows a curved motion path.
>
> **Whip pattern**   A movement pattern with the segments rotating in the same direction with maximal velocity. The end effector follows a curved motion path.

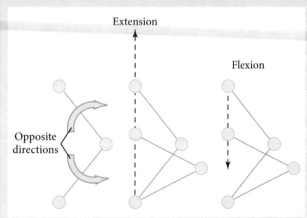

Figure 16.13	When the segments rotate in opposite directions, the endpoint moves in a linear path, extending when the origin and endpoint move further away and flexing when the origin and endpoint move closer together.

Extension

Flexion

Opposite directions

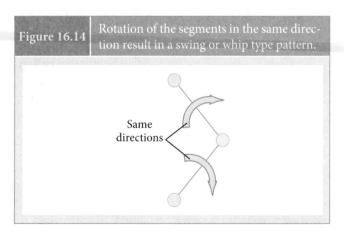

Figure 16.14	Rotation of the segments in the same direction result in a swing or whip type pattern.

Same directions

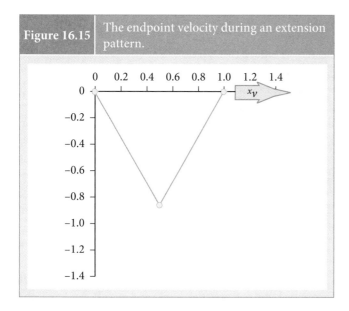

Figure 16.15 The endpoint velocity during an extension pattern.

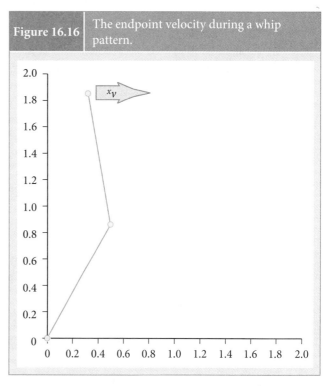

Figure 16.16 The endpoint velocity during a whip pattern.

The two basic movement patterns have their advantages and disadvantages. To illustrate them, compare throwing a baseball to throwing a dart. Throwing a baseball is usually done with a whip pattern, whereas throwing a dart is usually done with an extension pattern. Why?

In general, swing/whip patterns will lead to a greater end effector velocity than flexion/extension patterns. The standard two-link chain undergoing an extension pattern in the positive x direction is presented in Figure 16.15. Again, remember that these findings are only applicable to the chain in the example. Because the two segments have an equal length and the endpoint must move along a straight line, J2 must have an angular velocity that is twice that of J1. This means that there is a limit to how fast J1 can rotate (one half of J2). J2 will limit the velocity of the endpoint of the chain.

The same two-link chain has a whip pattern in Figure 16.16. The chain configuration is different because the goal is to maximize the velocity in the x direction. Here, there is no constraint on how fast J1 can rotate because the velocities in the y direction need not cancel. Even if we kept the magnitudes of the joint angular velocities the same, the whip pattern would still lead to greater endpoint velocities.

The relation between linear and angular velocity is

$$v = l_r\omega \qquad (16.19)$$

Where the length is a line from the joint to the endpoint, and linear velocity has a direction that is tangential to this line. The same can be done with chain: the length of the radius is just the length from the joint of interest to the endpoint (Figure 16.17). In the case of an extension pattern, the tangent to this line is also perpendicular to the direction of movement.

Therefore, the angular velocity of J1 is not directly contributing to the endpoint velocity. That is not to say that it is not important, or it does not play a role (it does). Without J1 rotating at half the velocity of J2, the end effector would not be traveling in a straight line during flexion/extension patterns. In contrast, during the whip pattern, a component of the tangent is contributing directly to the endpoint velocity. It is this component that is the difference between the endpoint velocities of extension and whip patterns.

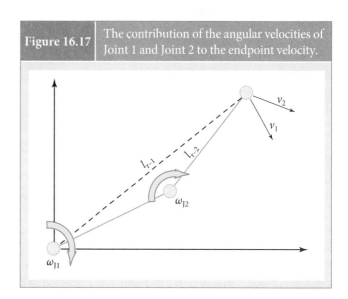

Figure 16.17 The contribution of the angular velocities of Joint 1 and Joint 2 to the endpoint velocity.

Although greater velocities are achieved with whip patterns, greater accuracy is achieved with flexion/extension patterns. With an extension pattern, the end effector is traveling along a rectilinear path toward the target. Subsequently, there is more time to correct for errors.[5] In addition, there is less velocity in the y direction. So if the angle to the target is off by a little bit, it will not result in as dramatic a change at the target position. With the whip pattern, the end effector is traveling with a curvilinear path. If the whip pattern was used to throw an object, there are very few points where the ball can be released and hit the target. So it requires greater precision. In addition, there are greater velocities in y direction, meaning that errors will be magnified if the angle is slightly off.

That is not to say that whip patterns cannot be accurate. Major league baseball players are highly accurate in their placement of the ball with whip patterns. They choose the whip pattern when throwing a baseball because they could not possibly achieve the high velocities necessary using an extension pattern. But it takes years of practice to skillfully place the ball. A dart is lighter than a ball, and the distance to the dartboard is much less than the distance from the mound to the plate. With a dart, you do not need to develop as high a velocity. Of course, the bull's-eye is much smaller than the strike zone, and thus darts require a high degree of accuracy. That is why you are better off using an extension pattern for darts.

Which pattern do you think a field athlete uses for the shot put? This question will be answered in the section on kinetics because there is a lot more to the story of movement patterns than just velocity and accuracy. But first, it is helpful to have a more precise way to quantify the movement between joints, particularly when they are not following the exact patterns described earlier.

16.1.3   Coordination and Coupling

Hopefully, it is obvious by now that most human activities require movement at more than one degree of freedom. Even the simple patterns of motion illustrated in the last section required two joints to be moving. Therefore, it is important to understand not only how each joint is moving independently, but also how they are interacting with each other. In this section, you will learn a few ways to do that.

First, it is important to understand a few ideas. In some situations, joints are compelled to move together. For example, stand up and keep your feet flat on the ground. Now, keeping your feet on the ground, try flexing your knees without moving your ankles or hips. You will find that you cannot do it: hip flexion and ankle dorsiflexion have

to occur with knee flexion. When movements have to occur together, they are said to be **coupled**.

> **Coupled motion**   Motions of two degrees of freedom that consistently occur together

Coupled motion is not restricted to movements between different joints. Coupled motion can also occur between different DOFs of the same joint. This happens when the axis of rotation is an oblique axis that is not coincident with any of the cardinal axes. For example, at the subtalar joint during weight-bearing, plantar flexion, adduction, and eversion are coupled during pronation. And in the spine, lateral flexion and rotation are coupled. More detailed treatments of joint mechanics cover these types of coupled motion.[7] In this text, you will mostly study coupled motion that occurs between joints.

There are also instances when joint motions are not compelled to occur together, but they usually do. In reaching for an object with your hand, shoulder flexion and elbow extension are not physically required to occur together, but they do as part of a coordinated movement. **Coordination** is the appropriate assembly and sequencing of degrees of

> **Coordination**   The appropriate assembly and sequencing of degrees of freedom

freedom. In other words, how many and which DOFs are used, and what is their appropriate sequence of motion in relation to each other?

There is probably a very fine distinction (if any) between joint coupling and joint coordination. Many authors use a "coupling angle" when talking about relative motion between joints or segments whether the relative motion is compulsory or not,[8] probably because it is difficult to tell if constraints placed on the movement make motions truly coupled. For the same reason, in this lesson coupling and coordination may be used interchangeably, unless otherwise noted.

Another issue is that relative motion can be discussed between absolute segment angle, joint angles, or a combination of the two. Here, the emphasis is on joint angles because they seem to be more intuitive. So although you should understand that there are many ways to view relative motion, this discussion will be limited to relative motion between joints.

Oftentimes, you will see joint angle data plotted as a function of time, as was done for two joints in Figure 16.18. Although you could make some educated guesses about what is happening between the two joints, some of the finer details may be lost. A better way to get a visual of the interaction would be to plot the two joint angles as a function of each

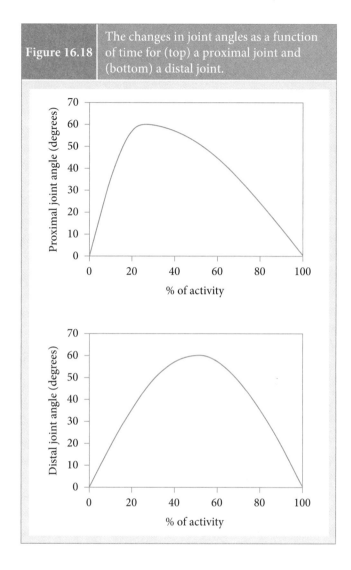

Figure 16.18 The changes in joint angles as a function of time for (top) a proximal joint and (bottom) a distal joint.

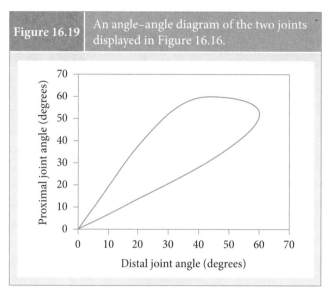

Figure 16.19 An angle–angle diagram of the two joints displayed in Figure 16.16.

Angle-angle diagram A plot of one joint motion as a function of another joint motion

other. The common practice is to put the more distal joint on the abscissa and the proximal joint on the ordinate (**Figure 16.19**). This is known as an **angle-angle diagram**.

There are some benefits and drawbacks to angle-angle diagrams. First, they provide a very nice picture of the interaction between the two joints. If both joints are rotating in the same direction, then the slope will be positive. If the joints are rotating in opposite directions, then the slope will be negative. The magnitude of the slope tells you something about the degree to which they are rotating. If the distal joint has two degrees of rotation for every one degree of rotation of the proximal joint (like the flexion/extension pattern), then the slope would be equal to 0.5. Conversely, if the proximal joint has a greater amount of rotation than the distal by

the same 2:1 ratio, then the slope would be equal to 2. If the slope is zero (a horizontal line), then the distal joint is rotating while the proximal joint is not. If the slope is undefined (a vertical line), then the proximal joint is rotating while the distal joint is not. The origin of the angle-angle diagram represents the neutral position for both joints. Some joints (such as the hip and shoulder) can have both positive and negative values, whereas joints that cannot extend beyond neutral (like the knee and elbow for many people) will only have values on one side of the zero line. If both joints return to their starting positions, the angle-angle diagram will end in the same place it started. If the motions follow the same, exact path during the return, the angle-angle diagram will be a straight line. If not, it will form a loop. A wider loop indicates a big difference between the departure and the return paths.

Two limitations of angle-angle diagrams are that you lose any temporal (timing) information and that they allow for only a qualitative and not a quantitative assessment of the activity. If you look at the angle-angle diagram in Figure 16.19, you can tell what Joint 1 is doing relative to Joint 2, but you do not know when that is occurring. This can be alleviated somewhat by annotating on the graph key events (such as initial contact and toe off during gait). In addition, you could put points on the angle-angle diagram and have those points represent a certain time scale. If the points are further apart, you know the joints have a greater angular velocity.

Biomechanicians (and scientists in general) love numbers. They generally eschew qualitative analysis in favor of putting numbers on things and then statistically comparing them. Various methods have been used to quantify interjoint coordination, but the one that is probably in most use today is

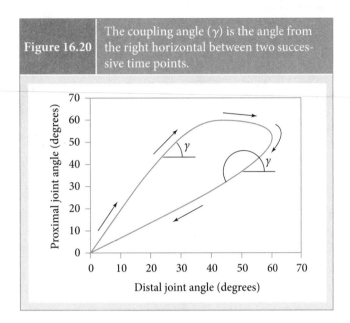

Figure 16.20 The coupling angle (γ) is the angle from the right horizontal between two successive time points.

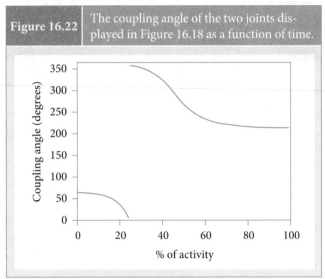

Figure 16.22 The coupling angle of the two joints displayed in Figure 16.18 as a function of time.

the coupling angle. With a coupling angle, you are finding the angle between a line connecting two successive time points and the right horizontal (**Figure 16.20**).[9] The coupling angle is given the symbol gamma, γ. The formula would be

$$\gamma' = \tan^{-1}\left(\frac{y'' - y'}{x'' - x'}\right) \qquad (16.20)$$

Quadrants are then assigned based on the coupling angle (**Figure 16.21**).[9] With respect to patterns of movement described earlier, anything in the first quadrant would represent a swing/whip pattern where both joints are rotating in the positive direction. Similarly, anything in the third quadrant would represent a swing/whip pattern where both joints are rotating in the negative direction. In the second quadrant, the proximal

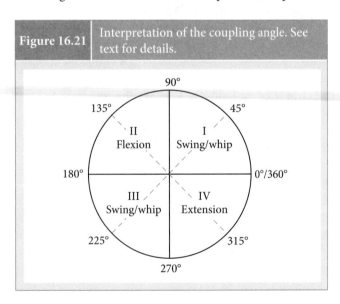

Figure 16.21 Interpretation of the coupling angle. See text for details.

segment is rotating in the positive direction, whereas the distal joint is rotating in the negative direction. This is a flexion pattern. Likewise, the proximal segment is rotating in the negative direction, and the distal joint is rotating in the positive direction in the fourth quadrant. This would be an extension pattern.

It should also be pointed out that, within the quadrants, any coupling angle that falls along one of the principal diagonals (45°, 135°, 225°, 315°) represents a one-for-one change in joint angles. Along the principal axes (0°, 90°, 180°, 270°) only one joint is moving. In between the diagonals and principal axes, one joint is rotating faster than the other.

For example, examine the first quadrant. At 45° there is 1° of positive rotation of the proximal joint for each 1° of positive rotation of the distal joint. At 90° the proximal joint is rotating in the positive direction while the distal joint is not moving. If the angle is between 0° and 45°, the distal joint is rotating faster than the proximal joint, whereas the reverse is true between 45° and 90°. Other quadrants would be interpreted in a similar fashion.

Coupling angles can be plotted as a function of time to show you how the joint coordination patterns are changing over time (**Figure 16.22**). These can sometimes be hard to interpret when they get disjointed because 0° and 360° are equivalent. I find it easier to interpret coupling angles as a function of time if they are plotted as a polar plot (**Figure 16.23**). With these plots, the coupling angles are the degrees on the circle, and the radius represents time. The time starts at the center of the circle. You can follow how the coupling angle changes with time if you start there and trace the plot with your finger as it extends outward.

The previous section looked at movement patterns from a qualitative perspective. Although it was useful in describing movement patterns in very general terms, it was not very precise. Angle-angle diagrams give you a visual representation

Figure 16.23	The same coupling angle-time curve in Figure 16.22 as a polar plot.

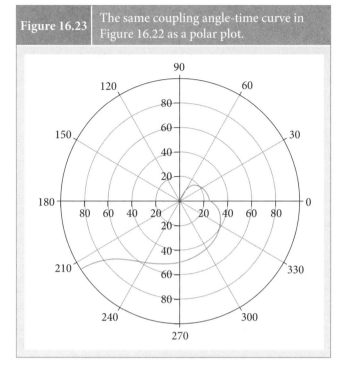

of the interaction between joints, but still provide a qualitative analysis. Coupling angles allow you to quantify motion between joints and then compare them statistically across trials or across subjects.

Section Question Answer

Overhead throwing, softball pitching, and darts are all "throwing" techniques, but they use different movement patterns. Overarm and underarm throwing both use whip patterns, but with overarm throwing, the shoulder is extending and the elbow is extending, whereas with underarm throwing the shoulder is flexing while the elbow is flexing. In contrast, darts use an extension pattern. The weight, speed, and accuracy requirements of the task all influence the decision as to which movement pattern is selected. Greater speed requirements use a whipping pattern, whereas greater accuracy requirements use an extension pattern. Force requirements also play a factor, and they will be discussed in the next section.

COMPETENCY CHECK

Remember:

1. Define the following terms: angle-angle diagram, chain configuration, compensatory motion, coordination, coupled motion, motion path, motor redundancy, movement pattern, and reach area.

Understand:

1. Assume a two-link chain each with a length of one. $J0 = 0$. Determine the endpoint location for the following joint angles (in degrees):

θ_{J1}	θ_{J2}	$^x p$	$^y p$
30	90		
60	120		
−40	45		

2. Describe the effect of the following restricted joint ranges of motion (in degrees) on the reach area:

θ_{J1}	θ_{J2}
20–90	10–110
10–110	20–90

3. Describe how motor redundancy and compensatory motion can be both helpful and harmful in completing tasks.
4. Describe the basic movement patterns and the benefits and limitations of each.
5. Construct angle-angle diagrams for each of the basic movement patterns.
6. Determine the coupling angle during the following motions (in degrees):

t	θ_{J1}	θ_{J2}	γ
1	30	60	
2	45	45	
3	90	60	

Apply:

1. Describe situations where coupled motion exists.
2. List activities where you would use a flexion/extension pattern instead of a swing/whip pattern, and vice versa.

16.2   KINETICS

Section Question

Jonathan has an injury to his elbow (Figure 16.24), but the athletic trainer told him it was caused by a deficiency at his shoulder and trunk. How can an injury to a more distal joint be caused by impairment at a more proximal joint?

Figure 16.24 How can an injury to a more distal joint be caused by impairment at a more proximal joint?

16.2.1 Interaction Torques

When examining single-joint motion in a gravitational field, the angular equivalent of Newton's second law is used:

$$I\alpha = T_{MTC} + T_g \qquad (16.21)$$

Realizing that inertial torque is defined as the product of the moment of inertia and angular acceleration, torque requirements of the MTC can be determined by rearranging the terms in Equation 16.21:

$$T_{MTC} = T_{inertial} + T_g \qquad (16.22)$$

The moment of inertia, I, is determined by both the mass of the segment and how that mass is distributed about the joint axis of rotation.

If you are looking at a chain, it is not quite as simple. First, the moment of inertia about a joint depends not only on the moment of inertia of its segment, but also on all the other segments and joints distal to it. In the two-link chain under consideration, the moment of inertia about J2 is determined in the usual way. The moment of inertia about J1 is determined by the mass and distribution of the mass of Segment 2, as well as the angle of J2. For example, consider flexing the hip, both with the knee extended (Figure 16.25a) and with the knee flexed (Figure 16.25b). Intuitively, you should know that it is more difficult (i.e., requires more torque about the hip for the same angular acceleration) to flex the hip with the knee extended. This is because the mass of the distal segment is further away from the hip joint in this chain configuration. The larger the mass of Segment 2, the further that center of

Figure 16.25 The moment of inertia is larger when the knee is extended than when the knee is flexed.

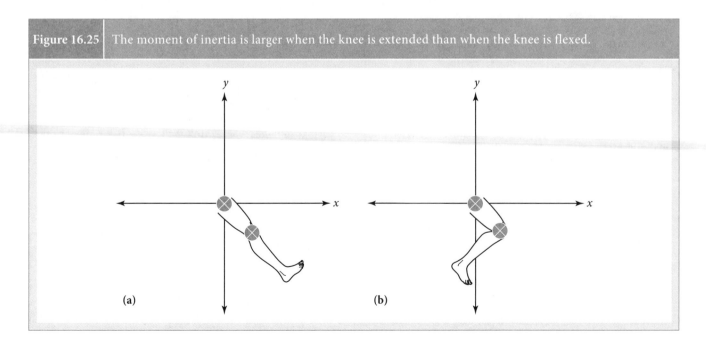

(a)   (b)

mass is from Segment 2, and a more extended J2 will all increase the moment of inertia about J1.

But the moment of inertia is not the only thing that is different with multijoint movement. Velocities and torques at one joint will affect the acceleration at every other joint.[10] To see a simple example of this,[11] position your upper extremity so that your shoulder and elbow are flexed and your wrist is relaxed. With the shoulder fixed, oscillate between flexion and extension of your elbow. You will find that your wrist is flexing and extending during the oscillations, even though there are no active MTC torques about the wrist joint. If you wished to keep the wrist in a neutral position, you would have to produce MTC torques to counteract the accelerations induced by the velocities and torques at the elbow.

The torques about a joint that are produced by other torques and velocities in the chain are referred to as interaction torques, and they can be quite large. Thus, with multijoint movement, Equation 16.22 becomes

Interaction torques
Torques created about a joint that result from the velocities and torques of other joints in a serial chain

$$T_{MTC} = T_{inertial} + T_{interaction} + T_g \qquad (16.23)$$

In addition, although it is easy to see how elbow flexion/extensor torques produce wrist flexion/extension torques, it is harder to visualize (but still present) that wrist flexion/extensor torques also produce elbow flexion/extension torques—and both produce torques at the shoulder. The mathematics involved with interaction torques can become quite complicated because all the torques and velocities at each joint must be determined simultaneously.[10] And the effects of these numerous torques depend on the chain configuration at that particular instant. The actual equations are beyond the scope of this lesson. What is important for you to know is that interaction torques can be quite large and are often exploited by skilled performers. How the CNS deals with all these interaction torques is a bit of a mystery.

16.2.2 Energy Transfers

Work-energy-power analyses are another way to examine movement. You already know that MTCs act as an energy source, generating energy and transferring it to the segments, and as an energy sink, absorbing energy from the segments. Here you will see how energy is transferred to different parts of the chain by two mechanisms. Two other mechanisms will be covered in the discussion on biarticular muscles.

The first case concerns muscle and single-joint concepts. When an MTC acts isometrically, it transfers energy from one segment to another.[12] A good example of this is heel raise

movement. Energy is generated by the plantar flexors and delivered to all the segments—raising their potential energy. MTCs of other key joints (such as the knees, hips, and spine) must act isometrically so that energy can be transferred up the chain. This type of energy can be thought of as *joint-to-link energy transfer* because energy generated by the MTCs changes the energy of remote links.[12]

The second type of energy transfer is a *link-to-link transfer.*[12] For this type of transfer to occur, segments must be rotating in the same direction. Link-to-link energy transfers cannot occur if the segments are rotating in opposite directions.[13] Whip/swing patterns can thus transfer energy between segments, but flexion/extension patterns cannot. Returning to the two-link example, imagine that Segment 1 is the thigh and Segment 2 is the shank, and the segments are rotating in the same direction like they would during the latter part of a kick. Note that the shank is rotating faster than the thigh, so the knee is extending. The joint velocity is the difference between the two segmental velocities:

$$\omega_{knee} = \omega_{shank} - \omega_{thigh} \qquad (16.24)$$

Assume there is a net extensor torque at the knee, which is equivalent to the torque produced by the MTC (i.e., no co-contraction). If you draw a diagram of each segment with the torques and angular velocities, this will be easier to visualize (**Figure 16.26a**). Power is equal to:

$$P = T\omega \qquad (16.25)$$

Therefore, at the thigh:

$$P_{MTC} = T_{MTC}\omega_{thigh} \qquad (16.26)$$

The thigh is rotating in the positive direction, but the knee extensor torque is negative on the thigh. This means that the thigh is losing energy at a rate of $T_{MTC}\omega_{thigh}$. Where is it going?

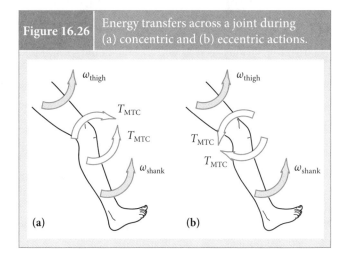

Figure 16.26 Energy transfers across a joint during (a) concentric and (b) eccentric actions.

It is going to the shank. At the shank, the torque and angular velocity are in the same direction:

$$P_{MTC} = T_{MTC}\omega_{shank} \qquad (16.27)$$

Therefore, the shank is gaining energy at a rate of $T_{MTC}\omega_{shank}$. The additional energy not received from the thigh is given to the shank by the MTC via a concentric muscle action:

$$P_{MTC} = T_{MTC}\omega_{shank} - T_{MTC}\omega_{thigh} \qquad (16.28)$$

which is precisely the definition of the net joint torque power:

$$P_{NJT} = T_{MTC}\omega_{shank} - T_{MTC}\omega_{thigh}$$
$$= T_{NJ}(\omega_{shank} - \omega_{thigh}) = T_{NJ}\omega_{knee} \qquad (16.29)$$

By a similar argument, if the torque at the knee were a flexor torque, then the net joint torque and the angular velocity of the thigh would be in the same direction. The thigh would be gaining energy at a rate of $T_{MTC}\omega_{thigh}$. Because the net joint torque and the shank angular velocity are in different directions, the shank would be losing energy at a rate of $T_{MTC}\omega_{shank}$. And the net joint torque would be absorbing energy at a rate of $T_{MTC}\omega_{knee}$. This situation is illustrated in Figure 16.26b.

16.2.3 The Role of Biarticular Muscles

In many ways biarticular MTCs (MTCs that cross two joints) are not conceptually different than single-joint MTCs: They can generate, absorb, and transfer energy. However, they do have some unique properties. After all, we would not have biarticular MTCs if they were not important. One neat thing is that, unlike a monoarticular MTC, a biarticular MTC may act on a joint opposite its anatomical classification.[14] This is due to the interaction torques. For example, the gastrocnemius can create an extensor torque at the knee if the interaction torque it creates at the ankle is larger than the torque it directly produces at the knee. This is a little beyond the scope of this lesson. Here you will concentrate on two other roles of biarticular muscles: energy transfer and directional control of endpoint forces.

In the previous section you learned about two different types of energy transfer: joint-to-link and link-to-link. Biarticular muscles are involved in two other types: *joint-to-joint* and *joint-to-link*.[12] Both types are very similar. The only difference is where the energy comes from. Because the joint-to-joint transfer is a little easier to understand, that is a good place to start.

Imagine a case where the lower extremity is going through an extension pattern, such as a squat movement. If the hip is extending, the rectus femoris would be stretched at its proximal insertion because it is a hip flexor. At the knee, the rectus femoris is an extensor and would be shortening at

Figure 16.27 If a biarticular muscle's length does not change, extension of J1 will cause an extension of J2. If a torque is extending J1, then it is a joint-to-joint transfer. If a torque is decelerating Segment 1, then it is a link-to-joint transfer.

its distal insertion. If the rectus femoris is not changing its length, it cannot generate or absorb energy (do work). It can, however, transfer some of the work done by the hip extensors to knee extension. In a similar fashion, the hamstrings can transfer energy of the knee extensors back up to the hip extension. And the gastrocnemius can transfer energy from the knee extensors to plantar flexors.

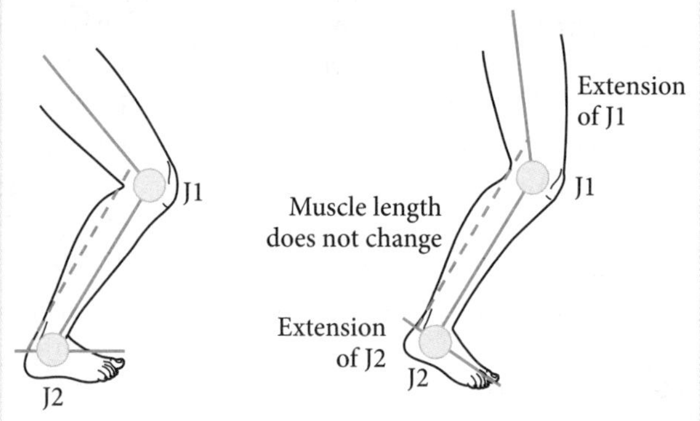

A link-to-joint transfer is very similar to a joint-to-joint transfer.[12] The difference is where the energy comes from. Imagine a vertical jump, which is a fast extension pattern. Near takeoff, the knee must stop accelerating or it will hyperextend. The two-joint gastrocnemius can transfer some of the rotational kinetic energy from the thigh (slowing knee extension) to the rotational kinetic energy of the foot. This allows the knee extensors to be active for a larger portion of the propulsive phase, thus increasing jump height. See Figure 16.27.

Biarticular muscles also play a role in directing the endpoint force vector. The magnitude and direction of the endpoint force is determined by the magnitude and direction of all joint torques in the chain. The effect of an isolated joint torque is theoretically straightforward.[15] In the two-joint chain, an isolated torque about Joint 1 will produce a force along a vector described by Joint 2 and the endpoint. Similarly, an isolated torque about Joint 2 will produce a force along a vector described by Joint 1 and the endpoint. See Figure 16.28.

Production of forces has to occur in all directions, not just these two. This requires coordination between *all* muscles. Biarticular muscles are not unique in this function, but they do differ from their monoarticular counterparts in the

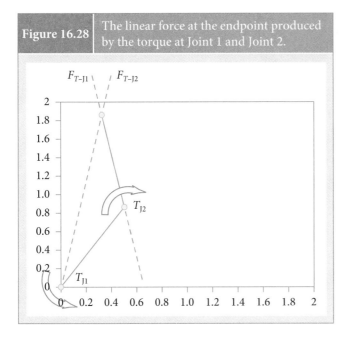

Figure 16.28 The linear force at the endpoint produced by the torque at Joint 1 and Joint 2.

principal direction at which they apply a force. Although the monoarticular muscles produce a force more in line with a pure torque at a single joint, the biarticular muscles create a significant transverse component because of the torques produced at both joints (Figure 16.29).[15] It is not that monoarticular muscles do not control the direction of forces, but the unique direction afforded by the biarticular muscles allows for a greater range and smoother control of forces than those provided by monoarticular muscles alone.

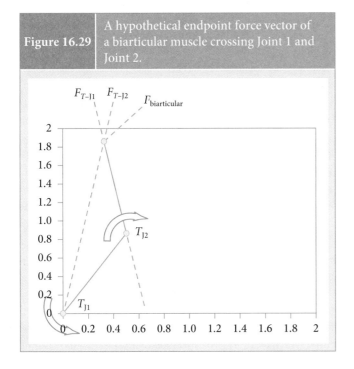

Figure 16.29 A hypothetical endpoint force vector of a biarticular muscle crossing Joint 1 and Joint 2.

16.2.4 Patterns of Movement Revisited

In the kinematics section, you learned that swing/whip patterns produced greater endpoint velocities, but that flexion/extension patterns were more accurate. The question was also posed: Which pattern would you choose for the shot put? If you analyze the shot put, you will see that those athletes use more of an extension pattern than the whip pattern seen with throwing a baseball (Figure 16.30). Why? A typical baseball weighs about 0.14 kilograms, whereas a shot put weighs 7.26 kilograms for men and 4 kilograms for women. The shot put used by men is 50 times heavier than a baseball. It is simply too heavy for the whipping pattern used to throw a baseball. Although the whipping pattern can generate a larger endpoint velocity, flexion/extension patterns can generate more force. Although this is also true when the chain is moving, it is easier to demonstrate via a static analysis.

Remember that whenever you push on something, it pushes back on you with an equal and opposite force in accordance with Newton's third law. We will use the same two-chain configurations in the velocity example (Figures 16.31 and 16.32). The masses of the segments will be ignored. Because the external force will be the same in each case, differences in torque requirements are due to differences in moment arms.

Figure 16.30 Why do shot putters use an extension pattern instead of a whip pattern?

© Jamie Roach/ShutterStock, Inc.

Figure 16.31	Relation between joint torques and endpoint force are due to the moment arm's length. Note in the extension pattern that there is no moment arm for Joint 1.

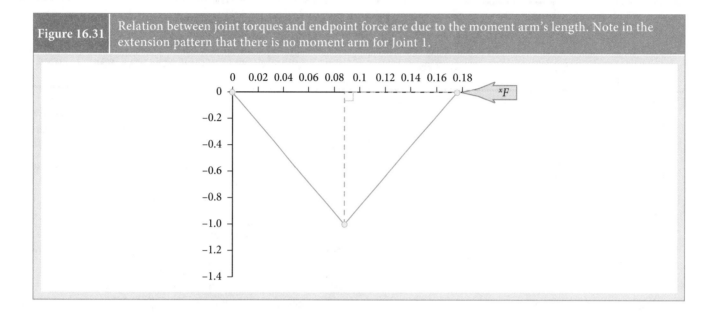

For both chain configurations, the torque about J2 caused by an endpoint force in the x direction is equal to:

$$T_{J2} = l_{\text{moment arm}} \times {}^{x}F$$
$$= -(l_{\text{segment2}} \times \sin(\theta_{J0} + \theta_{J1} + \theta_{J2})) \times {}^{x}F \quad (16.30)$$

The torque demands will be greatest when θJ_{0+1+2} is 90°. In the case of the extension pattern, this occurs when the J2 is

Figure 16.32	Relation between joint torques and endpoint force are due to the moment arm lengths. The moment arm for Joint 1 is quite large during the whipping pattern.

more flexed. The opposite occurs with the whip pattern. The moment arm is greater when J2 is extended. It is probably easier to see on the figure than it is to look at the equations.

The moment arm for the force in the x direction about J1 will include both the moment arm from the endpoint to J2 plus the moment arm from J2 to J1:

$$T_{J1} = l_{\text{moment arm}} \times {}^{x}F$$
$$= -(l_{\text{segment1}} \times \sin(\theta_{J0} + \theta_{J1})$$
$$+ l_{\text{segment2}} \times \sin(\theta_{J0} + \theta_{J1} + \theta_{J2})) \times {}^{x}F \quad (16.31)$$

In the case of the extension pattern, the moment arm would be zero about J1 because the two moment arms cancel. This is not the case with the whipping pattern. Once again, the more extended the chain is with the whipping pattern, the greater the torque demands are at the joints. Examine Figure 16.32.

A few points concerning the application of these patterns should be highlighted here. With the flexion/extension patterns, the demand is greater when the chain (both joints) is flexed. In a fully extended position, the demand is the least. You should be able to demonstrate this for yourself by simply standing. In a fully upright position, there is minimal muscular demand because the force is being transmitted across the bones without a need for muscular torque. You can stand in this position for quite some time. Now crouch down into a squat position (Figure 16.33). You will quickly realize how much greater demand there is on your musculature.

Also with the flexion/extension pattern, note how the torque demands on J1 were zero when the force vector passed through the joint. This is not necessarily a good thing because the (usually) larger, more proximal muscle mass is not contributing to the push or pull. If the end effector were offset from the joint

| Figure 16.33 | When standing upright, there is little muscular demand as the force is "taken up" by the skeletal system. In the semi–squat position, the force is still the same but the demand is "transferred" to the muscles. |

© Andrei Nekrassov/ShutterStock, Inc.

center, then there would be a torque about J1, and it could contribute to the movement. This is demonstrated in Figure 16.34. In this case, a small change in joint angles led to a difference in the torque demands across the two joints.

From a purely kinematic standpoint, to maximize endpoint velocity, all the joints should reach their maximum

| Figure 16.34 | Shifting the location of the endpoint point can increase the moment arm about Joint 1. |

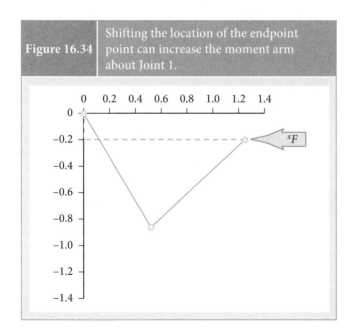

velocity at the same time during swing/whip patterns.[16] But this does not happen with human movement. Rather, joint rotations follow a proximal-to-distal sequence: a more distal segment's maximal velocity occurs after the next proximal sequence. If the human movement is different than what you would expect from the kinematics, then it means that there is something going on with the kinetics. (Remember that the takeoff angles of jumping and shot put are well below what you would expect kinematically, but are explained by the kinetics?)

There are probably a few explanations as to why this occurs.[17] First, the moment of inertia is less if the chain is initially flexed. For the same amount of torque, you can rotate the chain faster if the distal segments start flexed. Then, after the speed is increased, you start to increase the length of the chain by extending the joint. Second, energy transfers play an important part in human movement. To transfer energy from Segment 1 to Segment 2, Segment 2 must be rotating faster in the same direction as Segment 1. If both segments had the same velocity, the joint angle would not be changing, and energy could not be exchanged. The third explanation has to do with the force/velocity relations during concentric actions. You are at a disadvantage when producing MTC forces (which created torques about joints) at high speeds. Interaction torques are passive torques across a joint and thus are not subjected to the same force/velocity constraints. In addition, if the distal joint is flexing prior to extending, the MTCs will be put on stretch. This would allow the MTCs to take advantage of the stretch-shortening cycle and increased force (torque) output at high speeds. It is likely that all three mechanisms play a role in our adopting a proximal-to-distal sequencing during swing/whip patterns.

Flexion/extension patterns also tend to follow proximal-to-distal sequences during propulsive phases and distal-to-proximal sequences during braking phases. The major reasons for this sequencing include the energy transfers that occur with time as well as increasing the time/distance over which forces can be applied.

Section Question Answer

Many human movements require a high velocity of the endpoint of a chain. Proximal-to-distal sequencing is important in creating these high velocities because energy is transferred from the more proximal to the more distal joints, thus increasing their energy. Generating energy requires both torque and angular displacement (range of motion). Impairment of either decreases the energy generated by an MTC, which in turn decreases the endpoint velocity. In an attempt to "make up" for this loss of velocity, a person may try to increase the energy generated by the more distal joints by increasing the torque, ROM, or both. Distal joints tend to have lower ROMs and cannot generate as much torque. Thus, their capacity to generate energy is less. The increased demands placed on them by impairment at the more proximal joints may put them at risk.

SUMMARY

In this section, you learned about multijoint movements, and you saw how they were not simply the movement of multiple joints independently. The motions and torques about one joint affect the motions and torques at other joints. Although all muscles contribute to a movement, biarticular muscles have some unique properties that are used in transferring energy and directing the control of endpoint forces. Basic movement patterns include flexion/extension patterns and swing/whip patterns. Of course, these are basic patterns, but most movements are variations of one of these basic types. Within each movement pattern, the sequencing of motion is important. Proximal-to-distal sequencing is generally used when generating energy during a propulsive phase, whereas distal-to-proximal sequencing is generally used when absorbing energy during a braking phase. Now that you have these basic ideas down, it is time to apply them to a wide variety of activities that are typical in activities of daily living, work, and recreation.

REVIEW QUESTIONS

1. Define the following terms: angle-angle diagram, interaction torque, chain configuration, compensatory motion, coordination, coupled motion, motion path, motor redundancy, movement pattern, and reach area.

2. Explain how interaction torques occur.
3. List the different ways energy can be transferred in a multijoint system.
4. Trace the endpoint force vectors resulting from the different joint torques.
5. Explain the roles of biarticular muscles.
6. Describe situations where you would use one movement pattern over another and explain why you would choose to do so.
7. Describe situations where you would use a proximal-to-distal sequencing and situations where you would use a distal-to-proximal sequencing.

REFERENCES

1. Zatsiorsky VM. *Kinematics of Human Motion.* Champaign, IL: Human Kinetics; 1998.
2. Zatsiorsky VM. *Kinetics of Human Motion.* Champaign, IL: Human Kinetics; 2002.
3. Craig JJ. *Introduction to Robotics: Mechanics and Control.* Upper Saddle River, NJ: Pearson Prentice Hall; 2005.
4. Dodge S, Weibel R, Lautenschuetz A-K. Towards a taxonomy of movement patterns. *Information Visualization.* Fal-Win 2008;7(3–4):240–252.
5. Kreighbaum E, Barthels KM. *Biomechanics: A Qualitative Approach for Studying Human Movement.* 4th ed. Boston: Allyn and Bacon; 1996.
6. Luttgens K, Hamilton N. *Kinesiology: Scientific Basis of Human Motion.* 9th ed. Boston: WCB McGraw-Hill; 1997.
7. Neumann DA. *Kinesiology of the Musculoskeletal System.* 2nd ed. St. Louis: Mosby Elsevier; 2010.
8. Hamill J, Haddad JM, McDermott WJ. Issues in quantifying variability from a dynamical systems perspective. *Journal of Applied Biomechanics.* Nov 2000;16(4):407–418.
9. Chang R, Van Emmerik R, Hamill J. Quantifying rearfoot-forefoot coordination in human walking. *Journal of Biomechanics.* Oct 20 2008;41(14):3101–3105.
10. Hirashima M, Ohtsuki T. Exploring the mechanism of skilled overarm throwing. *Exercise and Sport Sciences Reviews.* Oct 2008;36(4):205–211.
11. Latash ML. *Synergy.* Oxford: Oxford University Press; 2008.
12. Zatsiorsky VM, Gregor RJ. Mechanical power and work in human movement. In: Sparrow WA, ed. *Energetics of Human Activity.* Champaign, IL: Human Kinetics; 2000.
13. Robertson DGE, Winter DA. Mechanical energy generation, absorption and transfer amongst segments during walking. *Journal of Biomechanics.* 1980;13(10):845–854.
14. Zajac FE. Muscle coordination of movement—a perspective. *Journal of Biomechanics.* 1993;26:109–124.
15. Hof AL. The force resulting from the action of mono- and biarticular muscles in a limb. *Journal of Biomechanics.* Aug 2001;34(8):1085–1089.
16. Putnam CA. Sequential motions of body segments in striking and throwing skills—descriptions and explanations. *Journal of Biomechanics.* 1993;26:125–135.
17. Blazevich A. *Sports Biomechanics. The Basics: Optimising Human Performance.* London: A&C Black; 2007.

PART V

Integrating the Levels

© Andrey Burmakin/ShutterStock, Inc.

Lesson 17 Putting It All Together

Lesson 17

Putting It All Together

© Maria Teijeiro/OJO Images/Getty Images

LEARNING OBJECTIVES

After finishing this lesson, you should be able to:

- Define the following terms: critical elements, intrinsic dynamics, and proficiency.
- List the four levels used in a biomechanical analysis.
- List the steps in analyzing movement.
- List the steps in constructing a hierarchical model.
- List the classes of movement and give examples of each.
- List the phases of movement.
- List the steps involved in determining the cause of a movement dysfunction.
- Compare and contrast the top-down and bottom-up approaches to analyzing movement. Give examples of when you would use each.
- Give examples of several types of constraints.
- Discuss the biomechanics involved with the following activities: standing balance, jumping, landing, cycling, wheelchair propulsion, striking, drinking from a glass, gait, and throwing.
- Describe the dysfunctional movement patterns of valgus collapse and lifting with a fully flexed spine.
- Complete a biomechanical analysis on an activity not discussed in this lesson.

INTRODUCTION

Everything you have learned up to this point was in preparation for this lesson. Although some lessons may have seemed like they were stand alone, this lesson attempts to integrate everything you have learned thus far. You will begin with a discussion of some of the considerations involved with analyzing human movement and conclude with the analysis of several tasks you see as part of the activities of daily living and sport. Consider the examples as not all-inclusive, set-in-stone, this-is-how-it is-done dogma, but rather some ideas on how you can perform a biomechanical analysis on any movement that interests you.

17.1 ANALYZING AND IMPROVING HUMAN MOVEMENT

There are many different ways to analyze human movement. The approach taken here is based on Newtonian physics. Throughout this text, we have taken a whole-part-whole approach. You learned about basic mechanical concepts at the whole body level by representing it as a point mass located at the center of mass (COM). You looked at the individual constituent parts that make up a joint. You also saw how each of these parts fits together into a single-joint system and then how single joints fit together into a multijoint system. The whole-part-whole approach is a nice way to learn mechanics, and it is a nice way to improve someone's mechanics. To begin, you should understand the difference between a top-down and a bottom-up approach (Figure 17.1).

Both approaches distinguish four levels: the whole body (or COM), total limb, joint, and tissue. At times, the total limb level may be equivalent to the whole body level. This may occur with a reaching task, but it would not be appropriate for throwing activities that include a substantial contribution from the lower extremities. The joint level refers to the movements and torques that occur at each degree of freedom. The

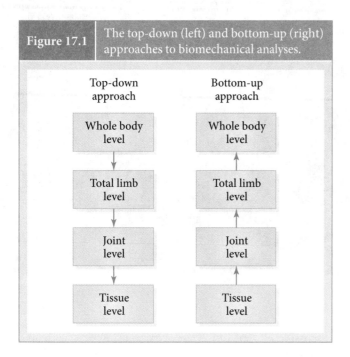

Figure 17.1 The top-down (left) and bottom-up (right) approaches to biomechanical analyses.

tissue level is the constituent parts—muscle–tendon complexes (MTCs), ligaments, joint capsule, articular surfaces, and so on.

Important Point!   There are four levels of biomechanical analysis: whole body, total limb, joint, and tissue.

With a top-down approach, you begin with the task objective and what you need the end effector (hand, foot, body's COM) to do. Then you see how the total limb (or limbs) functions to achieve that task. Next you examine the individual joints to determine how they contribute to the total limb function. Finally, you see how the individual joint motions stress the surrounding tissues. For the analysis of movement of most healthy individuals, I would recommend beginning with this approach. You will see many examples of this type of analysis in this lesson.

The reverse order is a bottom-up approach. You begin at the joint level and keep adding successive parts until you get to the whole movement. This approach is useful when studying pathomechanics and the effects of an intervention. For example, say you were interested in knowing how a patient with anterior cruciate ligament (ACL) reconstructive surgery, cerebral palsy, or stroke was coping with the injury. First, you would see what is different at the tissue level and then how these differences affect the joint(s) involved with the pathology. In the case of

Pathomechanics   The mechanics that are the result of an injury or illness

ACL reconstructive surgery, you would look for things like strength and range of motion at the knee. Assuming there were deficits, how did these deficits affect total limb function? And, again assuming there was an effect, how did it affect the ability to walk, run, or jump?

The bottom-up approach is also helpful in determining the effect of an intervention. Say you were interested in the effects of botulinum-A toxin (Botox) on the functioning of a child with cerebral palsy. You may begin by determining if the injection decreases the spasticity of the targeted and surrounding muscles. Next, you would be interested in finding out if the decreased spasticity (assuming it worked) leads to an increased range of motion at the joint(s). If there was an increased range of motion at the joint(s), you would then see how this changes the child's gait pattern. If there were changes at the total limb or whole body level that were not a result of changes at the joint or tissue level, then you would have to conclude that the effect was on something else (such as coordination).

Analysis of movement is an important part of changing someone's mechanics with the end goal being to either improve performance or decrease the risk of injury. A four-step process for doing so is outlined in Figure 17.2. Each of these steps is outlined next.

17.1.1   Step 1: Determine the Objective of the Movement

One of the *7 Habits of Highly Effective People* is to "begin with the end in mind."[1] It is also a good place to start when analyzing movement. The first question you should ask is: What is the goal of the task?[2] In other words, what is it that the person is actually trying to achieve with the movement? This is an important question because you cannot understand the mechanics of the movement until you understand the purpose of that movement.

Important Point!   You cannot understand the mechanics of the movement until you understand the purpose of that movement.

An analysis will be much better if the goal of the task is clearly defined. For example, the goal of a task may be to jump as high as possible or throw an object as far as possible. Those goals are very clear, but others are less so. These may lead to problems with the next step, which is identifying the mechanical correlates of the goal(s).

17.1.2   Step 2: Identify the Mechanical Correlates of Performance

Identifying the mechanical correlates of the goal shifts the focus to the performer, and what it is the performer is trying

| Figure 17.2 | A four-step approach to changing someone's movement mechanics. |

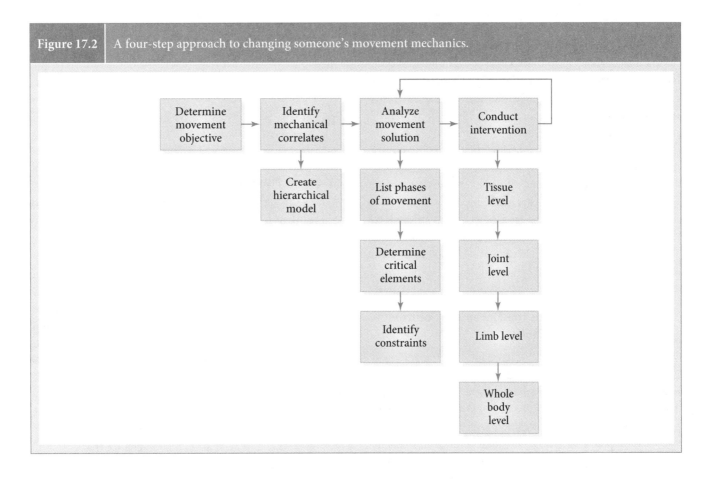

to optimize to achieve the desired goal.[2] During vertical jumping, the goal is to jump as high as possible. The height at the apex of the jump is determined by the height at takeoff and the velocity at takeoff, collectively referred to as the effective energy at takeoff.[3] You have clearly established what it is the performer needs to do to achieve maximal jump height in mechanical terms.

This step is relatively easy if there is a clear optimization criterion (i.e., the performer is trying to maximize or minimize some mechanical variable). It is harder when more than one factor influences the result. If the goal is to jump far instead of high, both the angle of takeoff and velocity of takeoff influence the distance. Although both of these factors influence the distance, they do not have an equal effect. Because the velocity term is squared, it will have a more profound effect than the angle of takeoff. In addition, velocity of takeoff is influenced by the angle of takeoff. So it is not a question of optimizing one factor over another, but rather the optimal interaction between the two.

This step is even more difficult when a criterion cannot be established. For example, what is the mechanical variable to be optimized when walking? The goal may be to get from

point A to point B, but in submaximal tasks such as these it is hard to say what should be maximized or minimized—if anything. For these cases, it may be best to analyze the demands and constraints placed on the task.

These difficulties notwithstanding, the best way to keep track of the mechanical factors affecting the result is by using a hierarchical model (Figure 17.3).[4,5] Hierarchical models were introduced previously, and you have seen them used throughout

| Figure 17.3 | The hierarchical model. |

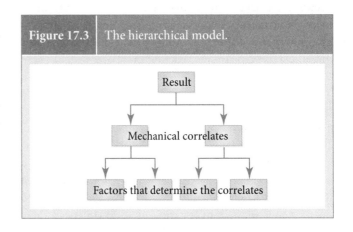

the text. To construct the model you put the result identified in Step 1 in the top box. In the box below, you put the mechanical correlates from Step 2. If you can identify other factors that determine one or more of these mechanical correlates, you put those in boxes below the previously identified factors. You continue this process for as long as you can identify additional factors. The key principle is that each factor should be completely determined by the factors linked to it from below.

Once all the factors are identified, go back and cross out those factors that cannot be changed.[5] This allows you to focus on the things that can be changed. For example, if the force due to gravity or air resistance is included in your model, you should not worry about them because you cannot do anything about them—unless of course you are designing equipment to alter the air resistance.

17.1.3   Step 3: Analyze the Movement Solution and Identify Faults

With the third step, the focus shifts from *what* the performer is trying to do to *how* she is doing it. What are the movement patterns that optimize the mechanical criterion stated earlier?[2] The mechanical criterion can be thought of as the movement *problem*. Here you identify the motor *solution*. How well the motor solution answers the movement problem is called proficiency.[6] So this step is really about determining how proficient someone's movement pattern is.

> **Proficiency**   How well a person performs a movement and achieves the goal of the task

Movements can be complex. It is usually helpful to break a complex movement into smaller, easier-to-analyze pieces. In addition, there may be some parts of the way a person moves that have a huge impact on the movement (called critical elements), whereas others have little to no impact. Because it may be time consuming (and not very fruitful) to try to keep track of every single degree of freedom (DOF), it is helpful to concentrate on the critical elements. Finally, not all options for performing a movement may be available due to certain constraints. Each of these areas is discussed in further detail later.

Classes of Movement

There are three broad classes of movement: discrete, serial, and cyclic. A discrete movement has a distinct beginning and end, without a movement repeating. An example of a discrete movement is the vertical jump. A serial movement has a distinct beginning and end, but links at least two distinct, discrete movements. A triple jump in athletics (where the athlete hops, bounds, and then jumps) would be a good example of a serial movement. A cyclic movement involves repeating a pattern over and over again. Walking, running, swimming, and cycling are all examples of cyclic movements.

> **Important Point!**   Three classes of movement include discrete, serial, and cyclic.

Phases of Movement

Knowing the class of movement is helpful in identifying key phases. Identifying phases are important because each phase contributes to the task performance. If the contribution of a phase to the overall mechanical criterion is less than optimal, performance may be less than optimal and/or more demand may be placed on the system in another phase.

Figure 17.4 shows the phases of a discrete task. A movement can usually be separated into three phases: preparation (countermovement), propulsion (action), and braking (recovery).[5] The purpose of the preparation phase is to put the body in an advantageous position for the propulsive phase, maximize the displacement (and thus energy generated) during the propulsive phase, and initiate the stretch-shortening cycle.[5] If the body is not in an optimal position to initiate the propulsive phase, then there is a good chance that either the movement will not be proficient or is potentially injurious. During the propulsion phase, the MTCs are generating energy and delivering it to the segments, usually following proximal-to-distal sequencing. During the braking phase, energy that was not transferred to an external object must be absorbed by the body (remember the conservation of energy). In addition, the body must be in a position so that it can perform a subsequent movement. Figure 17.4 emphasizes the point that these phases are not separate and distinct, but flow from one into another.

> **Important Point!**   The general phases of a discrete movement are preparation, propulsion, and braking.

During a serial task, each movement of the series may have its own preparation, propulsive, and braking phases.

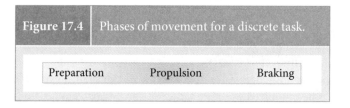

Figure 17.4	Phases of movement for a discrete task.
Preparation	Propulsion	Braking

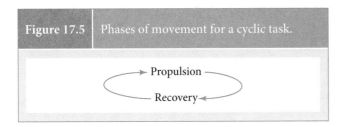

Figure 17.5 | Phases of movement for a cyclic task.

Propulsion
Recovery

Alternately, some phases may be combined. In the triple jump, the braking phase of one movement is the preparation phase of the next one. During cyclic tasks, there are also phases, usually involving propulsion and recovery (Figure 17.5):

- Swimming—hand in the water, and out of it.
- Cycling—pushing down on the pedal, and returning to the top.
- Gait—foot in contact with the ground, and foot in the air.

In addition, each phase may be further divided into subphases. This is usually a good idea when there are several important things happening in the same phase. You will see this when you analyze gait and throwing.

Critical Elements

One criticism of dividing movements into phases is that the division of the phases is somewhat arbitrary.[5,7] Due to the sequential action of multijoint movements, the lower body may be in the propulsive phase while the upper body is still in the preparation phase, or a more proximal joint may be decelerating while a more distal one is accelerating. An alternate method to identifying phases would be to determine "critical elements," or critical features, of the movement.[7] These are key aspects of the movement that are necessary for optimal performance.[8] For example, the depth of a countermovement is a critical element of a standing vertical jump. Obviously, you need a strong knowledge of the activity to determine the critical elements of the movement. Critical elements have been defined for some tasks, but are still missing for a large number of movements.

There are two additional things that you should understand about critical elements.[8] First, there is no correct "standard" for each critical element but rather a range of acceptable values. For example, the knee flexion angle of the vertical jump can be anywhere from 90° to 115°.[8] Second, the timing between critical elements is extremely important. As you have seen, the proximal-to-distal sequencing is an

Critical elements Aspects of a movement that are necessary for optimal performance

important principle of human movement. You will see an extension of this principle for many "upper body" movements such as throwing and striking: critical elements of the lower extremity occur before critical elements of the trunk, which occur before critical elements of the upper extremity. Some authors have referred to critical elements as "nodes."[9] I like this because it emphasizes the fact that these elements are linked through time. As with phases, a deficit in one critical element can increase the demands on another critical element to achieve the same output.

Important Point! The magnitudes of the critical elements are generally not discrete values, but fall in a range of acceptable values.

Important Point! The timing of the critical elements is as important as the critical elements themselves.

Constraints

In addition to examining the phases and/or critical elements that would be employed to achieve a particular mechanical objective, you must also consider the constraints that are imposed on the movement. There are three types of constraints: organismic, environmental, and task.[10] Organismic constraints can be thought of as the intrinsic dynamics of the person.[11] For this discussion, they will include the size and mass of the performer, as well as the range of motion and torque-producing capabilities of each degree of freedom. Environmental constraints are due to the physical surroundings. For example, walking on ice or uneven terrain would impose constraints on a person that are not ordinarily present in an ideal environment. Task constraints are due to the nature of the performance. If the movement is performed fast, you know that the torque production at the DOFs is decreased (at least during concentric actions). This could be thought of as a task constraint. Explicit task constraints can be the rules under which a sporting movement is performed. An often implicit task constraint is that the person does not get injured or lose balance while performing the task, so that he can perform it again.

Understanding constraints leads to some important observations. First, because no two people are "built" exactly alike, each will have a different set of intrinsic dynamics

Intrinsic dynamics The kinematic (range of motion) and kinetic (strength, power, endurance) capacities of each degree of freedom involved in a task (organismic constraints). This may mean that each one will choose a slightly different way to achieve the mechanical objective. Second, unless the task is performed under rather sterile conditions, there may be different environmental constraints that influence the movement solution. Third, constraints may change over time. An obvious long-term example would be an exercise or therapy intervention that increases the strength and/or ROM of the DOFs, while fatigue could change both in the short term. Both change the intrinsic dynamics of the performer, and thus the movement solution they employ.

Identifying Faults

One of the key reasons for performing a movement analysis is because the person is either performing a movement incorrectly or could be doing it better. In other words, they could be more proficient. Either that or they are performing the movement in such a way as to potentially injure themselves (which means they will not be proficient for very long). This means that they are either not doing something they should be doing (critical elements) or are doing something they should not be doing. This determination is not an easy one. It means that the performance must be evaluated against some criteria. Either an exemplary performance or a set of guiding principles can be used.[7] Caution must be used when a model performance is used, as each person has a different set of intrinsic dynamics and may vary their movement accordingly. There are also no universally agreed-upon set of principles that are used across a wide span of movements. Three important ones seem to be the use of the stretch-shortening cycle, sequencing of movements, and maximizing the distance/time over which a force is applied. Minimizing the external torque may be another important consideration in some cases.

17.1.4 Step 4: Determine the Cause and Conduct an Intervention

Many factors determine proficiency and injury risk beyond the movement pattern employed.[7] Thus, although it is not necessarily true that a "better" movement will increase performance or prevent injury, it is probably fair to say that performance will be limited and injury risk will be increased if a faulty movement pattern is employed. Therefore, the final step is to determine the causes of the faulty movement pattern and make the necessary corrections to fix them.

The process for doing so is outlined in **Figure 17.6**. The underlying assumption is that some sort of movement dysfunction exists. Otherwise, there would not be a need to correct something. Determining the presence of a movement dysfunction is done by examining the whole movement.

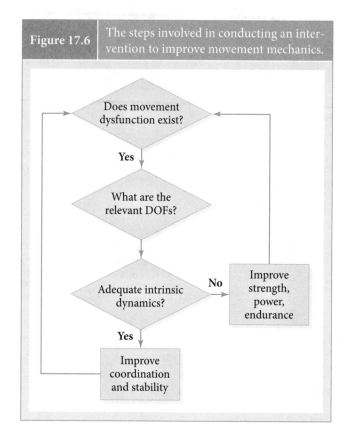

Figure 17.6 The steps involved in conducting an intervention to improve movement mechanics.

Assuming a movement function exists, the next step would be to identify the relevant DOFs involved in the movement. There are 244 DOFs in the body, so this could be a daunting task, but it is simplified a bit if you look for the critical elements of the movement.

Each task will impose certain demands on each DOF. These demands come in the form of range of motion, strength, power, and endurance. A person's intrinsic dynamics suggests that she has a certain capacity at each DOF. If the capacity exceeds the demands of the task, she should be able to perform it well. If the demands exceed the capacities, then impairment exists at one or more DOF, and performance will not be as proficient. Therefore, the next step is to examine the capacities of the individual DOFs. Although many people evaluate strength (the maximum force- or torque-producing capability), it is important to remember the force–velocity relation. Because force and torque output decrease with increased movement velocity (at least during concentric actions), it is important to evaluate power (i.e., torque- and force-producing capabilities at high movement speeds). Because an individual is rarely performing a task just one time, evaluating endurance at force- and velocity-specific conditions is also important.

If the individual parts do not have adequate capacities, then the intervention should focus on fixing them through an exercise program. It is time to evaluate the whole movement again. If a movement dysfunction persists despite adequate intrinsic dynamics, then the possibility exists that the cause lies with impairment of motor control. The intervention should focus on teaching or reteaching the motor skill. The movement would be evaluated again after the intervention.

Finally, performance could be poor even in the absence of impairments described earlier. In such cases, personal factors such as pain, fear, or motivation may be limiting performance. These may be more difficult to uncover and correct, but they are no less important.

In this section, you saw the general procedure used for the analysis of movement. Anything from very crude to very sophisticated measures can be taken for the third step. Obviously, more sophisticated techniques are generally more precise, but take more time, expensive equipment, and expertise to carry out. The fourth step goes beyond biomechanics and requires knowledge of exercise physiology, motor control, and maybe even psychology. Knowledge of biomechanics is but one part in your ability to improve someone's motor functioning. In the next section, we will conclude by examining some select motor activities and common movement faults.

COMPETENCY CHECK

Remember:

1. Define the following terms: critical elements, intrinsic dynamics, and proficiency.
2. List the four levels used in a biomechanical analysis.
3. List the steps in analyzing movement.
4. List the steps in constructing a hierarchical model.
5. List the classes of movement, and give examples of each.
6. List the phases of movement and the critical elements of each phase.
7. List the steps involved in determining the cause of a movement dysfunction.

Understand:

1. Compare and contrast the top-down and bottom-up approaches to analyzing movement. Give examples of when you would use each.
2. Give examples of several types of constraints.

Apply:

1. Describe an appropriate intervention for an individual who does not have adequate intrinsic dynamics.
2. Describe an appropriate intervention for an individual who has adequate intrinsic dynamics.

17.2 ANALYSES OF SELECT BASIC MOVEMENTS

In this section, you will examine some common motor tasks to identify the task objective, the mechanical correlates of the task, and some features of successful movement solutions. Entire volumes have been dedicated to some tasks, such as gait.[12] What follows here is not as much an in-depth analysis of each movement, but rather a discussion that highlights some of the salient features. In addition, two movement faults will be analyzed to show you what is considered dysfunctional movement patterns.

17.2.1 Standing Balance and Stability

You were introduced to some of the requirements for standing balance in the lesson on angular kinetics. There, you learned that the mechanical objective of the task was to maintain static equilibrium, which by definition requires that the sum of the forces are zero and the sum of the torques are zero. Two external forces are involved with standing: the person's weight (force due to gravity; F_g) and the ground reaction force. In this case, you can assume that the ground reaction force is only in the vertical (y) direction, $^yF_{GRF}$. These two forces also create two torques, each of which you can take about the ankle joint center (JC_a). The torques are a product of the forces and the perpendicular distances from the JC_a. The hierarchical model is presented in Figure 17.7.

In this model of standing balance, static equilibrium is maintained as long as the center of pressure (COP), representing the $^yF_{GRF}$ vector, is directly under the center of gravity (COG), representing the F_g vector. Stated another way, the $^\perp d_{COG}$ and $^\perp d_{COP}$ must be equal. If the $^\perp d_{COG}$ is longer in the anterior direction, it will create a net angular acceleration in the negative direction, and if it is longer in the posterior direction, it will create a net angular acceleration in the positive direction (Figure 17.8a). Angular accelerations in the opposite directions are induced if the $^\perp d_{COP}$ is longer (Figure 17.8b). One mechanical constraint is that the COP must be within the base of support (BOS)—you cannot have a ground reaction force outside the BOS. If the person were required to just stand, then a task constraint would be that she could not take a step.

Figure 17.7 | A hierarchical model for standing balance.

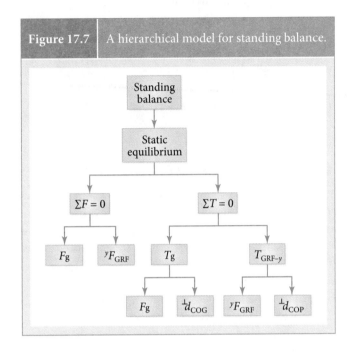

Figure 17.8 (a) If the COG is anterior to the COP, it will induce a forward angular acceleration. (b) If the COP is anterior to the COP, it will induce a backward angular acceleration.

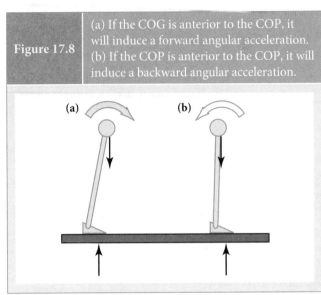

Important Point! To maintain static equilibrium while standing, the location of the center of gravity must be over the center of pressure.

If the COG moves away from the COP, it is not enough for the COP to return under the COG. If it did, there would still be an angular velocity in that particular direction—such a maneuver would only prevent a further increase in angular acceleration. To return the angular velocity to zero, there must be a net angular acceleration in the other direction. To illustrate this, consider the case where the COG moves anterior to the COP. This will induce a negative angular acceleration. If the COP were to return under the COG, there would still be an angular velocity in the negative direction, and she would still be at risk for falling. The COP must move anterior to the COG to induce a positive angular acceleration to return the angular velocity to zero.

If you think about it, the anterior movement of the COP must be pretty precise to create just enough of a positive angular acceleration to return the body to static equilibrium. That is probably unrealistic, and movement of the COP probably induces a positive angular acceleration that is larger than necessary. There is now a positive angular velocity that must be corrected for, and the COP must move posterior to the COG to correct it. Thus, there is probably a constant interplay with moving the COP anterior and posterior to the COG to keep it within the confines of the BOS.

It was originally hypothesized that small changes in the COP could be done using only the ankle joint. This was

known as an "ankle strategy."[13] A plantar flexor torque would move the COP forward and consequently "push back" the COG. Likewise, a dorsiflexor torque would "push forward" the COG. If the deviation was larger, a larger torque than what could be produced by the ankle would be necessary, and a "hip strategy" would be employed.[13] A hip extensor torque would "push back," whereas a hip flexor torque would "push forward." The knee joint was not thought to be involved with these corrections. More recent research suggests that the knee is involved in postural corrections, with a knee flexor torque involved in the "push-back" strategy and the knee extensor torque involved in the "push-forward" strategy.[14] From what you know about interaction torques from other lessons, it only makes sense that a knee torque would also be involved, even if its contributions are not as great as the other two joints. The push-back and push-forward strategies are illustrated in Figure 17.9.

If an even larger correction is necessary, a person may use their arms. If you did not think too hard about it, you would come to the erroneous conclusion that if your body was rotating in the positive direction, you would want to rotate your arms in the opposite direction. But if you were standing at the edge of a pool and your friend gave you a gentle shove, you would quite naturally rotate your arms in the same direction. How the arms affect the position of the COG is best understood in terms of the transfer of angular momentum. Remember that for every torque there is an equal and opposite reaction torque, and the same can be said for angular momentum. A positive rotation of the arms will create a negative rotation of the trunk. Because the trunk is much more

Figure 17.9	The push-back (a) and push-forward (b) movement strategies.

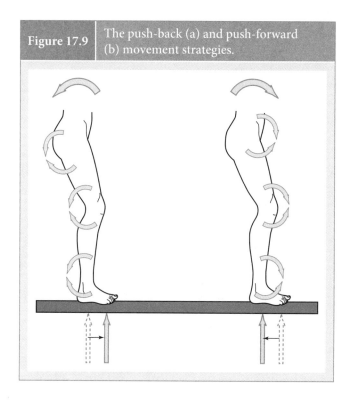

massive than the arms, the arms have to rotate very quickly to generate enough momentum to counteract a positive rotation of the trunk.

Important Point! To rotate the trunk, the arms should be swung in the opposite direction that you want the trunk to rotate.

The discussion thus far has concentrated on balance in the sagittal plane. Although less studied, the same principles apply to balance in the frontal plane (**Figure 17.10**). There are no degrees of freedom in the frontal plane for the knee or the talocrural joints. The net MTC torques that contribute to balance would come from the subtalar and hip joints. At the subtalar joint, eversion torques will accelerate the COM laterally, whereas inversion torques have the opposite effect. At the hip, an abductor torque will accelerate the COM laterally, and an adductor torque will accelerate it medially. Interestingly, the demand on these two joints would be incredibly high if they were the only ones controlling the COM in the frontal plane. A supple trunk that allows lateral flexion is necessary to overcome this difficulty.

Important Point! Subtalar eversion, hip abduction, and trunk lateral flexion work together to move the body's center of mass laterally.

Figure 17.10	Torques inducing a lateral (a) and medial (b) acceleration of the center of mass.

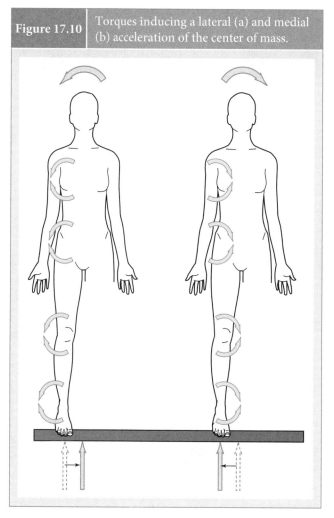

Case in Point

For this case, let us see how improving the intrinsic dynamics of an older adult can improve her balance. I mentioned before that to maintain static equilibrium, the COP must be directly under the COG. In addition, the COP must be within the confines of the BOS (a mechanical constraint). Thus, a theoretical limit for static equilibrium is the border of the BOS. But can that limit be achieved?

Consider a simple "model" with the center of mass on top of a rigid leg segment, a foot segment flat on the floor, and an "ankle" joint connecting the two segments. Let the angle between the vertical and leg segment be the angle of sway. **Figure 17.11** shows you the theoretical limit. Can you see that a larger angle of sway requires a greater torque at the ankle? If our hypothetical elderly woman does not have enough strength (an organismic constraint) to produce that torque, then her angle of sway must be less than the theoretical limit. Let us call that her actual limit. If her angle of sway

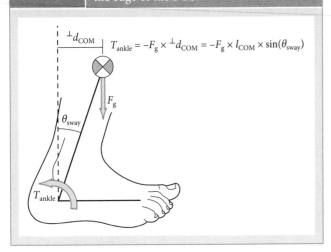

Figure 17.11 Increasing the angle of sway increases the magnitude of the torque at the ankle. The theoretical limit is where the COG is at the edge of the BOS.

$$T_{ankle} = -F_g \times {}^{\perp}d_{COM} = -F_g \times l_{COM} \times \sin(\theta_{sway})$$

exceeds her actual limit, she will fall if she cannot move her feet in time to establish a new base of support. Strengthening her muscles (in this case, the plantarflexors) will increase her actual limit. Provided she has adequate motor control, she will be able to have a larger angle of sway before she falls.

17.2.2 Jumping and Landing

Maximal-height vertical jumping is a very nice activity to study because there is a very clear task objective: jump as high as you can. There are also very clear mechanical correlates to jump height. If you wanted to think in terms of relative jump height (jump height minus standing height), performance is determined solely by the velocity at takeoff. If you wanted to think in terms of absolute jump height (jump height from the ground), performance is determined by both the velocity at takeoff and the height at takeoff.[3] This stems from the conservation of energy. To review the lesson on work–energy–power, energy is conserved while you are in the air:

$$0 = (E_{LK-2} - E_{LK-1}) + (E_{GP-2} - E_{GP-1}) \quad (17.1)$$

The kinetic energy would be zero at the apex of the jump, and the potential energy would be zero at takeoff, so Equation 17.1 becomes

$$E_{LK-1} = E_{GP-1} \quad (17.2)$$

And with the appropriate substitutions and solving for height, you get

$$h = \frac{v^2}{2a_g} \quad (17.3)$$

Important Point! The height of a jump is determined by the velocity of takeoff.

When constructing your hierarchical model, there are two approaches you can take. One uses impulse–momentum, and the other uses work–energy methods. Both of them will be presented here to show you how it is done, but often you will choose one over the other rather than doing both. With both, you make the assumption that, at the instant the propulsive phase begins, the velocity is zero and that the mass of the person does not change during the jump.

Important Point! You can analyze a jump using either impulse–momentum or work–energy principles.

With impulse–momentum, impulse is equal to the change in momentum:

$$F_{effective}\Delta t = m(\Delta v) \quad (17.4)$$

Because the initial velocity is zero, the final velocity ($v_{takeoff}$) is equal to the change in velocity. Rearranging Equation 17.4:

$$^{y}v_{takeoff} = \frac{^{y}F_{effective}\,\Delta t}{m} \quad (17.5)$$

The hierarchical model is presented in **Figure 17.12**.

With work–energy methods, you have two possibilities, depending on which side of the equation you want to include the gravity term. If gravity is considered a force that does work, then it is on the left-hand side of the question:

$$(^{y}F_{GRF} - F_g) \times {}^{y}d_{COM} = \Delta\left(\frac{1}{2}m\,{}^{y}v^2\right) \quad (17.6)$$

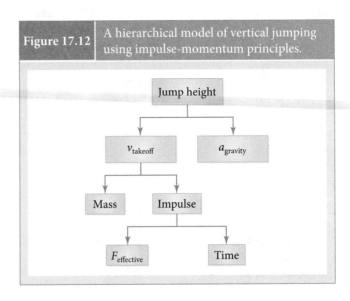

Figure 17.12 A hierarchical model of vertical jumping using impulse-momentum principles.

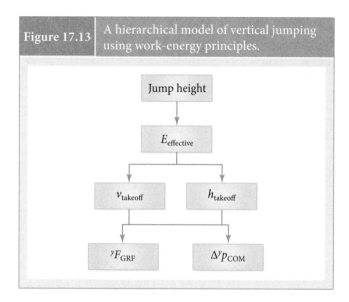

Figure 17.13 A hierarchical model of vertical jumping using work-energy principles.

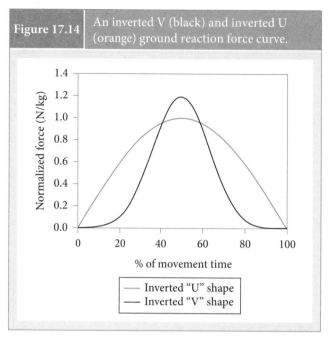

Figure 17.14 An inverted V (black) and inverted U (orange) ground reaction force curve.

In this case, the effective force performs work to change the velocity of the center of mass. If you include gravity on the right-hand side of the equation, then the force doing work is the ground reaction force in the y-direction, and it changes not only the kinetic energy of the COM but also its gravitational potential energy:

$$^{y}F_{GRF} \times {}^{y}d_{COM} = \Delta\left(\frac{1}{2}m\,{}^{y}v^2\right) + \Delta(ma_gh) \qquad (17.7)$$

This is the equation that is used to construct the hierarchical model in **Figure 17.13**.

> **Important Point!**   The force due to gravity can be included on either side of the work–energy equation, but it should not be included on both sides.

There are a few constraints to consider. One is that the projection of the center of mass is straight upward. Another is that the body must take off when the legs are fully extended. A final constraint is that the knee joint angular velocities in particular must be zero at takeoff, or the joints risk injury through hyperextension.

In general, the vertical ground reaction force curve could look like an inverted U or an inverted V (**Figure 17.14**).[15] In general, better jumpers tend to have a higher rate of force development and maintain the ground reaction force at a higher percentage of the maximum force than do lower performers. In other words, they have an inverted U for their vertical ground reaction force curve profile. This means the overall impulse is larger, but the peak is not. This leads to better performance and less potential for injury.

Phases of a jump can follow the basic pattern of preparation, propulsion, and braking. The preparation phase is the countermovement, where the body's COM is lowered. This phase begins with the initiation of the movement and ends when the body's COM is at its lowest point. The propulsive phase begins when the body's COM is at its lowest point and ends at the instant of takeoff. In the case of jumping, the braking phase is the landing.

The basic movement pattern of the jump is flexion–extension–flexion. During the countermovement, the hips and knees flex while the ankles dorsiflex. During the propulsive phase, the hips and knees extend while the ankles plantar flex. During the landing, the hips and knees flex and the ankles dorsiflex. There is a distal-to-proximal sequence during flexion and a proximal-to-distal sequence during extension. This sequencing allows the foot to stay in contact with the ground until nearly full extension is reached.

In general, there is an extensor torque required at all six joints during all three phases. An exception to this is seen at the knee, where there is a small flexor torque prior to takeoff. This flexor torque is thought to be a protective mechanism that prevents the knee from hyperextending. The extensor torques during the countermovement and landing are a result of eccentric MTC actions, whereas concentric MTC actions drive the propulsive phase. The net joint torque power is negative during the countermovement and landing phases and positive during the propulsive phase. Again, the exception to this is that there is a negative knee net joint torque power just prior to takeoff because the net joint torque is a flexor torque, whereas the knee is still extending.

The countermovement serves three important functions over and above increasing the distance (and time) over which the force is applied.[16] First, it activates the stretch reflex, which will lead to a more forceful concentric muscle action. Second, it will allow for the storage of elastic energy, which can be recouped during the propulsive phase. In addition, the shortening tendon is not subjected to the same force–velocity relation of the muscle fiber and can thus alter the force–velocity curve during the concentric muscle action. Third, the muscle is generating force during the descent so that it is "preforced" and starting higher on the force–time curve when the propulsion phase begins.

> **Important Point!**   The countermovement acts to elicit the stretch-shortening cycle, use elastic energy stored in the tendon, and preforce the muscle before the propulsion phase.

Several different energy transfers are thought to exist during the vertical jump.[17] The negative flexor power at the knee immediately prior to takeoff is necessary to prevent the knee from hyperextending. You would think that this has the effect of decreasing jump height. But the energy used to decelerate the thigh (and thus the knee) can be transferred to the ankle for plantar flexion via the biarticular gastrocnemius. This is a link-to-joint transfer. If the quadriceps continued to generate energy at this point, it is transferred by the gastrocnemius via a joint-to-joint transfer. Finally, the energy generated by the gastrocnemius to plantar flex the ankle results in an increase in the energy of all the body's segments (joint-to-link energy transfer).

A long jump is a jump for distance as opposed to height. The mechanics of a long jump are similar to those of a vertical jump. The major difference is that the COM rotates about the ankles in the direction of jump prior to the propulsive phase.[18] See Figure 17.15.

Energy is not added to the body while it is in the air, so the same amount of energy that is generated at takeoff must be absorbed at landing. In landing, the leg flexes in the opposite order of extension (from distal to proximal). Leg stiffness determines many aspects of the landing. If the leg is stiff, the COM will displace little, and consequently the vertical ground reaction force will be much larger. Because the joints are not displacing, little energy can be absorbed by the muscles and must therefore be absorbed by the passive structures of the body. If the leg is less stiff, the COM will undergo a greater displacement and the vertical ground reaction force will be smaller. The muscles will absorb a greater amount of energy

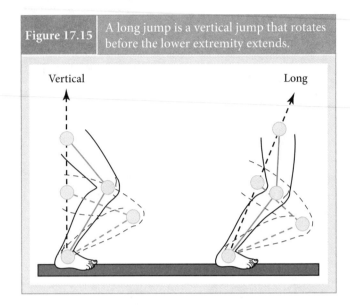

Figure 17.15   A long jump is a vertical jump that rotates before the lower extremity extends.

as the joints go through a greater range of motion. Somewhat ironically, the "softer" landing puts a greater demand on the muscles. See Figure 17.16.

> **Important Point!**   A softer landing is safer to perform because it decreases the magnitude of ground reaction force, yet it is more demanding because it increases the torque demand at each joint.

Case in Point

Landings can provide another example of how a person's intrinsic dynamics can affect a movement pattern. Consider a female adolescent volleyball player who must jump (and land) an estimated 200 times during a match. Let us focus on the landing, where her body must absorb energy.

Recall that the MTC has a huge capacity to absorb energy, but that it must be activated and lengthening to do so. At the joint level, this means that the MTC torque and the joint angular displacement must be in opposite directions. This is apparent during the landing, when there are extensor torques at the hip, knee, and ankle while those joints are flexing. Greater flexion angles require greater MTC torques, and she must have the requisite strength and ranges of motion to achieve those angles.

If she is deficient in either strength or ROM, then her MTCs will be limited in the amount of energy they can absorb. Energy not absorbed by the MTCs must be absorbed by other structures in the body (such as bones, cartilage, and ligaments). In addition, her MTCs must generate these torques while moving fairly rapidly—indicating that power is an important variable. And because she must do this repeatedly, power endurance is particularly important. As she fatigues,

The movements (a) and ground reaction forces (b) of "stiff" and "soft" landings.

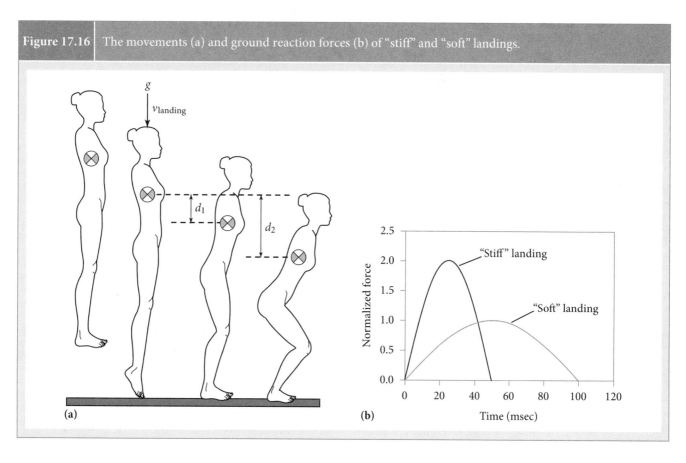

(a)

(b)

her flexion angles may decrease—indicating that less energy is absorbed by the MTCs. Her coach should make the appropriate substitutions before an injury occurs.

17.2.3   Cycling

Cycling is another multijoint activity that, like vertical jumping, occurs primarily in the sagittal plane. Unlike jumping, there is no clear performance criterion unless you are talking about maximum-speed cycling. In that case, the criterion is the maximum velocity of the bike, which is determined by the $^{\perp}F_{foot}$. Because the pedals are rotating fairly rapidly, it may be best to think in terms of the power delivered to the pedal rather than the force. Constraints on cycling include the fact that the feet stay in contact with the pedals, which follow a circular path. A mechanical constraint is that only the force perpendicular to the crank arm, $^{\perp}F_{foot}$, produces torque (Figure 17.17).

| Crank arm | The rigid body connecting the chain ring to the pedal |

Pedaling is an "in-between" movement pattern. While the foot is moving in a circular path in space like a swing or a whip pattern, the segments are rotating in opposite directions like a flexion/extension pattern. This is caused by the constraints of

the pedal. During the propulsion phase, the hip and knee are extending. Both joints are flexing during the recovery phase. The ankle stays somewhat neutral during the entire cycle, although some cyclists may try to plantar flex at the end of the propulsive phase.

There are extensor torques at all three joints (hip, knee, and ankle) during the propulsion phase. The kinetics of the recovery phase depends on whether or not toe clips are used. Toe clips can assist with the pedaling cycle by having the recovery leg "pull up" on the pedal. This would require flexor moments at

Only the component of the force of the foot that is perpendicular to the crank arm is effective in producing torque.

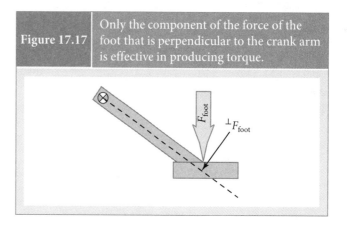

the joints. If toe clips are not used, the recovery leg could still lift itself up through hip and knee flexion, although none of that energy can be transferred to the pedal. Alternatively, the recovery leg could flex passively by the action of the propulsive leg pushing down on the one pedal causing the other pedal to rise up.

During the propulsive phase, the hip extensors and knee extensors cannot deliver their energy directly to the pedal. Rather, they deliver their energy to the lower leg (shank). The ankle plantar flexors transfer the energy from the shank to the pedal.[19] Stated another way, without the plantar flexors stiffening the ankle, the hip and knee extensor torques would drive the ankle into dorsiflexion. The plantar flexor torque counteracts this effect, allowing the hip and knee extensors to create force on the pedals (**Figure 17.18**).[20]

Important Point!   The ankle plantar flexors allow the energy generated by the hip and knee extensors to be delivered to the pedal.

Pedaling also highlights the role of biarticular muscles.[19] At the transition from the propulsive to the recovery phase, the monoarticular hip and knee extensors cannot deliver energy to the crank because of the position of the pedal (their contribution to $^{\perp}F_{foot}$ is minimal). However, the biarticular hamstrings can because of their large translational component. Similarly, during the transition from the recovery to the propulsive phases, the biarticular rectus femoris delivers energy to the crank that the monoarticular muscles cannot. This allows for a smooth transition between phases.

| **Figure 17.18** | Only if the ankle is "locked" in position can the force of the hip and knee be applied to the pedal. |

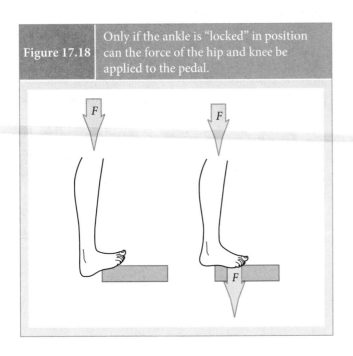

| **Figure 17.19** | A hierarchical model of wheelchair propulsion. |

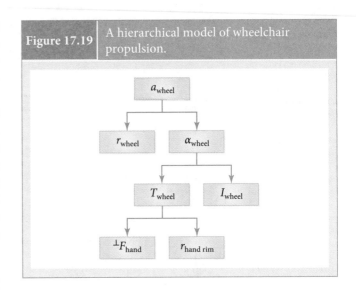

Important Point!   The biarticular muscles of the hip and knee allow for smooth transitions between propulsion and recovery and recovery and propulsion.

17.2.4   Wheelchair Propulsion

There has not been as much research on wheelchair propulsion as there has been on cycling, but many of the principles are the same. The hierarchical model shows that the acceleration of the wheel is determined by four factors: the radius of the wheel, the moment of inertia of the wheel, the radius of the hand rim, and the perpendicular force of the hand (**Figure 17.19**).

During the push phase, the shoulder joint is consistently flexing in the sagittal plane. The motions in the other planes are less consistent. During the first portion of the propulsion phase (where the hand is pulling up on the hand rim), the elbow is flexing, but it transitions to extending during the second portion (where the hand is pushing down on the hand rim). The relative amount of time in flexion compared to extension decreases with increased movement speed.[21] Thus, the propulsion phase can be further divided into a phase where there is a swing pattern (shoulder flexion–elbow flexion; commonly called a pull phase) and a phase where there is an extension pattern (shoulder flexion–elbow extension; commonly called a push phase). During the recovery phase, the hand can follow either a circular path or a linear path (**Figure 17.20**). The path of the recovery hand changes from a circular path to a linear path as the movement speed increases from 40% to 80% of maximum.[21]

The torque produced on the wheel is presented in **Figure 17.21**. Note that there are two peaks in the torque during the propulsion phase. The torque is determined

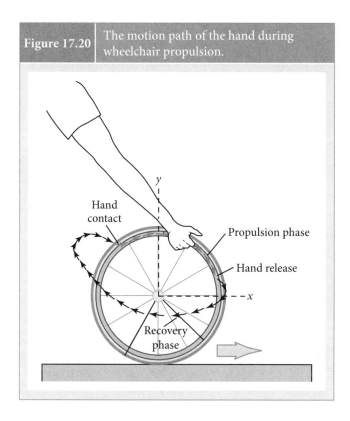

Figure 17.20 The motion path of the hand during wheelchair propulsion.

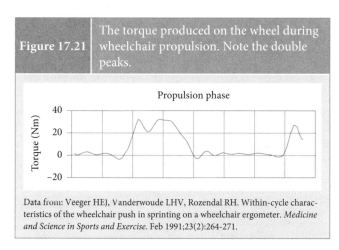

Figure 17.21 The torque produced on the wheel during wheelchair propulsion. Note the double peaks.

Data from: Veeger HEJ, Vanderwoude LHV, Rozendal RH. Within-cycle characteristics of the wheelchair push in sprinting on a wheelchair ergometer. *Medicine and Science in Sports and Exercise.* Feb 1991;23(2):264-271.

largely by the perpendicular force of the hand, which in turn is determined by the shoulder and elbow torques (**Figure 17.22**). (There could also be a considerable contribution from the trunk, but research is lacking.) At the shoulder joint, there is a flexor torque throughout the propulsion phase. At the elbow, there is initially a flexor torque followed by an extensor torque. Interestingly, the switch does not occur when the kinematics change from flexion to extension but when the flexor torque is no longer effective in delivering a perpendicular force to the wheel. This occurs

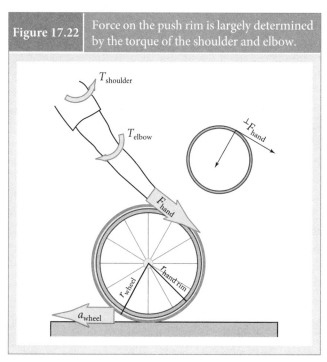

Figure 17.22 Force on the push rim is largely determined by the torque of the shoulder and elbow.

in the trough between the two peaks and happens well after the elbow starts flexing.[22]

Important Point! The elbow torque transitions from a flexor to an extensor during the propulsion phase when the flexor torque can no longer provide an effective force on the hand rim.

During the late recovery phase, there are shoulder and elbow flexor torques to decelerate and position the arm.[21] This should indicate to you that the stretch-shortening cycle is present, even in wheelchair propulsion, and provides further evidence that this is a fundamental principle of human movement.

Case in Point

Wheelchair propulsion demonstrates the interaction between anatomy and mechanics, or organismic and task constraints. Consider a new wheelchair user. If you were going to instruct her on how to use the chair using strictly mechanical principles, you try to get her to have the force of the hand applied in the tangential direction. However, this is not the most biomechanically efficient direction to apply the force.[23] More force can be applied in the tangential direction if the force is applied at an angle using the movement pattern described earlier. Human movement is subjected to mechanical, biological, and multisegment principles. Oftentimes, the biological principles supersede the other two.

17.2.5 Striking

When you think of striking, combat sports immediately come to mind. Striking includes more than just punching or kicking; any time you hit a ball with an implement, you are striking it. Baseball, softball, tennis, and golf are all activities that involve striking. Striking activities involve impacts and so are best studied using conservation of momentum principles. The objective of striking an object is slightly different in the combat sports than it is in the other activities listed. Neglecting the accuracy requirements, in both cases the mechanical criterion is the same: to impart as much momentum from your body (Body A) to the body you are striking (Body B). Certainly the mass and the velocity of Body B, as well as the coefficient of restitution, are important in determining the outcome of the impact. Assuming those things cannot be changed, the outcome is determined by the effective mass and velocity of Body A (**Figure 17.23**).

Important Point! Remember, in the mechanical sense a "body" is an object of interest. A ball is considered a body.

The phases involved in striking vary widely, depending on the tactical situation in which they are employed. For example, in a tennis serve there is a windup, which is the preparation phase. Such a windup would not be possible in a combat sport because it would "telegraph" the intention of the strike. However, a strike may be preceded by another one that can have the effect of a windup. A jab serves this function before a cross in boxing.

In general, strikes follow a preparation, propulsion, impact, follow-through, and recovery (**Figure 17.24**). During the propulsion phase, regardless of the movement pattern, almost all strikes follow a proximal-to-distal sequencing. In addition, the requirements for effective mass and velocity at impact seem to be at odds with one another: Increasing effective mass requires an increase in stiffness of the body, but increased stiffness decreases endpoint velocity. These two competing demands appear to be met by a "pulsing" or double activation, particularly of the more proximal musculature.[24]

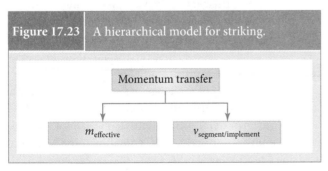

Figure 17.23 A hierarchical model for striking.

Momentum transfer

$m_{effective}$ $v_{segment/implement}$

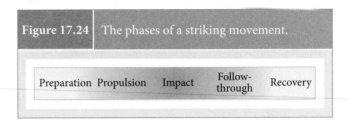

Figure 17.24 The phases of a striking movement.

Preparation Propulsion Impact Follow-through Recovery

Recall that peak velocity of a more proximal segment occurs before the velocity of the next, distal segment and that this process repeats down the chain. This would suggest that the more proximal muscle groups have a burst in activity to accelerate the proximal segment, and this burst occurs earlier in the movement than a burst in activity of the distal muscles. It appears as though the proximal muscles relax as the distal muscular increases. Then, at the instant before impact, the muscle activity bursts again to stiffen the body and increase effective mass at impact.

Important Point! During striking, the distal musculature relaxes after the initial acceleration and then acts again just prior to impact to stiffen the joints.

Just as the phases of a strike vary, so too do the movement patterns. A tennis stroke follows a whip pattern, whereas a boxing jab is an extension pattern. Certain martial arts kicks use a whip, and others use an extension pattern.

I mentioned earlier that phases are sometimes arbitrary, and the lower body may be in a propulsive phase while the upper body is still in a preparation phase. An alternative to this would be to break the movement up into critical elements that link the movements. A tennis serve is an example where this type of analysis is fruitful. In a tennis serve, the critical elements include[9]

1. Adequate knee flexion in cocking to knee extension at impact.
2. Hip and trunk rotation away from the court in cocking.
3. Scapular retraction and arm rotation in the scapular plane.
4. Weight transfer from back leg to front leg at ball impact.
5. Long axis rotation into ball impact and follow-through.

Notice that few of the critical features actually involve the upper extremity. One of the hallmarks of effective striking techniques is that energy is generated in the legs and trunk and then transferred to the arms. Indeed, the lower extremities and trunk contribute more energy than does the upper extremity for most striking skills. For examples, compare the relative estimated

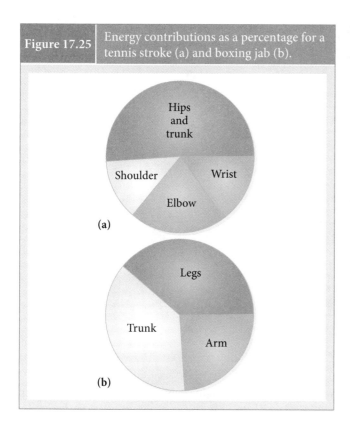

Figure 17.25 | Energy contributions as a percentage for a tennis stroke (a) and boxing jab (b).

Table 17.1 | The Phases for Drinking from a Glass

Phase	Start point	End point
Reaching and grasping	Hand begins movement	Hand begins to move toward mouth with glass
Forward transport	Hand begins to move toward mouth with glass	Drinking begins
Drinking	Drinking begins	Drinking ends
Backward transport	Drinking ends	Glass returned to table
Returning	Glass returned to table	Hand is back in initial resting position

Reaching and backward transport can be thought of as extension patterns, whereas forward transport is a flexion pattern. The angle-angle diagram for the shoulder and elbow is presented in Figure 17.26.[28] In addition to these motions, there is also considerable motion of the shoulder in the frontal plane (about 40° of abduction during the drinking phase).[28]

Object manipulation is part of the topic of angular kinetics. The hierarchical model, assuming a virtual finger, is represented in Figure 17.27. This model assumes that the object is held stationary in the hand, and rotating the cup in space is done by moving the wrist.

relative energy contributions of the various body segments during a tennis stroke[25] and a boxing punch (Figure 17.25).[26]

Case in Point

A tennis serve highlights the importance of identifying the critical elements of a movement. The first critical element identified earlier was adequate knee flexion during cocking. Consider a male tennis player who has restricted knee flexion. To achieve the same speed of the racquet head, he is going to have to generate energy from another source (another DOF). A likely candidate would be shoulder internal rotation torque. But by increasing shoulder internal rotation torque, he will also increase valgus loading at the elbow[27]—making him more susceptible to injury.

17.2.6   Drinking from a Glass

Drinking from a glass is something that you probably do several times a day without even thinking about it. The movement is usually smooth, coordinated, and done without error until there is an impairment that occurs after such events as an injury or stroke. Drinking involves several movements, including reaching for the glass and manipulating the glass once it is in your hand. Five phases have been identified with the movement (see Table 17.1).[28]

Figure 17.26 | A shoulder-elbow angle-angle diagram during a drinking task.

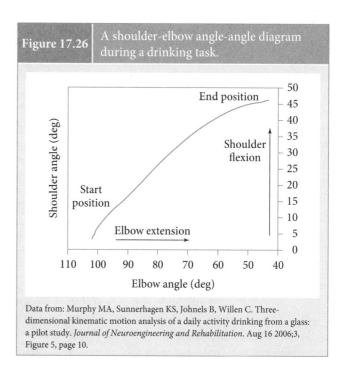

Data from: Murphy MA, Sunnerhagen KS, Johnels B, Willen C. Three-dimensional kinematic motion analysis of a daily activity drinking from a glass: a pilot study. *Journal of Neuroengineering and Rehabilitation*. Aug 16 2006;3, Figure 5, page 10.

Figure 17.27 | A hierarchical model for grasping.

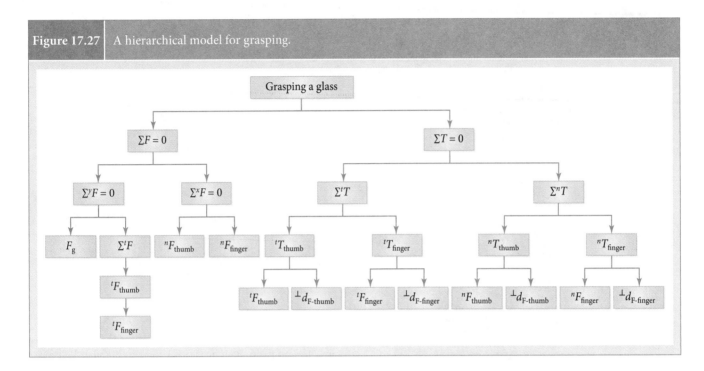

The forces in the hand must be sufficiently large to create a force in the normal direction that the force of friction will cancel the force due to gravity, but not so large as to crush the object that is being held. This force is created by the torque generated by the flexor digitorum superficialis and flexor digitorum profundus for the fingers. The thumb is a bit more complicated because the thumb motion is opposition. Torque is generated by the flexor, abductor, and opponens muscles of the thumb.

In addition, the forces generated by the fingers create torque about the center of mass of the object being held. To hold the object in static equilibrium, the net torque must also be zero. Titling a glass or writing with a pen requires precise control of these torques, which would not be zero during movement. You can begin to see how precise this control must be, and how a neurological impairment (such as a stroke or traumatic brain injury) can affect this functioning.

Case in Point

Patients with stroke provide a good example of compensatory motion during the reaching phase. If you are working with a patient with stroke, you will probably notice that he has a decreased range of motion at the elbow. To compensate for this, he will likely flex the trunk so that his hand can still reach the glass.[29] As a clinician, you now have a choice: You can either try to improve his impairment to make his intrinsic dynamics (and thus movement pattern) more "normal," or you could work to improve the capacity of the DOFs that he is using (i.e., the trunk) so that he is

more proficient with a different movement pattern. Which one would you choose?

17.2.7 Gait

Gait involves terrestrial locomotion. You may examine crawling in infants if you are interested in motor development, but here you will focus on typical adult human gait: walking and running. With maximum-speed running, like maximum-speed cycling, there is a clear performance objective—maximum average velocity. With walking and submaximal running, the choice of performance criterion is not as clear.

Walking

Walking is undoubtedly the most studied of all human activities. As mentioned earlier, there is no universally accepted criterion for walking. Gait velocity is determined by step length and step rate. Mathematically, you would want to maximize both of them if your goal was to maximize gait velocity. But it is important to remember that increasing stride length is only good up to a point. When you overstride, there is a very large braking impulse that is hard to overcome, which causes you to lose speed.

There are numerous ways to break the gait cycle up into phases. The simplest involves periods where the foot is in contact with the ground (stance) and periods when it is not (swing). The stance phase can be further broken down to periods of double-limb support (both feet flat in contact with the floor) and single-limb support (where one foot is in contact

Figure 17.28	The phases of walking gait.

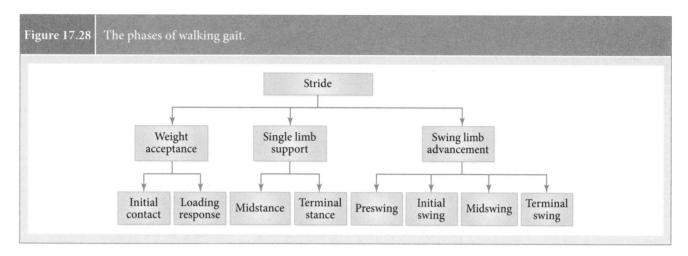

with the floor). However, a slightly different approach will be taken here. A stride will be looked at in terms of three functional tasks that need to be performed.[12] During a stride, you have to (1) accept weight onto the foot after it makes contact with the ground, (2) support your weight on a single leg, and (3) advance the swing limb in front of the body so that you can move forward. Within each of these tasks, there are several subphases (**Figure 17.28**). The demarcation of each phase, its objectives and critical elements are listed in **Tables 17.2** to **17.4**.[12]

Table 17.2	The Phases for Walking Gait

	Phase	Start point	End point
Weight acceptance	Initial contact	Foot just touches the floor	Foot is flat on the ground
	Loading response	Foot flat on the ground	Opposite foot lifts off of the ground
Single-limb support	Midstance	Opposite foot lifts off of the ground	COM is aligned over the foot
	Terminal stance	COM is aligned over the foot	Initial contact of the opposite foot
Swing limb advancement	Preswing	Initial contact of the opposite foot	Foot initially leaves the ground
	Initial swing	Foot initially leaves the ground	Swing foot is opposite stance foot
	Midswing	Swing foot is opposite stance foot	Shank is vertical
	Terminal swing	Shank is vertical	Foot just touches the floor

Table 17.3	The Objective for Each Phase of Walking Gait

Phase	Subphase	Objective
Weight acceptance	Initial contact	Position the limb for stance
	Loading response	Absorb shock
		Stabilize superincumbent body weight
		Preserve forward progression
Single-limb support	Midstance	Progress body over stationary foot
		Stabilize the supporting limb and trunk
	Terminal stance	Progress body beyond the stationary foot
Swing limb advancement	Preswing	Position the limb for swing
	Initial swing	Advance the limb from its trailing position
	Midswing	Clear the foot over the ground
		Advance the limb
	Terminal swing	Finish advancing the limb
		Prepare the limb for stance

Table 17.4	The Critical Elements for Each Phase of Walking Gait	
Phase	**Subphase**	**Critical elements**
Weight acceptance	Initial contact	Contact the floor with the heel
	Loading response	Stabilize the hip
		Restrain knee flexion
		Restrain ankle plantar flexion
Single-limb support	Midstance	Stabilize hip in the frontal plane
		Extend the knee
		Restrain ankle dorsiflexion
	Terminal stance	Forward free fall of the body
		Raise heel
Swing limb advancement	Preswing	Flex the knee
	Initial swing	Flex the hip
		Flex the knee
	Midswing	Flex the hip
		Dorsiflex the ankle
	Terminal swing	Decelerate hip flexion
		Decelerate knee extension
		Neutral position of the ankle

Important Point!   The three functional tasks of the gait cycle are weight acceptance, single-limb support, and swing limb advancement.

During the stance phase, the body acts like an inverted pendulum (**Figure 17.29**). Potential and kinetic energy of the COM is exchanged during the stance phase (**Figure 17.30**). To achieve this inverted pendulum-like motion at the COM, the stance leg has to be fairly stiff during walking. It is not perfectly rigid; there is some leg flexion to attenuate the ground reaction force during weight acceptance. The net muscle torques required to achieve this are presented in **Figure 17.31** and the resulting ground reaction forces in **Figure 17.32**. As you can see, the net muscle torque is highest at the ankle;

they are lower and more variable at the hip and knee.[30] Some energy is absorbed during weight acceptance, but most of the energy during single-limb support is transformed between kinetic and potential energy. Surprisingly little energy is generated by the hip or knee extensors. Energy is generated during swing limb advancement. As you might have guessed, the hip flexor torque and ankle plantar flexor torques are the

Figure 17.29	The body as an inverted pendulum during walking gait.

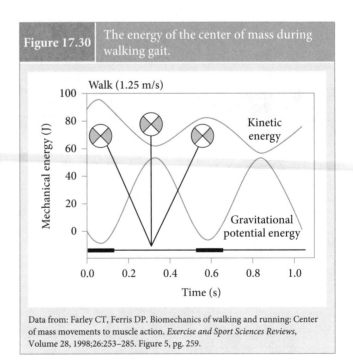

Figure 17.30	The energy of the center of mass during walking gait.

Data from: Farley CT, Ferris DP. Biomechanics of walking and running: Center of mass movements to muscle action. *Exercise and Sport Sciences Reviews*, Volume 28, 1998;26:253–285. Figure 5, pg. 259.

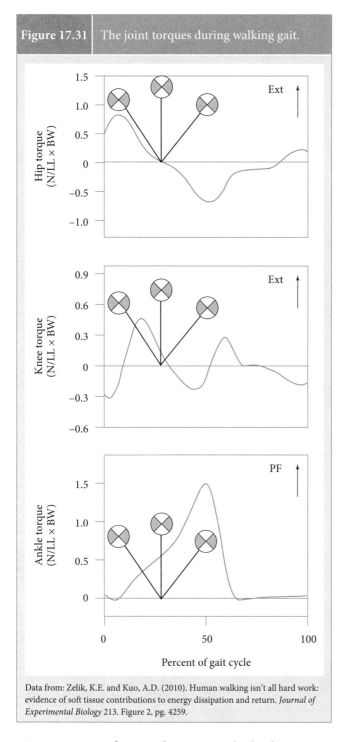

Figure 17.31 | The joint torques during walking gait.

Data from: Zelik, K.E. and Kuo, A.D. (2010). Human walking isn't all hard work: evidence of soft tissue contributions to energy dissipation and return. *Journal of Experimental Biology* 213. Figure 2, pg. 4259.

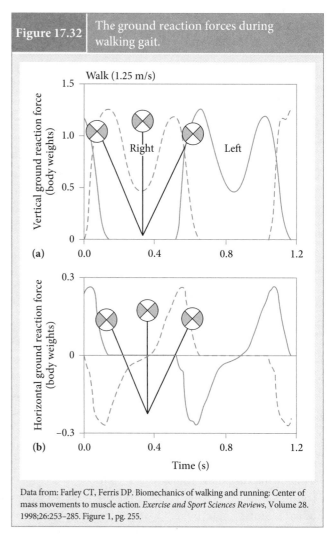

Figure 17.32 | The ground reaction forces during walking gait.

Data from: Farley CT, Ferris DP. Biomechanics of walking and running: Center of mass movements to muscle action. *Exercise and Sport Sciences Reviews*, Volume 28. 1998;26:253–285. Figure 1, pg. 255.

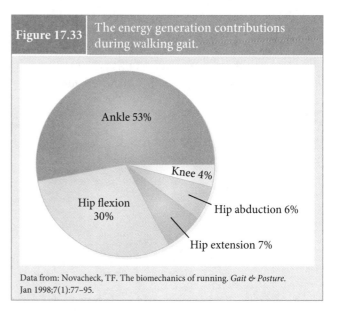

Figure 17.33 | The energy generation contributions during walking gait.

Data from: Novacheck, TF. The biomechanics of running. *Gait & Posture*. Jan 1998;7(1):77–95.

primary sources of energy during swing limb advancement (**Figure 17.33**).[31]

Running

Running, particularly maximum running, usually involves doing so for a prescribed distance. Running can therefore be analyzed in the macrocosm (entire length of the race) or the microcosm (one stride). To analyze a race in the macrocosm, critical elements include the top speed (instantaneous speed), how long it took the runners to get to the top speed (acceleration), how long they could hold their top speed, and what the difference was between top speed and final speed.

In the microcosm, it is usually instructive to examine running by comparing and contrasting it to walking. An obvious difference is that, in walking, there is always one foot in contact with the ground, and there are periods when both feet are in contact with the ground (double-limb support). In running, there is an airborne phase (where neither foot is in contact with the ground), and there are no periods of double-limb support.

> **Important Point!** The body acts like an inverted pendulum during walking and a spring mass during running.

Where walking gait is modeled as an inverted pendulum, running gait is modeled as a mass-spring. With the spring-mass model of the running gait, the leg is not as stiff as it is in walking. The leg flexes during the first half stance phase and extends during the second half of the stance phase (**Figure 17.34**). This explains why the COM is at its lowest height during midstance, which also corresponds to its lowest velocity. Hence, kinetic and potential energies are nearly in-phase and cannot be transformed from one to the other (**Figure 17.35**). The magnitudes of the net muscle torques are larger during running compared to walking.[30] In contrast to walking, the net muscle torques are the largest at the knee.[30] The net muscle torques and resulting ground reaction forces are presented in **Figures 17.36** and **17.37**. Unlike walking, an active leg extension during the second half of stance phase propels the body forward and into the air (this is achieved by the inverted pendulum during walking). It should not surprise you, then, that there is a much greater

| Figure 17.34 | The body as a spring mass during running. |

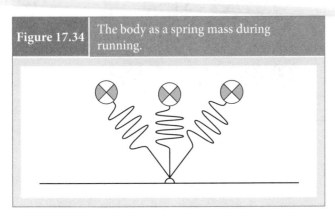

| Figure 17.35 | The energy of the center of mass during running. |

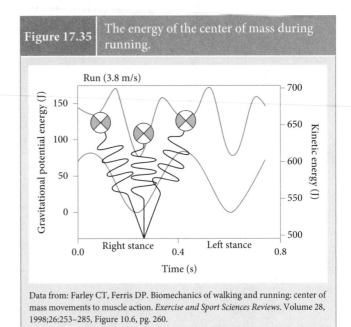

Data from: Farley CT, Ferris DP. Biomechanics of walking and running: center of mass movements to muscle action. *Exercise and Sport Sciences Reviews*. Volume 28, 1998;26:253–285, Figure 10.6, pg. 260.

energy contribution from hip and knee extension during running. The energy contributions of two running speeds are presented in **Figure 17.38**.[31]

Case in Point

A child with spastic hemiplegic cerebral palsy illustrates how people adapt their movement to their available intrinsic dynamics.[32] As I mentioned earlier, walking gait uses an inverted pendulum mechanism where most, but not

| Figure 17.36 | The joint torques during running. |

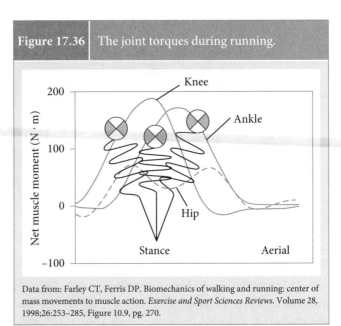

Data from: Farley CT, Ferris DP. Biomechanics of walking and running: center of mass movements to muscle action. *Exercise and Sport Sciences Reviews*. Volume 28, 1998;26:253–285, Figure 10.9, pg. 270.

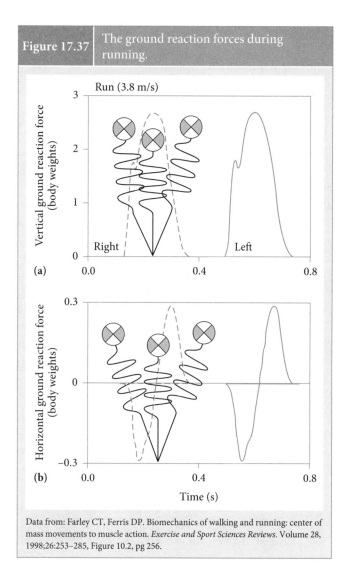

Figure 17.37 | The ground reaction forces during running.

Data from: Farley CT, Ferris DP. Biomechanics of walking and running: center of mass movements to muscle action. *Exercise and Sport Sciences Reviews*. Volume 28, 1998;26:253–285, Figure 10.2, pg 256.

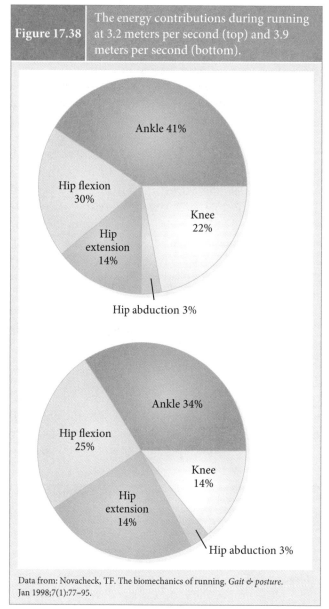

Figure 17.38 | The energy contributions during running at 3.2 meters per second (top) and 3.9 meters per second (bottom).

Data from: Novacheck, TF. The biomechanics of running. *Gait & posture.* Jan 1998;7(1):77–95.

all, energy is conserved. Some energy is lost during weight acceptance and must be replaced by the hip flexors and ankle plantar flexors during swing limb advancement. The child with cerebral palsy is not able to supply this energy with concentric muscle actions on their affected sides and therefore must adjust his movement pattern to make up for it. Typically, this is done by having the unaffected limb act like an inverted pendulum (as in normal walking) and the affected limb act like a mass spring (as in normal running).

On the unaffected side, the child will typically raise his COM higher than is seen with a normal child. This requires more work. The increased potential energy of the COM is then used to load the spring on the spastic, affected side when weight is transferred to it. The elastic energy on the spastic side is then used to propel the body back onto the unaffected side. Thus, instead of an exchange between

gravitational potential and kinetic energies, the gait of the child with cerebral palsy has an exchange between gravitational and strain (elastic) potential energies. Although this type of gait pattern requires more energy and thus more work to raise the COM, the efficiency is actually close to normal gait.[33] This suggests that the child found ways to exploit his unique intrinsic dynamics while walking.[32] As with the patient with stroke, as a clinician you have two options during rehabilitation: (1) try to improve his intrinsic dynamics and make his gait pattern more "normal," or (2) teach him to better use the intrinsic dynamics that are available to him.

17.2.8 Throwing

In a previous lesson, you learned about the different patterns used in "throwing" a shot put versus a baseball. Here the emphasis will be on overhand throwing using a whip pattern. Similar to the mechanical criterion in striking, the goal in throwing usually involves releasing the object with as much velocity as possible (in a particular direction, of course). To maximize the (linear) speed of the ball at release, you have to maximize the linear speed of the hand. To maximize the linear speed of the hand, you have to maximize the length of the hand, the angular velocity of the wrist, and the linear velocity of the wrist. To maximize the linear velocity of the wrist, you have to maximize the length of the forearm, angular velocity of the elbow, and linear velocity of the elbow. You would reiterate this process all the way through the non-throwing-side hip. You cannot change your segment lengths, so the velocity of a throw is largely determined by the linear speed of the lead hip and the angular velocities of all joints up to the throwing-side hand (**Figure 17.39**).

You now know that maximum velocity of the endpoint does not occur when each joint reaches its maximum velocity simultaneously, but follows a proximal-to-distal sequence that occurs in the context within the phases of a movement.[34] A pitch can be broken up into the three standard phases (preparation, propulsion, braking), sub-phases within each, and in some cases, sub-subphases (**Figure 17.40**). The start and ending points of each phase are listed in **Table 17.5**.

The stride is important because it increases the velocity of the lead hip. The stride also positions the lead hip, which is important for rotation of the pelvis. Arm cocking increases the external rotation and horizontal abduction of the shoulder, which puts the prime movers on stretch and increases the range of motion for the acceleration phase. It is estimated that the external rotation can be up to 180° during arm cocking.[35] During the acceleration phase, the shoulder internal rotation velocity can reach up to 7,000°/per second.[35] Although proximal-to-distal sequencing occurs for rotations in the sagittal plane, internal rotation of the shoulder usually happens after shoulder and elbow extension.[36] The large amount of kinetic energy generated during the propulsion phase must be absorbed by the body during the braking phase. During the braking phase, the shoulder external rotators must act eccentrically to control shoulder internal rotation.

Case in Point

The size and strength of the internal rotators are much smaller than external rotators and are therefore under tremendous stress during this phase. Therefore, it is imperative to increase the duration of the deceleration phase and involve more joints

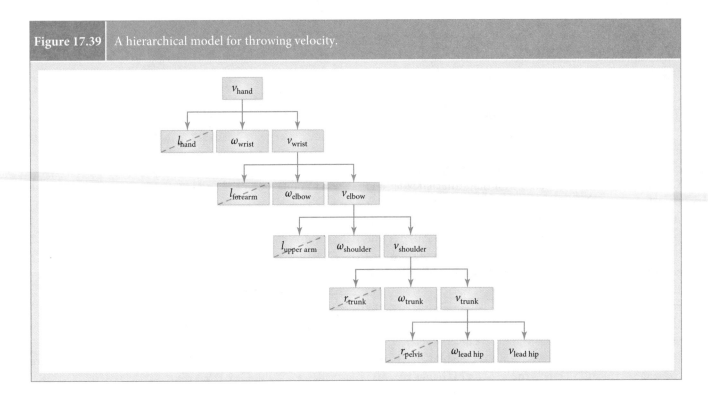

Figure 17.39 | A hierarchical model for throwing velocity.

Figure 17.40 | The phases of an overhead pitch.

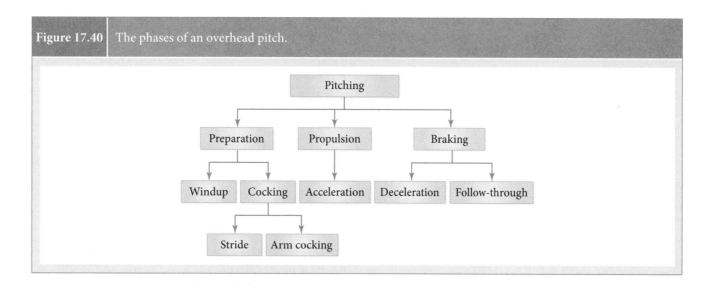

Table 17.5 | The Phases for an Overhead Pitch

Phase	Start point	End point
Windup	Initiation of leg movement	Lead knee reaches maximum height
Stride	Lead leg begins descent	Lead foot makes contact
Arm cocking	Lead foot makes contact	Maximum shoulder external rotation
Propulsion	Maximum shoulder external rotation	Ball release
Deceleration	Ball release	Maximum shoulder internal rotation
Follow-through	Maximum shoulder internal rotation	Return to balanced position

in the kinetic chain to absorb energy. A baseball pitcher who has a glenohumeral internal rotation deficit (called GIRD) in his throwing arm will not be able to increase the magnitude of glenohumeral internal rotation during the deceleration phase. This could decrease the amount of energy absorbed by the internal rotators. Energy not absorbed by them would be absorbed by other soft tissue structures, which could lead to injury of the shoulder.[37]

17.2.9   Some Dysfunctional Movement Patterns

Most of the time, when you are analyzing and correcting movement, you are looking for what it is that you want someone *to* do. Occasionally, you want to be looking for things that you *do not* want them to do. These are movement patterns that are considered dysfunctional because they either take away from performance or expose the body to potentially injurious stresses. Two such patterns will be explored here: valgus collapse and lifting with a fully flexed spine.

Valgus Collapse

A lower extremity flexion pattern is often used to decelerate the COM in many activities, such as landing from a jump or the braking phase in running. The overall objective here would be to absorb energy. The amount of energy absorbed at the whole body level depends on the angular displacement and torque produced at each DOF. Theoretically, most of that energy should be absorbed through dorsiflexion of the ankle and flexion at the hip and knee. However, pronation at the subtalar joint occurs when the foot hits the ground. Pronation is coupled with internal rotation of the tibia. Because there is minimal motion of the knee in the transverse plane, internal rotation of the tibia should be coupled with internal rotation and adduction of the hip. Thus, small amounts of energy are also absorbed at the subtalar joint and the hip in the frontal and transverse planes.

Excessive motion at the subtalar joint/hip can lead to a movement pattern referred to as valgus collapse (**Figure 17.41**). Obvious causes would be lack of strength/power/endurance of the subtalar invertors and hip external rotators, as these

Figure 17.41 | Valgus collapse.

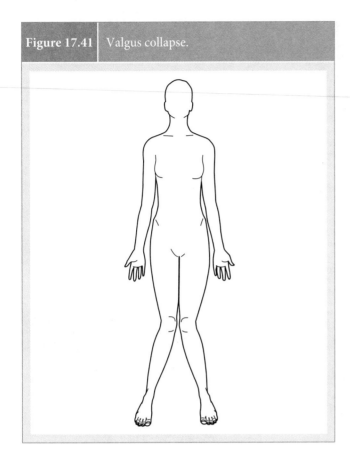

groups control the motions eccentrically during lower extremity flexion.[38-40] A less obvious cause may be restricted inversion at the subtalar joint or external rotation of the hip.[38-40] But if these joints are already starting in a position of eversion and/or internal rotation, they will go further into these motions if their total range of motion is the same. Finally, impairment in the sagittal plane (such as with ankle dorsiflexion) may also be a cause of valgus collapse.[40,41] This is because such an impairment would decrease the amount of energy that can be absorbed in the sagittal plane. In an effort to absorb more energy, other DOFs/planes of movement may be used.

Valgus collapse places increased demands at the knee. These increased demands can result in injury. Some of these injuries include anterior cruciate ligament (ACL) tears,[42] iliotibial (IT) band friction syndrome,[43] and patellofemoral pain syndrome.[44]

Lifting Mechanics

"Lifting" an object is a very generic term. Here, we will be referring to picking an object up from the ground. In such cases, there are two guiding principles for the movement. The first is to minimize the torque on the lumbar spine. This means that the object should be held as close to the trunk as possible. The second is to maximize the ability of the lumbar musculature to create a posterior shear force to counteract the anterior shear created by the load. To do this, full flexion of the spine should be avoided while lifting.[45]

As long as these two principles are followed, several different movement patterns can be adopted, including the squat, stoop, and golfer's pickup (Figure 17.42). You may think that the old adage to "bend the knees and lift with the legs is correct," and the squat technique should be employed in all cases. This is not necessarily the case. Each one may be appropriate, depending on the circumstances. The key is to ensure that the two aforementioned principles are followed.

Figure 17.42 | The squat (a), stoop (b), and golfer's pickup (c) for picking up an object.

Consider what would happen if you were lifting an object that was too large to be placed between your knees (an environmental constraint). Using the squat technique would mean that the load would have to be carried further away from your trunk (in violation of the first principle). In this case, it would be more appropriate if you used the stoop technique. Say you used the stoop technique, but you lacked the requisite hamstring flexibility. You may not be able to reach the object on the floor (an organismic constraint). To reach it, you may end up flexing the spine, which is in violation of the second principle. This is yet another example of how impairment in one area (hamstrings) can cause an injury in another area (the lumbar spine).

SUMMARY

You have gone through quite a long journey in biomechanics. Much like in the movie *The Matrix*, we live in a world that is governed by rules. Those rules are rules of physics. Unlike *The Matrix*, these rules cannot be broken, but when you understand them, you can use them to your advantage. The "code" of our world is mathematics, and it is this code on which biomechanics is based. Unlike dealing with inanimate objects, you must understand basic mechanics, biological characteristics of the neuromusculoskeletal system, and the interaction of multibody systems.[7] These concepts have been in development for a few hundred years, and you cannot expect to master them over a long weekend or even a semester. A journey of a thousand miles begins with a single step. I hope that this was an interesting first step, and I wish you well on your continued journey.

REVIEW QUESTIONS

Remember:

1. List the three functional tasks of the gait cycle.

Understand:

1. When would you use a push-back versus a push-forward strategy? Which muscular torques are involved with each?
2. Why would you swing your arms in the same direction that you are tipping to restore your balance?
3. Which joint torque are necessary to move the body's center of mass laterally? Medially? Why?
4. Explain the advantages of performing a countermovement before a jump.
5. Why is a "soft" landing safer, yet more demanding, to perform?
6. What is the role of the ankle musculature during cycling?
7. Why is there both a flexor and extensor torque at the elbow during wheelchair propulsion?
8. Describe the pulsing activation of the proximal musculature during striking activities and why it is important for performance.
9. Describe the force and torque requirements of holding a glass.
10. Explain the consequences of the total limb, ground reaction force, and joint torques for the body acting like an inverted pendulum during walking and a spring mass during running.
11. Describe why a valgus collapse is considered dysfunctional and its potential causes.
12. Describe why you should not lift with a fully flexed spine.

Apply:

1. Describe how you can improve performance or prevent injury for the following activities: standing balance, jumping, landing, cycling, wheelchair propulsion, striking, drinking from a glass, gait, and throwing.
2. How would you correct the dysfunctional movement patterns of valgus collapse and lifting with a fully flexed spine?
3. Complete a biomechanical analysis on an activity not discussed in this lesson.

REFERENCES

1. Covey SR. *The 7 Habits of Highly Effective People*. New York: RosettaBooks, LLC; 2004.
2. Caldwell GE, van Emmerik REA, Hamill J. Movement proficiency: incorporating task demands and constraints in assessing human movement. In: Sparrow WA, ed. *Energetics of Human Activity*. Champaign, IL: Human Kinetics; 2000.
3. Bobbert MF, Schenau GJV. Coordination in vertical jumping. *Journal of Biomechanics*. 1988;21(3):249–262.
4. Hay JG, Reid JG. *Anatomy, Mechanics, and Human Motion*. 2nd ed. Englewood Cliffs, NJ: Prentice Hall; 1988.
5. Bartlett R. *Sports Biomechanics: Reducing Injury and Improving Performance*. London: Spon Press; 1999.
6. Caldwell GE, Forrester LW. Estimates of mechanical work and energy transfers—demonstration of a rigid body power model of the recovery leg in gait. *Medicine and Science in Sports and Exercise*. Dec 1992;24(12):1396–1412.
7. Lees A. Technique analysis in sports: a critical review. *Journal of Sports Sciences*. Oct 2002;20(10):813–828.
8. Knudson DV, Morrison CS. *Qualitative Analysis of Human Movement*. Champaign, IL: Human Kinetics; 1997.
9. Lintner D, Noonan TJ, Ben Kibler W. Injury patterns and biomechanics of the athlete's shoulder. *Clinics in Sports Medicine*. Oct 2008;27(4):527+.
10. Davids K, Button C, Bennett S. *Dynamics of Skill Acquisition: A Constraints-Led Approach*. Champaign, IL: Human Kinetics; 2008.
11. Thelen E, Corbetta D, Kamm K, Spencer JP, Schneider K, Zernicke RF. The transition to reaching—mapping intention and intrinsic dynamics. *Child Development*. Aug 1993;64(4):1058–1098.

12. Perry J. *Gait Analysis: Normal and Pathological Function.* Thorofare: SLACK Incorporated; 1992.

13. Horak FB, Nashner LM. Central programming of postural movements—adaptation to altered support-surface configurations. *Journal of Neurophysiology.* Jun 1986;55(6):1369–1381.

14. Krishnamoorthy V, Goodman S, Zatsiorsky V, Latash ML. Muscle synergies during shifts of the center of pressure by standing persons: identification of muscle modes. *Biological Cybernetics.* Aug 2003;89(2):152–161.

15. Garhammer J, Gregor RJ. Propulsive forces as a function of intensity for weightlifting and vertical jumping. *Journal of Applied Sport Science Research.* 1992;6(3):129–134.

16. Zajac FE. Muscle coordination of movement—a perspective. *Journal of Biomechanics.* 1993;26:109–124.

17. Zatsiorsky VM, Gregor RJ. Mechanical power and work in human movement. In: Sparrow WA, ed. *Energetics of Human Activity.* Champaign, IL: Human Kinetics; 2000.

18. Ridderikhoff A, Batelaan JH, Bobbert MF. Jumping for distance: control of the external force in squat jumps. *Medicine and Science in Sports and Exercise.* Aug 1999;31(8):1196–1204.

19. Zajac FE. Understanding muscle coordination of the human leg with dynamical simulations. *Journal of Biomechanics.* Aug 2002;35(8):1011–1018.

20. Chen G. Comments on "Biomechanics and muscle coordination of human walking: Parts I and II." *Gait & Posture.* Apr 2004;19(2):206–207.

21. Vanlandewijck Y, Theisen D, Daly D. Wheelchair propulsion biomechanics—implications for wheelchair sports. *Sports Medicine.* 2001;31(5):339–367.

22. Veeger HEJ, Vanderwoude LHV, Rozendal RH. Within-cycle characteristics of the wheelchair push in sprinting on a wheelchair ergometer. *Medicine and Science in Sports and Exercise.* Feb 1991;23(2):264–271.

23. Rozendaal LA, Veeger HEJ, van der Woude LHV. The push force pattern in manual wheelchair propulsion as a balance between cost and effect. *Journal of Biomechanics.* Feb 2003;36(2):239–247.

24. McGill SM, Chaimberg JD, Frost DM, Fenwick CMJ. Evidence of a double peak in muscle activation to enhance strike speed and force: an example with elite mixed martial arts fighters. *Journal of Strength and Conditioning Research.* Feb 2010;24(2):348–357.

25. Kibler WB. Pathophysiology of overload injuries around the elbow. *Clinics in Sports Medicine.* Apr 1995;14(2):447–457.

26. Verkhoshansky Y. The dynamics of punching technique and speed-strength in young boxers. *Soviet Sports Review.* 1991;26(4):160–161.

27. Elliott BC, Fleisig GS, Nicholls R, Escamilla RF. Technique effects on upper limb loading in the tennis serve. *Journal of Science and Medicine in Sport.* 2003;6(1):76–87.

28. Murphy MA, Sunnerhagen KS, Johnels B, Willen C. Three-dimensional kinematic motion analysis of a daily activity drinking from a glass: a pilot study. *Journal of Neuroengineering and Rehabilitation.* Aug 16 2006;3.

29. Murphy MA, Willen C, Sunnerhagen KS. Kinematic variables quantifying upper-extremity performance after stroke during reaching and drinking from a glass. *Neurorehabilitation and Neural Repair.* Jan 2011;25(1):71–80.

30. Farley CT, Ferris DP. Biomechanics of walking and running: center of mass movements to muscle action. *Exercise and Sport Sciences Reviews.* 1998;26:253–285.

31. Novacheck TF. The biomechanics of running. *Gait & Posture.* Jan 1998;7(1):77–95.

32. Fonseca ST, Holt KG, Fetters L, Saltzman E. Dynamic resources used in ambulation by children with spastic hemiplegic cerebral palsy: Relationship to kinematics, energetics, and asymmetries. *Physical Therapy.* Apr 2004;84(4):344–354.

33. Detrembleur C, Dierick F, Stoquart G, Chantraine F, Lejeune T. Energy cost, mechanical work, and efficiency of hemiparetic walking. *Gait & Posture.* Oct 2003;18(2):47–55.

34. Alexander RM. Optimum timing of muscle activation for simple-models of throwing. *Journal of Theoretical Biology.* Jun 7 1991;150(3):349–372.

35. Fleisig GS, Barrentine SW, Escamilla RF, Andrews JR. Biomechanics of overhand throwing with implications for injuries. *Sports Medicine.* Jun 1996;21(6):421–437.

36. Hong DA, Cheung TK, Roberts EM. A three-dimensional, six-segment chain analysis of forceful overarm throwing. *Journal of Electromyography and Kinesiology.* Apr 2001;11(2):95–112.

37. Burkhart SS, Morgan CD, Ben Kibler W. The disabled throwing shoulder: Spectrum of pathology part 1: Pathoanatomy and biomechanics. *Arthroscopy—the Journal of Arthroscopic and Related Surgery.* Apr 2003;19(4):404–420.

38. Loudon JK, Jenkins W, Loudon KL. The relationship between static posture and ACL injury in female athletes. *Journal of Orthopaedic & Sports Physical Therapy.* Aug 1996;24(2):91–97.

39. Willson JD, Ireland ML, Davis I. Core strength and lower extremity alignment during single leg squats. *Medicine and Science in Sports and Exercise.* May 2006;38(5):945–952.

40. Sigward SM, Ota S, Powers CM. Predictors of frontal plane knee excursion during a drop land in young female soccer players. *Journal of Orthopaedic & Sports Physical Therapy.* Nov 2008;38(11):661–667.

41. Pollard CD, Sigward SM, Powers CM. Limited hip and knee flexion during landing is associated with increased frontal plane knee motion and moments. *Clinical Biomechanics.* Feb 2010;25(2):142–146.

42. Hewett TE, Myer GD, Ford KR, et al. Biomechanical measures of neuromuscular control and valgus loading of the knee predict anterior cruciate ligament injury risk in female athletes. *American Journal of Sports Medicine.* Apr 2005;33(4):492–501.

43. Noehren B, Davis I, Hamill J. ASB Clinical Biomechanics Award Winner 2006 Prospective study of the biomechanical factors associated with iliotibial band syndrome. *Clinical Biomechanics.* Nov 2007;22(9):951–956.

44. Powers CM, Ward SR, Fredericson M, Guillet M, Shellock FG. Patellofemoral kinematics during weight-bearing and non-weight-bearing knee extension in persons with lateral subluxation of the patella: a preliminary study. *Journal of Orthopaedic & Sports Physical Therapy.* Nov 2003;33(11):677–685.

45. McGill SM, Hughson RL, Parks K. Changes in lumbar lordosis modify the role of the extensor muscles. *Clinical Biomechanics.* Dec 2000;15(10):777–780.

Glossary

Abrasion wear—Interfacial wear that occurs as a result of a hard surface scraping a softer one

Abscissa—The horizontal axis on a two-dimensional graph

Absolute—Magnitude of the value, regardless of the sign

Acceleration—How rapidly something is changing velocity (speeding up or slowing down)

Accommodating resistance—A resistance that increases with the amount of force or torque applied to it

Acute injury—An injury that happens immediately

Adhesion wear—Interfacial wear that occurs as a result of surfaces sticking to one another and then tearing apart

Amortization phase—The time between the eccentric and concentric actions

Angle-angle diagram—A plot of one joint motion as a function of another joint motion

Angular acceleration—How quickly a body is speeding up or slowing down its rotation in a particular direction

Angular displacement—The change in orientation of a rigid body in reference to some axis

Angular position—The orientation of a rigid body in reference to some axis

Angular speed—How fast a body is rotating

Angular velocity—How fast a body is rotating in a particular direction

Anisotropic—Exhibiting different properties when measured in different directions

Antagonist—Muscles that have the opposite effect on a joint

Apex—The highest point of a trajectory

Area moment of inertia—A measure of a body's resistance to bending

Arthrokinematic motion—Motions at joint surfaces

Atrophy—A decrease in the physiological cross-sectional area

Average—A number representing the value of a quantity if that quantity did not change (was constant) throughout the period of interest

Axis—A straight line running through the origin specifying a direction from the origin

Axis of rotation—A fixed line about which a body rotates

Balance—The ability to control the body over its base of support

Ball bearing—A component that separates moving parts and takes a load

Bending—A load applied perpendicular to the longitudinal axis of a body, causing it to curve

Biomechanics—The study of the structure and function of biological systems by means of the methods of mechanics

Body—The object of analysis; it could be a whole person, a part of a person, or an inanimate object

Bone mineral content—The total amount of mineral in bone

Bone mineral density—The mineral content in an area or volume of bone

Boundary lubrication—The lubricating fluid prevents direct surface-to-surface contact

Braking force—A force that is causing a body to slow down

Braking torque—A torque that is decreasing the speed of rotation

Brittle—A characterization of an object that can only undergo very small deformations

Buckling—Deformation of a column

Cadence—The number of steps taken in a given period of time

Center of mass—A fictitious point where all the mass is considered to be concentrated

Chain configuration—The orientation of each link relative to each other and the global reference frame

Chronic injury—An injury that develops over time

Cocontraction—Activation of a prime mover and an antagonist simultaneously

Coefficient of restitution—The measure of elasticity of a collision between two objects

Cofunctional—Muscles that have the same effect on a joint or group of joints

Compensatory motion—Increased motion at one degree of freedom to make up for decreased range of motion at another degree of freedom

Compliance—The ratio of the change in deformation to the change in load; it is the opposite of stiffness

Components—Parts of a resultant vector, two or more vectors that are acting in different directions

Compression—A load that squeezes the parts of a body together

Concentric action—An action in which the muscle–tendon complex develops greater force than the external force acting on it and shortens; because the force and the displacement are in the same direction, the MTC is doing positive work—increasing the energy of the skeletal system

Contact force—The force created when two bodies are touching each other

Coordination—The appropriate assembly and sequencing of degrees of freedom

Coupled motion—Motions of two degrees of freedom that consistently occur together

Couple—Two forces that are equal in magnitude, opposite in direction, and in the same plane; the effect of a couple is pure rotation with no translation

Crank arm—The rigid body connecting the chain ring to the pedal

Creep—An increase in strain when the stress is held constant for a period of time

Critical elements—Aspects of a movement that are necessary for optimal performance

Decouple—Allow joined subsystems to operate independently

Deformation—A change in dimensions of a body

Degrees of freedom—The number of movements available. Movement must occur in two directions to equal one DOF

Derived variable—A variable that is formed by multiplying or dividing it by other variables

Direction—A pointing toward something, determined by its orientation and sense

Displacement—A change in position

Distance—How far a body has traveled

Ductile—A characterization of an object that can undergo very large deformations

Dynamic—Moving

Eccentric action—An action in which the muscle–tendon complex develops less force than the external force acting on it and lengthens; because the force and the displacement are in the opposite direction, the MTC is doing negative work—decreasing the energy of the skeletal system

Economy—The amount of energy required to perform a certain amount of work (MEE)

Effective force—The vector ma_{COM}, which is the effect of the net sum of all force vectors acting on a body

Effective mass—The portion of a body's mass that is involved with a collision

Effective torque—The vector $I_{COM}\alpha_{COM}$, which is the effect of the net sum of all torque vectors acting on a body

Efficiency—The amount of mechanical energy that can be expended with a given amount of energy

Elastic collision—A collision where two objects bounce off each other without any deformation or loss of heat

Elastic deformation—A deformation in which the object returns to its original dimensions after the deformation

Elastic modulus—The ratio of stress to strain

Elastohydrodynamic lubrication—The amount of surface area is increased by the pressure of the fluid

Electromechanical delay—The time between the onset of electrical activity at the muscle and production of measurable force

Energy—The state of matter that makes things change, or has the potential to make things change

Extension pattern—A movement pattern with the segments rotating in opposite directions. The end effector follows a linear motion path as it moves away from the origin of the chain.

External (locomotor) work—Energy required to change the motion or location of the center of mass

Failure tolerance—The stress level above which failure will occur

Fatigue—Any reduction in the force-generating capacity of the total neuromuscular system, regardless of the force required in any given situation

Fatigue wear—Wear that is the result of microdamage

Flexion pattern—A movement pattern with the segments rotating in opposite directions. The end effector follows a linear motion path as it moves toward the origin of the chain.

Fluid film lubrication—Movement increases the amount of fluid between articulating surfaces, thus increasing their separation

Force—A push or pull by one body on another

Fracture—The breaking of a bone

Frame of reference—The perspective from which the movement is described

Functional spinal unit—Two adjacent vertebrae and the intervertebral disk

Gait—Locomotion over land

Gravitational potential energy—The potential energy that a body has due to its position

Ground reaction force—The equal and opposite force the ground applies back on the person

Hydrodynamic lubrication—The amount of separation between articulating surfaces is increased by the fluid when a wave of fluid is created by the two surfaces moving tangentially to one another

Hyperplasia—An increase in the number of muscle fibers

Hypertrophy—An increase in the size of the muscle fibers

Hysteresis—The amount of energy lost to heat between loading and unloading

Impulse—The product of average force and time that force is applied; it is equal to the change in momentum

Inelastic collision—A collision in which two objects stick together after they collide

Inertia—A resistance to change in motion, specifically a resistance to change in a body's velocity

Instantaneous—The value of a quantity at a particular moment in time

Interaction torques—Torques created about a joint that result from the velocities and torques of other joints in a serial chain

Interfacial wear—Wear that occurs when two surfaces come in direct contact

Intra-abdominal pressure—Pressure inside the abdominal cavity

Intrinsic dynamics—The kinematic (range of motion) and kinetic (strength, power, endurance) capacities of each degree of freedom involved in a task

Isoform—A different form of the same protein

Isometric action—An action in which the muscle–tendon complex develops a force that is equal to the external force acting on it and does not change its length; because there is no displacement, the MTC is doing no work—the energy of the skeletal system remains unchanged during isometric actions

Kinematics—The study of motion without considering what is causing the motion

Kinetic energy—The energy that a body has due to its motion

Kinetics—The study of the forces that cause motion

Lever—A rigid body that is used in conjunction with a pivot point to multiply the force or speed applied to another body

Load—An externally applied force

Margin of safety—The difference between the failure tolerance and actual stress applied to a body

Mass—The amount of matter in an object

Material failure—A breaking apart of the material

Mechanical energy expenditure—The amount of mechanical energy necessary for a prescribed motion

Mechanics—The study of forces and their effects

Mechanopathology—The mechanics that result in an injury

Moment of force—("Moment" for short), is synonymous with torque; it is the turning effect of a force

Moment of inertia—The angular equivalent of mass, gives an indication of how difficult it will be to rotate an object

Momentum—A resistance to change in velocity of a moving body

Motion path—The trajectory of movement by any point of interest

Motor redundancy—The ability to perform the same task in multiple ways

Motor unit—A motor neuron and all the muscle fibers it innervates

Movement pattern—Any recognizable spatial and temporal regularity, or any interesting relation, between moving bodies

Net—The total value after summing all the individual values

Neutral axis—The line along which there is neither compressive nor tensile loading on a body during loading

Ordinate—The vertical axis on a two-dimensional graph

Orientation—A particular reference line

Origin—The place where the frame of reference begins

Osteoarthritis—The progressive degeneration of the articular cartilage and the bone deep to it

Osteokinematic motion—Rotations of bones

Osteopenia—Lower-than-normal bone mineral density

Osteoporosis—Severe decrease in bone mineral density

Parabola—A type of plane curve

Pathomechanics—The mechanics that are the result of an injury or illness

Pelvic rotation—Movement of a pelvic bone in relation to the sacrum

Pelvic tilt—Movement of the entire pelvic girdle with respect to the hips and spine

Performance—How large a deviation, and how quickly a system returns to its intended position or trajectory following a disturbance

Perturbation—A force causing a disturbance or deviation in a system

Plane—A smooth flat space defined by two axes

Plastic deformation—A deformation in which the object does not return to its original dimensions after the deformation

Plasticity—The ability to adapt or change

Point—A way of representing a body that has no dimensions

Polar moment of inertia—The resistance to torsional loading

Position—Location in a reference frame

Potential energy—The energy a body has that has the potential to change something, but it is not currently changing anything

Power—The time rate of doing work; alternatively, how quickly energy is entering or leaving the system, or how much force can be produced while moving quickly

Prehension—The act of grasping or seizing something

Proficiency—How well a person performs a movement and achieves the goal of the task

Projectile—An object in the air that is only subject to the force of gravity and wind resistance after it leaves the ground

Propulsive force—A force that is causing a body to speed up

Propulsive torque—A torque that is increasing the speed of rotation

Qualitative—Subjectively describing something using words rather than measuring it

Range of motion—The amount of movement available at a joint in a given direction

Range—The horizontal displacement of a projectile

Rate—How quickly a value is increasing or decreasing with time

Reach area—The space in which the end effector can be positioned

Reaction force—As a consequence of Newton's third law, for every force created by body A on body B, there is a force of equal magnitude and opposite in direction created by body B acting on body A

Relative height—The difference between the vertical position at takeoff and the vertical position at landing

Relative impulse—The amount of impulse that can be generated relative to body mass

Relative Velocity—How fast one body is moving in relation to another body

Resultant—A vector that is equivalent to the combined effect of two or more vectors

Rigid body—A body that maintains a constant shape

Robustness—The magnitude of the disturbance that a system could tolerate and still return to its intended position or trajectory

Sarcopenia—Age-related decrease in muscle mass

Scalar—A quantity that only has a magnitude

Sense—Specified by two points; going from point B to A has the opposite sense (and opposite direction) of going from point A to B on the same line

Shank—The part of the leg from the knee to the ankle

Shear—A load that causes one part of a body to move parallel past another part

Shoulder elevation—Raising the arm overhead in any plane

Slope—The incline of a line on a graph from the horizontal axis

Specific tension—The amount of force that can be produced per cross-sectional area

Speed—How fast a body is moving

Sprain—An injury to a ligament that occurs when it is stretched beyond its capacity

Squeeze film lubrication—The amount of separation between articulating surfaces is increased by the fluid when the two surfaces are compressed together

Stability—The ability of a system to produce a reference position or trajectory in the presence of a disturbance

Static equilibrium—A special case where both the linear and angular accelerations are zero, and thus the sum of the external forces and sum of the external torques are zero

Static—Not moving

Step—The period from the initial contact of the one foot to the initial contact of the other foot

Stiffness—A resistance to change in deformation; an object with greater stiffness requires greater forces for the same amount of deformation

Strain energy density—Relative amount of energy stored by the material

Strain potential energy—The energy a body has due to its deformation

Strain–rate dependency—When mechanical properties are dependent on the rate of change of strain

Strain—The change in dimension normalized to the original dimension

Strength curve—A plot of the maximum torque produced about a joint as a function of joint angle

Strength (muscle)—The ability to produce force

Strength (material)—The amount of loading an object can withstand before failure

Stress-relaxation—A decrease in stress when the strain is held constant for a given period of time

Stress—The way a force is distributed within a body

Stretch-shortening cycle—A concentric action immediately after an eccentric action; energy stored during the eccentric action contributes to the movement during the concentric action

Stride—The period from initial contact of one foot to the next initial contact of that same foot

Strut—A structural unit designed to resist compressive forces

Superincumbent body weight—The body weight that is lying on top of the segment(s) of interest

Swing pattern—A movement pattern with the segments rotating in the same direction with a submaximal velocity. The end effector follows a curved motion path.

Synergist—Muscles that work together to produce a movement

System—The object of analysis that is made up of two or more bodies

Task failure—The inability to continue or complete a desired action

Tension—A load that pulls the parts of a body apart

Tie rod—A slender structural unit that is designed to resist tensile forces

Torque—The turning effect of a force

Torsion—The type of loading that exists when there is a twist around the neutral axis

Toughness—The amount of energy that can be absorbed by a body before failure

Trajectory—The path of a projectile

Truss—A structure composed of one or more triangular units

Vector—A quantity that has both a magnitude and a spatial direction

Velocity—How fast something is moving in a particular direction

Wear—Surface material is deformed and removed by frictional forces

Weight—The force due to gravity; weight always acts in the downward, vertical direction and has a magnitude of 9.81 m/sec^2 times the body's mass

Whip pattern—A movement pattern with the segments rotating in the same direction with maximal velocity. The end effector follows a curved motion path.

Work—The process of changing the amount of energy in a system

Yield point—The amount of deformation that marks the transition from elastic to plastic deformations, and deformation beyond this point results in a permanent deformation

Young's modulus—The ratio of stress to strain

Index